COMMON CONVERSION FACTORS, BRITISH ENGINEERING UNITS TO METRIC

British	×	= Metric
inches	25.4	millimeters
feet	0.3048	meters
yards	0.9144	meters
miles	1.609	kilometers
square inches	6.452	square centimeters
square feet	0.0929	square meters
acres	0.405	hectares
square mile	2.590	square kilometers
ounce, mass	28.35	grams
pound, mass	0.4536	kilograms
ton, mass	0.9072	metric tons
pound, force	4.448	newtons
ton, force	8.896	kilonewtons
psf	47.88	newtons/meter^2
psi	6.895	kilonewtons/meter^2 or kilopascals, kPa
psi	6.895×10^{-3}	megapascals, MPa
ton/square foot	95.76	kPa
ton/square foot	1.024	kilograms/meter^2
1 atmosphere, STP	101.3	kPa
1 bar	100	kPa
fluid ounce	30	milliliters
quart	0.95	liters
gallon	3.8	liters
pound, force per cubic foot	16.025	kilograms/meter^3
mile per hour	1.609	kilometers/hour
feet per year	0.966×10^{-6}	centimeters/second
BTU	252	calories
kilowatt-hour	860,421	calories

Metric to British conversion factors are the reciprocals of the conversion values shown above.

Geology
Applied to Engineering

TERRY R. WEST

PURDUE UNIVERSITY

Geology
Applied to Engineering

 Prentice Hall, Englewood Cliffs, New Jersey 07632

In all things of nature there is something of the marvelous.

Aristotle 388–322BC

LIBRARY OF CONGRESS CATALOGING-IN-PUBLICATION DATA

West, Terry R.
    Geology applied to engineering / Terry R. West.
        p. cm.
    Includes index.
    ISBN 0-02-425881-4
    1. Engineering geology.   I. Title.
TA705.W47 1995
550′.24624—dc20

93-48184
CIP

Editor: Robert A. McConnin
Production Supervisor: Spectrum Publisher Services, Inc.
Production Manager: Aliza Greenblatt
Text Designer: Robert Freese
Cover Designer: Heather Scott

© 1995 by Prentice Hall, Inc.
Simon/Schuster Company
Englewood Cliffs, New Jersey 07632

Printed in the United States of America

10 9 8 7 6 5 4 3 2 1

ISBN 0-02-425881-4

Prentice-Hall International (UK) Limited, *London*
Prentice-Hall of Australia Pty. Limited, *Sydney*
Prentice-Hall Canada Inc., *Toronto*
Prentice-Hall Hispanoamericana, S.A., *Mexico*
Prentice-Hall of India Private Limited, *New Delhi*
Prentice-Hall of Japan, Inc., *Tokyo*
Simon & Schuster Asia Pts. Ltd., *Singapore*
Editora Prentice-Hall do Brasil, Ltda., *Rio de Janeiro*

Preface

This text, developed over a number of years, has been used by the author in two distinctly different courses, Geology for Engineers and Engineering Geology. Geology for Engineers, a first geology course for engineering students, relates the physical aspects of geology to civil engineering construction. Typically, no prerequisites are required and the course is populated by sophomore and junior engineers. It is a laboratory course with the end-of-chapter exercises forming a basis for laboratory studies on minerals, rocks, maps, geologic processes, and applied geology. Additional problems are included with each chapter to provide opportunities for engineering computations. These are tied to example problems presented in the text.

Of the 20 chapters in the book, only Chapter 7, Elements of Soil Mechanics, has not been assigned in its entirety for engineering students. However, portions of Chapter 7 can be referenced to emphasize specific concepts from other chapters. For example, the section on effective stress can be assigned in conjunction with ground-water geology or engineering and environmental geology.

Engineering Geology is the second course in which the text has been used. This is a division of the geologic sciences that involves geologic principles and engineering fundamentals. The objective of the course is to assure that the geologic factors affecting the location, design, construction, and maintenance of engineering projects are recognized and properly provided for. At the working level, it involves engineering applications for geologists and forms the counterpart to geology for engineers.

Engineering Geology is an upper-division course for undergraduate geology majors and graduate students, with an emphasis on construction or environmental concerns. Also enrolled are civil engineers with a background in geology and those who are interested in geologic applications to construction. The course builds on a basic knowledge of geology by relating geology to engineering fundamentals and construction concepts.

Chapter 7, Elements of Soil Mechanics, provides basic information for the Engineering Geology course. Also used in the course are Chapter 6, Engineering Properties of Rock; Chapter 13, Physiographic Provinces and Engineering Considerations; Chapter 14, Landslides, Subsidence, and Slope Stability; Chapter 15, Ground-Water Geology; Chapter 18, Earthquakes and Geophysics; Chapter 19, Subsurface Investigation and Site Selection; and Chapter 20, Engineering Geology and Environmental Geology.

A major advantage accrues to both courses as a result of the inclusion of the entire collection of subject matter. Although students in Engineering Geology have had course work on rocks and minerals, they will find the summary sections on engineering problems for specific rock use (Chapters 2 through 5) of considerable interest. Such details are not included in conventional physical geology books. Also, for the diligent student in civil engineering, Chapter 7, on soil mechanics, will help breach the gap between civil engineering and applied geology.

Students enrolled in engineering geology courses typically have diverse backgrounds in terms of their geologic training. Therefore, it serves an important purpose to have chapters on basic geology, such as glacial geology, stream processes, and structural geology, available for review purposes.

Problem sets provided at the end of each chapter serve as home problems for student work. Weekly assignment of problems yields a systematic way to progress through the course material. Numerous solved problems are included in the text to illustrate the procedure for performing engineering calculations.

The text is liberally illustrated both with photographs and line drawings. In a beginning text particularly, illustrations provide the extra detail required to clarify new concepts and discussions. Care has been taken to explain in a clear fashion both the geologic and engineering details. Discussions are based on the geology and engineering experience obtained by the author through teaching, research, and consulting. Working as both a registered professional engineer and an applied geologist has provided the author with a wide range of experience in the related fields of engineering and geology. Degrees in geology, geological engineering, and civil engineering also set the stage.

English and metric units are used simultaneously in the text, and a detailed conversion table is provided in Chapter 7, Elements of Soil Mechanics. Endpapers inside the front and back covers also provide conversions and standard equations for soils engineering. The purpose of the dual listing is twofold. Metrication appears to be inevitable in the United States, yet the persistent use of English units still prevails in many engineering offices and in common application in the business environment. Second, today's science and engineering students must be able to convert from metric to English units with ease and also perform complete calculations based on either system.

A detailed index is provided at the back of the textbook. This index should prove helpful to the student in finding definitions and locating discussions on related subjects. Because of the wide diversity of geology and engineering, this index serves as an aid in associating the different areas of study.

The Geology for Engineers and Engineering Geology courses approach the boundary between geology and civil engineering from opposite directions. Knowledge of both geology and engineering is needed to operate successfully within the construction geology specialty and many challenging, technical problems fall in this boundary zone. The field is dynamic because new developments in engineering and geologic studies continue to impact the specialty. It is the purpose of this text, *Geology Applied to Engineering*, to provide training for students so they can bridge this gap between geology and civil engineering.

ACKNOWLEDGMENTS

My sincere thanks go to the students, colleagues, teachers, staff, family, and friends who contributed valuable ideas on teaching basic and applied geology to students at the university level and/or who provided the needed moral support to complete this extensive project. Also acknowledged are those individuals who directly assisted with the preparation of the manuscript and the illustrations for this book.

I wish to thank the reviewers of the manuscript—Glenn R. Brown, University of Toronto; W.B. Fegusson, Villanova University; Nels F. Forsman, University of North Dakota; Ronald E. Gallagher, University of Toledo; Charles E. Glass, University of Arizona; Donald M. Keady, Mississippi State University; Charles J. Ritter, University of Dayton; John D. Rockaway, University of Missouri–Rolla; and N. Luanne Vanderpool, University of Illinois–Chicago Circle—who provided helpful suggestions for the text. I also want to thank Dr. Henry O. A. Meyer for reading Chapter 1 on the Origin and Development of the Earth and Mrs. Richard Gelsleichter who read the text as a nongeology or engineering specialist to test its general interest and readability.

The photographs from outside sources are acknowledged as they appear in the text. The others, which comprise most of the total, are photos from my collection, taken on numerous geology and engineering field investigations over the years. These photographs have served as visual aids for numerous classes on applied geology.

In deep gratitude for her encouragement and support during the preparation of this text, the book is dedicated to my wife, Shirley Mueller West.

T. R. W.

Brief Contents

Contents

Geology
Applied to Engineering

1

Origin and Development of the Earth

THE EXPANSE OF SPACE

The universe extends in all directions as far as we can detect matter. It includes all of the cosmos or all the matter that exists. With a 200-in. telescope stars can be seen about 2 billion light-years[1] away, and with a radio telescope a distance of more than 3 billion light-years can be scanned. From this it can be seen that the universe is immense beyond human comprehension, with unit distances measured in light-years and with total dimensions extending for billions of light-years.

The points of light we see in the sky at night are mostly stars. Only a few are planets moving visibly across the sky, and occasionally there are hazy patches of light that a telescope is unable to resolve sharply into a single point. These are nebulae, the other galaxies beyond the Milky Way, the galaxy in which Earth orbits. Astronomers now estimate that about a billion galaxies exist in the universe.

[1]One light-year is the distance light travels in a year = 9.3×10^{12} km = 5.8×10^{12} miles. The speed of light, $c = 2.998 \times 10^5$ km/sec = 1.863×10^5 miles/sec.

Galaxies, the building blocks of the universe, are concentrations of billions of stars. They vary in size and form, but many are spiral-shaped with curved arms trailing from their centers. When viewed along their spiral diameter, they appear as a disk that bulges outward at the center. The nearest galaxy to the Milky Way is Andromeda, or the Great Nebula, located 2.2 million light-years away.

The Milky Way is home to all the stars that we can see clearly without a telescope; they number about 10^{11} or 100 billion stars. In the Milky Way, stars are more or less clustered together to form the constellations, which were named centuries ago with reference to animals, mythical gods, famous people, or common objects. About 90 constellations have been named based on the forms they seemed to resemble. Today constellations are used as navigational aids and as a convenience to orient and describe locations in the sky.

This chapter discusses the Earth's origin and development, beginning with the universe and focusing on the details of the Earth itself. Information about the solar system and the universe, with some ideas on their origin, is included in the discussion.

MODERN THEORIES
OF THE UNIVERSE

The extent of the universe is so great that the past is revealed to us on a delayed basis; indeed, we can compare the past directly with the present. Light that began its journey hundreds, thousands, millions, and even billions of years ago is now being received on Earth from different locations in the universe. One aspect learned by comparing these delayed messages shows that energy and matter in the distant past functioned much as they do today. This reinforces the hypothesis proposed by geologists that uniform physical laws prevailed to drive the geologic processes throughout the Earth's history.

We know enough about the cosmos or about cosmology (the structure of the universe) to construct theories about the origin of the universe. As with most theory development, one must start with a list of known facts and attempt to build a theory consistent with those basic data. Such facts are as follows:

- The universe is vast and extends as far as humans can measure.
- Mass is concentrated into collections of stars known as galaxies.
- Stars undergo stages of development from red giants to white dwarfs.
- The universe is expanding, with stars at the periphery of the universe moving away the fastest.

This last entry deserves further consideration. Expansion of the universe is indicated by the Doppler effect. This phenomenon shows that the frequency of light or sound is lowered as an object accelerates away from the observer. For example, the pitch (frequency) of an automobile horn decreases as the car accelerates away from a point of reference. Because light emitted from the farthest stars is shifted or displaced toward the red end of the visible spectrum, the stars must be accelerating away from the observer. The light's wavelength is increased toward red, and therefore its frequency is reduced ($c = \lambda \upsilon$ where c is the speed of light, a constant; λ is the wavelength (in units of length); and υ is the frequency of the waves or 1/time).

The Evolving Universe Theory

If matter is accelerating away from a central location and the rate of acceleration and the distance can be estimated, one can estimate how long ago matter left the common source by working backward. This forms the basis of the *Evolving Universe* or *Big Bang theory*.

According to the proponents of this theory, George Lamaitre and George Gamow, all the matter of the universe was once tightly packed into an extremely dense region with a radius of about 160 million km (100 million miles). The initial mass was shattered by a cataclysmic explosion, hurling matter and radiant energy into space and bringing about the creation of the present chemical elements. This event happened between 7 and 15 billion years ago. As expansion occurs, the density of matter becomes less (i.e., volume increases; mass remains constant; mass/volume, or density, decreases).

The initial central mass consisted of subatomic particles not yet organized into elements. The elements formed when protons, neutrons, and electrons united during the explosive episodes. The elements were built in turn from these building blocks, with neutrons and protons grouping together by fusion to yield elements of greater and greater atomic number.

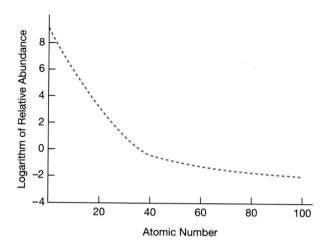

FIGURE **1.1** Relative abundances of the elements in the universe plotted against atomic number.

Hence, the abundance (frequency) of the elements in the universe decreases with increasing atomic number. This relationship is illustrated in Figure 1.1. Several problems violate the otherwise smooth curve that might be expected to develop; there are no stable atoms of mass 5 or 8 and the abundance of iron is greater than the smooth curve would suggest.

Several questions have been raised about the authenticity of this hypothesis, the first having to do with the relative abundance of elements and iron just mentioned. The second is concerned with the effects of the cataclysmic explosion. If such an immense explosion occurred, why is there no mark on the universe as a result of it? This question may have been answered at least in part by the 1978 Nobel prize winners in astronomy, Dr. Arno Penzias and Dr. R. W. Wilson, who discovered a faint glow of radiation throughout the universe that could be the aftereffects of the Big Bang.

Other questions posed are how the initial mass of matter formed and what took place before its existence. In answer to these queries, the expansion-contraction universe theory is proposed. This theory claims that the universe expands and contracts repeatedly, perhaps an infinite number of times. Following expansion, the universe will reverse its direction and contract into the central mass to begin another cycle.

The Steady-State Theory

A second theory on the origin of the universe is not concerned with the beginning or end of the universe. In this theory, matter is continually replenished in one place, while it is being destroyed in others. Expansion of the universe is accepted as fact but new matter is formed to keep pace with expansion, and a steady density of matter is maintained. If steady-state density is maintained, a sequence of evolving stars is required.

Stages of Star Development

The stages of development for stars are an observed fact. Gas clouds comprise a large percentage of the matter in the universe. A rotating volume of gas will take on a flattened shape and spiral form as a direct consequence of its rotation. As the gas cloud rotates, masses tend to grow by gravitational attraction and, eventually, large masses of about equal size (protostars) will appear. These protostars are not true stars in the sense that they do not give off light. Condensation finally produces the dense bodies that begin to glow with light as temperatures reach 1 million °C and their hydrogen undergoes fusion. An initial volume of gas must be compressed (reduced) about a million times to become a star, so that stars in a typical galaxy are separated by great distances. In the Milky Way, the average distance between stars is about 4.3 light-years.

Stars can be classified according to two variables, surface temperature (color) and absolute visual magnitude or absolute brightness. The majority of stars plot along a band of decreasing temperature and brightness known as the *main star sequence*.

The brightest stars, blue giants, are very hot, blue stars with surface temperatures between 11,000 and 25,000 K. (K = degrees Kelvin; zero K = –273 °C). Stars like the Sun are yellow stars with an intermediate temperature, between 5000 and 6000 K. White dwarfs are small, very faint stars. The bright red stars, the giants and super giants, have surface temperatures less than 5000 K.

The majority of stars in the universe are the size or mass of the Sun or smaller. However, on a scale from smallest to largest, the Sun is considered a lightweight star. The sequence of events that occurs in the evolution of different stars is dependent on their mass during the developing stages.

LIGHTWEIGHT STARS Lightweight stars, such as the Sun, turn into red giants when their hydrogen is consumed and a helium core begins to expand. If the mass of this remaining core is less than 1.4 solar masses when the nuclear reaction ceases, it will shrink back into a white dwarf. Because of electron degeneration that counteracts the gravity force, the body reduces to a volume about that of the Earth and one teaspoon of matter equals about 5 tons (4500 kg). The white dwarf cools eventually to become a black dwarf, a completely burned-out star.

MIDDLEWEIGHT STARS   Middleweight stars (4 to 8 solar masses in size when on the main star sequence) evolve into red supergiants, which are much brighter than red giants, when their hydrogen is consumed. Helium in the core is burned in fusion, expanding the star, which then shrinks back when all of the helium is used. When the carbon core is burned next, the star eventually explodes in a great burst known as a *supernova*. Finally the star collapses back into a white dwarf or to a neutron star, depending on its central mass. If the remainder is less than 1.4 solar masses, it turns into a white dwarf in the same way as did the lightweight star. But if it is between 1.4 and about 3 solar masses, it collapses to become a neutron star. The collapse is so great because of this sequence of events that one teaspoon of matter equals about 1 billion tons (9.1×10^{11} kg).

HEAVYWEIGHT STARS   Heavyweight stars contain from 8 to 60 or more solar masses when on the main sequence. In their evolutionary process they go through the expansion and contraction sequences as helium, carbon and possibly magnesium cores are consumed by nuclear reaction. If, when the final collapse begins, following the supernova occurrence, the remaining portion is less than 3 solar masses (but greater than 1.4), it will eventually become a neutron star in the same manner as the larger, middleweight stars described earlier. If, however, this final portion is greater than 3 solar masses, it will continue to collapse on itself, going through electron and neutron degeneration, and finally form a black hole. The collapse is so great there is no known force in the universe that will stop it, and apparently the density approaches infinity. This is a star that has been compressed to such a great extent that radiation (including light) can no longer escape from it. Indeed it has become absolutely black in nature (thus the term *black hole*) consuming all radiation it encounters, giving none of it back.

Materials from supernova explosions pour into space where they mix with the elemental hydrogen already present. Here construction of new galaxies and stars can take place.

In the Steady-State theory the heavier elements are not produced during an initial cataclysm; in fact, no such big bang is required. These elements are created instead in the central regions of red giants and when supernovas explode. Whether the hydrogen consumed in the fusion processes of stars is created from energy in regions of rarified space is debatable.

The rate of occurrence of supernovas appears to be great enough to provide a suitable source of heavy elements for the universe. This adds credence to the Steady-State theory. In fact, several supernovas have been observed in the Milky Way during recorded time—for example, in 1054 AD, 1572 AD, and 1604 AD—and each for a time was much brighter than any planet. No similar events have occurred in the Milky Way since the invention of the telescope.

Several difficulties are raised by the Steady-State theory. For one, the well-documented degeneration of stars (red giant to white dwarf) does not prove the transformation of energy to mass or the fusion of lighter elements to form heavy ones. The two need not be related. In another evaluation, a highly considered astronomer, Dr. Martin Ryle, evaluated two volumes of space that existed at different times (different distances from the Earth). The first was 1 billion light-years away, and the second was 0.5 billion light-years away. His calculations showed that the two did not have the same density or the same mass per unit volume. Recall that the Steady-State theory claims that density is a constant throughout the universe. A rebuttal to Ryle's finding is that his samples were not sufficiently large or statistically sound to evaluate this density adequately.

A further argument against the Steady-State theory is the finding mentioned previously that strongly favors the Evolving Universe theory. This involves the faint glow of radiation throughout the universe discovered by Dr. Penzias and Dr. Wilson, which could be the aftereffects of the Big Bang.

ORIGIN OF THE SOLAR SYSTEM

The Sun has a family of nine planets that revolve around it in a counterclockwise direction as viewed from the north. The distance of the various planets from the Sun ranges from 0.4 AU to almost 40 AU and their periods of revolution around the Sun are as short as 88 days (Mercury) and as long as 248.4 years (Pluto).[2]

A growing body of information now exists based on recent studies and space probes that provides a detailed description of the solar system. As before, in our discussion on the origin of the universe, we begin by listing the known facts about our object of study, now the solar system. Any hypothesis must be consistent and explain the facts known about the solar system. These facts are as follows:

1. All the planets revolve around the Sun in the same direction, in elliptical but almost circular orbits, that lie nearly in the same plane, and most of the moons revolve in the same direction.

2. All planets except for Uranus rotate in the same direction as they revolve around the Sun, which is counterclockwise as one views the Earth from the north. Uranus rotates in a clockwise direction.

3. The planets are spaced at regular intervals from the Sun, a discovery known as the Titius-Bode rule, and it yields a geometric progression in distances. Stated verbally, the progression is obtained by writing 0, then 3, and continuing the progression by doubling the previous number to obtain 0, 3, 6, 12, 24, 48, 96, and so forth. Adding 4 to each number in the series yields 4, 7, 10, 16, 28, 52, 100, and so forth. Finally dividing each number by 10, we get 0.4, 0.7, 1.0, 1.6, 2.8, 5.2, 10, and so forth. This is the approximate distance from the Sun in astronomical units for the planetary orbits. However, the outermost two planets, Neptune and Pluto, deviate from this sequence. Pluto has a very eccentric orbit around the Sun and it is a tiny body only 1/5 the diameter of the Earth.

4. Ninety-nine percent of the mass of the solar system is in the Sun but 98% of the angular momentum is in the planets (angular momentum = mass × angular velocity = mass × velocity × distance from center of rotation). Angular velocity is the rate of movement along a path measured as an angle from the center.

5. The terrestrial, or inner, planets (Mercury, Venus, Earth, Mars) are small, dense planets (4 to 5.5 times that of water) and the outer planets (gas or giant planets) have low densities (0.7 to 1.7 times that of water) and are more similar to the Sun than to the terrestrial planets.

Development of Hypotheses

In general, the various hypotheses that have been proposed to explain the origin of the solar system fall into two groups. The first requires a catastrophic event, the accidental intervention of another star, in its attempt to explain the observed facts. These are the so-called "second-body" or "collision" hypotheses. The other group, known collectively as "single-body theories," requires no influences beyond the Sun or solar system in attempting to explain the observed facts.

Chronologically, theories on the origin of the solar system were proposed by some scientists and then disputed by others on the basis of the theory's inability to answer the established facts known at the time. An early proposal was the second-body hypothesis of Georges-Louis Buffon in 1749 known as the *dynamic encounter theory*. It suggested that a passing star pulled off two filaments of hot gases from the Sun. This gaseous material condensed to form the planets. The primary problem with this proposal was that 1) the gases would disperse into space, and 2) no difference in composition would develop between the inner and outer planets. Other shortcomings would be suggested later in other second-body hypotheses.

Immanuel Kant, a German philosopher in 1755, and Pierre-Simon Laplace, the great French mathematician, in 1796 independently proposed a theory for which they are now collectively given credit. Known today as the *nebular* or *Kant-Laplace hypothesis*, it was the first of the single-body proposals. It suggested that a mass of gas cooled and began to contract. As it did, its rotational speed increased, much like skaters can increase their speed when spinning by drawing their arms inward. This is a consequence of the law of conservation of angular momentum. Rings of gases were spun off from the center and these condensed to form the planets. It was more than 100 years before this hypothesis was discounted because the rings contained too little mass to form planets, but mainly because the Sun would contain most of the angular momentum of the system under these conditions.

With the nebular, single-body proposal in disrepute, Jeans and Jefferies in 1917 attempted to patch

---

[2]An astronomical unit (AU) is the average distance from the Earth to the Sun, equal to 1.5×10^8 km or 9.3×10^7 miles.

up the second-body hypothesis of Buffon, by introducing their *tidal hypothesis*. According to this hypothesis, the matter that was pulled from the Sun in cigar-shaped masses as the second star went by was solid material. This would yield more angular momentum to the planets and was an attempt to explain the shortcomings of the nebular hypothesis, because it would prevent the material from being lost into space as in Buffon's dynamic encounter hypothesis. The shortcomings of this proposal were 1) no difference in composition of the inner and outer planets would develop, and 2) the close appearance of a second star in the vastness of space is a highly improbable occurrence. More recently, astronomers have pointed out that most of the material pulled from the Sun would come from its interior and would be so hot (1 million °C) that gases would be dispersed through space with explosive violence rather than condensing into planets. This weakness in the theory, together with the low probability of the two-star encounter and the reassertion of the revised nebular hypothesis, has now apparently discounted entirely the theory of the second-body origin of the solar system.

In 1944, van Weizsacher proposed the turbulent hypothesis and soon thereafter a series of other hypotheses were developing and being discounted in close succession as specifics were questioned and changes were made. All were single-body hypotheses and many explained the origin of the Sun as well as the planets. This is an important issue if the origin of both the Sun and the planets is to be considered. The problem of angular momentum and distribution of mass between the Sun and the planets is easily explained if one begins with a nonrevolving Sun and rotating gases and dust around it. The serious question is how did this stationary star and whirling dust cloud come into existence? How did the Sun stop whirling? The recent group of hypotheses, sometimes referred to as the *dust-cloud hypotheses*, considers the complete story of the origin of the solar system.

The Dust-Cloud Hypothesis

Interstellar matter is widely distributed throughout space. In this vast region, rarified matter exists that is about 99% gas and 1% dust. The gases are the ever-present hydrogen and helium, and the dust-size particles have compositions similar to terrestrial material. These include silicon compounds (silicates),

iron oxides, ice crystals, and a host of other small molecules and compounds, including some organic ones.

Light pressure from the stars supposedly causes this matter to pack together to form a cloud about 9600 billion km in diameter. As the cloud packs together it begins to rotate faster, similar to the Kant-Laplace hypothesis. Its rotation continues to increase as it collapses slowly under its own gravity forces until it reaches a diameter of about 5900 million km and then collapses rapidly from there, perhaps in a few hundred years. The increased pressure of the contracting cloud greatly increases the temperature until it reaches about 1 million °C and nuclear burning (or fusion) begins. The Sun begins to radiate as a star and the planets and satellites are derived from minor dust streams in the original cloud before the last stages of collapse occur.

The dust-cloud hypothesis leaves two important features of the solar system unexplained, the spacing of the planets from the Sun and the angular momentum of the Sun and the planets. From research conducted by Fred Hoyle in 1960 and other related studies, the process of magnetic coupling has been presented to explain the angular momentum problem.

Magnetic Coupling

The contribution of magnetic coupling is explained in the following way. Stars are known to have magnetic fields that extend into the surrounding space. During formation of the solar system, the more rapidly rotating Sun dragged the less rapidly rotating disk of dust with it. The linkage between the Sun and rotating dust provided a rotating solar magnetic field. Interaction occurred with the gases whose particles served as tiny magnets. The Sun's angular momentum was transferred to the rotating disk of dust. This increased rotation hurled the gaseous portion to the outer reaches of the solar system where they condensed into the great planets. With this outflow of gases, the remaining particles near the Sun consisted mostly of the heavier materials, iron, silicon, and magnesium oxides. These formed into the small terrestrial planets by collision and gravitational attraction.

Evaluation of the Dust-Cloud Hypothesis

The dust-cloud hypothesis and magnetic coupling still do not explain the spacing of the planets around the Sun (Titius-Bode rule). This is left for others to

FIGURE **1.2**   (a) Sketch of the Milky Way from the side showing the location of the
     solar system.

discover. There is also some question about whether magnetic coupling completely answers the complicated aspects of how the Sun slowed its rotation and how the planets incorporated the additional angular momentum. These details require further study.

A subject introduced some years ago to remedy the weaknesses of the second-body hypothesis provides a point of interest in this discussion. This involved the appearance of two stars (that is, a double star) approaching the Sun, to initiate the solar system. As the theory goes, one of the double stars is destroyed by the pull of the other two stars on it. Debris from the destroyed star would supply materials for the planets around the Sun. Subsequent calculations showed that the debris could go into orbit around the Sun, but would not condense into planets. The hypothesis was discarded along with all the other two-body explanations of the origin of the solar system.

The interesting aspect of this hypothesis is that a double star approaching the Sun is a more likely occurrence than the approach of a single star. About 1 star in 100, such as the Sun, is a single star. Also, it is dynamically unlikely that double stars are capable of holding planets.

In the Milky Way, our galaxy of 10^{11} stars, it is estimated that the number of single stars is 10^9.

Considering the needs of heat supply, light, size, age, and so forth, the stars with planets similar to the Earth would range from 1 in 1000 to 1 in 1 million. These statistics suggest that in the Milky Way alone, there may be 1000 to 1 million planets on which humanlike forms could exist. In addition, there are about 1 billion galaxies in the universe.

THE MILKY WAY

The Milky Way is a spiral-shaped galaxy with two spiral arms extending from its central mass. The diameter of the nebula is about 100,000 light-years and the Sun is located about 3/5 the distance from the center (30,000 light-years) in one of the spiral arms. If viewed along its central plane, the galaxy would appear as a flattened disc with a central bulge. A sketch of the galaxy is shown in Figure 1.2(a). The central bulge is about 30,000 light-years thick. In the spiral arms, that dimension reduces to about 10,000 light-years. A photograph of a spiral galaxy found in Ursa Major (the Big Dipper) and located about 4.2 million light-years from Earth, is shown in Figure 1.2(b).

The nearest stars to the Sun are in the Proxima Centauri, which can be observed in the southern

FIGURE **1.2**   (b) Photograph of a spiral galaxy in Ursa Major, distance about 4.2 million light-years. Kitts Peak National Observatory, photo courtesy of National Optical Astronomy Observatory (NOAO).

hemisphere. They are about 4.3 light-years away, which is the average distance between stars in the Milky Way. Stars in the galaxy rotate slowly about the central mass, and it takes the Sun and its planets about 200 million years to complete one rotation.

On a clear night, the Milky Way can be seen in a band that stretches across the sky. During the summer our view from Earth is toward the star Sagittarius and the center of the galaxy. In the winter, the view is away from the center and toward the outer portions, along the spiral arm of the Milky Way.

A generally accepted view today holds that the lighter elements were formed by the Big Bang, which initiated the expanding universe, and the heaviest elements, such as uranium, were synthesized in a supernova when it imploded to form a neutron star, blasting its remains into space. Nearly 80 bright supernovas have been observed by telescope beyond the Milky Way since 1885. The abundance and range of heavy elements and related isotopes found on Earth suggests that the matter comprising our solar system has been through at least one supernovation.

The Sun, then, is apparently a second- or third-generation star.

THE LIKELIHOOD OF LIFE ON OTHER PLANETS

As previously stated, a single as opposed to a double star represents only about 1% of the total population. Although only single stars can hold planets in orbit, in the Milky Way there are from 1000 to 1 million planets on which humanlike forms could exist.

Despite the apparent abundance of planets that could conceivably support life, we must remember that distances in space are extremely vast. The minimum distance between such planetary systems is measured in distances of tens of light-years or in units of billions of kilometers or miles. These distances are obviously much beyond the capability of human travel in the twentieth century.

A number of conditions are necessary before a planet can support life forms as advanced as humans. The primary requirements include the proper temperature, amount of light, gravity, atmospheric composition and pressure, and water. Other requirements, perhaps of lesser significance, involve other life forms present, wind velocity, dust, and radioactivity. Some other conditions that might make a planet uninhabitable are excessive meteorite bombardment, extensive volcanic eruptions, high frequency of earthquakes, and possibly an extreme level of electrical activity (lightning). Consequently, planets must meet some major requirements in order for them to be inhabitable by humans:

Mass of planet: Must be greater than 0.4 earth mass[3] to produce and retain a breathable atmosphere and less than 2.35 earth mass since surface gravity must be less than 1.5g.

Period of rotation: Must be less than about 96 hours (4 earth days) to prevent excessively high daytime temperatures and excessively low nighttime temperatures.

Age of the planet and star about which it orbits: Must be greater than 3 billion years to allow for appearance of complex life forms and the production of a breathable atmosphere.

Axial inclination or inclination of equator to the plane of orbit and level of illumination from its sun: These determine the temperature patterns on the surface. Illumination at low inclinations should lie between 0.65 and 1.35 times the Earth's norm (between 10 and 20 lumens/cm^2). However, certain combinations of illumination up to 1.9 times the Earth's norm and inclinations up to 81° are compatible under marginal conditions.

Orbital eccentricity: Must be less than about 0.2 because greater amounts of eccentricity produce unacceptably extreme temperature effects on the planetary surface.

Mass of the star: Must be less than 1.43 solar mass,[4] because residence time on the main sequence of stars must be more than 3 billion years. Mass also must be more than 0.72 solar mass because smaller stars yield an incompatibility between acceptable illumination levels and tidal retardation of a planet's rotation. For the rare class of

planets with extremely large or close satellites, the lower range of the star's mass is extended to 0.35 solar mass.

Binary star system: If planets orbit this system, the two stars must be quite close together or very far apart to prevent instability of planetary orbits and not produce a level of illumination on the planet that is too variable.

When all of these requirements are met, there is a very good possibility that the planet will be inhabitable.

There are 14 stars within 22 light-years of the Earth that qualify as promising candidates for having planets that humans could inhabit. The closest are Alpha Centauri A and Alpha Centauri B, which lie within 4.3 light-years. Their combined probability for containing at least one planet inhabitable by humans is 10.7%. For the 14 most promising candidates combined, the probability that at least one planet is inhabitable is about 43%. Unfortunately, these range from a distance of 4.3 to 22 light-years away from Earth, which is far beyond the distance that humans will reach in the foreseeable future.

THE FATE OF LIFE ON EARTH

The ultimate fate of life on Earth is tied to the evolution of the Sun. It is a fairly old star, about 5.5 to 6 billion years in age. It will continue as a normal or main sequence star for about another 2 billion years at which time it will begin to heat up and expand. This is a consequence of having exhausted much of its hydrogen through the fusion process; thus, it will undergo expansion prior to its demise. The temperature will increase tremendously, boiling the oceans on Earth, killing all life. The Sun may even expand sufficiently to engulf the Earth. Following this red giant stage the Sun will cool to form a burned-out star.

Major changes in the configuration of the Earth's surface could occur before the Sun begins to heat up and expand. Radioactive decay in the Earth's interior provides the flow of heat to form mountains and move the continents around on the planet. This is discussed later in the chapter in a discussion on plate tectonics.

Radioactive decay will continue to diminish with time as the radioactive elements are consumed in the reaction process. Eventually, heat flow will diminish,

[3]The mass of the earth $\simeq 6 \times 10^{24}$ kg.
[4]Solar mass is the mass of the Sun = 3.35×10^5 earth mass = 2.01×10^{30} kg.

bringing an end to mountain building and volcanic eruptions. The relentless pounding of the seas against coastal regions over geologic time will flatten the continents, filling the ocean basins with sedimentary debris. The final effect will be a flat continental platform, greatly diminished in size, located at or near sea level. It would cause a major impact to most forms of life on Earth. This is estimated to occur in about 2 billion years.

It was once thought that the Moon had emerged from the Earth well after the onset of planetary development. Moving in an elliptical path away from the Earth, it was suggested that the Moon would eventually return to Earth yielding a catastrophic collision and very likely destroying all advanced forms of life. This was estimated to occur in about 2 billion years.

Studies of lunar rocks have subsequently shown, based on rock and mineral analysis, that the Moon has a somewhat different composition than the Earth. There is no history of intense oxidation or effects of free water in lunar rocks, both of which are so prevalent in rocks on Earth. It is now concluded that a collision between the Earth and Moon will not occur as previously proposed. The currently supported theory on lunar origin is that the Moon separated from the Earth when an extremely large mass impacted the growing earth mass, early during planetary accretion stage, more than 5 billion years ago.

A question sometimes raised is whether life can be rejuvenated on Earth after the Sun cools down from the red giant stage and before it becomes a burned-out star. Based on the history of the Earth it takes about 3 billion years under proper conditions for advanced life forms to develop. Oceans would have to form anew along with a proper atmosphere. It would seem unlikely that all these constraints could be met, but simple forms of life could develop before the Sun cools completely.

CONFIGURATION OF THE EARTH'S SURFACE

Early Developments

The planet Earth was formed by the accretion of planetesimals during the formation of the solar system about 4.7 billion years ago. Composed of silicon compounds (silicates), iron and magnesium oxides, and lesser amounts of other elements, the planet started out as a cold mass. It began to heat up as three mechanisms contributed heat to the system. The energy of motion of the infalling planetesimals was converted to heat and the resulting compression contributed more heat to the planet. These two events accounted for an initial temperature of about 1000 °C within the first million years of Earth's existence.

Melting of the Earth

The third mechanism that contributed to the heating of the Earth was radioactive decay. Uranium, thorium, potassium, and the other radioactive elements eventually contributed enough heat (reaching temperatures of 2000 °C or more) to melt the iron in the Earth, causing the so-called "iron catastrophe" to occur. Iron, being heavier than the other common substances, sank toward the center of the Earth and displaced lighter materials. This marked an event of catastrophic proportions because as the iron migrated toward the center of the Earth, it released large amounts of gravitational energy that were also converted to heat. This additional heat produced a temperature rise on the order of 2000 °C, which melted much of the remaining portion of the Earth.

This differentiation process caused first by the melting of the iron, followed by the additional melting of most other materials, converted a generally homogeneous planet into a zoned body. The resulting configuration consisted of a dense iron core, a surface crust composed of lighter materials that had a lower melting temperature, and an intermediate zone comprising the mantle. The differentiation process also most likely triggered the escape of gases from the interior and eventually led to the formation of the atmosphere and the oceans. Differentiation is thought by many earth scientists to have occurred about 4 billion years ago.

Formation of the Oceans and Atmosphere

The water on the Earth's surface today most likely came from chemically bound hydroxyl groups (OH^- groups) attached to minerals in the crust. Such minerals as the micas and amphiboles contain hydrogen and oxygen linked as hydroxyls. As the Earth warmed, water vapor was carried to the surface dissolved in magma. This outgassing is a consequence of differentiation of the Earth that yielded the zones of different composition.

The volcanic gases consisted mainly of water vapor, carbon dioxide, hydrogen, hydrogen chloride, carbon monoxide, and nitrogen. It is likely that much of the outgassing occurred early in the Earth's history, between 4 and 3.8 billion years ago, and coincided with extensive volcanic flows during that time interval. Periodically since this early time, many volcanic eruptions of short duration have followed this initial, lengthy eruptive phase. The oceans were formed from the early outgassing as was the atmosphere and only small amounts of juvenile or new water have been added to the oceans after the initial period. Most of the water delivered to the atmosphere by volcanic eruptions today is recycled, meteoric water.

The early atmosphere did not contain free oxygen in any of the following three forms: normal molecular atmospheric oxygen (O_2), ozone (O_3), or rare atomic oxygen (O). There are several reasons why this is indicated: 1) No plausible source of free gaseous oxygen for the early atmosphere has been proposed, 2) the composition of the early atmosphere generally agreed on by earth scientists involves gases that would combine with and remove free oxygen, and 3) free oxygen in the early atmosphere would inhibit the origin of life and the fossil record shows that such life existed billions of years ago.

The composition of the early atmosphere was likely made up of carbon monoxide, carbon dioxide, nitrogen, and water vapor. Another hypothesis, no longer seriously considered, suggests that the early atmosphere was primarily methane and ammonia. Geologic evidence does not support the presence of a methane-ammonia–rich atmosphere.

The oldest known sedimentary rocks are 3.8 billion years of age, but those greater than 2.6 billion years old contain relatively small amounts of limestone and are, instead, rich in chemically precipitated silica. An ammonia-rich atmosphere would have favored the deposition of limestone (or dolomite), while greatly limiting the precipitation of silica. Therefore, if an initial methane-ammonia atmosphere existed, it had evolved into an atmosphere dominated by carbon monoxide, carbon dioxide, nitrogen, and water vapor by about 3.8 billion years ago.

These oldest sediments also supply additional information about the past. The small amounts of dolomite found in these rocks contain unoxidized iron compounds, further supporting the position that little or no free oxygen existed in the atmosphere at the time. Abundant Precambrian iron stones suggest a reducing atmosphere, free from oxygen. Also, the presence of dolomite, any at all, suggests a salty ocean, because precipitation of dolomite requires a saline solution. The sea was certainly fully salty by 2 billion years ago as evidenced by extensive formations of dolomite of that age in southern Africa.

A third item of information shown by the sedimentary column is the continuous existence of liquid water on the Earth for the last 3.8 billion years. This also suggests surface temperatures between the freezing and boiling point of water, which in turn indicates that either 1) the Sun supplied an extensive amount of heat even at an early time, 2) the Earth was able to retain heat at that level, or 3) loss of heat from the interior kept the temperature in balance.

The production and accumulation of free oxygen apparently came only after life had evolved to at least the level of green algae. It and higher forms of plant life use sunlight to convert carbon dioxide and water into organic matter and oxygen, through photosynthesis. Not until the oxygen produced had satisfied the demands of other atmospheric gases and combined with them, did free oxygen begin to accumulate in the atmosphere. This would also remove any remnants of methane or ammonia that were still present in the atmosphere.

Much of the carbon monoxide and carbon dioxide from the early atmosphere became locked in the carbonate rocks of the geologic column, that is, in the limestones and dolomites. Some hydrogen, carbon dioxide, oxygen, and other elements formed coal and petroleum, which are also stored in the Earth's crust. The small amount of CO_2 still in the atmosphere (a few hundredths of 1%) is quite important because of its role in photosynthesis.

ARCHITECTURE OF THE EARTH'S SURFACE TODAY

Many schoolchildren, while studying the globe or maps of the Earth, observe that the continents would fit together if they were merely moved around. South

America could be tucked under Africa, North America against Eurasia, with Antarctica and Australia drawn up from below to fit under the others. This very idea was proposed scientifically by the German meteorologist Alfred Wegener in 1915. Many scientists in the United States scoffed at his proposal of continental drift and it was generally discounted except in other parts of the world, until intensive studies following World War II provided positive evidence of its occurrence. Today the concepts of seafloor spreading and plate tectonics are firmly established as the likely mechanisms that have continued to move the continents apart during the past 200 million years.

Prior to the accumulation of this geophysical evidence, only similarities between rock types and ages of rocks (including fossil evidence) across the continental boundaries, which previously would have been in contact, were advanced in support of continental drift. This evidence was not convincing enough to change the status quo and no consensus favoring Wegener's proposal ever developed.

Seafloor Spreading

The spreading of the seafloor from midocean ridges has been established by oceanographic studies and age determinations of rock. Seafloor geophysics has also provided convincing evidence. Variations in the direction of the magnetic field for the basaltic rocks on either side of the spreading ridges indicate the strong symmetry of the units moving away from the center. During the period of seafloor spreading, a number of reversals in polarity of the Earth's magnetic field have occurred; that is, the pole attracting the compass needle has switched directions from north to south a number of times during the 200 million years of spreading. Sequences of magnetic variations on opposite sides of the spreading ridge match each other like opposite fingers on the left and right hands. This symmetry across the spreading ridge is illustrated in Figure 1.3.

The rates of spreading on either side of the ridge range from 1 to 6 cm/year. This suggests a total spreading rate of from 2 to 12 cm/year. At this rate it would have taken between 36 and 288 million years for the present Atlantic Ocean to attain its present width. This is generally consistent with the 200 million year estimate for the duration of seafloor spreading.

An interesting feature about the midocean spreading ridges is that they are sites for both volcanic and

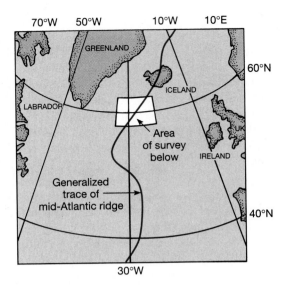

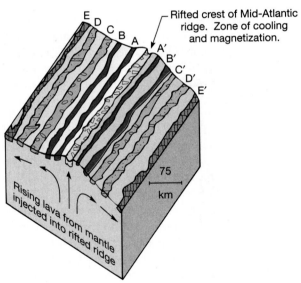

FIGURE 1.3 Seafloor spreading along the Mid-Atlantic ridge south of Iceland.

earthquake activity. Iceland, for example, with its well-known volcanic and thermal activities, sits exactly on the crest of the Mid-Atlantic ridge. Other spreading centers in the oceans that are also locations for earthquakes include the east Pacific ridge (or rise), the mid–Indian Ocean ridge and the ridges or rises that encircle Antarctica. These can be observed in Figure 1.4. The earthquakes that occur along these spreading centers are mostly of shallow depth and relatively small magnitude.

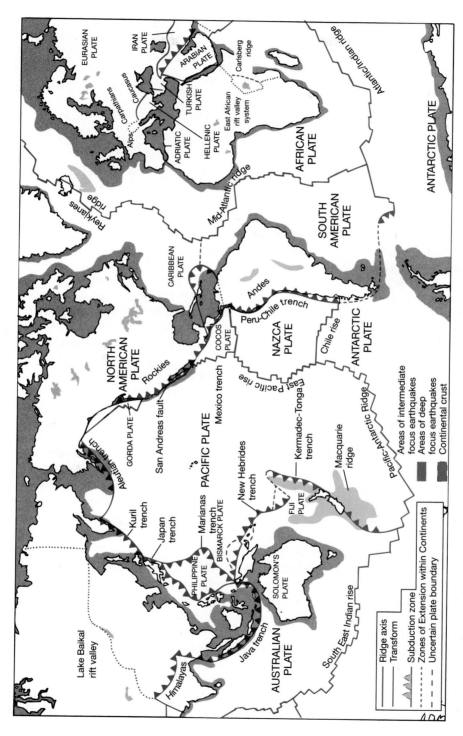

FIGURE 1.4 Plates and plate boundaries of the Earth.

Because these are locations where heat is welling up from the interior of the Earth, the rocks below are too hot and plastic to accumulate high levels of stress. Therefore, only smaller magnitude earthquakes occur along these ridges.

Plate Tectonics

Plate tectonics is a unifying theory that answers many questions about the Earth. Included are such aspects as varieties and distribution of rocks, history of sedimentary rock sequences, position and nature of volcanoes, earthquake belts, mountain systems, deep-sea trenches and ocean basins.

It is known that the surfaces along which earthquake foci align are the boundaries of blocks that move independently of each other. These blocks are referred to as *plates* and their movement as *plate tectonics.* Plates are typically thousands of kilometers across but only 100 to 200 km thick (60 to 120 miles). Both oceanic and continental crust can cap these plates, so they are sometimes referred to individually as *continental plates* and *oceanic plates.* The continent is actually imbedded in the moving plate and is carried passively by it. In fact, many of the plates extend well beyond the continental outline itself; for example, the African plate (Figure 1.4) is nearly twice the area of the African continent. Hence, it is about half continental crust and half oceanic crust at the surface. The Pacific plate by contrast is nearly all oceanic material except for a sliver of California along the San Andreas fault. The Eurasia plate is really two ancient plates now joined or sutured together along the Ural Mountain and it includes both continental areas plus much of the North Atlantic and Arctic oceans.

The plates are rigid slabs consisting of a continental and/or oceanic crustal cap plus part of the underlying mantle. The combined crust and upper mantle is referred to as the *lithosphere.* These lithospheric plates ride on a weak, plastic zone below known as the *asthenosphere.* A cross section depicting these layers is shown in Figure 1.5.

The Earth's surface today is divided into about eight large rigid plates and a dozen lesser ones (Figure 1.4). All the crustal plates seem to be moving relative to each other except for Africa, which apparently has remained relatively fixed in position for several tens of million of years. Well inside the boundaries of some plates are located chains of volcanoes, such as the Hawaiian Islands situated near

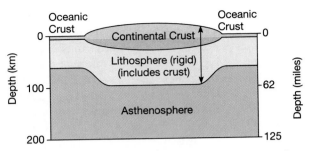

FIGURE 1.5 Cross section of the upper portion of the Earth, showing the continental crust, oceanic crust, lithosphere, and partially molten asthenosphere.

the center of the Pacific plate. These are thought to mark the locations of fixed hot spots beneath the moving plates and can be used to show rates and directions of movement.

There are three basic types of plate boundaries or margins: 1) the zones of divergence or spreading, that is, the typical ocean ridges; 2) the transform margins where plates slide sideways past each other; and 3) the zones of convergences where the plates move directly toward each other. The deep-seated earthquakes of the world occur mostly along the zones of convergence and to a lesser extent along the transform margins.

The ocean spreading ridges are also known as *accretional margins.* New basaltic ocean crust is formed where heat and volcanic flows well up from the interior of the Earth. The Mid-Atlantic ridge, a most prominent spreading center, is also the site for volcanic eruptions and shallow focused earthquakes.

Transform margins or transform fault zones occur where the plates slide sideways past each other. The San Andreas fault is the boundary where the Pacific plate is moving northward relative to the southward-moving North American plate. Another transform fault lies between the North American plate and the Caribbean plate.

The zones of convergence are essentially of two types, consuming margins and collisional margins. At the consuming margins, the dense oceanic rock dives either beneath the lighter continental crust or beneath a somewhat thinner oceanic crust as in deep-sea trenches. The descending plate is consumed by warming and melting at depth. This melting produces molten rock material, which ascends and provides volcanoes and igneous intrusions above the leading edge of the descending plate. Along the North American coast from northern California to British Columbia the Pacific plate is plunging under the North American plate.

Where moving oceanic plates of normal thickness underride the thinner oceanic sequences adjacent to continents, arcuate, deep-sea trenches with paralleling island arcs develop. These island arcs are prominent around the northern and western Pacific and Indonesia. Some of the Caribbean Islands are also examples.

In situations where the moving continent overrides the seafloor, the deep-sea trenches formed have a straight alignment, and linear chains of volcanoes rise through the continents as in the Andes of South America. Not only is the ocean floor consumed in such cases along with scraped-off islands and seamounts, but also pieces of the adjacent continent are pushed into the trenches and are subsequently carried beneath the advancing continent along with the oceanic rocks. These down-thrusted rocks are eventually carried to great depths to mix with molten rocks of continental origin and to rise eventually as melts of intermediate composition. Ore minerals may be associated with this volcanism in the continental margins because this process concentrates the metals that were finely dispersed throughout the deep-sea sediments and continental crust.

Collisional margins are the final type of plate convergence. Where oceans are drastically narrowed or completely closed by overriding and converging continents, the relatively light, high-standing continental crust is compressed and folded to produce the linear trending mountain ranges such as the Himalayas, the Alps, and the Appalachians. The latter were formed when the ancestral Atlantic Ocean closed about 320 million years ago when the African plate collided with the North American plate. Sediments that had accumulated on the margins of the North American continent were compressed, yielding the folded Appalachians and associated structural features. The Atlantic Ocean reopened when the two plates began moving apart some 200 million years ago.

Plate tectonics provides an explanation for many aspects of a global significance. These include volcanoes, earthquakes, folded mountains, island arcs, deep-sea trenches, continents, and ocean basins, in addition to the youthfulness of the oceanic rocks and the great age of their continental counterparts.

The zoned layering of the Earth formed early in its history has provided the necessary life-giving oceans and atmosphere. The Earth's internal heat fueled by radioactive decay provides the driving mechanism to propel the rigid plates across the plastic asthenosphere. The Earth is a dynamic body undergoing change. In subsequent chapters, we concentrate on the materials that compose the Earth and on the processes that shape it today.

EXERCISES ON THE ORIGIN AND DEVELOPMENT OF THE EARTH

1. What are the dimensions of the universe in light-years, kilometers, and miles? By what means is this determined?

2. How are galaxies and constellations related to the subdivision of the universe? Approximately how many galaxies are there? Name the galaxy nearest to the Earth. What is its distance?

3. Name our own galaxy. About how many stars does it contain? Based on this, approximately how many stars exist in the universe?

4. What are the two primary theories on the origin of the universe? Which one is favored today and why?

5. Is the Sun the same as, smaller, or larger than an average star? What will likely happen to it in the future? Explain.

6. How are black holes formed? Is it likely that they are extremely common? Explain.

7. On the origin of the solar system, what two groups of hypotheses have been proposed? Which is favored today? Why is it so difficult to account for the relationship of angular momentum of the Sun and planets? How is it best explained?

8. What is the Titius-Bode law? It can be written as $d_p = 0.4 + 0.3(2)^x$ where $x = -\infty$ for $p = 1$ (mercury) and $x = (p - 2)$ for $p \geq 2$, where p is the number of the planet achieved by counting outward from the Sun and d is the distance of the planet from the Sun in astronomical units. Based on this equation complete the table on page 16. Note that the percent error is great for Pluto. Explain.

9. How far away is Proxima Centauri, the closest star to the Sun? If a spaceship traveled at twice the escape velocity of the Earth[5] how long would it take to reach this star?

[5]The Earth's escape velocity = 11 km/sec or 25,000 miles/hr.

Planet	Actual Mean Distance from Sun (km × 10^7)	Actual Mean Distance from Sun (AU)	Distance by Titius-Bode Law (AU)	Percent Error
Mercury	5.85			
Venus	10.80			
Earth	15.00	1.00	1.0	0
Mars	22.80			
Asteroids	33.0–50.0			
Jupiter	78.00			
Saturn	143.1			
Uranus	288.0			
Neptune	459			
Pluto	591			

10. What are the limitations concerning gravitational attraction of a planet if it is to be hospitable to humans? What is meant by the statement that the combined probability of finding a habitable planet in the Centauri A and B stars is 0.107? Also within 22 light-years of the solar system the probability of finding a habitable planet is 0.43. What is meant by this statement?

11. At speeds of 1/10 the speed of light, how long would it take to travel to Centauri A or B? How far could humans travel at this velocity in a 20-year round-trip mission?

12. It is likely that oceans could be observed on planets at 7 AU whereas forests could be seen at 3 AU from an approaching space rocket. How many years travel away from the surface would this be at 1/10 the speed of light and at 25,000 miles/hr? Large cities may be visible from 5 million miles away. How many astronomical units does that equal?

13. Why is the melting of the Earth to form a zoned interior so important in the development of the Earth as we know it?

14. What is the evidence that the early atmosphere of the Earth contained no free oxygen? Why is the presence of methane and ammonia not likely in the range of 3.8 to 2.6 billion years ago?

15. Why is it likely that less carbon dioxide is present in the Earth's atmosphere today than it was 3 billion years ago? What is the important role of carbon dioxide in the atmosphere today? Explain.

16. Why is it likely that much of the oceans were formed early in the Earth's history? If this is so, what is the origin for all the water released by volcanoes today? Explain.

17. Why was continental drift discounted by most scientists following Wegener's proposal of this in 1915? What evidence had been advanced prior to the post–WWII studies of the ocean basins?

18. By what means is seafloor spreading indicated? How does it relate to plate tectonics?

19. Do all plates consist either of all continental or all oceanic crust? Explain this relationship.

20. What is meant by the asthenosphere and the lithosphere? Draw a sketch, and label it properly.

21. Name the different types of plate boundaries. Give examples of each type of plate boundary based on the specific plates and movements on the Earth. See the appropriate figure in the text for assistance.

22. What are the major aspects of global significance that have been explained by plate tectonics?

23. Could there have been plate tectonics on the Earth without zoning of the interior? Could it have occurred if the Earth had no oceans? Explain.

24. What evidence is there that plate tectonics on Earth had occurred even prior to the onset of the current movement some 200 million years ago? Explain.

Additional Readings

BATES, D.R., 1964, *The Planet Earth*, Pergamon Press, New York.

CLOUD, P., 1978, *Cosmos, Earth and Man*, Yale University Press, New Haven, CT.

DOLE, S.H., 1970, *Habitable Planets for Man*, 2nd ed., American Elsevier Publishing Co., New York.

PASACHOFF, J.M., and KUTNER, M.L., 1978, *University Astronomy*, W.B. Saunders Co., Philadelphia.

PRESS, F., and SIEVER, R., 1986, *The Earth*, 4th ed., W.H. Freeman and Co., New York.

2 SKIM

Minerals

Geology is concerned with the study of the Earth—its surface expression, composition, structure, and internal activity, plus the processes by which it was formed and those that mold it today. This includes the history of the Earth in both the physical and biological sense. In this text, the physical nature of the Earth prevailing at the present time provides most of the material for study.

The term *rock* applies to the solid materials that form the outer rocky shell or crust of the Earth. Three broad groups are considered based on origin: igneous rocks are those that have cooled from a molten state; sedimentary rocks are those deposited from a fluid medium, usually water, and typically as products of weathering of other rocks; and metamor-

phic rocks are formed from preexisting rocks by the action of heat and pressure.

Rocks typically are assemblages of one or more minerals. The mineral composition and the rock texture, or nature and pattern of the assemblage, are useful descriptive features that aid in rock identification. For this reason, the study of minerals is the proper place to begin our discussion of Earth's materials.

OCCURRENCE OF MINERALS

Minerals are the naturally occurring elements or chemical compounds that comprise the soil and rock materials of the Earth. They are found in all geologic

FIGURE 2.1 Calcite, exhibiting rhombohedral cleavage. Note the incipient cleavage planes inside the individual calcite samples.

environments including the alluvial sands along a river bed, soils of a plowed field, and bedrock exposures in a mountainous region. To appreciate the engineering significance and properties of rocks and soils, a knowledge of minerals plus an ability to identify the most common examples are necessary requirements.

Although more than 2000 minerals occur in nature, only about 100 are common. These 100 plus minerals form the basis for study in a typical college course in mineralogy. Included in that collection are metallic and nonmetallic ore minerals and the less common members of the carbonate, sulfate, silicate, and other fundament mineral groups. Calcite, a common carbonate mineral, is illustrated in Figure 2.1.

Despite the hundreds of known minerals, only 25 or so make up the common, rock-forming varieties. These abundant minerals provide the primary influence on engineering constructions, whether founded on soil or on bedrock. Thus, for many engineering projects, a knowledge of these 25 minerals will supply sufficient mineralogical detail. These few minerals can usually be identified by careful study of their physical properties without sophisticated equipment. Also, the ability to identify these minerals can be

achieved to a reasonable degree during a single, well-supervised laboratory session.

In an exercise on mineral identification, evaluations of the diagnostic physical and chemical properties are used to determine the appropriate mineral name with the aid of a mineral identification chart. With some practice, the key physical properties of the common minerals are easily remembered so that identification is soon possible without the table of reference. In a beginning exercise, large specimens are provided that depict the physical properties of the minerals in a prominent fashion. Once physical properties are recognized in these select samples, identification of typical specimens with less prominent features can be accomplished.

APPLICATION TO ENGINEERING

Mineral identification can be of great importance to the field engineer. Recognition of gypsum in a limestone that lies along a proposed tunnel centerline will provide information on possible problems of swelling in the presence of water and subsequent deterioration of the final concrete lining. The presence of pyrite in a dark shale can suggest ultimate problems of deterioration from acidic waters. Swelling clays in a shale can provide a warning that slope stability may be a problem or a heaving condition may develop on wetting of the clay. Associated problems with various mineral substances may be fully appreciated by the engineer, but first the troublesome mineral must be recognized if the problem is to be predicted.

The usual progression of study proceeds from mineral to rock identification. Because most rocks are assemblages of one or more minerals, rock identification evolves directly from mineral study. With addition of the descriptive feature, texture, the procedure for rock identification is established. Rocks are classified first into geologic groups based on origin—igneous, sedimentary, and metamorphic—and then supplied with a specific rock name from within that group. Mineral identification is an important step in classifying rocks and therefore a knowledge of minerals is a necessity. Engineering classifications of rock are typically based on the geologic rock name and on several, significant strength properties of the rock. Mineral identification leads to rock identification, then subsequent modifiers can be added to provide engineering importance.

FIGURE 2.2 Cluster of quartz crystals.

MINERAL DEFINITION

A mineral is formally defined as a naturally occurring chemical element or compound, formed by inorganic processes, with an ordered internal arrangement or pattern for its atoms, that possesses a definite, chemical composition or range of composition. This is a restrictive definition; oil and coal do not qualify nor does volcanic glass (which lacks the ordered internal structure) or manufactured glass. Blast furnace slag, a by-product of steel making, which is commonly used as road base material and as concrete aggregate, does not qualify because it is glassy and not naturally occurring.

An ordered internal arrangement of atoms in a mineral is known as the mineral structure or crystalline structure. In contrast, a nonordered internal arrangement is termed *amorphous,* or without form, and it occurs in liquids or supercooled liquids such as glass. Significantly, minerals are crystalline and glass is amorphous. Therefore, each mineral has a specific composition, or specific range of composition, plus a specific crystalline structure. The structure refers to a specific repetition of the atomic array, that is, a repeating relationship of constituent atoms comprising the structure. Thus a mineral name such as quartz entails not only the composition SiO_2 but also the specific atomic array of silicon and oxygen that comprises the quartz structure. Calcite, in the same manner, involves the composition $CaCO_3$ but also the calcite structure in which calcium and the carbonate radical are arranged in a specific fashion. A cluster of quartz crystals is shown in Figure 2.2.

Some minerals have identical compositions but different crystalline structure. Pyrite and marcasite, both FeS_2, are examples of this and both qualify as separate minerals. In pyrite, the iron atoms are equally spaced in all directions, but in marcasite their spacing differs depending on crystallographic direction. Such pairs of minerals are called *polymorphs.* Diamond and graphite (C), calcite and aragonite ($CaCO_3$), and quartz and cristobolite (SiO_2) are other common examples.

MINERAL FORMATION

When minerals solidify from the liquid state or form in other ways, they yield an internally ordered, solid material. This process, termed *crystallization,* occurs when many crystals form independently, increase in size, and finally grow together to give a mosaic pattern of interlocking crystals.

Crystallization occurs in nature by solidification of silicate melts on cooling, by precipitation of crystals from a water solution, by sublimation from the vapor phase, and from other solids (recrystallization) during the alteration and metamorphic processes. Silicate melts crystallize when the temperature is lowered to the point of fusion (known as freezing in the case of water) and solid crystals begin to form. Ice crystals forming in water under freezing conditions are common examples of crystallization by fusion. Magma, or molten rock within the Earth's crust, begins to crystallize when its temperature falls below the melting point of certain common silicate minerals. This process of silicate crystallization is more complicated than that of ice from water because one phase or mineral will crystallize out of the melt at a certain temperature and subsequently react with the remaining liquid as cooling proceeds.

Precipitation from solution can occur after evaporation of some or all of the liquid. A reduction in the liquid volume yields less of the dissolving agent to hold the ions in solution, and precipitation occurs once saturation is reached. A reduction in temperature of the liquid may also yield a reduction in solubility of the constituents, thereby triggering precipitation. A change in pH (hydrogen ion concentration) of the liquid may also reduce solubility and cause certain solids to precipitate. Also, a reduction of pressure in the liquid, such as when water flows onto the Earth's surface from within a rock mass, may trigger the precipitation of certain minerals. Biological activity along with chemical reactions between a mineralizing solution and the host rock (as in ore deposit formation) also cause minerals to precipitate from a fluid medium.

MINERAL IDENTIFICATION

The properties of minerals used for identification are dictated by their composition and structure, that is, the elemental building blocks and the way they are assembled. Chemical properties aid in some determinations, but by and large for common rock-forming minerals, physical properties are used for identifica-

tion. Physical properties can be recognized either on sight or after applying several simple tests. These properties include color, streak, luster, hardness, specific gravity, cleavage, fracture, crystal form, magnetism, tenacity, diaphaneity, presence of striations, and reaction to acid.

Color as a distinguishing feature of minerals can either be deceptive, nondefinitive, or in limited situations quite diagnostic. Some minerals are found in several different colors, so color identification for these should be used cautiously. For other minerals, such as olivine, color is a diagnostic feature. Therefore, color is always noted in mineral identification but rather than used alone is considered with other properties in the identification process.

The term *streak* refers to the color of the finely powdered mineral. By grinding a mineral to powder and then examining its color (easily accomplished by scratching on an unglazed porcelain surface), a common base is obtained for comparison. Surface texture and irregularities on mineral surfaces can produce colors that are considerably different from its streak. Certain minerals have a distinctive streak. For example, hematite (Fe_2O_3) yields a deep-red streak despite the various colors portrayed in sample specimens.

Luster is the overall appearance of the surface of a mineral in reflected light. Luster is divided into two groups for convenience: metallic and nonmetallic. Minerals that show a high degree of reflection from an opaque surface like that of a metal are appropriately said to have a metallic luster. Nonmetallic luster includes all other lusters that do not show this extreme degree of opaque-surface reflection. The nonmetallic lusters are further subdivided according to their particular appearance. Common descriptions of nonmetallic lusters are:

Adamantine	Brilliant luster displayed by gemstones such as diamond
Vitreous	Glassy luster, like that of broken glass
Pearly	Irridescent luster of a pearl
Greasy	Luster such that the surface appears to be coated with a film of oily liquid
Silky	Luster like silk caused by a fine fibrous appearance such as that of silky gypsum (satin spar)
Earthy	Dull luster like that of dry soil.

TABLE 2.1   Mohs' Scale of Hardness

Minerals	Level of Hardness	Useful Tools
Talc	1	
Gypsum	2	
		Fingernail 2.5
Calcite	3	Copper penny 3
Fluorite	4	
Apatite	5	
		Steel knife blade or glass plate 5.5
Orthoclase	6	
Quartz	7	Unglazed porcelain plate (streak plate) 7
Topaz	8	
Corundum	9	
Diamond	10	

Hardness is the resistance that a mineral shows to being scratched. It is governed by the elements in the mineral and their arrangement plus the strength of the bonds that bind them and resist scratching. Graphite and diamond, two polymorphs of carbon, differ in internal structure and bond strength and hence have greatly different hardness values.

In geologic studies Mohs' hardness scale (Table 2.1) is used to compare mineral hardness. The scale ranges, as shown in Table 2.1, from 1 to 10 with a representative mineral for each integer value. The increments from 1 to 9 are about equal in amount but the difference between 9 and 10 is considerably greater, estimated to be perhaps 30 times larger than the other units.

Several simple tools are used to determine the physical properties of minerals, which leads to mineral identification. Included are those used to test hardness: a steel knife with hardness of about 5.5, a glass plate of a similar hardness, a copper penny with a hardness of 3, and fingernails with a hardness of 2.5 (Table 2.1). Other useful tools are an unglazed porcelain steak plate, with a hardness of 7, used to streak minerals; a small bar magnet; and a 10× hand lens, which is useful when searching for striations or small crystal forms. Added to these items is a small bottle of dilute hydrochloric acid. In determining hardness, an attempt is made to scratch the mineral with the tool

and, in turn, to scratch the tool with the mineral. A comparison of these effects should tell which is harder.

The *specific gravity* of a mineral is a dimensionless number expressing the ratio between the mass (or weight) of the mineral and that of an equal volume of water. Specific gravity (SpG) is a function of the chemical composition and packing as 1) elements with high atomic weight increase the specific gravity, and 2) differences in mineral structure, because the closer the elements are packed together, the higher the specific gravity.

Specific gravity is measured with the following equation:

$$SpG = \frac{\text{Mass (or weight) of mineral in air}}{\text{Mass (or weight) of equal volume in water}}$$

$$= \frac{\text{Mass (or weight) of mineral in air}}{\text{Mass (or weight) in air} - \text{Mass (or weight) in water}}$$

This is obtained by weighing the mineral in air and then in water. The difference between the two is the buoyancy force of the water, and it equals the mass of water with the same volume as that of the mineral specimen.

The common rock-forming minerals (quartz, feldspar, calcite, etc.) as a group have a specific gravity near 2.7, which forces the average for common rocks toward this value. Metallic minerals, notably the sulfides, have specific gravities higher than 2.7, more nearly 5.5 or greater. Mercury has a specific gravity of 13.6 and pure gold of 19.3.

Specific gravity is used in a relative sense as a simple physical test. By working with a piece of mineral about the same size each time, for example, about the size of a sugar cube, the heft can be compared to a similar size piece of quartz or feldspar. Low values (around 2.3), average values (nearly 2.7 as in quartz or feldspar), or high values can be indicated by this simple comparison.

Specific gravity of a mineral or other material, multiplied by the unit weight of water (or the mass density of water), yields the unit weight of the material (or the mass density of the material, respectively). The unit weight of water is 62.4 lbs/ft^3 and its mass density is 1 g/cm^3 or 1000 kg/m^3. Under the international system for measurements (the SI system) mass density in kg/m^3 is used.

As an example, a solid piece of calcite, SpG 2.72, would weigh 2.72(62.4) = 169.7 lbs/ft^3 and would have

FIGURE 2.3 Biotite, showing one plane of cleavage. (Hamblin, WK, *The Earth's Dynamic System.* 2nd Ed., Macmillan, New York.)

a mass density of 2.72(1000) = 2720 kg/m³. In the cgs system, still used in some laboratory studies, the mass density would be 2.72(1 g/cm³) = 2.72 g/cm³.

Cleavage is the ability of a mineral to break along smooth parallel planes. This property is also due to the arrangement of the atoms and the types of bonds between them. Cleavage in a mineral indicates a direction of preferred weakness in bond strength between atoms. As the structure or atomic arrangement is repeated in space, this weakness will be repeated as well, so that cleavage occurs in families of planes all parallel to each other.

Cleavage is described according to its degree of development as good, moderate, or poor depending on how readily the parallel breaks occur. It is also described according to how many planes of cleavage with different orientation occur, or by the shape of the geometric form enclosed by three or more planes. One plane of cleavage or two planes at right angles are common descriptions used for cleavage. Regarding geometric forms, table salt, or halite (NaCl), has three planes of cleavage at right angles, which is termed *cubic cleavage* because of the resulting cube-shape form. Calcite exhibits cleavage in three directions, but not at right angles, which yields a rhombohedral shape known appropriately as rhombohedral cleavage. This is illustrated in Figure 2.1. Biotite, a type of mica, has a single prominent plane of cleavage, as shown in Figure 2.3.

A *fracture* is an irregular break that occurs in the absence of cleavage and is the usual consequence when minerals are broken. Some types of fracture can be used for identification purposes. Conchoidal frac-

ture, or fracture that yields curved surfaces, is particularly useful in recognizing quartz. Fibrous fracture and hackly fracture, derived by breaking along irregular, sharp edges, can be used for identification as well.

Crystal form involves the display of well-formed crystal faces by a mineral. Crystal faces are an external manifestation of the internal arrangement of atoms. Some minerals exhibit their crystal forms more readily than others. Among the common rock-forming minerals, quartz and garnet exhibit crystals most frequently. Crystal forms should be distinguished from cleavage faces. Crystal faces form during the crystallization process, not when the mineral piece breaks, so crystal faces will be destroyed if the specimen is broken. Cleavage faces can be obtained again and again as parallel faces break off when the mineral is struck. Incipient parallel breaks can commonly be seen within the cleavage pieces, indicating a readiness to form these new faces (Figure 2.1). Crystal forms, however, lack such parallel faces within the specimen that would emerge on breaking. This is why crystals are valued more highly as specimens than are cleavage fragments, because they are a unique form in themselves, which will be lost if breakage occurs.

When a specific mineral grows freely in an open space it takes on its own characteristic shape because the angles between adjacent crystal faces are constant for similar crystals of that mineral. The faces may be different in size owing to different growth rates in two directions, thus yielding a deformed crystal, but the interfacial angles will still remain the same. Crystal faces in minerals occur along the directions that intersect the maximum number of atoms and thus are directly related to the atomic array.

For convenience, the faces are defined in reference to crystallographic axes, three or four in number, which are imaginary lines intersecting at the center of the crystal. Minerals are classified into six crystal systems based on the symmetry demonstrated by their crystal forms. Planes of symmetry and axes of symmetry are used to describe these symmetry relationships. A plane of symmetry divides an object into two halves, each the mirror image of the other. These symmetry planes also contain one or more crystallographic axes.

An axis of symmetry is a line through the crystal about which rotation yields a repeat of the same pattern or crystal face. (For example, a square has a

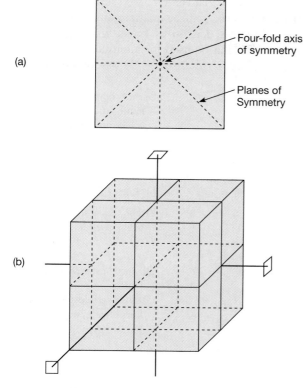

(a)

Four-fold axis
of symmetry

Planes of
Symmetry

(b)

FIGURE 2.4 Planes of symmetry and axes of symmetry for a square and a cube. (a) Square showing basic symmetry elements. (b) Cube showing basic symmetry elements.

fourfold axis of symmetry through its center, it also has four planes of symmetry.) These details are illustrated in Figure 2.4. In Figure 2.4(a) rotation about the axis each 90° yields the next side of the square. Four turns of 90° are needed to return to the beginning position, and this yields a fourfold axis of symmetry or of rotation. A cube is shown in Figure 2.4(b). It has three axes of fourfold symmetry, one through each of the opposing faces. They are each marked with a four-sided form. The cube also has nine planes of symmetry, the three shown plus two diagonal planes each across the three pairs of faces as illustrated in Figure 2.4(a).

The cube demonstrates the highest degree of symmetry according to these symmetry elements. In all, 32 different classes of symmetry are possible. Based on the ways these classes can be grouped

relative to the crystallographic axes, they are collected into the six crystal systems. Information about the crystal systems is given in Table 2.2.

The crystallographic axes for the six crystal systems are shown in Figure 2.5. A number of crystal forms associated with various minerals are also illustrated. Note for beryl that an *a* axis can be placed through the center of (perpendicular to) each vertical face with the *c* axis vertical or through the horizontal faces at the bottom and top, yielding the basic hexagonal system for reference. This method of relating crystal faces to the basic crystallographic axes, that is, recognizing where the axes and planes of symmetry occur, is the way in which crystals are assigned to crystal classes. This overall subject, known as *crystallography*, is typically included as part of a first course in mineralogy.

Magnetism is the attraction of a mineral to a magnet. The most notably magnetic mineral is magnetite, which is strongly attracted even to a simple bar magnet. Several other less magnetic minerals exist in nature, for example, pyrrhotite ($Fe_{1-x}S$), but they are less common than magnetite.

Tenacity is the resistance a mineral shows to various destructive mechanisms such as crushing, breaking, bending, tearing, and so forth. Minerals can be described according to the following terms.

Brittle	Breaks or shatters easily.
Malleable	Can be hammered into thin sheets.
Sectile	Can be cut into thin shavings with a knife.
Ductile	Can be drawn into a thin wire.
Elastic	Bends but resumes its original shape when released.
Flexible	Bends but does not resume its original shape on release.

Diaphaneity is the ability of a mineral to transmit light. The terms used to express diaphaneity are:

Transparent	A clear outline of an object can be seen through the mineral.
Translucent	Light is transmitted through the mineral but objects cannot be discerned.
Opaque	No light is transmitted through the mineral even on its thinnest edges.

TABLE 2.2   Details on Crystal Systems of Minerals

System	Axes	Planes of Symmetry (Max.)	Mineral Examples°
Cubic (or isometric)	Three equal axes at right angles to one another	9	Garnet, fluorite, diamond, halite (rocksalt), galena, sphalerite, pyrite, magnetite
Hexagonal	Four axes: three equal and horizontal and spaced at 60° intervals; one vertical axis	7	Beryl, apatite, tourmaline, calcite, quartz
Tetragonal	Three axes at right angles: two equal and horizontal, one vertical axis longer or shorter than the others	5	Zircon, cassiterite, chalcopyrite
Orthorhombic	Three axes at right angles, all unequal	3	Olivine, enstatite, topaz, barite
Monoclinic	Three unequal axes: the vertical axis (*c*) and one horizontal axis (*b*) at right angles, the third axis (*a*) inclined in the plane normal to *b*	1	Orthoclase feldspar, hornblende, augite, biotite, gypsum
Triclinic	Three unequal axes, no two at right angles	None	Plagioclase feldspar, turquoise

°Also see mineral descriptions in Appendix A.

Striations are fine parallel lines that appear to be scribed onto cleavage faces of the feldspar, plagioclase. They occur because of twinning of the feldspar crystal, that is, the reversal of direction of component parts of the crystal with respect to each other. Striations are useful in telling orthoclase from plagioclase because orthoclase lacks striations whereas plagioclase typically shows them.

Reaction to acid is another test. Although not a physical property, effervescence in acid is used in conjunction with the other tests to distinguish minerals. Calcite is easily identified by the application of HCl because it effervesces freely. Dolomite reacts much more slowly and this feature forms a basis for distinguishing those two carbonates.

The table for identifying the common rock-forming minerals that are most likely to be encountered during engineering investigation and construction is presented in Appendix A, Table A.1. Students should become very familiar with the most common

minerals because that knowledge forms the basis for rock identification. Table A.3 in Appendix A contains the most common minerals of economic significance. Minerals from this second group are not likely to occur commonly in rocks encountered in most engineering studies but knowledge of them may prove of interest to many engineers. They do illustrate well the range of physical properties discussed previously. If additional knowledge of minerals is desired, a course in mineralogy is recommended for engineers, particularly those specializing in aggregates for construction and in highway materials.

ROCK-FORMING MINERALS

Silicates

Silicates, as the most important group of rock-forming minerals, comprise the majority of the common minerals. In the classification used in mineralogy, subgroups of the silicates are developed on the basis of

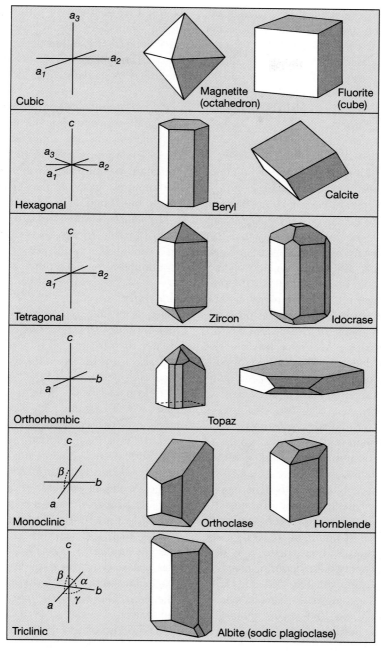

FIGURE **2.5**   Mineral crystal systems.

their silicon-to-oxygen ratio, which controls the crystal structure. This ratio strongly influences the properties and weathering characteristics of silicates. As an example, olivine has a ratio of 1 to 4 (silicon to oxygen), whereas for quartz the ratio is 1 to 2. These two minerals are vastly different; olivine is a high-temperature ferromagnesian that weathers rapidly, whereas quartz is a low-temperature mineral highly resistant to weathering. The subgroups of silicates are not detailed here because that task is beyond the scope of a basic applied geology text. It is significant to note that the silicon-to-oxygen ratio of silicates drastically affects their behavior.

Another means for classifying silicates is to subdivide according to the presence of certain prominent elements. This yields the following grouping: ferromagnesians and nonferromagnesians (feldspars, quartz, muscovite, and garnet).

Ferromagnesians

Ferromagnesians are an important group of common silicates and include those that contain iron and magnesium and in general are dark in color. Because iron and magnesium are heavy ions, ferromagnesians tend to have a slightly higher specific gravity than 2.7, the average specific gravity of quartz, feldspars, and calcite. Ferromagnesians collectively include olivine, pyroxenes, amphiboles, and biotite.

Olivine is a light-green-colored mineral with a characteristic sugary texture caused by the presence of anhedral or rounded grains. Its hardness of 6.5 to 7 is rather deceptive because the grains break easily, suggesting a lower hardness. The chemical composition is $(MgFe)_2SiO_4$ and the specific gravity can range from 3.2 to 4.2—as the iron content increases, so does the specific gravity. The diagnostic features of olivine are its green color and sugary texture.

Augite is the common member of a group of silicates known as pyroxenes. It is a dark green or black mineral that exhibits two planes of cleavage, nearly at right angles to each other. This right angular cleavage and dark color are diagnostic of augite. In rocks augite tends to appear as short stubby crystals.

Hornblende is the most common mineral of the amphibole group of silicates. Like augite, it is dark green to black but displays two directions or planes of cleavage at angles of approximately 56° and 124°. This difference in cleavage angles distinguishes hornblende from augite (56° and 124° versus about 90°, respectively). In rocks, hornblende tends to appear more needle shaped than does augite. Idocrase, shown in Figure 3.5, is a complex silicate mineral, usually brown in color, that forms in limestones during contact metamorphosis.

Biotite is a silicate belonging to the mineral group known as micas. Layers of micas can be peeled off or cleaved easily, yielding sheets of the platy material depicting its one prominent plane of cleavage. This is shown in Figure 2.3. The plates are flexible and elastic; that is, they will spring back to their original shape when bent and released. Biotite is the dark mica appearing as either dark green, black, or brown, and has a specific gravity of 2.7 to 3.2. It is best identified by its dark color and platy cleavage.

Nonferromagnesians

The nonferromagnesian silicates do not contain iron and magnesium in combination but usually have calcium, sodium, or potassium instead. Muscovite is the nearly colorless mica counterpart of biotite. It has the same properties of platy cleavage and elasticity as biotite but is lighter in color. Its specific gravity is 2.8 to 3.1 and its hardness of 2 to 3 compares well with that of biotite. In thin sheets it is transparent, taking on a yellow color in thicker sheets. Some substitution of aluminum for silicon may occur.

Feldspars

Taken as a whole, the feldspars are the most common rock-forming silicates. Collectively they comprise greater than 50% of all minerals in the Earth's crust and also form the basis for igneous rock names and their classification.

Feldspars can be divided into two types: orthoclase, which contains potassium ($KAlSi_3O_8$), and plagioclase, which contains varying amounts of calcium and sodium. Plagioclase forms a continuous series, with sodium and calcium substituting for each other in the silicate structure. The cation composition ranges from all sodium to all calcium with every possible proportion in between. If greater than 50% sodium is present, the mineral is sodic plagioclase; if greater than 50% calcium is present, calcic plagioclase is the term used. In many cases, the sodic plagioclase is assumed to be albite and the calcic plagioclase to be labradorite for hand specimen identification. These are the most common plagioclase minerals.

The feldspars as a group have a vitreous luster, a hardness of 6 to 6.5, and a specific gravity of 2.5 to 2.7. They also have two planes of cleavage at nearly right angles to each other.

After concluding that a mineral is a feldspar, further subdivision is needed to identify the specific feldspar. Commonly, orthoclase ranges from white to pink in color, whereas plagioclase ranges from white to black. As the calcium percentage increases in plagioclase, the mineral becomes darker in color.

Consequently, albite is a white plagioclase and labradorite is a dark-gray plagioclase. The property other than color that separates the two feldspars is the presence or absence of striations. Striations are the parallel lines on a cleavage face that result from twinning in the crystal. Orthoclase does not exhibit striations. Therefore, striated feldspars are assumed to be plagioclase. A problem arises when white plagioclase does not exhibit striations visible to the naked eye or with a hand lens. These minerals are difficult to identify specifically without the aid of a polarizing or petrographic microscope.

Quarts Minerals
Quartz (SiO_2) is a common rock-forming mineral that is second only to the feldspars in abundance. Quartz can occur in a variety of colors depending on the presence of minor impurities. This yields such varieties as pink-colored rose quartz, purple-colored amethyst quartz, dark-colored smoky quartz, and cloudy-looking milky quartz. Quartz is best identified by its hardness of 7 and conchoidal fracture. When crystals are present, their six-sided shape is diagnostic of quartz.

Chert is a cryptocrystalline or finely crystalline variety of quartz that occurs primarily in sedimentary rocks. Many geologists refer to the dark variety of chert as flint. Chert is of particular interest to the engineer because it presents certain problems when used as an aggregate material. Therefore, the engineer must be able to identify it. It too has a hardness of 7 (although weathering may suggest a lower value), has a dense appearance, and shows conchoidal fracture.

Chert may be a troublesome material when used in concrete for mechanical and chemical reasons. Lightweight or weathered cherts have a tendency to break or pop out when exposed to freezing and thawing temperatures on a concrete surface. Weak cherts also tend to reduce the strength of concrete if they comprise a high percentage of the aggregate.

Some cherts have been known to react with high-alkali cements to cause cracking and expansion of the concrete. A reduction in the Na_2O and K_2O percentage in the cement below 0.6% usually alleviates this problem. This is called the alkali-silica problem.

Other Silicates
Garnet is an accessory mineral found in igneous and metamorphic rocks. There are a number of different types of garnet that form a solid solution group as do many silicates. The most abundant garnet is red in color, has a hardness of 6.5 to 7.5, and may appear as well-formed multifaced crystal forms belonging to the cubic crystal system. Color and hardness are the most diagnostic features for garnet.

Two common silicates that appear in metamorphic rocks are talc and chlorite. Talc is a soft mineral with the hardness of 1 and is usually white in color with a greasy feel. It has one good cleavage, yielding tiny platy masses with a pearly luster. Rocks with abundant talc are referred to as soapstone by well drillers, construction personnel, and many geologists. Chlorite is a green platy mineral with a hardness value from 2 to 2.5, which is slightly greater than talc. It has a vitreous luster and its plates are flexible but not elastic.

Oxide Minerals
The oxides of importance in regard to common rock-forming minerals are the iron oxides. Hematite (Fe_2O_3) is recognized most easily by its red streak as is limonite ($FeO \cdot H_2O$) by its brown streak. Limonite may occur in gravel deposits in significant quantities yielding several percent by weight. It makes a poor concrete aggregate because it causes staining and popouts on the concrete surface after only several cycles of freezing and thawing.

Magnetite (Fe_3O_4) occurs less commonly than the other iron oxides. It is highly magnetic and can be identified readily by this property. It also has a metallic luster and a specific gravity of 5.2.

Sulfide Minerals
The most common sulfide mineral is pyrite (FeS_2). Pyrite, or fool's gold, occurs in igneous, metamorphic, and sedimentary rocks to some degree. It is most easily recognized by its brassy color and metallic luster, hardness of 6 to 6.5, and cubic crystal shape. A similar mineral, chalcopyrite ($CuFeS_2$), is more yellow colored and has a hardness of 4.5. Chalcopyrite is a valuable source of copper and appears as a primary copper ore in many mining areas.

Pyrite is considered a nuisance mineral if present in gravel pits used for concrete construction. Staining of the surface of concrete may occur following oxidation. Certain varieties of pyrite yield sulfate ions in such quantities that the sulfate is able to attack concrete. If pyrite is present in facing stone for buildings, unsightly stains can occur after weathering.

For these reasons, it is necessary to identify this mineral and to determine its percentage in rock considered for engineering purposes.

Carbonate Minerals

The important carbonates are calcite and dolomite. Calcite ($CaCO_3$) is distinguished by its effervescence in dilute hydrochloric acid, its rhombohedral cleavage, and its relatively low hardness of 3. Dolomite [$CaMg(CO_3)_2$] effervesces only slowly in dilute HCl and is slightly harder than calcite with a hardness of 4. Dolomite will effervesce only when powdered, whereas calcite effervesces actively in large pieces and in the powdered form as well.

Sulfate Minerals

The sulfate minerals, though widely represented in nature, have only two members that are reasonably common. These are gypsum ($CaSO_4 \cdot 2H_2O$) and anhydrite ($CaSO_4$). Gypsum is distinguished by its low hardness of 2 and usually a white color. It may appear in three forms: the massive alabaster, as satin spar with its silky luster, or in a transparent variety, selenite. It also has a relatively low specific gravity of 2.2 to 2.4. Gypsum occurs as a chemical precipitate from seawater and when present in thick deposits of several feet or more is of economic value. Gypsum deposits are quite soluble in ground water and may be cavernous if located near the Earth's surface. In this respect, the same problems must be considered as with cavernous limestones. These problems are discussed in Chapter 15 on Ground-Water Geology.

Anhydrite is the water-free form of $CaSO_4$. Its hardness is greater than gypsum, nearly 3, and usually occurs as a white to blue, massive, vitreous to dull material. It is distinguished from calcite and dolomite by its lack of effervescence in acid. A powdered form should be used for this comparison because dolomite effervesces to a limited extent.

It is important that anhydrite be recognized in nature because it has the property of swelling when wet and converting to gypsum. Accompanying this change is an increase in volume with disastrous effects if present in the foundation of an engineering structure or in a tunnel. Certain cases in which anhydrite was present in the abutment of a dam have resulted in undesirable aftereffects as the anhydrite absorbed water, expanded, and caused cracks in the structure.

Clay Minerals

The clay minerals are a group of fine-grained mineral varieties that are of major significance to the engineer. They comprise an essential portion of the soil and therefore yield a strong influence on soil behavior.

The term *clay* in the geologic literature has been used to pertain either to a size fraction or to a group of minerals. This is an important distinction because in some soil classifications the clay size can include nonclay minerals. Details concerning clay size are provided in the discussion of soil classifications in Chapter 8. For now, we can assume that clay is that fine portion of the particle size distribution that is smaller than 1/256 mm according to the Wentworth scale and of a reasonably similar size for the other important soil classifications.

Types of Clay Minerals

The clay minerals consist of five major groups: kaolinite, halloysite, illite, vermiculite, and smectite (or montmorillonite). They are divided into two categories, those with two layers or sheets in their structural repeat unit and those with three. Kaolinite and halloysite are two-layer clays; illite, vermiculite, and smectite are three-layer clays.

Clay Mineral Structures

Two kinds of layers comprise the clay minerals, a silica tetrahedral sheet and an octahedral sheet. The silica sheet consists of a series of tetrahedra of four oxygens (with a silicon atom in the center) that extends in two directions, yielding the sheet structure. The octahedral sheet consists of octahedrons formed by six oxygens or hydroxls with a magnesium or aluminum atom in the center. It extends in two dimensions to provide the sheet structure. These are illustrated in Figures 2.6 and 2.7.

These two distinct layers can be joined, yielding a repeat unit of one tetrahedral layer and one octahedral layer or a *t-o structure,* or they can be combined to yield an octahedral layer sandwiched between a tetrahedral layer top and bottom, known as a *t-o-t structure.* The t-o structure is a two-layer clay whereas the t-o-t structure is a three-layer clay. The repeat units are stacked one on top of another, yielding the clay platelets that make up the particles of clay in nature. The two-layer and three-layer structures are illustrated in Figure 2.8.

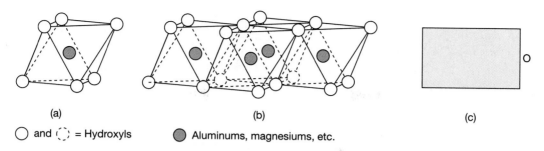

○ and ⟨ ⟩ = Hydroxyls   ● Aluminums, magnesiums, etc.

FIGURE **2.6**   Diagrammatic sketch showing (a) single octahedral unit, (b) the sheet structure of the octahedral units, and (c) the schematic of the octahedral sheet.

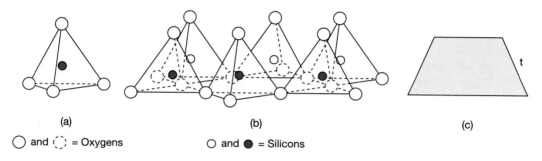

○ and ⟨ ⟩ = Oxygens   ○ and ● = Silicons

FIGURE **2.7**   Diagrammatic sketch showing (a) single silica tetrahedron, (b) the sheet structure of tetrahedrons, and (c) the schematic of a tetrahedral sheet.

Chlorite occurs as a green-colored clay mineral in its megascopic form as in chlorite schist. It is also relatively common in soils, usually as a minor constituent along with other clay minerals. It is a mixed-layer clay consisting of the three-layer t-o-t structure plus an octohedral layer before the t-o-t portion is repeated. This is illustrated in Figure 2.9.

Two-layer clays are held together by ionic bonds between the sheets with no charge imbalance. They are nonswelling clays; kaolinite and halloysite belong to this category.

Three-layer sheets have some charge imbalance because of substitution of ions in the octahedral and tetrahedral sheets. In illite this is neutralized by potassium ions between the three-layer units, which hold together strongly enough that illite is not a swelling clay. In smectite and vermiculite, the substitution in the tetrahedral and octahedral layers is more extensive and somewhat random so that an orderly balancing by potassium ions at the unit boundaries is not possible. Other cations, typically Ca^{++}, Mg^{++}, and Na^{++} are present as well. This allows for water to gain access between the clay units, yielding the swelling clays. Smectite swells more than does vermiculite, the specific amount depending on the cations present. Expansive clays are a major geologic hazard in the United States. They are landslide prone and disrupt building foundations, roads, and other structures when they undergo expansion.

Base Exchange Capacity

Base exchange capacity is a property ascribed to most clay minerals. It is defined as the ability to attract cations to its surfaces and to exchange them stoichiometrically (valence for valence) for each other. Therefore, a plus-two cation (Ca^{++}, for example) can replace 2 plus-one cations (Na^+) on the crystal lattice. Negative charges on the clay surfaces provide the charge deficiencies that are satisfied by these cations. The sources of the negative charges on the lattice are crystal terminations and substitutions in the tetrahedral and octahedral sheets of the silicate structure.

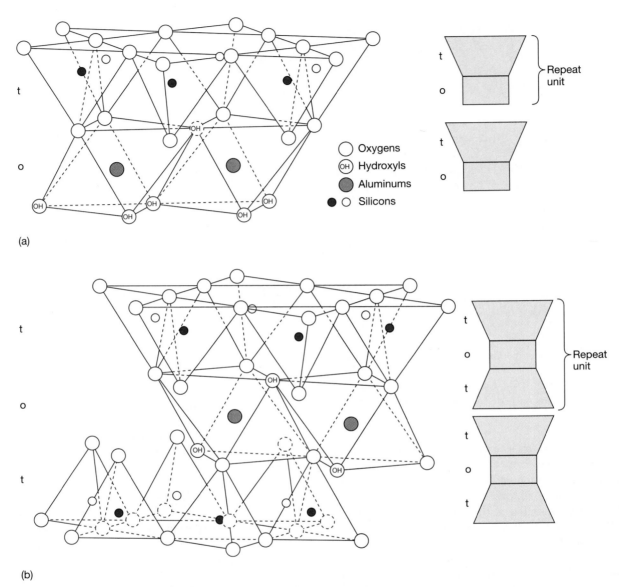

FIGURE 2.8 (a) Two-layer or t-o structure clay. (b) Three-layer t-o-t structure clay.

Crystal terminations increase as the particle size decreases so that finer grained particles will have a greater base exchange capacity than coarser grained ones. Clays with a greater degree of substitution in their tetrahedral and octahedral sheets will also have a greater base exchange capacity. Smectites are both extremely fine grained and prone to major substitu-tion within their silicate structure, which is why these minerals have a high base exchange capacity.

Smectites have the highest base exchange capacity and kaolinite has the lowest. The other clays lie somewhere between these extremes. The subject of base exchange capacity is discussed further in Chapter 7, Elements of Soil Mechanics.

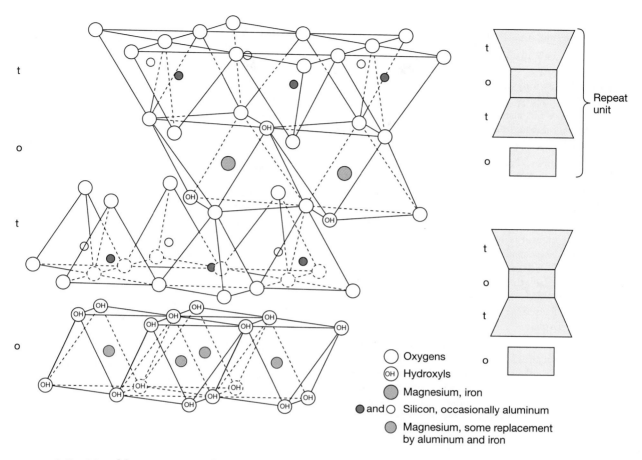

FIGURE 2.9   Mixed-layer structure for chlorite.

Zeolites

Zeolites are silicate minerals related generally to the feldspars, but they have major substitutions in their silicate structures that yield a high density of negative charges on their mineral surfaces. For this reason they are used for water softening, which is the removal of Ca^{++} and Mg^{++} from a water supply. Na-zeolite is obtained by adding NaCl to the zeolite material. The hard water is allowed to pass over the Na-zeolite so that Ca^{++} and Mg^{++} replace the Na^+ on the zeolite. Ca^{++} and Mg^{++} in water form an insoluble residue when mixed with soap suds, resulting in bathtub scum, which is a well-known nuisance. Na^+ forms no such residue so by using it to replace Ca^{++} and Mg^{++}, the water becomes softened.

EXERCISES ON MINERALS

1. Name one or more diagnostic properties and the specific value or information for those properties that would be used to differentiate between the following mineral pairs:
   (a) calcite and quartz
   (b) feldspar and quartz
   (c) calcite and feldspar
   (d) pyrite and limonite
   (e) amphibole (hornblende) and pyroxene (augite)
   (f) mica and clay minerals
   (g) orthoclase and sodic plagioclase
   (h) calcite and dolomite

(i) calcic plagioclase and pyroxene (augite)

(j) biotite and amphibole (hornblende)

2. A mineral with a hardness of 7 appears in two different forms. One is a six-sided barrel-shaped piece with a pyramid at its top, and it is white in color and transparent. The other is pink and translucent and has no smooth sides but instead has rounded fracture surfaces. What mineral would this be and how do you account for the greatly different appearance?

3. A white mineral has a hardness of 6, two planes of cleavage nearly 90° apart, and shows striations on two of the cleavage faces. What specific mineral would this be?

4. Limestone (which consists mostly of calcite) is being quarried in one location and quartzite (which consists mostly of quartz) is being quarried in another. Which of the two rock materials would cause greater wear on the shovels, trucks, and crushing equipment used to load and size the material for construction use? Explain. Which material would be more resistant to abrasion if used as a road base for a highway or railroad? Explain.

5. Both weathered chert and sound quartz pieces are common constituents in stream gravels. Both have the chemical formula SiO_2. Yet the weathered chert is an unsound material in concrete when subjected to freezing and thawing cycles, whereas quartz does not have this problem. Explain why this is the case.

6. Pyrite (FeS_2) is a common accessory mineral in coal-bearing rocks. What chemical is developed when water comes in contact with pyrite and what is the environmental effect of this?

7. Large fissures and caves develop in rocks composed of calcite. Relate this geologic condition to the properties of calcite studied in this exercise. Would caves also form in dolomite? Explain.

8. Montmorillonite (smectite) is a clay mineral that expands extensively on wetting. What would be the effect of locating a footing for a building in this material if wetting and drying of the clay can occur with seasonal changes?

9. Zeolites are a group of silicates that form from weathering or near-surface processes and appear as secondary fillings and veins in preexisting rocks. They have an expansive quality similar to that of some clay minerals. What adverse effects could occur if zeolite-bearing rocks (a) are used as an aggregate for concrete or (b) form the abutments (or side walls) for a concrete dam? Explain.

Additional Readings

BERRY, L.G., MASON, B., and DIETRICH, R.V., 1983, *Mineralogy*, 2nd ed., W.H. Freeman and Co., New York.

HERMAN, C., ZIM, H.S., and SHAFFER, P.R., 1963, *Rocks and Minerals*, Golden Press, New York.

KLEIN, C., and HURLBUT, C.S., JR., 1985, *Manual of Mineralogy*, 20th ed., John Wiley & Sons, New York.

SINKANKAS, J., 1966, *Mineralogy, A First Course*, D. Van Nostrand Co., Princeton, NJ.

3

Igneous Rocks

THE NATURE OF ROCKS

Geologic Definition

At the outset the reader should recognize that geologists and engineers do not always use the same definition for the term *rock*. To a geologist, rocks are essential units of the Earth's crust whose origin, classification, history, and spatial aspects are impor-

tant. Physical properties such as massiveness or durability of the material are not directly involved, and, geologically speaking, glacial ice, sand, marble, coal, and basalt are all rocks.

A working definition for *rock* that is sometimes used in beginning geology courses states that they are common assemblages of one or more minerals. This is true generally, but there are exceptions if most of the material is nonmineral. For example, coal consists mostly of nonminerals because it has an organic origin, and volcanic glass (obsidian, etc.) is nonmineral because of its amorphous or noncrystalline nature. Yet both are rocks. Therefore, a more acceptable geologic definition for rock is that just provided: Rock is an essential part of the Earth's crust, and involves details of origin, classification, history, and spatial relationships.

Engineering Definition

Engineering definitions of rock differ from those used in geology; engineers consider rock to be a hard, durable material. From an excavation point of view, rock is any material that cannot be excavated without blasting. All other earth materials would be termed *earth* or *soil*. Another definition for rock indicates that it is earth material that does not slake when soaked in water. Bedrock is the hard continuous rock mass, lying below the soil or at the Earth's surface, that must be excavated by blasting. Soil, by contrast, is the loose, unconsolidated material, lying above bedrock, that can be excavated by conventional means.

Confusion between the geologic and engineering definitions for soil can occur because some materials classified as rock by geologists can be easily excavated. For example, thick loess deposits (wind-deposited silt) and extensive gravel and sand deposits persisting over large areas would be categorized as rock units by geologists but would be considered soil by engineers. In this text, with its emphasis on engineering applications, the engineering definition is generally followed. A discussion of rock versus soil is considered further in Chapter 8, Rock Weathering and Soils.

IGNEOUS ROCKS

Rocks are divided into three main types depending on their origin: igneous, sedimentary, or metamorphic. Igneous rocks are those cooled from a molten state; sedimentary rocks are deposited in a fluid medium, usually water; and metamorphic rocks are formed from preexisting rocks by the action of heat and pressure.

Igneous rocks include the deep-seated, intrusive rocks that cool slowly within the Earth as well as the extrusive lavas and pyroclastics that form at the surface and are quick to cool. Pyroclastics are those materials ejected from volcanoes that settle out in the form of ash deposits and coarser grained rock units.

Rocks formed at intermediate depth within the Earth, such as dikes and sills, also cool at an intermediate rate. They comprise another important group of igneous rock bodies.

Each continent of the world has extensive areas in which rocks of predominantly igneous origin are exposed. Igneous rocks make their appearance on the landscape in the centers of mountain belts, in eroded plateaus marking previous mountains, as isolated igneous intrusions, and as extensive outpourings of molten rock. Although sedimentary rocks occur at the surface or below the soil zone in many areas of the globe, they comprise only about 5% by volume, whereas some 70% of the volume in the outer 10 miles of the Earth is composed of igneous rocks.

A construction engineer, if involved with projects in many portions of the western United States or near the Appalachian region in the east, will by necessity encounter igneous rock as building foundation materials, in excavations, as sources for highway construction, and in various other engineering projects. The particular problems associated with igneous rocks and their wide diversity of type and nature are a worthy subject of study for engineers who plan to work with earth materials.

FORMATION OF IGNEOUS ROCKS

Those igneous rocks that cool at the Earth's surface are the product of volcanic activity. When eruptions of molten rock, generated within or below the Earth's crust, reach the surface, the opening is enlarged by explosions and slumping of molten and semisolid material. This crater is located at the top of a mound of accumulated lava and pyroclastic debris. The growing mound of material capped by this crater is the well-known volcano.

Lava extruded from a volcano flows over the adjacent land surface, eventually cooling and hardening into solid rock. This is one variety of volcanic or extrusive rock. Intermittent between the lava flows, most volcanoes have periods of explosive eruptions in which pieces of hardened volcanic rock are thrown into the air only to settle out as ash deposits and coarser sized materials. These pyroclastics are commonly interbedded with the lava flows.

At depths within the Earth, the molten material or magma after intruding older rocks cools slowly and hardens without reaching the surface. Magma differs from the lava flowing on the surface typically in its greater content of dissolved fluids, which tend to escape when reaching the Earth's surface. Rocks that crystallize inside the Earth are called intrusive or plutonic igneous rocks. Since intrusive rocks cool very slowly compared to extrusive or volcanic rocks, their mineral grains are much larger than those comprising volcanic rocks.

Igneous bodies formed at intermediate depth, that is, below the surface but not located as deep as plutons, are called *hypabyssal rocks*. Typically these rocks have textures intermediate in size between coarse-grained intrusive rocks and fine-grained extrusive ones. They may even have a porphyritic texture, one with coarse crystals surrounded by a fine-grained groundmass.

Magmas are hot, viscous, siliceous melts containing dissolved gases. The most abundant elements present are silicon and oxygen and the primary metal ions are potassium, sodium, calcium, magnesium, aluminum, and iron. Dissolved gases typically are H_2O, CO_2, and SO_2. Igneous rocks form as the magma cools and the composition can vary greatly, depending on the original composition of the melt and the history of cooling and implacement.

The silica (SiO_2) content of igneous rocks varies from about 40% to more than 80%. Magmas and rocks that contain abundant silica were originally termed *acidic* because of the early notion that rocks were formed from a silicic acid through replacement of the hydrogens by metal cations, as in the formation of salts from acids. Therefore, the more silica in the rock, the more silicic acid that was present originally. Today it is known that, instead, minerals form from silicate melts through a complex path of crystallization, but the term *acidic* persists to some extent in the literature as a descriptive term. Note, however, that *high-silica igneous rock* is the preferred terminology.

Correspondingly, low-silica rocks were termed *basic* and those with extremely low silica content (less than 50%) were ultrabasic (also ultramafic, meaning extremely dark). These terms describe an important fact about igneous rocks: their extreme range in silica content.

Low-silica magmas are less viscous (more fluid) than high-silica ones and in fact there is considerable question as to whether the high-silica melts could flow as a lava from a volcanic vent. Instead, high-silica extrusive rocks are more likely pyroclastic in origin. For low-silica lavas, temperatures of about 1000 °C have been measured in volcanoes. Fluxes such as dissolved gases are commonly present, which lowers the melting temperature of the magma.

Volcanoes and their conduits provide the passageways between bodies of magma in the crust and the Earth's surface. The morphology and origin of volcanoes are considered in the next section.

TYPES OF ERUPTIONS

Several types of volcanoes or volcanic eruptions can be distinguished as follows:

1. Fissure eruptions in which lava exits quietly along lines of fracture at the Earth's surface with the emission of only small quantities of gas
2. Shield volcanoes with large flat lava cones forming gentle slopes (Hawaiian type)
3. Cinder cones with steep slopes formed by pyroclastic eruptions
4. Composite cones or stratocones formed by alternating lava flows and pyroclastic eruptions, the latter accompanied by emission of much gas and sometimes by violent explosions.

The last three types, shield volcanoes, cinder cones, and composite cones, are collectively known as central eruptions.

Fissure Eruptions

Fissure eruptions occur quietly along linear fractures in the Earth as low-viscosity basalts issue forth and spread rapidly over large areas. In the geologic past, floods of basalt have poured out at different locations throughout the world and have been attributed to fissure eruptions. These features are known as plateau basalts or flood basalts because of the tendency to form extensive plateaus of layered basalt flows.

The only documented lava flood of historic time occurred in Iceland in 1783 when the Laki fissure was nearly 20 miles (32 km) long from which lava spread over an area of about 260 mi^2 (660 km^2). Iceland itself is a remnant of vast lava floods that accumulated for the last 50 million years and blanketed an area of more than 200 million mi^2 (500 million km^2). The hardened lava is believed to be 9000 ft (2800 m) thick.

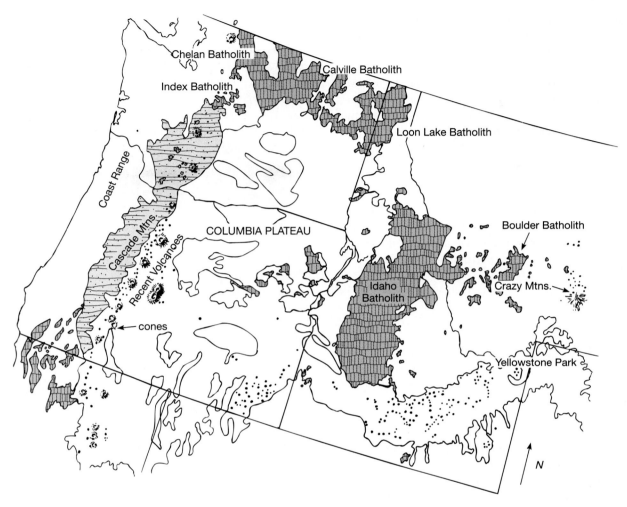

FIGURE 3.1 Distribution of volcanoes, lava flows (stippled), and igneous intrusions (cross hatched) in northwestern United States.

The plateau basalts of the Columbia Plateau in Washington, Oregon, Idaho, and northern California cover 50,000 mi^2 (130,000 km^2)—see Figure 3.1. In some areas more than 5000 ft (1500 m) of extrusive rock have been built by a series of fissure eruptions. Individual eruptions deposited layers from 10 ft (3 m) to 300 ft (100 m) thick. The total volume of rock poured out during this Miocene volcanic episode has been estimated as more than 25,000 mi^3 (100,000 km^3).

Other extensive areas of the world with major plateau basalt accumulations are the Deccan Plateau of India with an area of 400,000 mi^2 (1,000,000 km^2) and a thickness of more than 5500 ft (1800 m), the Parana of Brazil and Paraguay with an area of 300,000 mi^2 (750,000 km^2), north-central Siberia, the Ethiopian Plateau near Victoria Falls on the Zambezi River, Greenland, Antarctica, and northern Ireland. Much older flood basalts (Precambrian in age) are found in northern Michigan and in the Piedmont region of the eastern United States. The basalt flows of the world are estimated to cover more than 1 million mi^2 (2.5 million km^2).

Shield Volcanoes

Shield volcanoes are large flat cones consisting almost entirely of lava flows of basaltic composition, which poured out in quiet eruptions from a central vent or closely related fissures. The cones are much broader than their height with slopes seldom steeper than 10° at the summit and 2° at the base.

Mauna Loa is a shield volcano on the island of Hawaii and is the largest of its kind in the world. It rises 2.5 miles (4 km) above sea level but measured from the seafloor, it stands 6 miles (10 km) high with a diameter at its base of 60 miles (100 km). This yields an estimated volume of nearly 6000 mi^3 (24,000 km^3) of basalt, all accumulated from individual lava flows only tens of feet (a few meters) thick. The island of Hawaii actually consists of five volcanoes that have built up from the seafloor and have coalesced to form the island.

Cinder Cones

Cinder cones are small mountains consisting primarily of fragmental (or pyroclastic) debris ejected from the central vent. They achieve slopes equal to the angle of repose of the clastic debris, or about 30° to 40°, and they seldom exceed 1600 ft (500 m) in height. Many cinder cones have flows of basalt issuing from their base. Paricutín in Mexico erupted in a farmer's field in 1943 and built a cinder cone 500 feet (150 m) high during its first week of activity. It is currently a dormant volcano about 1300 feet (400 m) high.

Composite Cones

Composite cones or stratocones are built up by alternating lava flows and beds of pyroclastics. This is the most common form of large continental volcano and they are characterized by slopes up to 30° at the summit tapering to 5° near the base. Composite cones are represented by such famous volcanoes as Fujiyama, Vesuvius, Stromboli, and Mayon on Luzon in the Philippines. Mount Shasta in northern California, a stratocone in the Cascade Range, is shown in Figure 3.2.

If the central vent of the volcano becomes too high or gets plugged, the lava may find an easier route through the cone to open a subsidiary vent on the volcano's flank. Subsidiary vents commonly occur on major volcanoes.

MORPHOLOGY OF VOLCANOES

The features associated with volcanoes and other details of eruptions are illustrated in Figure 3.3. The

FIGURE **3.2** Mount Shasta, northern California, is a composite cone in the Cascade Range. (Ward's National Science Establishment, Inc.)

large mound of lava and pyroclastic debris formed at the Earth's surface is the volcanic cone and the depression at its summit, the crater. Enlarged craters are known as calderas, most of which are formed by collapse, but some occur by explosion. The passageway to the crater is the volcanic pipe or conduit. Satellite cones typically form on the flanks fed by the main conduit.

In some cases, particularly for cinder cones, the volcanic debris can erode, leaving the lava-filled pipe, or volcanic neck or plug, exposed as a resistant column of rock. Dikes feeding away from the center of the neck may be preserved as well. Ship Rock, New Mexico, and Devils Tower, Wyoming, are examples of volcanic necks and are shown in Figure 3.4. Columnar jointing in Ship Rock is seen to be vertical near the top but closer to horizontal near the base. These joints tend to form perpendicular to the cooling surface (or surface of the cone).

Fissure eruptions and plateau basalts are shown in Figure 3.5, along with the different types of cones.

VOLCANIC DEPOSITS

Lava pours from volcanoes or fissure eruptions moving down the slope to fill low areas in the terrain. Basalt is highly fluid, and can flow at high velocities, 60 mi/hr (100 km/hr) and for long distances, 30 miles (50 km). Successive flows build individual layers from tens of feet (or few meters) to 300 feet (100 m) thick.

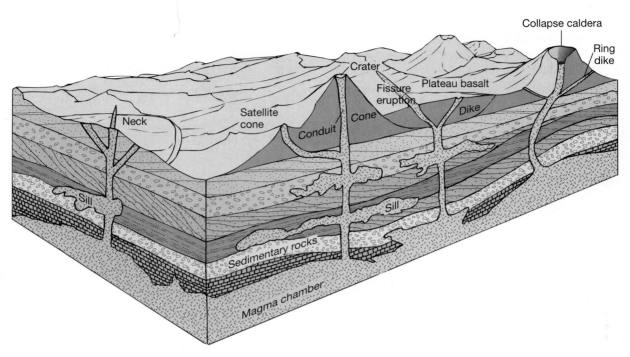

FIGURE 3.3 Cross section of volcanoes and related features.

Basaltic flows on Hawaii give rise to two primary surface forms: pahoehoe (pa-hoy-hoy) and aa (ah-ah). Pahoehoe forms ropy features on the surface and, therefore, yields a rough blocky surface. Aa shows a relatively smooth surface. This illustrates that aa is the more fluid of the two lavas and it yields thinner individual flows.

As lava cools and contracts, polygonal shrinkage cracks form perpendicular to the cooling surface. Such features are called columnar joints. In thick flows the surface hardens well before the interior and if the pressure is sufficiently great, lava will break through the sloping crust, extruding molten rock and leaving a lava tube or cave behind. In many cases the pressure pushes up the surface of the flow to yield a pressure ridge, an elongated blister with a central fracture through which the gas escapes. Therefore, lava flows may contain openings that easily transmit water or they can collapse under construction loads.

Pillow lavas are rounded masses about an inch (a few centimeters) to 6 ft (2 m) across that resemble a stack of pillows with their rounded side up. They occur when lava flows into seawater or into freshwater lakes and ponds. The pillows generally have glassy borders. The rounded shape of the pillows can be used to tell the original top from bottom of a deformed sequence because the pillows are concave downward when formed. Pillows are also used to indicate subaqueous flows.

The glassy material in the pillow lavas may alter to a yellowish clay-type mineraloid called palagonite. Through alteration or weathering, palagonite can weather further to clay. Palagonite-rich basalts are prone to deterioration in the presence of water when such materials are used as a road aggregate. These materials in the northwestern United States have led to serious highway pavement problems in the past.

Lavas can be glassy, finely crystalline, or combinations of both depending on their cooling rate. When water vapor and other gases are released, gas cavities or vesicles can form. A common rock formed in this way is vesicular basalt.

In addition to lava flows, extrusive igneous rocks are formed as pyroclastic deposits. These consist of fragments of volcanic material blown out of the vent, and collectively such fragments are called tephra. These ejected particles can consist of rocks, minerals, or glass and are classified according to size into dust (the finest), ash (up to about 100 cm), and larger elliptical-shaped pieces known as volcanic bombs.

FIGURE 3.4 Erosional remnants of volcanic cones at Ship Rock, New Mexico (above), and Devil's Tower, Wyoming (below). (Ward's National Science Establishment, Inc.)

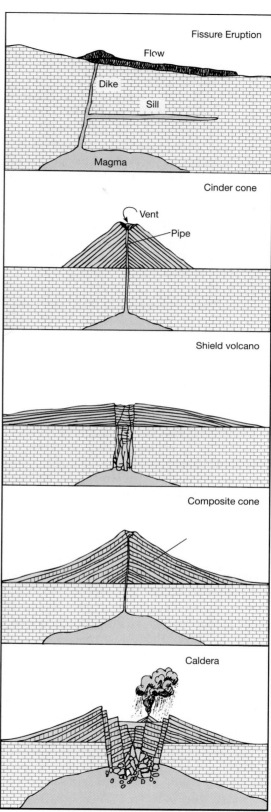

FIGURE 3.5 Various types of volcanic eruptions.

Commonly, bombs are ejected in a plastic state and harden before reaching the ground. Large size fragments of solid rock blown into the air are known as blocks and these can be thrown 6 mi (10 km) or more. Volcanic ash and dust can be carried great distances by volcanic explosions. In the eruption of Krakatoa in 1883 dust reached the upper levels of the atmosphere and was dispersed around the world.

MOUNT SAINT HELENS ERUPTION

The eruption of Mount Saint Helens in the spring of 1980 provides an example of a major pyroclastic explosion. A detailed account of the eruption is presented here based on reports by the U.S. Geological Survey (USGS).

Mount Saint Helens in southwestern Washington is one of the major volcanoes in the Cascade Range, which parallels the Pacific Coast from northern California to southern British Columbia. It is a stratovolcano as are most of the other volcanic peaks in the Cascades, consisting of alternating layers of lava, pumice, and other pyroclastics, plus mudflow and landslide debris. According to plate tectonics the Cascades formed as a consequence of the Pacific Plate plunging below the North American Plate along a convergence boundary. Melting of the oceanic plate occurs as it is forced deeper and, as melting occurs, magma rises along the leading edge to form the chain of volcanic mountains.

Much of the Cascade Range was formed in an active eruptive cycle between 40,000 and 2500 years ago. The individual peak forming Mount Saint Helens accumulated during the past 2500 years with much of the upper portion formed in the past few hundred years. Geologic studies show Mount Saint Helens has had many periods of explosive activity in the past 4500 years interspersed with quiet periods up to 500 years in length.

Around 1800 a major eruption deposited pumice and ash several feet thick in an area extending for many miles to the northeast of the crater. Indians, early explorers, and fur trappers provided vivid stories of the eruptions they witnessed during the first half of the nineteenth century. Activity was greatest during the 1830s and 1840s, but after 1857 the eruptive cycle came to an end.

Geologists of the USGS performing geologic hazard studies predicted in 1975 that Mount Saint Helens was likely to erupt very soon geologically, possibly before the end of the twentieth century. They predicted this eruption would yield great volumes of pumice, ash, and pyroclastic flows. Huge mudflows and subsequent flooding of adjacent river valleys were also anticipated. Many of their worst fears were realized in early 1980.

On March 27, 1980, Mount Saint Helens ended its 123-year quiescent period with an eruption that threw ash and steam high into the air after opening a summit crater 250 ft (75 m) wide. This was preceded on March 20 by an earthquake measuring 4.1 on the Richter scale. Earthquakes grew stronger after March 20 and people were told to keep a safe distance from the general area. After the March 27 eruption residents, campers, loggers, and tourists staying in local commercial lodges were evacuated within a distance of 15 miles (25 km) from the summit.

Earthquakes and small eruptions continued, opening a second crater that eventually combined with the first. By early April the crater was 1500 ft (450 m) wide and growing. Harmonic tremors (earth vibrations of a constant nature in contrast to the stronger short-duration earthquakes) were detected, signifying that magma was moving upward from below. Scientists set up their instruments and waited for the expected eruption. Sightseers rushed to the area and roadblocks had to be erected to keep them outside the danger zone. Only scientists and public safety officials had access to areas near the mountain.

In early May the north side of the mountain began to bulge, swelling as much as 3 to 6 ft (1 to 2 m) in a day's time. Small earthquakes triggered occasional avalanches of rock and ice, which tumbled down the mountainside.

The big explosion occurred Sunday, May 18, at 8:32 A.M. When an earthquake of Richter magnitude 5 occurred, the growing bulge gave way in a gigantic avalanche of rock, mud, and ice followed by major explosions directed both vertically and horizontally. Nearly 0.75 mi³ (3 km³) of the cone was blasted loose, sending the lighter material high into the air and the heavier portion out and downward. This powerful blast of gas, ash, and rock fragments flattened virtually everything in its path for a distance extending 10 to 16 miles (16 to 25 km) to the northeast, north, and northwest of the mountain. People were killed as far as 16 miles (25 km) away from the crater.

Trees were sheared off or uprooted, stripped of their limbs, and charred by the blast. Vehicles were tumbled by the explosion, then smashed and broken when they hit the rocks and trees. Plastic melted and metal twisted. Everything was devastated in the path of the air blast.

A giant avalanche of rock, ice, and water raced down the north slope of the volcano, spilling into the Toutle River. Spirit Lake at the base of the mountain was partially filled by the avalanche, which sloshed water up the ridges, raising the lake level about 200 feet (60 m). Pyroclastic flows of hot gases, ash, and pumice with temperatures as high as 1600 °F (870 °C) roared down the mountain at nearly 100 mi/hr (160 km/hr) wiping out everything in its path.

Water from overflowing lakes and from melted snow and glaciers mixed with the avalanche and volcanic debris to create massive mudflows. These cascaded down the surrounding streams and river valleys causing floods by damming stream beds and overflowing the channels. Floodwaters swept away thousands of logs, heavy logging equipment, and even large bridges. Heated water and mud killed essentially all the fish in the neighboring Toutle River.

Rescue teams including members of the county sheriff's office, the U.S. Forest Service, and the military were dispatched immediately. The work was dangerous and difficult with limited visibility because of the ash, steam, and smoke from the forest fires. Within 36 hours, 170 people had been rescued but millions of dollars worth of damage had occurred and many lives were lost. At last count 34 fatalities were verified and another 28 people were unaccounted for and presumed to be dead.

Floodwaters carried great volumes of sediment down the Toutle River to the Cowitz River and finally into the Columbia River. In all, 55 million yd^3 (42 million m^3) of debris were deposited, which blocked the navigational channel of the Columbia River to ocean-going ships for a distance of 9.5 miles (15 km). The Columbia River contains the navigational channel for the port of Portland, Oregon, which is located about 30 miles (50 km) up river from the Pacific Ocean. The Cowitz River enters the Columbia River some 15 miles (25 km) upstream from the ocean and

15 miles (25 km) downstream from Portland. When the Columbia River was blocked, the normal draft of 40 ft (12 m) was reduced to 14 ft (4 m) in a number of places. Ships upstream were stranded for days until the Army Corps of Engineers could get the channel dredged and reopened to shipping.

Clouds of ash traveled east-northeast across Washington blocking out the Sun for hours. Fields and row crops were buried by ash in the central and eastern parts of the state. Ash fell in Washington, Idaho, and western Montana bringing transportation to a halt until visibility cleared. Four days later ash had crossed the United States and had reached the East Coast. The ash cloud was eventually tracked around the world.

PYROCLASTIC DEPOSITS

Pyroclastic rocks are formed when the fragments become cemented together or lithified. Volcanic tuffs consist of ash and dust size pieces; volcanic breccias consist primarily of larger sized pieces, those above 4 mm in size.

A spectacular and potentially disastrous type of eruption associated with high-silica magmas is known as an ash flow or nuee ardente. It occurs when hot ash, dust, and gases are ejected en masse in a glowing cloud that moves downslope at high speed. The solid fragments are buoyed up by the hot gases yielding minimal frictional resistance to this incandescent avalanche. Velocities up to 60 mi/hr (100 km/hr) are possible because the expanding gas reduces frictional resistance by forcing the grains apart.

In 1902 an ash flow occurred on Mount Pelee on Martinique in the Caribbean. It engulfed the town of St. Pierre in one minute, killing 28,000 people. Deposits formed by a nuee ardente are poorly sorted and lack bedding.

Typically the fragments remain hot after the ash flow comes to rest. After settlement and compaction, the fragments are fused by volcanic glass to yield a welded tuff or ignimbrite. Ash flows can be quite vast, forming deposits more than 300 ft (100 m) thick and more than 40,000 mi^2 (100,000 km^2) in extent. Many geologists today consider that rock units previously thought to be rhyolite flows are actually ignimbrite sheets because of the high viscosity of silica-rich melts.

FUMAROLES, GEYSERS, AND HOT SPRINGS

In areas of former or dormant volcanic activity, steam and gases (CO_2, HCl, H_2S, HF) at high temperature may be emitted from gas vents (or fumaroles) for a long period of time. Geysers are springs of boiling water and steam, supplied from thermal zones below, which erupt on a periodic basis. Geysers differ from hot springs in that hot springs flow constantly and usually have a lower temperature than that of geysers. Yellowstone National Park in northwestern Wyoming contains numerous geysers, fumaroles, and hot springs. Old Faithful, the famous geyser and tourist attraction in Yellowstone, erupts about every hour shooting water to a height of nearly 200 ft (60 m).

Iceland, a land of numerous hot springs and geysers, is a location where these geothermal sources are used for heating and cooking purposes. Areas in southern California and northern Mexico are currently being exploited for geothermal power generation. Cerro Puerta, southwest of Mexicali, Mexico, is the site of an electrical generating station using the high-temperature source (steam) of the subsurface. Thirty miles (50 km) to the north near El Centro, California, several American companies have constructed geothermal power plants to evaluate electrical generation.

INTRUSIVE ROCK BODIES

Igneous bodies of major areal extent that have solidified inside the Earth are called plutons. Typically they are classified according to their size, shape, and relationship to the rock they have invaded (known as the country rock). Plutons include batholiths, stocks, dikes, sills, and laccoliths. Easily the most common are those tabular igneous intrusions, dikes and sills, which feed from other plutons. These various intrusive bodies are illustrated in Figure 3.6. Topographic expression of extrusive igneous formations is also depicted.

Magma rises through the crust in several ways. It can work its way upward by stoping, that is, by breaking off blocks of country rock and assimilating them into the melt. A partially digested piece of country rock in a pluton is called a xenolith. By giving off heat, magma can elevate temperatures in the adjacent rocks and make them flow plastically aside. It can also forcibly bow the rock mass upward, causing tension cracks to develop, which are then exploited by upward migration of the intrusion.

Batholiths are the largest plutons and are discordant in nature, that is, they cut across the bedding in the country rock. By definition they have a cross-sectional area (or map area) of 40 mi^2 (100 km^2) or more. They have steep boundaries and extend to great depths, estimated at from 6 to 18 miles (10 to 30 km) into the crust. These enormous igneous bodies, typically composed of granite or granodiorite, crystallized several miles (thousands of meters) below the Earth's surface. Uplift and extensive erosion are required to expose these bodies to the Earth's surface.

Batholiths are found in the centers of mountain belts and in the shield areas of continents such as in the Canadian Shield of North America. They are considered to be the roots of mountains, and the axis of the batholith typically follows that of the mountain range. The Sierra Nevada of eastern California are such an example. The Idaho Batholith is another extensive exposure of these immense plutons; it covers an area of nearly 16,000 mi^2 (41,000 km^2).

Stocks are discordant plutons, similar to batholiths, but smaller. They must have a map exposure of less than 40 mi^2 (100 km^2), but usually occur more in the range of 1 to 10 mi^2 (a few square kilometers to tens of square kilometers). Stocks may grade upward into volcanic necks and many have porphyritic textures, that is, visible crystals in a fine-grained groundmass.

Dikes and sills are tabular bodies that extend outward from other igneous intrusions. Dikes, perhaps the most common intrusive bodies, cut across the bedding planes or the foliation of the country rock (and thus are discordant). Typically they are from 1 to 10 ft (0.3 to a few meters) thick, but their thickness can range all the way from 1 in. to 1000 ft (a few centimeters to hundreds of meters). The largest known dike in the world is the Great Dike of Rhodesia (Zimbabwe), which runs cross country for 375 mi (600 km) with an average width of 6 mi (10 km).

Dike implacement is related to the fracture pattern in the country rock. Magma is forced out through these fractures from the central conduit, stock, or other body. Dikes radiating from volcanic necks are common and are illustrated in Figure 3.4

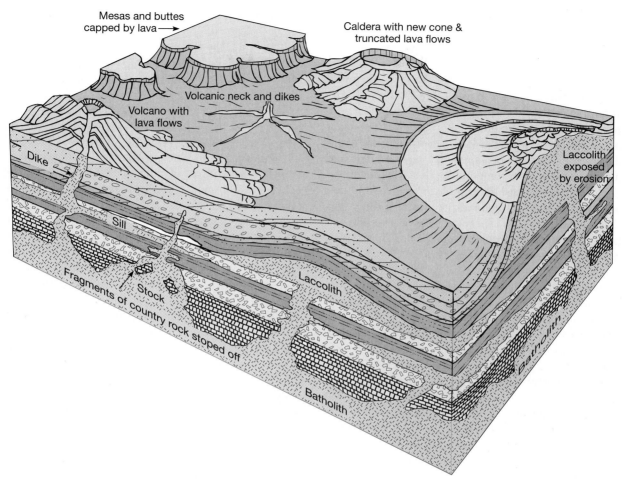

FIGURE 3.6 Igneous intrusive bodies.

for Ship Rock, New Mexico. In some locations dikes can radiate outward to form circular features called ring dikes. Large ring dikes yield circular features up to 15 mi (25 km) in diameter. This is illustrated in Figure 3.7. Dike swarms are parallel or radiating dikes consisting of numerous clustered dikes. They may form as fissure fillings or along tension fractures that developed during magma implacement.

Dikes can be either more resistant or less resistant to weathering than the country rock, which yields either wall-like features cutting across the countryside or elongated narrow trenches.

Sills are tabular bodies formed when magma is injected along bedding planes of layered rocks;

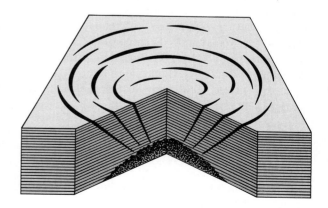

FIGURE 3.7 Three-dimensional view of ring dikes.

hence, they are concordant features. They range in thickness from 1 in. to 1000 ft (a few centimeters to hundreds of meters) and can extend over broad areas. The Palisades Sill along the Hudson River in New York and New Jersey is a well-known example in the eastern United States. It is 1000 ft (300 m) thick. The Whin Sill in the north of England is about 100 ft (30 m) thick, has a diorite composition, and has an extent of nearly 1600 mi^2 (4000 km^2). It provides an excellent quality roadstone material with high strength and good durability.

Sills are distinguished from flows of extrusive rock in that sills lack the common features of these lava flows, namely, ropy or blocky surfaces or vesicular and pillow structures in the hardened lava. Sills are typically coarser grained than lava flows because they cool more slowly, and the country rock above and below shows thermal effects such as baking. Sills protected by surrounding rock also lack signs of weathering, which may occur in lava flows. Sills can contain pieces of the country rock through which they were injected.

Sills are not necessarily horizontal, but merely concordant to the bedding of the enclosing rock. Tilted beds may also contain sills. Sills are fed from local dikes, volcanic necks, or come directly from stocks and batholiths.

Laccoliths are formed when the injection of magma along bedding planes forces the sediments upward in a dome or mushroom shape. Like sills, the intrusive pressure of the magma is able to overcome the weight of the overlying strata (or the overburden pressure), which causes uplift and makes room for the magma. This suggests a limit in the depth of rock cover during placement, such that the lithostatic pressures can be overcome. Laccoliths have a flat floor and an arched roof. They can be a mile or so thick (thousands of meters) and a mile or so (several kilometers) in diameter. Typically they have porphyritic textures. Laccoliths were first described in the Henry Mountains of southern Utah, a portion of the Colorado Plateau.

IGNEOUS ROCK TEXTURES

Igneous rocks are classified on the basis of two characteristics: mineral composition and texture. Mineral composition is shown indirectly by rock color,

a particularly useful feature for fine-grained materials where individual minerals cannot be discerned. Rock texture is a measure of the overall size, shape, and orientation of the constituents. It involves a description of the individual grains and the mutual relationship between them.

If individual constituents are sufficiently large to be discerned by the naked eye, the texture is said to be coarse grained or phaneritic. The lower limit of this size is about 0.5 mm in diameter. For phaneritic grains, identification of individual mineral constituents is possible and rock identification as a result is made easier.

Extremely large grains, those greater than about 10 mm, deserve special attention because of their size. Such texture is called pegmatitic and the rock type is known as a pegmatite. The mineral composition is used to supply the modifying term. For example, when quartz and orthoclase are present, the rock is termed a granite pegmatite. Extremely coarse-grained pegmatites do occur in nature wherein orthoclase crystals 1 ft (0.3 m) in length can be found. Such deposits are formed from the last magmatic liquid, after most of the intrusive body has solidified. Pegmatites crystallize by slow cooling in the presence of volatiles and rare elements. Crystals are able to grow to considerable size without terminating against a fast-growing neighbor.

Fine-grained rocks are those whose grains or crystals cannot be discerned with the naked eye. Such rocks also are termed aphanitic. They have a dull massive appearance and show subdued conchoidal fracturing. As one would expect, different degrees of fineness occur with regard to these textures.

Aplitic textures, those that occur in the dike rocks known as aplites, are fine sugary textures with crystals just barely discernable by the naked eye. The grains are anhedral, showing a poor degree of crystal outline. Aplites are usually light in color and contain quartz and orthoclase. However, even when using a 10× hand lens, these mineral crystals are difficult to observe.

The typical aphanitic texture is represented by a basaltic rock with a nearly equal, fine grain size. These rocks look massive and have a rather dull appearance because they do not have any large cleavage faces to reflect the light. No individual grains can be discerned, so specific mineral identification is not possible without the use of a petrographic (polarizing) microscope.

A rock with a texture composed of coarse-grained crystals in a fine or aphanitic groundmass is termed a porphyry. The coarse phaneritic crystals are called phenocrysts and their mineral composition can be determined as easily as can be coarse-grained constituents in a phaneritic rock. Porphyritic textures occur in extrusive lava flows, as dikes and hypabyssal rocks, or even in marginal portions of deep-seated plutons.

A number of other specific textural terms exist that are used by geologists to describe the spectrum of textural possibilities. One of particular interest is the diabasic texture. In this case, two minerals of nearly equal size, and usually in the finer range of the phaneritic designation, are intergrown to form a rather unique pattern. The minerals involved are plagioclase feldspar in lath or rectangular shapes associated with equidimensional pyroxenes. The interlock is particularly strong, therefore the rock is highly durable and massive. Diabase commonly occurs in dikes and sills as in the Palisades Sill along the Hudson River in New Jersey.

Rocks with glassy textures, for example, obsidian, represent the extreme case of rapid cooling. These rocks are noncrystalline or amorphous. Conchoidal fracture is an obvious feature that accompanies the vitreous or glassy luster. Such rocks form on top of or within lava flows where rapid cooling is provided by direct contact with the air. The most common rock textures of igneous rocks are shown in Figure 3.8.

The relationship between cooling rate and texture should be apparent from the previous discussion. Simply stated, rocks that cool more slowly form larger crystals. Phaneritic rocks cool within the Earth, whereas aphanitic rocks occur as surface flows. Porphyries show a combined history of slow cooling to form the phenocrysts, followed by an accelerated phase that yields the fine groundmass. Glassy textures indicate the extremely high rate of cooling in which crystals are allowed no time to form. Therefore, the texture of an igneous rock is used to infer its origin or mode of formation.

Pyroclastic textures are those composed of clasts or pieces of material that are of volcanic origin and are held together by some cohesive or cementing material. The clastic texture can be determined by the ease with which pieces of material are broken from the rock. However, there are also sedimentary rocks with a clastic texture. Mineral content and appearance must be used to distinguish between clastic extrusive rocks and clastic sedimentary ones.

Some textures of volcanic rocks are due in part to the escape of gasses near the surface of the lava flow. This is why pumice, the frothy top of a lava flow, is filled with air bubbles. Some basalts also contain these holes or vesicles. As mentioned earlier, they are called vesicular basalts and occur commonly in volcanic areas.

In some instances the vesicles become filled with secondary minerals such as calcite or zeolite at a time subsequent to the actual lava flow. These fillings are called amygdules and a basalt so filled with secondary minerals is termed an amygdaloidal basalt.

Care should be taken to distinguish between amygdules and phenocrysts, both for academic and practical purposes. As previously noted, zeolites have a tendency to dissolve when in contact with water, which makes them undesirable constituents in an aggregate material. The shape of amygdules is helpful in distinguishing them from phenocrysts, because amygdules are usually rounded or misshapened much like a compressed air void. Also a preferred orientation or alignment of the cavity fillings is commonly exhibited, which is related to flow directions in the basalt. Phenocrysts typically do not show such preferred alignments.

MINERALOGY OF IGNEOUS ROCKS

Igneous rocks are composed primarily of silicate minerals plus only minor amounts of oxides, sulfides, and other mineral groups. Most silicates are relatively hard, with values on Mohs' hardness scale ranging from 5 to 8. This is why igneous rocks as a group are hard and rather resistant in the unweathered state.

It is risky to consider certain minerals to be exclusively igneous, metamorphic, or sedimentary because many minerals can originate in different ways or persist from one rock group to another. Yet it is meaningful to recognize that certain minerals com-

FIGURE 3.8   Different textures of igneous rocks: (a) very coarse grained, (b) coarse grained or phaneritic, (c) fine grained or aphanitic, (d) porphyritic, (e) pumaceous, and (f) glassy.

monly occur primarily in igneous, metamorphic, or sedimentary rocks and, therefore, their presence may be indicative of a certain origin. More worthwhile, however, is the combined use of texture, mineralogy, and field relationships to ascertain the rock origin.

The common silicate minerals in igneous rocks are feldspars (orthoclase and plagioclase), pyroxenes (in-cluding augite), amphiboles (hornblende included), quartz, olivine, and the micas (muscovite and biotite). Other accessory minerals are pyrite, feldspathoids, zeolites, and other less abundant silicates. The common minerals just listed are often found in igneous rocks but are also present in certain metamorphic rocks. In some cases many will persist as clastic grains in sedimentary rocks, so the student is cautioned not to use constituent minerals exclusively to determine rock names.

CLASSIFICATION OF IGNEOUS ROCKS

As mentioned earlier, igneous rock classifications are based on two properties: texture and mineral composition (mineralogy). Rock mineralogy is sometimes difficult to determine, particularly in fine-grained or aphanitic rocks. Fortunately, a general relationship between color and mineral composition can be used. Dark-colored minerals, that is, the green, black, brown, and dark-gray colors, are most commonly ferromagnesians. These minerals are low in silica and crystallize early as the magma cools. Calcic plagioclase is also dark in color, crystallizes early from the magma, and is associated with the ferromagnesian minerals. By contrast, the light-colored minerals (white, light gray, pink, red) are high in silica and crystallize late in the magmatic process. Quartz, orthoclase, and sodic plagioclase are light-colored minerals with a high silica content.

The classification of igneous rocks appears in Appendix A, Figure A.1, in chart form. In this diagram the igneous classification is presented graphically to enable a better understanding and easier identification of igneous rocks. The lower portion indicates the mineral composition of the rock in question. After this is determined, a vertical line is extended upward in the diagram. Based on the textural description a horizontal line is extended from the right side of the diagram until it intersects the vertical line and a decision is reached on the appropriate rock name.

As discussed previously, texture is also indicative of origin. This relationship can be seen by comparing the left and right sides of the igneous rock classification (in Appendix A, Figure A.1). Equal-sized, coarse-grained texture is indicative of the slow cooling at depth for intrusive rock. Fine-grained texture suggests the opposite: rapid cooling at the Earth's surface; and a porphyry typically denotes two stages of cooling rate, first slowly at depth to form the phenocrysts and then rapidly at the surface to form the fine-grained ground mass.

Glassy texture is listed below the fine-grained entry in Figure A.1 and it denotes a rock mass devoid of crystal form. Cooling has been so rapid that no nucleation of crystals could occur. In the glassy texture, color can be deceiving because only a minor amount of ferromagnesian minerals will yield a dark appearance. Such is the case for obsidian, which contains only a small amount of dark minerals, which provide a black color; yet obsidian is a high-silica rock.

Pyroclastic textures are also listed on the rock classification. Pyroclastic rocks comprise an important portion of igneous rocks and are included to provide a complete picture. Pyroclastics are subdivided according to the grain size of their constituent pieces. In a more detailed rock description, the composition of the pyroclastic is added to the fine-grained term in the form of a prefix, yielding such terms as rhyolite tuff or andesite tuff. This degree of description is not typically pursued in a beginning course on applied geology.

IGNEOUS ROCK DESCRIPTIONS

Granite is a coarse-grained igneous rock composed of orthoclase and quartz. Minor amounts of sodic plagioclase, hornblende, biotite, and muscovite may also be present. *Syenite* is also a coarse-grained igneous rock similar to granite except that it contains no quartz. Orthoclase is the essential mineral for the rock. *Gabbro* is another coarse-grained igneous rock containing calcic plagioclase and olivine or pyroxene (for example, augite). It is typically dark gray or black in color.

Diorite is a coarse-grained igneous rock intermediate in composition between granite and gabbro. It is usually lighter in color than gabbro because of the presence of light-gray plagioclase (sodic plagioclase). The rock typically has a salt and pepper appearance.

Rhyolite is the fine-grained equivalent of granite, that is, it contains orthoclase and quartz but has an aphanitic texture. This aphanitic texture makes it impossible to identify individual mineral grains, so the overall color is used instead for identification. *Trachyte* is the fine-grained equivalent of syenite having an aphanitic texture and containing abundant orthoclase but no quartz. The term *felsite* is used to refer collectively to the fine-grained high-silica rocks including both rhyolite and trachyte.

Basalt, the fine-grained equivalent of gabbro, is dark gray to black in color and contains calcic plagioclase. As the most common extrusive igneous rock, it is found in many areas of the world. Of slightly coarser texture than basalt but having a similar mineral composition is the rock diabase. It normally forms in a near-surface dike or sill. Diabase, basalt, and sometimes other dark fine-grained igneous rocks are designated by the term *trap* or *trap rock.*

Andesite is the fine-grained equivalent of diorite and is composed of sodic plagioclase, pyroxene, and usually amphibole. Andesites are lighter in color than basalts, showing a light gray to green color. Borderline cases between basalt and andesite are difficult to determine without the use of a polarizing microscope.

Ultrabasic rocks (or ultramafic) are those lower in silica than the basaltic composition. These rocks, which contain only minor amounts of calcic plagioclase, are almost entirely composed of olivine, pyroxene, or combinations of the two. Dunite is a fairly common ultrabasic rock that consists almost entirely of olivine. Peridotite consists of both olivine and pyroxene.

Obsidian is a high-silica glass that is dark in color and exhibits obvious conchoidal fracture. Its distinctive dark color is due to minor amounts of magnetite finely dispersed through the glass. Obsidian forms during extremely rapid cooling on the surface usually above or adjacent to a volcanic vent.

Pumice is the frothy, low-density rock that forms atop a lava flow because of the accumulation of gas bubbles at the surface. The small, numerous holes reduce the bulk specific gravity to a level even below that of water. Pumice is usually white to light gray in color with a fairly high silica content.

Scoria is a coarse, frothy rock of basalt composition and is the low-silica equivalent of pumice. It is usually red to black in color and forms on the surface of basalt flows.

Porphyries are named on the basis of the mineral composition and the relative percentage of phenocrysts versus groundmass. For example, a rock of granite composition is termed a granite porphyry if the phenocrysts make up more than 50% of the rock, and in like manner it is termed a rhyolite porphyry if the groundmass predominates.

Volcanic tuff is a pyroclastic rock composed of ash and dust derived from volcanic explosions. Tuffs include fragments less than 4 mm in size. They may be pieces of pumice, volcanic glass, or fragments of lava that were thrown into the air. Welded tuffs are pyroclastic rocks that become cemented or welded together by the incandescent heat of the deposits.Welded tuffs typically are difficult to tell from rhyolite or andesite.

Volcanic breccia is a pyroclastic rock composed primarily of fragments greater than 4 mm in size. Fragments of lava comprise most of the rock with lesser amounts of glass and pumice. Scoria may be present as well.

A simplified classification (Figure A.1) and tabulation sheet for igneous rock identification are supplied in Appendix A. Proceeding from left to right, the student can supply information that, when used along with Table A.1 on minerals, can indicate the rock name and information on origin.

ENGINEERING CONSIDERATIONS OF IGNEOUS ROCKS

Some of the engineering problems associated with igneous rocks were mentioned in the previous chapter on mineral identification. A more complete summary is included here for systematic review.

1. The use of igneous rocks as aggregates in Portland cement concrete can cause problems. In some instances fine-grained siliceous materials have caused volume expansion. The alkali-silica reaction problem can be alleviated by using low-alkali cements or nonreactive aggregates, or by adding pozzolans to the concrete mix. Pozzolans are natural rock materials or fly ash from the smokestacks of coal-burning power plants, which contain finely ground silica. They yield a nonexpansive reaction product when united with the alkali of the cement.

   The reactive igneous rocks include those that contain volcanic glass with a composition ranging from rhyolite through andesite. Basaltic glass contains too little silica to be reactive. Pyroclastic rocks containing glass with a high silica composition also can be reactive. This includes most tuff, volcanic breccia, obsidian, and pumice.

   Other siliceous materials in addition to these igneous varieties have proven reactive with high-alkali cement. Opal, or $SiO_2.nH_2O$, is considered to be the most active mineral constituent in this regard. Chalcedony, a fibrous variety of SiO_2 that occurs as the common agate, may also be reactive. Some cherts have proven troublesome but many observers suggest this is due to the presence of opal, and supposedly nonopaline cherts would not be reactive. Some schists, shales, and other rocks have also caused expansion in connection with the alkali-silica problem.

2. Very coarse-grained igneous rocks are undesirable for use as aggregates for construction. With increasing grain size, abrasion resistance is reduced, and the rock is less suitable for use as a base course (road base), concrete aggregate, or source of riprap (large stone used for slope protection along rivers and seacoasts).

3. The presence of certain minerals in igneous rocks makes the rock undesirable for some engineering uses. Zeolite minerals, which are relatively soluble in water, are undesirable in aggregates that will be exposed to the weathering process.

4. In foundations for engineering structures such as dams, bridge piers, and underground installations, weathered igneous rock or any other weathered rock is to be avoided. Excavation must extend through this material into sound rock.

5. Dimension stone includes rock used for tombstones and monuments plus facing stone for buildings. Igneous rocks are commonly used for this purpose because of their resistance to weathering. Rock that undergoes staining, fracturing, and spalling of the surface must be avoided when selecting the proper building stone. Strong, fresh, and unaltered igneous rocks yield the most suitable materials. Common, unweathered and unaltered phaneritic (coarse-grained) varieties are selected for building stone. These range from granite to gabbro.

EXERCISES ON IGNEOUS ROCKS

1. What is the difference between a rock and a mineral? Are all rocks composed only of minerals? (*Hint:* Consider volcanic glass and coal.)

2. Why are aphanitic igneous rocks more difficult to identify than are phaneritic ones? How is identification accomplished?

3. How does granite differ from diorite? How does basalt differ from obsidian? How does rhyolite differ from tuff?

4. Dikes can be aphanitic, porphyritic, or phaneritic. How is this possible? Explain. Dikes are typically steeply dipping (from the horizontal). How do you think dikes gain access through the existing rock? Explain.

5. Volcanic rocks can vary enormously in strength and in their other physical properties. Relate this to the two different varieties, lava rocks and pyroclastics. Which would be stronger and why?

6. Intrusive rocks are coarse-grained. Why are those of finer crystal size typically stronger than those with larger crystals? Consider rocks of different composition and then of the same composition. Explain.

7. What would be the maximum unit weight for a piece of pumice that floats in water? If the specific gravity of the material that comprises the pumice is 2.65, what is the porosity of the rock? Recall that the unit weight of water = 62.4 lbs/ft^3 and porosity = volume of the voids/total volume.

8. Volcanic ash deposits sometimes contain expansive clays such as smectite (montmorillonite) and commonly these deposits are water bearing. What engineering problems can be anticipated for such materials?

9. What are the properties of igneous rocks that make them good sources for gravestone monuments and facing stone for buildings? Why would dunite and zeolite-rich basalt not make good facing stone? (*Note:* Olivine weathers rapidly in a humid climate.)

10. Refer to the discussion on the Mount Saint Helens eruption. What mass of material was exploded from the mountain on May 18, 1980? How much is this in terms of cubic yards, cubic feet, and cubic meters? What are the individual hazards that developed as a consequence of this eruption?

11. Mount Rainier, Mount Hood, and Mount Lassen are other volcanic cones in the Cascades. What cities are they near? What major geologic hazards could they cause? Explain.

12. Based on discussions in the previous chapters, how are flood basalts probably related to plate tectonics?

Additional Readings
BARTH, T.F.W., 1962, *Theoretical Petrology*, 2nd ed., John Wiley & Sons, New York.

ERNST, W.G., 1969, *Earth Materials*, Prentice Hall, Englewood Cliffs, NJ.

MASON, B., 1960, *Principles of Geochemistry*, 2nd ed., John Wiley & Sons, New York.

WILLIAMS, H., TURNER, F.J., and GILBERT, C.M., 1954, *Petrography*, W.H. Freeman and Co., San Francisco.

4

Sedimentary Rocks

The focus in this chapter is on the classification, origin, and identification of sedimentary rocks. It is appropriate to recall that 95% of the outer 10 mi (16 km) of the Earth's crust is composed of igneous and metamorphic rocks. However, the emphasis here is that 75% of all rocks exposed at the Earth's surface are sedimentary. These rocks form the outer veneer above the more abundant igneous and metamorphic rocks below. For this reason it is likely that many engineers will work more with sedimentary rocks than with the other rock types.

Sedimentary rocks are those rocks composed of mineral grains or crystals that have been deposited in a fluid medium and subsequently lithified to form rock. This fluid medium in the broadest sense can be either water or air. However, pyroclastic rocks, which comprise a major portion of those sedimented in air, are included in this text in the igneous rock classification. Other authors have chosen to include pyroclastics with sedimentary rocks, because both can display clastic particles and bedding characteristic. In this text, however, pyroclastics are considered to be igneous rocks because both the extrusive rock varieties, that is, lava flows and pyroclastics, have a volcanic origin and both are found in the same geographic location.

51

RELATIVE ABUNDANCE

The three principal types of sedimentary rocks comprise more than 99% of the total: shale, sandstone, and limestone make up 46%, 32%, and 22%, respectively. The great abundance of shale in nature is easily overlooked. Shales form slopes rather than prominent cliffs and do not provide good bedrock exposures, so they are observed less commonly by the casual observer than are the more resistant rocks. Shale slopes weather readily into soil, which supports plant life that soon covers the shale outcrop.

SEDIMENTS VERSUS SEDIMENTARY ROCKS

A distinction must be made between sediments and sedimentary rocks. Sediments are a product of mechanical and chemical weathering. They are pieces of loose debris that have not been lithified, that is, have not been hardened into a rock material. In the engineering sense such sediments are considered to be soil consisting of a combination of gravel, sand, silt, and clay. Sediments are found in stream bottoms, wide floodplains, deltas, and alluvial fans where deposition by flowing water has occurred. Sedimentary rocks are held together by various types of cementing agents, such as calcite, quartz, or iron oxide, or by the compaction of their mineral grains into an indurated mass.

LITHIFICATION

Lithification is accomplished by several different processes that may act singly or in combination to change sediments into sedimentary rocks. The prominent contributors are compaction, cementation, and crystallization.

In compaction the pore space between individual grains is gradually reduced by the pressure of the overlying sediments. This is accomplished by the rearrangement of grains and a reduction in water volume. Coarse deposits of sand and gravel experience some compaction but fine-grained sediments containing silt and clay undergo a much greater reduction in volume. As compaction continues, the particles are forced closer together and the deposit becomes more dense and coherent. Estimates indicate that clay-rich deposits buried to depths of 3000 ft (1000 m) have been compacted some 60% from their original volume, whereas sand units are reduced only 30%.

It is also significant that the term *compaction* has a different meaning in engineering geology or geotechnical engineering than it does in a strict geologic sense when applied to lithification. These contrasting definitions are explored in detail in Chapter 14. The subject of subsidence is also discussed in that chapter.

Cementation involves the filling by a binding agent of the spaces between individual particles in a sediment. As mentioned previously, calcite, quartz, and iron oxide are the most common cements but less abundant ones include opal, chalcedony, anhydrite, and pyrite. The cementing material precipitates and crystallizes from the water circulating through the sediment. Coarse-grained deposits with their larger interstitial pore spaces are more easily cemented together than are finer ones. For this reason it is extremely difficult for clay-rich rocks to become well cemented and most are held together by compaction with only a minor contribution from cementation.

Crystallization occurs in sediments some time after deposition, and in some cases it happens long after lithification is completed. New minerals crystallize or crystals of existing minerals increase in size. Such minerals crystallize from amorphous, colloidal substances in the fine mud. This contributes to lithification in finer grained sediments.

FORMATION OF SEDIMENTARY ROCKS

The constituents that comprise sedimentary rocks are supplied in two ways, the first from broken pieces of preexisting rock. These sediments, produced by rock weathering and known as detrital (Latin for "broken down") material, are transported into the sedimentation basin. The second contribution to sedimentation is chemical precipitation and shell deposition. Its immediate source is the depositional basin, although dissolved solids were supplied originally by the landmass.

Particles of gravel, sand, silt, and clay derived from erosion of the land surface are examples of detrital sediments. They are typically carried to the oceans by streams and deposited some distance offshore. Chemical deposition is usually accomplished by precipitation of dissolved materials from the water. This may occur by direct inorganic precipitation or through the action of plants or animals that extract chemicals from the water.

Many organisms such as corals and clams extract calcium carbonate from seawater and use it to build their shells. Other organisms, for example, a variety of algae known as diatoms, secrete a microscopic skeleton (or test) composed of silica. This is a major source of sedimentary chert. When the animals and plants die, their skeletons settle to the bottom, becoming part of the accumulating sediment. This provides the biochemical portion of the deposit. Precipitation of calcite around detrital and chemically deposited constituents yields lithification of the sediments to form sedimentary rocks. Individual rocks typically contain portions of both constituents although most are predominantly either detrital or chemical in nature.

The sedimentation process controls how sediments are laid down or deposited. Factors that effect sedimentation can be enumerated by starting with the basin of deposition and working backward to the landmass. The source of the sediment, how it was transported to the basin, and the environment of deposition all come into play.

The source of detrital material is a preexisting rock mass located on the land surface. The nature of this rock—its composition and texture—influences what form the rock fragments will take. Rock weathering plays an important role in how the constituent mineral grains are changed into disaggregated particles. Weathering is discussed in Chapter 8. Regarding the sedimentary process we note that both the solid particles (detrital material) and the dissolved constituents are dictated by the original rock and the weathering process.

Transportation, both type and duration, influences how these weathering products will change before arriving at the depositional basin. This is controlled by the geologic agents of transport: stream flow, ground water, glaciers, ocean currents, wind, and gravity. Each of these agents is discussed in later chapters.

Sedimentation occurs when a geologic agent is no longer able to transport material any further. Suspended particles are deposited at a point of sudden energy loss, such as a decrease in stream velocity. Dissolved solids will precipitate when environmental changes are sufficient to trigger this action, which may be chemical or biochemical in nature. After a subsequent change in conditions, sediment transport may be resumed but the detention time between travel episodes may consist of weeks, years, or millenium. The final destination can be a stream valley, lake bottom, or zone in the ocean. This final sedimentation zone provides additional influence owing to its own chemical and physical makeup. In all we see that a sedimentary rock is a product of its rock origin plus the geologic developments from weathering to deposition and lithification.

SEDIMENTARY ROCK CLASSIFICATION

Sedimentary rocks are subdivided on the basis of texture into four major groups: 1) clastic, 2) coarsely crystalline or interlocking grains, 3) fine grained or cryptocrystalline, and 4) whole fossils or their alteration products. Each of these comprises a major group of sedimentary rocks.

Clastic Sedimentary Rocks

Clastic sedimentary rocks are composed of preexisting mineral grains or rock pieces that settled out of water (or air) and were subsequently cemented or compressed together to form rock. Those held together by compaction are some of the weakest sedimentary rocks and tend to slake readily when soaked in water. Compaction shales and compaction siltstones are examples of these low-strength clastic rocks.

Clastics are further subdivided based on the size of the clasts or grains that comprise the rock. Also, the rock name is determined by the size of the most prevalent grains. Table 4.1 shows the relationships between particle size and clastic sedimentary rocks. The Wentworth scale of particle size, commonly used by geologists, is followed in the table.

Conglomerates and Breccias

Conglomerates and breccias are composed of gravel-sized pieces, the distinction being that the grains comprising a conglomerate are rounded whereas those in a breccia are angular. Rounded particles

TABLE 4.1   Sediments and Clastic Sedimentary Rocks

Sediment	Size	Sedimentary Rock
Gravel	>2 mm	Conglomerate (if particles rounded)
		Breccia (if angular)
Sand	2 to 1/16 mm	Sandstone
Silt	1/16 to 1/256 mm	Siltstone (commonly called shale)
Clay	< 1/256 mm	Claystone (commonly called shale)
Silt and clay		Shale

suggest the common erosion process in which rock pieces are rounded during transport by stream flow some distance before reaching the sedimentation basin. Angular pieces denote a short transport distance with quick burial, typical of areas with rapid sedimentation and subsidence of the basin. This denotes an active area of deposition that receives major amounts of sediment as it subsides.

Sandstones

Many quartz- and feldspar-rich sandstones are only partially cemented and hence contain a high degree of interconnected pores. If the sand grains are loose and can be removed by rubbing the rock surface, the expression *friable sandstone* is applied. This porosity of sandstones is conducive to high permeability (the ability to transmit fluids) and thus many sandstones are good reservoir rocks for water, petroleum, and natural gas. For the same reason, such sandstones below the ground-water table can yield great volumes of water when tunnels are driven through them. This can be a major problem that significantly reduces the rate of tunneling and reduces worker safety. Special construction techniques are needed to minimize ground-water inflow.

As previously mentioned, the term *sandstone* pertains to a clastic sedimentary rock composed of sand-sized grains. When the grains are essentially all quartz, the rock is called a quartz sandstone. A sandstone composed of feldspar fragments (usually orthoclase), or pieces of preexisting rocks which themselves are rich in feldspar, is called an arkose or arkosic sandstone. Dark rock particles from basalt or other low-silica igneous rocks comprise those sandstones known as graywackes. They contain clay minerals, including chlorite in the matrix, which sur-

rounds the rock fragments and holds the rock together. Graywackes are common in mountainous regions and in plateaus associated with mountains. They are typically interbedded with dark shales and possibly chert beds and subaqueous lava flows, as in the California coastal ranges.

Generally, graywackes are more strongly cemented than are friable, quartz sandstones but not as strongly as limestones or related sedimentary rocks. In some regions, materials engineers have used graywackes for concrete aggregate and as base course (road base) materials for highways. Problems with degradation (or reduction in the size of the aggregate pieces) have occurred when these materials were used. When used as concrete aggregates some graywackes have proven to be alkali-silica reactive, causing expansion and extensive cracking of the concrete.

Some limestones consist of sand-sized fragments of broken fossils or of rounded carbonate grains. These rocks are termed *clastic limestones* because they are composed of individual grains of calcite that are cemented together. Geologists and engineers should refrain from referring to clastic limestones as a sandstone or calcareous sandstone on both drilling logs and in engineering geology reports. The term clastic limestone is more appropriate and less confusing. In many areas of the United States, sandstone carries with it the connotation of quartz sandstone, because most sandstones of the midcontinent region are indeed quartz-rich. Hence, use of the term *calcareous sandstone* on a drilling log suggests a sandstone that is mostly quartz with some calcite perhaps as the cement, rather than a rock composed primarily of sand-sized pieces of calcite. Clastic limestones can consist of broken fossils, oolites (small rounded grains of calcite), or other calcite fragments.

Shales

Shales are clastic, fine-grained sedimentary rocks composed of silt and clay. An additional feature usually associated with shales is that they show fissility or the prominent, closely spaced bedding planes along which the rock will break. Fissility indicates a certain strength caused by compaction of the material.

The term *shale*, however, is commonly applied in a more general way to include all clastic sedimentary rocks that are finer grained than sandstone. This is one of the reasons why shale has the reputation of being such a varied material from one geologic unit to another. As a general group, shales vary greatly and, in some cases, other terms such as siltstone, mudstone, or claystone are more appropriate.

Rocks lacking fissility, strictly speaking, are not shales. Mudstones, like shales, consist of silt and clay, but they lack fissility. Siltstones consist predominantly of silt-sized pieces and therefore do not exhibit a plastic nature when cut or rubbed. Instead, the gritty nature of the silt grains can be realized by grinding a small piece of the rock between one's teeth. Claystones consist predominantly of clay minerals, lacking the grittiness of siltstones and the fissility of shales.

For purposes of identification in a first geology course for engineers, only two groups of fine-grained clastic sedimentary rocks are commonly considered. These are 1) siltstone, which shows a recognizable grittiness and lack of clay, and 2) shale, which has both silt and clay present or predominantly clay and might or might not illustrate fissility.

An engineering concern regarding shales and related rocks is whether or not they will require blasting for excavation. Earth moving can be accomplished in some fine clastic sedimentaries either by wheeled scrappers, dozer blades, or by ripper teeth attached to the back of a dozer and pulled across the rock surface. The rippability of the shales and siltstones is a function of fissility, cemented strength, and type of interbedding of the units. Thin alternating beds of soft shale and sandstone are easily ripped, massive siltstones are not, with the various other fine clastic rocks ranging somewhere between these extremes. The seismic velocity of geologic units (sound wave or P wave velocity) has been used to estimate rippability. This technique is considered more fully in a subsequent discussion in Chapter 18, Earthquakes and Geophysics.

Argillites are slightly metamorphosed shales or shale-like rocks. They commonly have greater strength than shales and are used in areas of limited aggregate supply for base course and concrete aggregates. The problems involved with their use are similar to those associated with graywacke sandstones.

Coarsely Crystalline Sedimentary Rocks

Coarsely crystalline sedimentary rocks, or those with large interlocking crystals, are an important group of nonclastic rocks. The interlocking nature of the crystal grains is similar to that of a phaneritic igneous rock. Instead of consisting of silicates, these rocks are composed mostly of carbonate, sulfate, and chloride minerals.

Sedimentary rocks displaying interlocking crystal textures are formed by precipitation from water or by recrystallization of a finer material during and after lithification. Some limestones show a recrystallized mosaic of calcite crystals in which the individual crystals are of a coarse sand size.

Mineral constituents will precipitate from water when the solubility is exceeded for that particular chemical compound. Solutions can become saturated (or oversaturated) with respect to a certain compound either by changes in temperature (usually a decrease), by changes in pH of the solution, or by a reduction of the liquid volume by evaporation. Turbulence in the water may also encourage precipitation if the water is nearly saturated relative to a particular chemical.

Mineral crystals become intergrown or intermeshed as they grow larger in size and abut each other. Because of this, the interlocking texture of precipitated and recrystallized sedimentary rocks can be distinguished from clastic rocks with their rounded grains or pieces held together by a cementing agent.

Carbonates

The two most common carbonate rocks are limestone, composed mostly of the mineral calcite ($CaCO_3$), and dolomite or dolostone, which is composed primarily of the mineral dolomite [$CaMg(CO_3)_2$] plus varying amounts of calcite. Some limestones are precipitated directly from seawater.

Precipitated limestones commonly are fine grained with a smooth texture. (Normally, recrystallization is required for limestones to develop phaneritic crystals.)

Commonly these rocks are called lithographic lime-stones, because such limestones of a fine-grained, smooth variety were once used for lithographic plates in the early printing industry. Because of their fine texture, lithographic or dense limestones are not classified with coarsely crystalline ones even though most other sedimentary rocks formed by precipitation are included in that group. This illustrates a problem of deciding rock origins based on descriptive features. A consistent relationship between origin and texture does not necessarily hold true when extended to a large group of rocks such as chemically deposited sedimentary rocks. Relating the origin and texture for specific rocks still provides the best means for rock study and classification.

A considerable amount of calcium carbonate is known to precipitate directly from solution in the open, warm oceans of the tropics and subtropics. When it is deposited as a continuous, fine-grained material, lithographic limestone results, but in other cases the precipitated calcite serves as a cementing agent for other constituents. Dolomite, however, is only rarely deposited directly from the ocean. The solubility of Mg^{++} is sufficiently great that primary dolomite (directly precipitated from seawater) is fairly rare and is generally formed only when seawater evaporates in an enclosed basin.

Dolomites are formed, most commonly, by the replacement of Mg^{++} for some of the Ca^{++} in lime-stones long after the rock has lithified. Time is an important factor in this replacement process: Older sedimentary sequences have a higher dolomite percentage in the carbonates than do younger sequences with a generally similar geologic setting. In rocks of the Silurian age or older (400 million years or more) the chances are high that the carbonate rocks will be dolomite rather than limestone. In the replacement process, many dolomites take on an interlocking texture of visible crystals. Those that do this are classified into the coarsely crystalline category. Other dolomites have such small crystals that their outline cannot be seen with the naked eye. These are classified under the fine-grained, dense textural group that also contains lithographic limestone.

Evaporites

Evaporites are sedimentary rocks derived when sea-water is evaporated in an isolated portion of the ocean. Minerals precipitate sequentially with decreasing water volume and the order of crystallization is dictated by the relative solubilities of the mineral constituents.

Seawater contains about 3.5% dissolved solids in the form of cations, anions, and complex molecules. Nearly 78% of the solids are represented by Na^+ and Cl^-. When seawater is reduced to about one-half its original volume, calcite and some iron oxide are precipitated. With additional reduction in water volume to about one-fifth the original, gypsum ($CaSO_4 \cdot 2H_2O$) is precipitated. Another sulfate mineral, anhydrite or $CaSO_4$, also occurs commonly in sedimentary rock sequences.

Since anhydrite is more abundant in the subsurface than gypsum, it is generally assumed, based on field evidence, that gypsum is formed from anhydrite by descending ground water. Anhydrite is likely formed at depth under the heat and pressure of a growing rock column in a sedimentary basin. The apparent sequence is as follows: Gypsum is precipitated from seawater in an evaporite deposit; gypsum alters to anhydrite over geologic time during burial; and gypsum is formed again when erosion brings the evaporite deposit close enough to the surface that ground water comes in contact with it.

Gypsum has economic importance because it is used in large volumes to make gypsum wall board (plasterboard) and also as an additive in the manufacture of Portland cement. Anhydrite, which does not hydrate easily in industrial processes, is not as valuable as gypsum but can be used in the manufacture of Portland cement. It is also harder than gypsum (4 versus 2 on Mohs' hardness scale) causing greater wear on crushing and handling equipment.

With 10% of the original water volume remaining, NaCl begins to crystallize. Magnesium salts (sulfates and chlorides) are formed on further evaporation and the last to crystallize are NaBr and KCl. All of these salts, which form after NaCl, precipitate with about 2% of the original water volume remaining.

Commonly the rock name for evaporite deposits is based on the most prevalent mineral, for example, anhydrite rock and gypsum rock. For NaCl-rich rocks, the mineral name halite is not applied but instead the common name, rock salt, is used. Bittern salts refer to potassium- and magnesium-rich salts and are so named because of their bitter taste.

Fine-Grained or Cryptocrystalline Rocks

Fine-grained, cryptocrystalline, or amorphous features are those sedimentary textures that are too fine to discern with the naked eye. Rocks belonging to this category that were previously discussed are lithographic or dense limestone and dense dolomite. The other prominent member of the group is chert, including flint and opal.

Chert

Chert is a dense, hard, siliceous material (when unweathered) that occurs as rounded to irregular nodules or as persistent layers in sedimentary rocks. It is located chiefly in limestones, but also in graywacke sequences. Cherts are also widely found in stream gravels, which contain a significant portion of sedimentary rock constituents. These may be rounded, polished pieces showing all degrees of weathering from very little (still smooth and dense) to deeply weathered (white, soft, and porous).

Chert in its unweathered condition has many of the characteristics of quartz: a hardness of 7, conchoidal fracture, and high resistance to chemical weathering. In fact, the primary constituent of chert is cryptocrystalline or finely crystalline quartz. The dark-gray or black variety of chert is referred to as flint in many areas of the United States; in some localities, however, gray, black, or brown varieties are termed flint. Flint, chert, and volcanic glass (obsidian) were used by Stone Age humans to make arrow points, spear heads, and cutting knives. White chert, which develops a weathered, brown surface coating, is the common variety present in stream gravels.

Chert is formed either through primary deposition of silica in the oceans in a colloidal form or as a secondary replacement in carbonate rocks after deposition. Examples of both primary and secondary chert deposition are widespread. Secondary chert may be formed when minute silica or fragments of siliceous shells distributed throughout a rock mass are first dissolved and then concentrated by precipitation to form a large nodule near the center of the contributing rock.

ENGINEERING PROBLEMS OF CHERT   As mentioned previously in Chapter 2, cherts are subject to two problems when used as concrete aggregates. Deeply weathered chert develops surface pop-outs when used in concrete that undergoes freezing because of the high porosity of weathered chert. The second concern is that certain cherts undergo an alkali-silica reaction with high-alkali cements. This reaction leads to cracking and expansion of concrete and, ultimately, to failure of the material.

The constituents in the cherts that lead to the alkali-silica reaction are chalcedony and opal. Cherts can consist of three different constituents: opal and chalcedony as just mentioned, along with fine or cryptocrystalline quartz. Opal is an amorphous variety of SiO_2 containing some water of hydration, and chalcedony is SiO_2 with a fine, fibrous texture, which can only be discerned using a petrographic microscope. Opal is highly reactive with alkalis—only 0.25% of the aggregate by weight is needed to cause considerable expansion in alkali-rich concrete. For chalcedony a percentage of about 3% of the coarse aggregate is needed to cause this expansion.

To prevent the alkali-silica reaction problem, three procedures can be followed: 1) Use low-alkali cement (<0.6% alkali as Na_2O), 2) use nonreactive aggregate or add nonreactive aggregate to reduce the reactive portion by dilution to a level below the percentages mentioned in the previous paragraph, or 3) add pozzolans, which include certain volcanic rocks and fly ash or ash collected from smokestacks where coal is burned. These finely ground high-silica materials yield a nonexpansive reaction product with the alkali in the cement.

To solve the pop-out problem of weak cherts, the percentage of these low-density materials (SpG less than 2.45) is kept at a minimum. This is accomplished by selectively choosing gravel sources and by reducing the maximum allowable size of the coarse aggregate. This latter procedure is effective because large, weak pieces are more prone to freeze-thaw damage than are smaller ones, because the larger pieces are unable to expel all the water prior to freezing and must absorb the entire volume increase when ice crystals form.

Whole Fossil Rocks

Whole fossils or their alteration products yield the fourth and final textural group for sedimentary rocks. These textures are formed from calcite fossils or from materials with a high organic content, including coal.

Fossiliferous Limestone

Fossiliferous limestones consist of abundant calcite fossils of sand size or bigger, many of which are still intact or unbroken. The fossils are cemented together by minor amounts of calcite, which hold the pieces together where they touch or fill the interstices between the fossils. In most fossiliferous limestones, the grains are large enough that they can be abraded with ease. Rocks of this type commonly are too weak to be used for aggregates in concrete or bituminous mixes or as a support for road construction (base course).

If the coarse fossil limestone has its interstices filled with calcite cement, the rock is simply termed a fossiliferous limestone. If instead only the points of contact between fossil fragments are held by calcite, the rock is known as coquina. Because the shells in much of the coquina are gravel sized, the term *shell conglomerate* is sometimes used.

Another calcite-rich rock of marine origin is the familiar material chalk. It is a special variety composed of minute shells or shell fragments cemented lightly together with calcite. Foraminifera, tiny protozoans (small marine animals), have shells, or tests, composed of calcium carbonate that comprise the chalk. These shells are deposited in the deep water beyond the depositional zone of clastic sediments derived from the land.

Rocks Rich in Organic Matter

Organic matter is present in nearly all sedimentary rocks. In marine sediments, the range is about 0.5% for the lower limit to about 10% for the upper limit, the latter occurring in black, pyrite-bearing shales. In general, fine-grained rocks tend to contain more organic debris than do coarse-grained ones.

Sedimentary deposits also exist that contain up to 100% organic material. These are the freshwater organic accumulations. One of the most important of these is the coal group, which includes peat, brown coal or lignite, bituminous coal, and anthracite coal. A second variety of organic debris comprises the petroleum-rich rocks.

Peat is associated with freshwater swamps. Organic remains are prevented from complete decay by the smothering action of the water, which covers the dead trees and mosses. Organic residues may comprise 70% to 90% of the total accumulation. Coal is formed through burial of the peat with an attendant heat and pressure increase. Water is removed through burial and volume reduction, and an increase in carbon content is also accomplished. The progression through the coal series is accomplished by increased vertical pressure. Anthracite coal is formed during intense folding of sedimentary rocks and is classified as a metamorphic rock.

The petroleum series of organics have as their probable origin compounds that collect on the bottom of ocean basins, lagoons, and estuaries. This accumulation is richer in fatty and protein substances than are peats. Normally they are mixed with clay and silt, yielding a lower organic composition than that present in peat beds. With burial and increases in pressure from the overburden, oil shale is formed first and then eventually petroleum. Oil shale is a black shale containing considerable amounts of hydrocarbons mixed with the silt and clay. Research is currently under way to obtain petroleum from oil shales at a cost competitive with other energy sources. Extensive areas of oil shale are present in Colorado and adjacent states, with deposits of lesser extent in many other states of the United States.

SEDIMENTARY ROCK IDENTIFICATION

Along with information presented in this chapter, sedimentary rocks can be identified by means of Table A.5 in Appendix A. The sedimentary rock classification is based first on texture and then on additional physical characteristics of the rock.

The first step in the identification process is to determine the characteristics of the texture: 1) clastic, 2) coarsely crystalline, 3) fine grained or dense, or 4) composed of whole fossil or their alternation products. In the case of clastic sedimentaries, the grain size is used to further subdivide the groups. Then the shape of the grains (conglomerate versus breccia) or the nature of the grains (siltstone versus shale; quartz sandstone versus arkose) is used to provide the rock name.

For the whole fossil group, the nature of the fossil or organic material is used to determine the rock name. Peat, lignite, and coal are identified in this way. For the fine-grained or dense rocks, the mineralogy is determined in order to name the rock. The amount of chert, calcite, or dolomite present determines what rock name will apply.

In the case of limestones the procedure may be reversed. After the sedimentary rock is observed to effervesce freely in cold hydrochloric acid, the rock is recognized as a limestone. The texture is used to give the rock its complete name. Textures may be clastic (composed of broken pieces of calcite), coarsely crystalline, fine grained or cryptocrystalline, or consisting of whole fossils, with two extremes for the last case: large gravel-sized shells (coquina) or tiny microscopic ones (chalk). These limestones typically are deposited in marine environments.

FEATURES OF SEDIMENTARY ROCKS

Stratification, or bedding, of sedimentary rocks is commonly the most significant rock feature. From an engineering point of view, bedding provides the extremes in directional properties of sedimentary rocks and also, to a large extent, the anisotropic nature of these materials. Extreme differences in strength, permeability, electrical conductivity, and seismic velocity are introduced by stratification. These properties are examined in Chapter 6, Engineering Properties of Rocks. Horizontal bedding planes are shown in a limestone quarry in Figure 4.1.

Other features of sedimentary rocks that merit discussion are the primary structures of bedding planes (cross bedding, ripple marks, mud cracks, basal conglomerates, graded bedding and fossil position) and the secondary structures of these rocks (concretions, nodules, geodes, and stylolites).

A valuable characteristic of the primary bedding plane features is that commonly they can be used to tell the original top from the original bottom of the sedimentary sequence. These sedimentary features are examined following a further discussion of bedding.

Bedding

The layering or bedding of sedimentary rocks marks the solid-water interface between the sediment and the water medium at the time of deposition. As additional sediment is added the previous boundary is preserved by minute differences in sedimentation.

Bedding planes are manifested by changes in the depositional materials. They are characterized by

FIGURE 4.1 Horizontal bedding planes in a limestone quarry.

slight changes in the following properties: color, mineral content, or texture. Reduction in water content and eventual lithification reduces the sediment thickness but under most circumstances the subtle changes of bedding are preserved in the rock.

Partial weathering of a sedimentary rock enhances these bedding features, making them appear more obvious and sometimes promoting fracture along these planes. Fresh rock cores of shale may show only subtle indications of bedding, but if the cores are exposed to wetting and drying conditions (left lying out in the weather for a period of time) they will form thin wafers a few millimeters thick. Sedimentary rocks used as dimension stone on the exterior facing of buildings or for stone walls should possess only subtle bedding effects because these planes will be exploited by weathering. In most situations, bedding planes are placed in the horizontal rather than vertical position to minimize the ingress of water into the dimension stone.

Bedding planes for most sedimentary rocks were originally oriented in essentially the horizontal position. This occurs because most sedimentary rocks have a shallow marine origin and these rocks are formed on the continental shelf where the ocean bottom closely approaches the horizontal. A basic concept of sedimentary rocks known as the *principle of original horizontality* states this horizontal condition. This is shown in Figure 4.1.

There are some exceptions to this original horizontality of beds. Reef deposits of limestone or dolomite provide an example of such an exception. In this situation the ocean bottom slopes away from the reef and different animal and plant assemblages thrive at different water depths, ranging from the surf zone to as deep as 300 ft (100 m). Figure 4.2(a) is a sketch of a limestone reef forming in the ocean and 4.2(b) is a photograph of a stone quarry in a dolomite reef that was uplifted after lithification. In Figure 4.2(a) the fossil reef continues to accumulate shells as the biological life cycles occur and organisms die. If the reef sinks under its own weight it can continue to grow while utilizing the same water depth, and perhaps persist for spans of millions of years. The Silurian period (about 400 million years ago) at various locations in the United States is marked by the development of extensive fossil reefs with compo-

sitions that have changed from limestone to dolomite over geologic time. Much of the original porosity of these reefs has been preserved despite the dolomitization process.

Bedding Plane Features

Bedding associated with dunes developed in sand is commonly inclined to the base of the dune. This feature, called cross bedding or false bedding, occurs when sand dunes migrate in the direction of current movement. Three types of beds comprise a single dune, the bottom set, foreset, and topset beds, as illustrated in Figure 4.3. The foreset beds, which represent the sloping face of the dune, may be inclined at angles from 5° to 30°. When a change of current direction occurs, the topset beds erode, truncating the foreset beds. New topset beds are deposited that intersect the old foreset beds at an acute angle, whereas the foreset bed below retains its smooth curve, which becomes tangent to the base. This provides a means of telling the original bottom (tangent portion) from the original top of the bed (truncated), regardless of how much the sequence has been tilted subsequently by rock folding. In Figure 4.3(c) the original beds have been overturned and now lie upside down. For vertical beds one can also determine the original top from the original bottom in this manner. A photograph of an ancient sand dune formation in Zion National Park, Utah, is shown in the lower portion of Figure 4.3.

Ripple marks are undulations along the bedding plane of a sandstone preserved through the lithification process. They develop on sand dunes, beaches, or stream bottoms. An example is shown in plan view in Figure 4.4: a close-up is shown in the upper photo and a view of the ripple marks on a rock face in the lower one.

Ripple marks are of two varieties: symmetrical ones with pointed cusps and asymmetrical ones showing rounded troughs and ridges. Symmetrical ripple marks are formed by the back and forth motion of water along the seacoast but beyond the surf zone. These are oscillation ripple marks and can be used to tell the original top from original bottom of the sedimentary sequence. The cusps point toward the original top of the unit. Figure 4.4 shows symmetrical ripple marks.

Asymmetrical ripple marks depict current ripple effects or those with a persistent direction of move-

(a)

(b)

FIGURE 4.2   (a) Formation of a limestone reef. (b) Stone quarry in a dolomite reef.

ment. Because of their rounded troughs and ridges, no difference is indicated between the original top direction and that of the original bottom in the sequence. Whether it is the portion below the rippled surface or the imprint of that surface being observed cannot be discerned. If the top of the bed is ascertained by another means, the direction of the current movement during deposition can be found based on the sloping surface.

A basal conglomerate sometimes comprises the first layers formed above an erosion surface or when an abrupt change in sedimentation occurs. It consists of rock pebbles or cobbles surrounded by sand, which grades upward into a typical sandstone. The pebbles mark the residual weathered rock supplied to the basin when the new sedimentation sequence began. A basal sandstone therefore can be used to tell the original top from the original bottom of the sequence; obviously the conglomerate is at the bottom. This is illustrated in Figure 4.5.

Unfortunately, conglomerates do not always mark the beginning of a sandstone unit because intraformational (within the formation) conglomerates also occur. If these are repeated within a sandstone unit because of changes in the supply of detrital sediment, the top cannot be discerned from the bottom based on these conglomerates.

Graded bedding is indicated when the particles in a sedimentary bed vary from coarse at the bottom to fine at the top. This feature therefore can be used to discern the original top of the bed. Graded bedding is formed by the rapid deposition of particles from turbid water carrying a range of different sizes. It is generally agreed that these are deposited by turbidity or density currents that sweep down steep slopes in an ocean or large lake. They were responsible for breaking the transatlantic cable in 1929 and it is suspected that they occasionally transport large volumes of sediment from the upstream areas toward the dam in human-made reservoirs. Graywackes

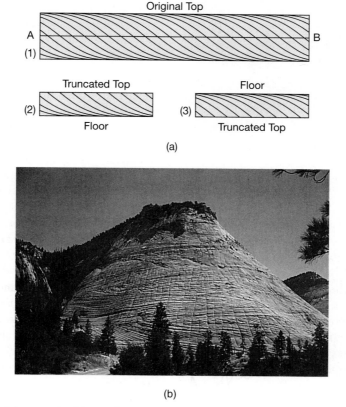

FIGURE 4.3 (a) Cross bedding, showing (1) original complete set of beds, (2) top portion truncated, along A–B, and (3) these beds overturned from original position. (b) Cross bedding in sandstone at Zion National Park, Utah.

(a)

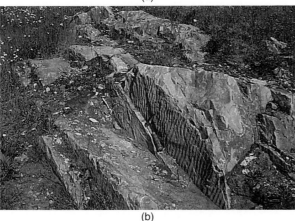

(b)

FIGURE 4.4 Ripple marks. (a) Close-up of ripple-marked surface. (b) Ripple marks on a vertical bed.

sometimes show graded bedding and consequently it is believed that such rocks are deposited in deep marine environments by turbidity currents.

Mud cracks and position of fossils are the final primary features of bedding planes considered in this discussion. Mud cracks occur when silt or clay dries out and shrinks to yield polygonal cracks. If the cracks are filled by sediment, particularly sand, the mud cracks may be preserved after the rock is lithified. The presence of mud cracks in ancient rocks shows that the sequence was exposed to the air and that the mud cracks pinching out with depth indicate the original top of the sedimentary sequence. These details of mud cracks are illustrated in Figure 4.6.

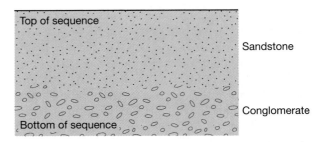

FIGURE 4.5 Basal conglomerate in sandstone sequence.

When large fossils such as the valves (or the half shell) of a clam come to rest on the ocean floor, shifting currents will ensure that the shell lies concave downward. Therefore, the original bottom of the sedimentary sequence is in the concave direction of the shells. This can be observed in the cross section of the sedimentary rock sequence of Figure 4.7. Raindrop impact impressions and animal tracks can sometimes be observed along sedimentary bedding planes. These can be useful features concerning the environment of deposition.

Secondary Structures

Many sedimentary rocks contain structures that were formed after the sediment was deposited and the rock lithified. These are secondary structures and, unlike the primary features, do not indicate conditions prevalent during deposition. They do document the sedimentary origin of rocks that contain them and many prove helpful during identification and study of geologic history. The features of interest are nodules, concretions, geodes, and stylolites.

A nodule is an irregular, knobby body of material that differs in composition from the sedimentary rock that contains it. Nodules are usually aligned parallel to the bedding of the rock and several may be joined to form a continuous bed. Nodules are typically 1 to 4 in. (2.5 to 10 cm) in diameter. Chert or flint comprises most common nodules and they typically occur in limestone or dolomite, but also in shale. Nodules probably form when silica distributed throughout the sediment is concentrated in a central location during and after lithification. A nodule is shown in Figure 4.8.

Concretions are local concentrations of material in a sedimentary rock that have a rounded, smooth surface and range in size from less than 1/4 in. (1 cm)

FIGURE **4.6** Mud cracks in (a) recent sediments and (b) in lithified rock.

to 3 ft (1 m) in diameter. They typically consist of the same cementing material that lithified the rock and therefore are probably composed of calcite, iron oxide, or silica, which comprise the common cements for sedimentary rocks. Concretions, typically are

more strongly held together than is the host rock; a large concentration is illustrated in Figure 4.8(c).

Geodes, attractive objects commonly available at rock and mineral shops, are roughly spherical in shape with a hollow interior partially filled with mineral crystals. They typically range from about fist size to 1 ft (30 cm) in diameter. A layer of banded chalcedony normally forms the outside layer, and crystals of quartz, calcite, or dolomite project inward toward the center. Various sulfide minerals and other less common varieties are also observed on an infrequent basis. A quartz-filled geode is illustrated in Figure 4.8(b).

Most geodes are found in limestones and typically are most abundant within a specific zone of a geologic unit. Some geodes may be nearly solid and very dense if the crystals inside have grown together. Geodes released from their geologic units by weathering or erosion are sometimes found along stream beds in limestone terrain.

The formation of hollow geodes in a solid rock is an intriguing question for geology students and mineral collectors alike. It seems reasonable that a simple opening would close as a result of pressure from the overlying rocks. First a water-filled cavity develops in the sediments, probably when plant remains decay within a buried deposit. As the sediment consolidates during the rock-forming process, a wall of silica, colloidal in nature, forms around the water cavity, isolating it from the surrounding material. With time, freshwater washes through the sediments expelling the saltwater, but the protected cavity retains its highly saline water. This sets up an osmotic effect as the higher concentration inside the void tries to equalize with the water outside through the semipermeable membrane, the silica wall. As long as osmosis continues, pressure is exerted outward from the cav-

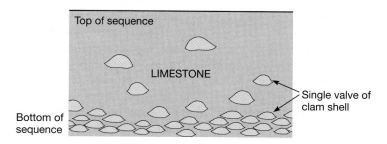

FIGURE **4.7** Fossil orientation in sedimentary sequence. Shells lie with concave portion downward.

FIGURE 4.8 Nodules, geodes, and concretions. (a) Chert nodules in limestone with pieces of chert weathered from rocks. (b) Quartz-filled geode. (c) Large septarion concretion showing cracks filled with secondary quartz.

ity, which expands little by little until the salt concentrations are equalized. At this juncture, osmosis stops, the outward pressure dissipates, and the cavity no longer expands. With time, the silica wall dries, chalcedony crystallizes from the colloidal silica, and the wall contracts, yielding cracks. At some later time ground water seeps through these cracks and crystals begin to grow inward, toward the center, from the chalcedonic lining. As a consequence, a crystal-lined geode is formed. A significant point is that in geodes crystals grow inward, but in concretions they grow outward.

Stylolites are serrated planes or surfaces that occur parallel to bedding in limestones. They are formed by the dissolving action of water as it moves along the bedding planes followed by deposition of impurities. In a cross section perpendicular to bedding they appear as irregular lines along selected bedding planes, and are common features in polished limestone and marble slabs used in building interiors. In some instances people have mistaken stylolites for fossils, worm burrows, and primary sedimentary features. Stylolites are illustrated in Figure 4.9.

FIGURE **4.9** Stylolites in limestone appear as wavy lines and a dark surface.

ENVIRONMENTS OF DEPOSITION

Sedimentary rocks are formed in many different locations or sites in the geologic environment. These areas of sediment accumulation and subsequent lithification may be referred to as sedimentation basins, indicating areas of low elevation that receive sediments from streams flowing into them. The environments of deposition can occur at any intermediate point between the source of detrital material and the ocean basins. The oceans of course provide the ultimate, low elevation toward which particles move by gravity. When lithification is accomplished after the sedimentation process, sedimentary rocks are formed. The various environments are shown in Table 4.2.

TABLE **4.2** Environments of Sediment Deposition

Continental	Mixed Continental and Marine	Marine
Desert	Littoral	Shallow sea
Glacial	Lagoon	Intermediate sea
Fluvial	Estuary	Deep sea
Piedmont	Delta	
Valley flat		
Lake		
Swamp		
Cave		

Continental Environment

Sedimentary rocks of continental origin are formed when sediments accumulated in continental environments undergo lithification. Examples are lithified sand dunes from desert environments (Figure 4.3) or sandstones formed from continental stream deposits. Both are common occurrences in the geologic past as indicated by the rock column.

Fluvial materials are those deposited by flowing water. They can originate in piedmont areas (intermountain basins that accumulate debris at the edge of the mountain front) where the sudden change in stream gradient causes coarse particles to be deposited. Valley flat or flood plain deposits, also known as alluvium, can also be lithified into sedimentary rocks.

Sediment deposited by glaciers or sediment in lakes, swamps, or caves can undergo lithification as well to yield sedimentary rocks. Swamp environments provide lignite and coal after compaction and burial.

Mixed Environments

The mixed continental and marine environments are the important locations along the boundary of the landmass and the ocean. Deltaic deposits may provide extensive rock units, for example, in the Catskill area of upstate New York. Littoral deposits, those formed by ocean currents paralleling the coast, are important reservoir rocks for petroleum production. Locations of the coastlines in the geologic past provide valuable information regarding oil migration and accumulation. Rock sequences are studied in great detail by oil geologists in an area prior to subsurface drilling.

Marine Environment

For marine environments the shallow sea category (continental shelf) provides the most significant depositional area. In Figure 4.10 the continental shelf is shown in relation to the other marine environments. The major portion of all sedimentary rocks of the continental United States was formed in large inland seas, which lapped up on the continent at various times in the geologic past. The common limestones, shales, and sandstones so prevalent in the central United States were formed on the continental shelf. These are referred to as the rocks of the stable shelf or stable interior. The rivers of the world drop most of their sediment on the continental shelves where marine rocks are forming today.

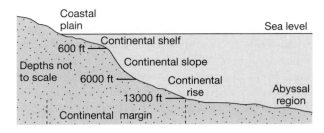

FIGURE 4.10 General features of continental margins and ocean basins.

The intermediate sea zone receives some sediments from the continental landmass but only in selected areas of the world. These rocks may occur in great volumes in certain areas. Such basins include the depositional troughs of global proportions, which will eventually give rise to new mountain chains. Sedimentation basins and mountain building are discussed in Chapter 10.

The deep sea receives very small quantities of sediment from the landmass, primarily because of the great distance sediment must travel in the oceans to reach the abyssal depths. For water depths in excess of 6000 ft (1800 m) the pressure is so great that $CaCO_3$ will redissolve before reaching the ocean bottom. Animal life is sparse because of the lack of nutrients, which are supplied by rivers entering the ocean at the coastline. Therefore, the only sedimentation that occurs, and it accumulates very slowly, is provided by fallout from the atmosphere.

Shallow Marine Environment

An important characteristic of shallow marine sedimentary rocks is the definite pattern and sequence of rock types that parallel the ancient coastline. This can be used to establish the geologic history of an area and to explain the details of the rock sequence.

One of the commonly used examples to illustrate shallow marine deposition is a pattern of sandstone, shale, and limestone formed on the gently sloping continental shelf. The basis for this distribution of rock types is that the coarsest detrital material would settle first, causing sandstones to form a strip closest to shore; the finer detrital material would settle next, yielding a band of shale beyond the sandstone; and limestone in the absence of detrital material would form the third band from shore through precipitation

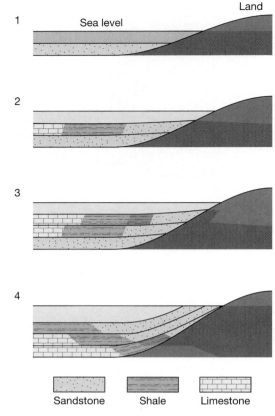

FIGURE 4.11 Diagrams showing progressive on-lap onto the land mass, followed by off-lap.

of $CaCO_3$ from seawater. This is illustrated in a series of cross sections in Figure 4.11.

To continue this example, we now observe what occurs as the shoreline adjusts to sea level changes. In the second and third diagrams of Figure 4.11 the sea has advanced (transgression or on-lap), yielding deeper water for all locations, and the zones of deposition are displaced to the right (toward the landmass). In the fourth diagram the water level has dropped dramatically, showing retreat (regression or off-lap), and the zones of deposition have been displaced seaward. In cross section this yields wedge-shaped deposits of sandstone, shale, and limestone.

This example provides the detail needed to determine the change in sea level indicated by the occurrence of different sedimentary rocks. A stratigraphic column that illustrates the sea level changes is pre-

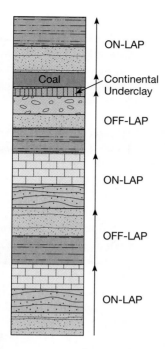

FIGURE **4.12** Deposition by an advancing and retreating sea.

sented in Figure 4.12. This is based on what is observed in a core hole or boring obtained at a single location. Beginning at the bottom (which represents the event that occurred first) an increase in water depth is shown indicating on-lap or an advancing sea. The next three units show off-lap, then on-lap again, and finally a sustained off-lap, yielding continental deposits and coal formation. Then on-lap occurs again.

Additional Details of Carbonate Deposition

Several varieties of calcite-rich materials are deposited in continental environments rather than under marine conditions. Examples are tufa, caliche, marl, and travertine. Tufa is a calcareous material found where hot or cold springs and seeps emerge at the ground surface. $CaCO_3$ precipitates because of evaporation of the water that previously contained the dissolved materials. Tufa is usually porous, spongy, and rather fragile because of its mode of formation. It may contain twigs, soil, or other debris, which is incorporated in the evaporating moisture. By contrast, siliceous sinter or geyserite is a silica-rich deposit formed by evaporation of water from hot springs

and geysers. Its origin is somewhat similar to tufa but, of course, differs in composition.

Caliche is a calcareous deposit that forms within the soil profile in semiarid regions. Two schools of thought exist concerning its origin. In one explanation, the calcium carbonate is deposited at the base of the soil profile by water percolating downward from the surface. The second possibility is the deposition by capillary action of water moving upward from below. Caliche may consist of a rather indurated deposit up to 6 in. (15 cm) thick and found at some depth in the soil profile.

Marl is a term that has been applied to several types of sedimentary deposits that are mixtures of calcium carbonate plus mud and/or sand. Marl also may refer to impure limestones found in freshwater lakes or to marine muds composed primarily of calcium carbonate. These are supposedly formed near the outer margins of shale deposition. The term has also been applied to certain calcareous green sands, for example, those found in New Jersey.

Travertine is a deposit of calcium carbonate commonly found in limestone caves. It is also referred to by such familiar terms as dripstone, flowstone, or Mexican onyx. Travertine is precipitated from downward percolating waters, which seep into the cave system through cracks and joints in the limestone. Stalactites, stalagmites, pillars, and curtains of calcium carbonate are the well-known wonders composed of travertine. Formation of caves and travertine deposits are also discussed in Chapter 15.

Dolomite deserves further explanation because of its significance as a common sedimentary rock. It is composed mostly of dolomite mineral [$CaMg(CO_3)_2$], plus varying amounts of calcite. The term *dolostone* has been proposed for this rock to remove the confusion caused by the use of the term *dolomite* for both a rock and a mineral. In many areas of applied geology the term dolostone has not been readily accepted so that dolomite still prevails as the commonly used rock term.

Dolomites can contain considerable amounts of noncarbonate minerals, typically quartz or clay minerals—as much as 25%, with 10% not being uncommon. Calcite can also be present in large percentages in a dolomite, up to one-half of the carbonate fraction. These would be termed calcic dolomites; in like manner a limestone that contains appreciable amounts of dolomite is called a dolomitic limestone.

A problem that develops in some Portland cement concrete is caused by the alkali-carbonate reaction. Cracking and an expansion of the concrete occurs, eventually causing failure of the structure. Fine-grained calcic dolomites that are rich in clay comprise the reactive rocks. Nearly equal amounts of dolomite mineral and calcite plus 10% clay or more in the fine textured rock depict this problem aggregate. High-alkali cement (greater than 0.6% Na_2O) is also needed to trigger the reaction. The procedures for preventing this reaction are to 1) avoid dolomites with this composition and texture, 2) reduce the amount of the troublesome aggregate by adding nonreactive rock materials, or 3) use low-alkali cements.

ENGINEERING CONSIDERATIONS OF SEDIMENTARY ROCKS

1. The alkali-silica reaction problem in Portland cement concrete was discussed previously in the chapter on igneous rocks. Sedimentary rocks that can be involved in this reaction are chert (including opal and chalcedony) and graywacke. Metamorphic rocks that might be reactive include argillite, phyllite, impure quartzite, and granite gneiss.
2. The alkali carbonate reaction problem is related to a specific carbonate composition and texture as discussed in the previous section.
3. Limestone and dolomite provide the best sedimentary aggregates for construction materials. Siltstone, shale, quartz sandstone, and conglomerate are generally not acceptable; graywacke is marginal.
4. Stream and terrace gravels commonly contain weak rock pieces that yield nondurable aggregates in concrete. Weathered chert, shale, siltstone, clayey carbonates, iron oxide nodules, and friable sandstones can cause pop-outs at the concrete surface after a number of freeze-thaw cycles.
5. Coarse-grained limestones abrade too severely to be used for aggregates for construction. Such rock particles lose gradation owing to a reduction in particle size.

6. Sedimentary rocks used as dimension stone for the facing of buildings should be nonstaining and resistant to weathering effects. High-purity, clastic limestones have proven durable; clayey carbonates and quartz sandstones may be subject to spalling. Quartz sandstones are used for flagstone walls.
7. Shales and siltstones can provide a suitable foundation for buildings, dams, and bridges. A deep weathering profile is one concern, and the slabbing loose of rock due to stress release after excavation is another. Placement of a concrete slab over the shale and possibly the use of rock bolts to restrain the rock are common practices.
8. Limestones, dolomites, and evaporite deposits can exhibit an irregular soil-rock interface in their weathering profiles. Pinnacles (of rock) and pipes (of soil) are common. Care must be taken to ensure that heavy structures are founded completely on solid rock.
9. Sinkholes and underground conduits in limestone and dolomites must be recognized and properly dealt with when founding buildings in these terrains. Changes in the existing ground surface or of subsurface drainage by construction should be evaluated with great care.
10. When water is impounded behind a dam, if limestone lies at the rim or within the reservoir area, careful consideration is required. The presence of solution channels in the limestone that extend to another surface drainage area will lead to leakage unless the channels are filled by grouting.
11. Compaction shales have caused problems when used as rock fills in highway embankments. When such shales become wetted in the embankment, slaking occurs, this material fills the voids between the rock pieces, and surface subsidence and slope instability occur. To solve the problem, these shales need to be broken during placement and compacted into a solid mass in the same manner used for soil materials.
12. Conglomerates are basically weak sedimentary rocks because they are poorly cemented and highly porous. Water movement through this rock removes the cement and increases permeability. When encountered in dam abutments and foundations, conglomerates require special treat-

ment to increase their strength and reduce permeability.

13. Sedimentary rocks containing anhydrite are troublesome to engineering structures such as dams, highways, and tunnels because the mineral will alter to gypsum in the presence of water, yielding an increase in volume and considerable stress on the structure adjacent to it. The presence of anhydrite must be recognized and steps taken to reduce its effect.

EXERCISES ON SEDIMENTARY ROCKS

1. Bedding is one of the most prominent features of sedimentary rocks. Would you expect a difference in strength parallel and perpendicular to bedding? In what way? How would permeability be affected? Explain.

2. Limestones can be dark in color. What is the likely cause of this coloration? How would a block of fine-grained limestone be distinguished from basalt? Which would be stronger and why?

3. Clastic limestones are composed of broken fossil fragments; fossiliferous limestones are composed of whole fossils. How do you think the environment of deposition differed between the areas where these rocks were formed? (Assume both are marine in origin.) Explain.

4. Quartz sandstones commonly are only partly cemented and contain a high volume of pore space. Would quartz sandstone be a good supplier of ground water? Shale also has a high porosity; would it have a high permeability? How does shale differ from sandstone as a supplier of ground water?

5. A ledge of rock in a quarry is described as a cherty, argillaceous (clay-rich), fine-grained dolomitic limestone. The chert is deeply weathered. What problems might develop if this rock were used as an aggregate for Portland cement concrete?

6. Large pieces of limestone are being considered for use as slope protection stone for a harbor area. The apparent specific gravity of the rock was found to be 2.52; how much would a large rectangular piece 3 × 4 × 5 ft weigh in pounds? What would be its mass in kilograms per cubic meter? If diabase were used instead, with an apparent specific gravity of 2.96, what would be its weight for a piece of the same size as the limestone? What would be the difference in effect of the two pieces relative to resisting the energy of ocean waves? Explain.

7. A volcanic breccia is a coarse-grained pyroclastic rock; a sedimentary breccia was discussed previously in this section. How can these two rocks be distinguished from each other? Explain in detail and give information on the origins of the two rocks.

8. Fine-grained dolomite might not effervesce in cold, dilute hydrochloric acid and chert of course will not. What tests should be performed to tell these two rocks apart? What might be done to make the acid test more helpful? Explain.

9. Why are fossils less common in shale than they are in limestone? Consider the origin of the materials that comprise these rocks.

Additional Readings

LAPORTE, L.F., 1979, *Ancient Environments,* 2nd ed., Prentice Hall, Englewood Cliffs, NJ.

PETTIJOHN, F.J., 1957, *Sedimentary Rocks,* 2nd ed., Harper and Row, New York.

WILLIAMS, H., TURNER, J.F., and GILBERT, C.M., 1954, *Petrography,* W.H. Freeman and Co., San Francisco.

5

Metamorphic Rocks

$\mathbf{M}$etamorphic rocks are formed within the Earth's crust from preexisting rocks through the action of heat and pressure. The rocks retain their solid form as metamorphism proceeds. This is accomplished by the growth of existing minerals and rearrangement of the overall chemical compositions to form new mineral species. If complete melting occurs, the process is igneous rather than metamorphic. In some cases, hot fluids are supplied to the rock mass so that the overall composition changes somewhat during metamorphism. The original rocks can be igneous, sedimentary, or metamorphic.

As with intrusive igneous rocks, metamorphic rocks may be found at the Earth's surface when the overlying rocks are removed by erosion. For those rocks formed by regional metamorphism, which are discussed in the next section, this indicates the removal of at least 3000 ft (about 1000 m) of overlying rock.

METAMORPHIC PROCESSES

The two main processes that yield metamorphic rocks are contact metamorphism and regional metamor-

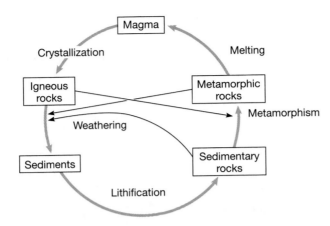

FIGURE **5.1**   Geologic rock cycle.

phism. When hot magma intrudes into surrounding rock, a reaction occurs at the zone of contact. The altered zone varies in thickness from about an inch (a few centimeters) to 100 ft (about 30 m) and even as great as 1.25 miles (2 km). This zone, containing slightly enlarged crystals, new minerals, and occasionally some ores such as sulfides or oxides, surrounds a batholith, stock, or other large, intrusive igneous body. Examples of rock types formed along igneous contacts are skarns and hornfels, both fine-grained massive rocks. Because these rocks contribute only a minor portion to the total volume of metamorphic rocks and are not easily distinguished from country rock, they typically are not included for rock identification in a beginning geology course.

Regional metamorphism occurs over extremely large areas, involving thousands of square miles (thousands of square kilometers) and thicknesses of 3000 ft (1000 m) or more. These rocks are best exposed in the roots of old, folded mountains and in Precambrian continental shields. When most minerals of preexisting rocks are subjected to the heat and pressure of regional metamorphism, they are extended beyond the stability field in which they were formed. The increased heat weakens chemical bonds and accelerates the rate of chemical reaction, thus encouraging new minerals to form that are stable under these extreme conditions. The pressure encourages growing crystals to align themselves in a manner that will minimize pressure effects. The final outcome is the formation of new minerals, of increased size, oriented in a preferred direction.

Regional metamorphism occurs in sedimentary basins where a thickness of 30,000 to 40,000 ft (9000 to 12,000 m) of sedimentary rocks has accumulated. For such rocks to be exposed at the surface, extensive uplift and erosion of thousands of feet of overlying rock must have occurred. Typically, surface outcrops of regionally metamorphosed rocks are extremely old, commonly Precambrian in age, because of the great expanse of time needed for the sedimentation, metamorphism, and erosion processes to occur.

GEOLOGIC ROCK CYCLE

Definite relationships exist among igneous, sedimentary, and metamorphic rocks. Geologic processes change one rock type into another. The way in which rock types and processes are related is shown in Figure 5.1 in a diagram known as the rock cycle.

Beginning with magma, crystallization yields igneous rock, which can be either intrusive or extrusive. These rocks can weather, yielding broken rock fragments and other weathering products known as sediments. When sediments become lithified, they yield sedimentary rocks. Metamorphism of these rocks produces metamorphic rocks, which, in turn, on melting provide magma once more. Shortcuts across the circle are shown; igneous rocks can be metamorphosed to metamorphic rocks and sedimentary or metamorphic rocks can weather to sediments.

The intricate history of a rock unit can be illustrated by following it around the rock cycle (Figure 5.1). For example, quartzite is a metamorphic rock composed of metamorphosed quartz sandstone. The quartz grains in the sandstone existed as sediments before the rock became lithified. At some point in its geologic history, the quartz sand weathered from a coarse-grained igneous rock, most likely a granite. Quartz in the granite probably comprised from 10% to 20% of the rock.

For example, a rock unit such as the Baraboo Quartzite, which is about 1000 ft (300 m) thick and likely once covered much of the state of Wisconsin, was derived from 5 to 10 times this volume of igneous rock that had to be eroded to provide the quartz. All of this took place prior to formation of the Baraboo Quartzite, which occurred about 800 million years ago. Wisconsin has an area of about 54,000 mi^2 (139,000 km^2) and if the quartzite averaged 1000 ft thick, a rock volume of 10,000 mi^3 (41,700 km^3) is obtained for the Baraboo Quartzite. Five times this volume gives 50,000 mi^3 (208,500 km^3) for the original igneous rock if it contained about 20% quartz and none of the quartz was chemically destroyed during the weathering process.

TEXTURES OF METAMORPHIC ROCKS

Most rocks that have undergone the extremes of heat and deforming pressure during regional metamorphism contain minerals arranged in parallel, flat layers or elongated grains in segregated bands of color. This feature of metamorphic rocks is known as foliation. Based on this property, metamorphic rocks can be subdivided into two groups, foliated and nonfoliated. The four distinct types of foliation are slaty, phyllitic, schistose, and gneissic, in order of increasing metamorphic grade or level of metamorphism. Foliated rocks also have a tendency to break along planes of weakness associated with the foliation. This property is known as rock cleavage and should not be confused with the cleavage of individual mineral grains discussed in Chapter 2.

Slaty cleavage consists of fractures along smooth planes in a rock that are separated by distances measuring less than a millimeter. This feature is much more pronounced than breaks along bedding in a sedimentary rock.

Phyllitic cleavage consists of parallel fractures that are closely spaced but generally farther apart than slaty cleavage. It occurs in rocks with a visible sheen of muscovite on the cleavages and these sometimes appear as crenulations or wavy undulations forming the cleavage surfaces.

Schistose cleavage consists of flakes or plates that are visible to the naked eye. The plates are formed by mica sheets or by needle-shaped minerals such as hornblende.

Gneissic structure consists of alternating bands of coarsely crystalline minerals a few millimeters or centimeters thick. Bands commonly consist of light-colored silicate minerals with thinner bands of dark mineral. These bands may or may not exhibit rock cleavage or actual foliation. With increases in platy minerals, the gneissic structure tends more and more toward schistose foliation. Photographs of a garnet mica schist, a gneiss, and an argillite are provided in Figure 5.2. Argillites, discussed in Chapter 4, are slightly metamorphosed shales that have greater strength and integrity than a typical shale.

TYPES OF METAMORPHIC ROCKS

Names for foliated metamorphic rocks are typically based on their texture. For this reason, when the texture of these rocks is recognized the naming of the rock is accomplished as well. In the case of schists, the minerals present provide the modifying term for the rock such as hornblende schist or garnet, muscovite schist. For nonfoliated metamorphics, mineral composition forms the basis for the rock name.

Slate is a rock formed from shale through low-grade metamorphism. It is fine grained and its slaty cleavage is a result of the alignment of microscopic flakes of muscovite, formed from the clay minerals of the shale. It occurs in various colors including black, gray, green, and red. The black color is caused by carbonaceous material or fine pyrite and the green and red by the ferrous and ferric states of iron oxide, respectively.

Phyllite is a metamorphic rock with a composition similar to slate but with larger pieces and greater quantities of muscovite present. This provides a sheen to the cleavage surfaces. Crenulations of the cleavage surfaces are sometimes present, indicating a slightly higher grade of metamorphism than slate.

Schist is the most abundant rock formed by regional metamorphism. Typical of schists are the

(a)

(b)

(c)

FIGURE 5.2 Hand specimens of (a) a garnet mica schist (b) a banded gneiss, and (c) an argillite.

FIGURE 5.3 Highway cut through strongly foliated metamorphic rock, a banded gneiss.

needle-shaped minerals. Garnet, magnetite, or other accessory minerals may also occur. A highway road cut through strongly foliated rock, mostly gneiss, is shown in Figure 5.3.

Gneiss is a coarse-grained metamorphic rock with a banded appearance caused by alternating layers of silicate minerals. It is formed during high-grade, regional metamorphism. Although gneiss does exhibit rock cleavage to a certain extent, its cleavage is much less pronounced than that of a schist. As the bands of mica increase in a gneissic structure, rock cleavage becomes more developed. Gneisses formed from coarse-grained igneous rocks (e.g., granite and gabbro) consist of alternating bands of quartz, feldspar, and hornblende or augite, whereas those derived from graywacke or other clayey sedimentary rocks contain muscovite and chlorite in addition to the other minerals. Those platy minerals tend to increase rock cleavage characteristics.

Marble is a common metamorphic rock composed almost entirely of calcite or dolomite. It is coarse grained and is formed by either contact or regional metamorphism, although the latter occurrence is more common. Marble, which is derived from limestone or dolomite, differs from these sedimentary counterparts by its larger grains and absence of sedimentary features, such as fossils and bedding planes, which are destroyed during metamorphism.

In the metamorphic process, mineral grains enlarge and tend to align themselves in a preferred

clearly visible flakes of platy minerals, for example, muscovite, biotite, talc, chlorite, or hematite. Needle-shaped minerals such as hornblende may also dominate the composition. The prominent minerals present provide modifiers for the rock name, for example, muscovite schist or hornblende schist. Considerable amounts of quartz and feldspar may be present but in lesser amounts than are the platy and

direction, but the rock shows no foliation because only a single mineral species is present. Viewed under a petrographic microscope, the preferred orientation of the carbonate crystals can be observed.

Marble is an important dimension stone used for interior decoration for walls, panels, and stairways of buildings but also as an exterior facing for buildings. Its various colors and structures result from impurities in the original sedimentary rock and from geologic occurrences such as faulting and brecciation with subsequent void infilling. Common colors are black, green, red, and brown plus various shades of gray.

Quartzite, a common, nonfoliated metamorphic rock, is formed by the metamorphism of quartz sandstone. The grains of quartz in the sandstone are firmly bonded by silica, which fills the pore spaces during metamorphism. It is distinguished from quartz sandstone by its more massive, dense nature; lack of clastic particles, which are easily abraded loose; and the way it breaks *through* the constituent sand grains rather than around them. Quartzite is distinguished from marble by its much greater hardness and the effervescence in acid evidenced by marble but not quartzite.

Details of the textural features of quartzite can be discerned only through use of a petrographic microscope. Both the original quartz grains and the overgrowths of quartz yielding an interlocking texture can be seen under the microscope.

Quartzites are white in color when no impurities are present in the rock. Several other colors occur as a consequence of iron or other minor constituents being present. Pink, red, green, and purple are common colors for quartzite.

METAMORPHIC GRADE

In the metamorphic process a sequence of changes occurs as the intensity of heat and pressure increases. This is a measure of metamorphic grade, low-grade rocks forming first; medium-grade and high-grade rocks thereafter if the degree of metamorphism continues to increase. Based on this concept, a sequence of changes can be noted with increasing extent of metamorphism. The changes that occur as shale and basalt are metamorphosed are as follows:

Shale → Argillite → Slate → Phyllite → Mica schist → Feldspar schist → Migmatite → Granite

Basalt → Greenstone → Hornblende schist → Feldspar amphibolite → Hornblende migmatite → Granodiorite

In these two examples, feldspar schist and feldspar amphibolite are high-grade metamorphic rocks, whereas slate and greenstone are low grade. Hornblende is a type of amphibole, so that feldspar amphibolite is a banded rock of abundant feldspar and hornblende. Two concepts are involved in the development of these two sequences: migmitization and granitization.

Migmatites

Migmatites are mixtures of rocks, typically gneiss and granite, which are found in extremely high-grade metamorphic terrains typically in the centers of eroded mountain chains. These rocks combine the high-grade metamorphic rock, gneiss, with the apparently igneous rock granite. In this case, however, the granite is thought to be formed by extremely intense metamorphism where the minerals are recombined into a nonfoliated rock.

Granitization

The process by which granite is formed by intense metamorphism rather than by igneous implacement is known as granitization. Certain field conditions support this concept, such as discovery of areas in which sedimentary formations (such as shale) grade into schists and then into migmatites. Most geologists agree that some granites are likely formed by igneous intrusions and some by granitization. They disagree on the percentage of each that make up the total volume of granite in the Earth's crust. Geologists who favor the igneous origin suggest that perhaps 15% of all granites form by granitization and 85% by igneous intrusions. Those who favor granitization might say that the two percentages are reversed.

There is another problem for intrusive granites in addition to the field evidence of metamorphic mountain belts mentioned previously. This is the so-called "space problem," that is, what happened to the great volume of rock that was displaced by the intruding magma? It could not just be pushed aside or domed upward because the rock cover is too great and the volume of rock represented by the granite pluton would be too vast. Granitization, a replacement process, seems more likely as the origin of the largest batholiths such as the Sierra Nevada. However, portions of this batholith, as field evidence shows, were formed by igneous intrusions.

METAMORPHIC ROCK IDENTIFICATION

Identification of metamorphic rocks is accomplished by use of Table A.7, which is supplied in Appendix A. The first step is to recognize the type of foliation shown by the unknown rock. This foliation determines the rock name and only a modifier based on the mineral content needs to be added. For the nonfoliated rocks, the mineral content is used to determine quartzite from marble. This is based on the distinction between quartz and calcite or quartz and dolomite; these differences were described in Chapter 2.

In Appendix A a tabulation sheet for metamorphic rock identification is also presented (Table A.8). The information indicated by the table headings should be found for each sample. This leads directly to the proper identification of metamorphic rocks.

ROCK IDENTIFICATION FOR ALL VARIETIES

In the discussions on igneous and sedimentary rocks and again for metamorphic rocks, each genetic type was considered independently of the other two. After identifying rocks with this approach, one is presented with the challenge of distinguishing individual rocks from a collection comprised of all three rock types. This requires that, when assigning a name to an unknown specimen, a distinction be made between all the common rocks. This clearly is more difficult than when only one genetic group of rocks is considered at a time. To aid in this identification process of the total group of rocks, the following is presented for consideration.

Several general rules of thumb are used to isolate one genetic rock type from the other two. These are based on the following considerations: hardness of rock; diagnostic minerals; directional features such as bedding, foliation, and orientation of grains; and other features including presence of fossils, clastic texture, sedimentary structures, and organics.

Hardness

Igneous and metamorphic rocks are generally considered to be hard, whereas sedimentary rocks are soft. This is a result of the predominance of silicate minerals in the first two rock types whereas sedimentary rocks include carbonates and evaporites, which are inherently soft. The silicates measure 6 or more on Mohs' hardness scale compared to carbonate and sulfates, which are in the range of 3 to 4. Also contributing to the soft aspect of sedimentary rocks is the clastic texture that many of these rocks possess. The fragments are only loosely cemented together so pieces are easily abraded from the rocks. Even though the grains themselves may be hard, as in a quartz sandstone, the friable nature of the rock gives it an appearance of being soft. Exceptions exist as would be expected, such as hard sedimentaries (chert), soft igneous rocks (tuff), and soft metamorphics (marble). However, these exceptions are easily remembered simply because they do not conform to the general rule.

Diagnostic Minerals

Certain minerals occur predominantly in only one of the genetic rock types. Garnet, muscovite, graphite, chlorite, and talc are more common in metamorphic rocks. Olivine occurs in igneous rocks whereas calcite, clay minerals, chlorides, sulfates, and chert most commonly occur in sedimentaries. Quartz and feldspar can occur in all three types.

Directional Features

Directional features are most prominently displayed as bedding in sedimentary rocks and as foliation in metamorphic rocks. Bedding planes are indicated by color and textural changes, by separation in the rock, and by the overall shape of rock pieces. Most sedimentary rocks show at least some indication of bedding, particularly those with clastic textures. Pyroclastics may show layering and some lava rocks show flow banding. Both are typically less prominent than is bedding in a sedimentary rock. Other details such as mineral composition can be used to tell layered igneous rocks from bedded sedimentaries.

Foliation in metamorphic rocks is more prominent than is bedding in most sedimentary rocks. The individual types of foliation—slaty cleavage, schistosity, and gneissic structure—are typically easy to identify, leading directly to rock identification.

Other Features

Other features include the presence of fossils, indicating a sedimentary origin; a clastic texture, suggesting either sedimentary rocks or pyroclastics (with rounded clastic pieces suggesting sedimentaries); the presence of organics depicting sedimentary rocks; certain features unique to specific rock types such as small dikes running through igneous or metamorphic rocks; and the sedimentary features—stylolites, geodes, and nodules, for example. The absence of

certain features can also be used for identification purposes. Lack of directional features may suggest an igneous origin, and coarse calcite grains without fossil evidence may suggest a metamorphic rock, marble.

By applying these concepts collectively, a major step toward rock identification is accomplished. Specific ways to distinguish between two similar rock types are easily learned. Hardness and reaction to acid separate quartzite from marble, mineral content suggests the difference between tuff and graywacke, and degree of crystal interlock between granite and sedimentary breccia. Table 5.1 presents a summary of rock identification aids based on genetic rock type.

TABLE 5.1 Identification Aids for Determining Genetic Rock Types

Common Minerals	Textures	Structures in the Field	Other Features
Igneous Rocks			
Olivine°	Nondirectional, coarse	Columnar jointing	Strong crystal interlock
Augite	to fine	Flow banding	Nonfossiliferous
Hornblende	Porphyritic		Usually hard rocks
Plagioclase	Vesicular		Pyroclastics composed
Orthoclase	Amygdoloidal		of angular, igneous pieces
Biotite, muscovite, quartz	Pyroclastic		
Sedimentary Rocks			
Quartz	Layered texture	Bedding	May be fossiliferous
Calcite	Clastic particles usually	Concretions	Usually soft
Clay minerals+	rounded	Cross bedding	Mud cracks
Chemical precipitates+	Cement or matrix present	Ripple marks	
Halite	May be friable	Mud cracks	
Gypsum		Stylolites	
Organic material+		Geodes	
Coal, peat			
Chert+			
Metamorphic Rocks			
Garnet×	Commonly foliated: slaty,	Rock cleavage	Fossils absent
Muscovite, abundant×	phyllitic, schistose,	Minor folds and faults,	Large crystals
Hornblende	gneissic	dikes	Nonclastic
Chlorite, abundant×			
Graphite×			
Talc×			
Quartz			
Orthoclase			

° Good indicator of igneous rocks.
+ Good indicator of sedimentary rocks.
× Good indicator of metamorphic rocks.

ENGINEERING CONSIDERATIONS OF METAMORPHIC ROCKS

1. Foliated metamorphic rocks commonly yield rock pieces with elongated shapes when crushed. Gneiss, argillite, and phyllite may be used as concrete aggregates. Flat and elongated pieces cause mixing problems in fresh concrete and directional properties in hardened concrete.

2. Gneiss containing abundant mica can yield problems when used as a concrete aggregate. Freeze-thaw and wetting-drying effects cause micas to fail along their prominent cleavage plane. Flaking of aggregates from the concrete surface results. Schists typically are not used as aggregates because of the abundant mica present.

3. Foliated rocks possess prominent directional properties. Strength, permeability, thermal conductivity, and seismic velocity are affected by the direction of foliation. Care should be taken that loads (from bridges, dams, building foundations) are not transferred to foliated rock masses in a direction closely parallel to the foliation.

4. Metamorphic rocks may be deeply weathered in the eastern United States (Piedmont Plateau) and the depth to bedrock is quite variable. Saprolite, partially weathered rock, is extensive. Care must be taken to found heavy structures, or to locate tunnel alignments, in sound rock whenever possible.

5. Coarse-grained gneisses, like granites of a similar size, abrade severely when used as aggregates for construction. These rocks lose gradation by abrasion, resulting in a reduction of particle size.

6. Slate, schist, and phyllite are subject to rock overbreak during blasting of rock cuts or tunnels because of their pronounced rock cleavage. High stress concentrations in tunnels may occur for the same reason.

7. The stability of rock slopes is greatly affected by the attitude of foliation with respect to the rock slope direction. When foliation dips steeply into an opening, rock slides commonly occur. Rock bolts or tendons may be needed to prevent such failures.

8. As mentioned in the sedimentary rock discussion, phyllite and argillite can yield alkali-silica reactive aggregates.

9. Marble is subject to the same problems as limestone. Solution cavities and channels may develop, resulting in similar problems of leakage of reservoirs and collapse of newly formed sinkholes.

10. Quartzite is a massive, hard rock and has a major abrasive effect on crushing and sizing equipment. It is more expensive to process than most rocks because of this property.

EXERCISES ON METAMORPHIC ROCKS

1. Quartzite can be confused with basalt. What do they have in common? How would you distinguish between the two?

2. Prepare a list of the rocks and minerals listed in the identification tables for common minerals and rocks supplied in Appendix A. Put them in alphabetical order or in some other sequence that mixes the terms so they are no longer grouped into minerals and individual genetic groups of igneous, sedimentary, and metamorphic rocks. Place the correct symbol adjacent to each term, M for mineral, I for igneous, S for sedimentary, and X for metamorphic. Review the answers to ensure that you can accomplish this from memory.

3. Metamorphic rocks are found in the roots of old mountain systems and in continental shields. Describe what is meant by these two locations. Where in eastern North America would such rocks be found? Why are there metamorphic rocks in the stream gravels of the upper Midwest?

4. Coarse crystals of calcite can exist in limestone and they also occur in marble. How would you tell these two rocks apart? Describe how their origins differ. Be specific.

5. In a certain geologic terrain quartzite and granite are found in contact. Assume that both are essentially the same age and both were subjected to equal extremes of heat and pressure during the metamorphic process. Why has the quartz sandstone been metamorphosed to quartzite but the granite has not been changed to gneiss?

6. If a conglomerate is metamorphosed, what does it become? How would it be distinguished from the original conglomerate?

7. When shale is metamorphosed it changes into different rocks depending on the extent of metamorphism. Show the steps that occur with higher grades of metamorphism. What minerals form in this process?

8. Along a certain steep-sided stream valley slate dips into the hillside on one side of the stream and into the valley on the other. Draw this in cross section. A road is to be placed along the base of the valley wall by removing rock in a side hill cut. On which side of the stream should the cut be placed? Explain. Why is this so? In addition to the foliation, what other rock weaknesses would be of concern? How would this affect the road cut?

9. A schist sample has an unconfined compressive strength (strength at failure) of 17,000 psi. If the sample tested was 2 in. in diameter, what total load was applied to the sample to cause failure? What is the value of the compressive strength in tons per square foot, kilograms per square centimeter, and kilograms per square meter?

10. Review the listings under the engineering considerations section (near the end of each chapter) for igneous rocks (Chapter 3), sedimentary rocks (Chapter 4), and metamorphic rocks (this chapter). Condense each entry into a single phrase or short sentence. A total of 28 entries is involved.

Additional Readings

BEST, M.G., 1982, *Igneous and Metamorphic Petrology*, W.H. Freeman and Co., New York.

ERNST, W.G., 1969, *Earth Materials*, Prentice Hall, Englewood Cliffs, NJ.

TURNER, F.J., 1981, *Metamorphic Petrology*, 2nd ed., McGraw-Hill, New York.

WINKLER, H.G.F., 1974, *Petrogenesis of Metamorphic Rocks*, 5th ed., Springer-Verlag, New York.

6

Engineering Properties of Rocks

F rom an engineering point of view, rocks are significant for two major reasons: 1) They are an important building material with numerous applications to construction use and 2) many engineering structures are founded on rock; their safety depends on the stability of the rock foundation and the adjacent rock mass.

The physical properties of rocks determine their behavior as construction materials and as a structural foundation. Two classes or measures of these properties exist: the first, called "rock properties," are measured on small samples in the laboratory; the second

are those known as "rock mass properties," which require a large mass of rock to determine the overall behavior for a large volume. Therefore, testing of rock mass properties is, by necessity, performed in the field. Typically, rock mass properties are controlled by weakness planes in the rock rather than by the properties of the intact material itself. Tests on intact rock provide values for rock properties.

Rock and *stone* are terms with specific, engineering meanings. Sometimes the terms are used interchangeably by nonspecialists, but this is incorrect. *Rock* refers to a geologic formation in its natural location, that is, still in place as part of the bedrock mass. *Stone* refers to blocks or fragments excavated from quarry ledges that have been prepared for construction use. Consequently, stone has undergone human processing, which increases its cost and presumably its value.

Rock properties in most cases where engineering construction is involved are determined by specified testing procedures. In the following sections standard rock properties are discussed first, then those properties measured on aggregates for concrete and for other types of construction are considered, followed by a brief overview of rock mass properties.

ROCK PROPERTIES

When rock is to be used as a foundation for engineering structures, the relevant properties are mass density, strength, and compressibility. By comparison, for rocks used as construction materials, the primary concerns are mass density to some extent, but mainly strength and durability. Because mass density is common to both lists, this topic is discussed first.

Specific Gravity, Mass Density (Weight Density), and Absorption

Specific gravity (SpG), discussed in Chapter 2, is directly related to mass density. For massive solids the mass density is equal to the specific gravity multiplied by the density of water; that is, by 1 gm/cm^3 or 1000 kg/m^3, which yields the maximum mass per unit volume the solid can exhibit as it assumes that no pores are present. A common method to determine specific gravity (presented also in Chapter 2) is to weigh a solid in air and in water to obtain the values:

$$SpG = \frac{\text{Mass of solid in air}}{\text{Mass of equal volume of water}}$$

$$= \frac{\text{Mass of solid in air}}{\text{Mass of solid in air} - \text{Mass of solid in water}}$$

This holds true because the volume of a solid is equal to the volume of water it displaces. In like manner, the buoyancy force is equal to the volume displaced multiplied by the density of water, which equals the mass of the solid in air minus the mass of the solid in water.

Unit weight is used in many calculations in the United States rather than mass density. Unit weight can be obtained from specific gravity by multiplying the value by the unit weight of water, that is, by 62.4 lbs/ft^3.

Unit weights for common varieties of rock are given in Table 6.1, which also lists other physical properties of rocks.

Most rocks, however, have some porosity so their density is somewhat less than the maximum indicated in the preceding equation. To obtain a specific gravity that also accounts for porosity, the mass is determined in air in a dried condition, determined in air in a saturated condition, and determined in water in a saturated condition. These masses can be variously combined to give three different specific gravities.

$$SpG_d = \frac{A}{B-C} \quad \text{or bulk specific gravity}$$

$$SpG_s = \frac{B}{B-C} \quad \text{or bulk specific gravity, saturated surface dried}$$

$$SpG_a = \frac{A}{A-C} \quad \text{or apparent specific gravity}$$

where

A = mass of solid in air, dried for 24 hr in an oven

B = mass of solid in air, saturated surface dried

C = mass of solid in water, saturated.

Saturated surface dried means the rock sample has been saturated by soaking in water for 24 hr and the excess water at the surface removed by drying.

The water absorption of the rock is calculated from these measurements by the following:

$$\% \text{ absorption} = \frac{B-A}{A} \times 100$$

Regarding aggregates for concrete, the bulk specific gravity, saturated surface dried (SpG_s), is the term typically used; it is also preferred for most other engineering applications. Equations are available that show the relationship among these three measures of specific gravity. If absorption and one of the three specific gravities are known, the other two can be calculated.

Rock Strength

In general, a rock can be subjected to three primary types of stress: compressive, shear, and tensile. Compressive stress tends to decrease the volume of the rock by forces that act inward and directly opposite to each other. Shear stress is caused by two equal forces acting in opposite directions as a couple, and tensile forces tend to pull a substance apart by outward-acting, equally opposing forces. Compressive, tensile, and shear stresses are illustrated in Figure 6.1.

TABLE 6.1 Physical Properties of Rocks

Rock	Compressive Strength (psi) q_u	Shear Strength (psi) S_0	Tensile Strength (psi) T	Modulus of Elasticity × 10^6 psi E	Angle of Shearing Resistance φ	Poisson's Ratio μ	Unit Weight (pcf) γ	Porosity n%
Igneous								
Granite	13,900–34,720	1950–6940	970–3470	2.77–8.3	45–60	0.15–0.24	162–175	0.5–1.5
Coarse	7630–10,460	1500–2000			48–56			
Pegmatitic	6190	1040			58			
Fine	31,900	2700			70			
Slightly altered	9460	1420			58			
Syenite				8.33–11.1			162–170	0.5–1.5
Quartz monzonite	30,500	3650			63	0.17		
Monzonite porphyry	18,100	2390			59			
Diorite	25,000–41,700	2010	2080–4170	9.31–13.9	54		168–175	0.1–0.5
Diabase (dolerite)	27,800–48,600	3470–8330	2080–4860	11.1–15.3	55–60		168–190	0.1–0.5
Gabbro	25,000–41,700		2080–4170	9.31–15.3			175–193	0.1–0.2
Basalt	20,800–41,700	2780–8330	1390–4170	8.33–13.9	50–55		175–181	0.1–1.0
Andesite	18,700–19,100						137–144	10–15
Tuff	530					0.11		
Metamorphic								
Gneiss	7000–27,800		700–2800	2.7–8.3	48–73	0.11	175–187	0.5–1.5
Massive granite gneiss	32,390	4500			66			
Granite gneiss	7630–12,000	1500–1800			48–56			
Schistose	12,000	1800			73			
Weathered schistose	7700–13,400	1800–2200			59			
Quartzite	20,830–41,660	2780–8330	1390–4170	5.7–8.3	50–60		162–168	0.1–0.5
Marble	7150–34,720	2080–4170	970–2800		35–50	0.25–0.38	162–168	0.5–2.0
Slate	12,110–34,720		970–2800				162–168	0.5–2.0
Schist	1160–17,000			0.6–2.8		0.08–0.20		
Biotite	7750–12,000							
Biotite-chlorite	5300–17,000							
Sedimentary								
Sandstone	2780–23,600	1100–5560	560–3470	0.69–11.1	35–50	0.17	120–161	0.5–26
Graywacke	7900	1700			47			
Shale, general	1390–13,900	417–4170		1.4–4.9	15–30			
Clayshale	180–1040	40–160						
Siltstone	4120–7290	750–1000			57–64			
Mudstone				2.8–6.9				
Coal	700–7000		280–700	1.4–2.8				
Limestone	4170–34,700	1100–6940	700–3470	1.4–11	35–50	0.16–0.23	137–162	5–20
Chalk	750	60			23			
Dolomite	11,100–34,700		2080–3470	5.5–11.6	50–65		156–162	1–5

Note: Compressive strength, English units, 1 psi = 6.895 kN/m^2 or 6.895 kPa in SI units; 1 megapascal = 10^6 Pa = 10^3 kPa ≈ 145 psi.
Unit weight, 1 pcf = 16.02 kg/m^3 density.

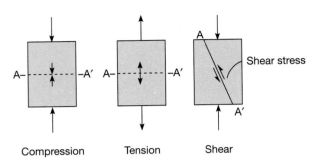

FIGURE **6.1** Compression, tension, and shear stress diagrams.

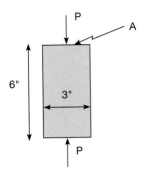

FIGURE **6.2** Loading for uniaxial compression testing example.

The compressive strength is the compressive stress required to break the specimen. It is measured in pounds per square inch (psi) in the British units used by most engineers in the United States or in newtons per square meter in SI units (1 psi = 6.895 kN/m² = 6.895 kPa).[1] The unconfined compressive strength (or uniaxial compressive strength) pertains to rocks unconfined at their sides while the load is applied vertically until failure occurs. The strength, σ, is shown by the formula:

$$\sigma = P/A$$

where P is the failure load in pounds (or newtons), A is the cross-sectional area of the sample in square inches (or square meters), and σ is the strength or pressure in psi (or in newtons per square meter).

The unconfined compressive strength of rocks ranges from a few hundred psi (about 1000 kPa or one MPa) for weak shales to more than 40,000 psi (280,000 kPa or 280 MPa) for diabases and some basalts and quartzites. Representative values for compressive strengths of individual rock types are given in Table 6.1. An example problem is provided next.

EXAMPLE PROBLEM: UNIAXIAL COMPRESSION TESTING

A rock core of limestone is 3 in. in diameter and 6 in. long (Figure 6.2). It is loaded to failure in an unconfined compression testing machine. If the failure load was 62,150 lb, what is the unconfined compression strength σ of the limestone sample?

ANSWER

$$\sigma = \frac{P}{A} = \frac{62{,}150 \text{ lb}}{\frac{\pi d^2}{4} \text{ in.}^2} = \frac{62150}{\frac{\pi 3^2}{4}} = 8792.4 \text{ psi}$$

Triaxial compression tests are performed on rock specimens to determine how the rock will behave under different confining pressures. In this test the sample, usually a rock core with a length twice its diameter, is placed in a heavy steel cylinder. A rubber jacket is placed around the rock core, with fluid filling the space between the rock and the cylinder wall. The ends of the sample are subjected to increments of increasing load, while confining pressure is applied to the fluid and held constant. Failure occurs when the rock fractures, yielding displacement. The confining pressure and maximum axial load are recorded.

The axial load required to cause failure of a rock increases with increasing confining pressure for this test. By performing two or three tests at different confining pressures, the relationship between axial compressive strength and confining pressure can be determined for that rock material. Data are plotted on a shear stress–compressive stress diagram known as a Mohr diagram. This is shown in Figure 6.3(a) for triaxial testing and in 6.3(b) for unconfined (or uniaxial) compression testing conditions. Details on the Mohr circle procedure for representing stresses are provided in Chapter 7, which discusses soil mechanics.

The parameters shown in the diagram are σ, τ, S_0, and ϕ; and A and B mark the points of failure, indicating the shear stress at failure for those condi-

---

[1] The unit kPa stands for kilopascal or 1000 pascals, MPa stands for megapascal or 1 million pascals. A pascal is a N/m² and a kPa is a kN/m², 1 psi = 6.895 kPa, and 1 MPa ≈ 145 psi.

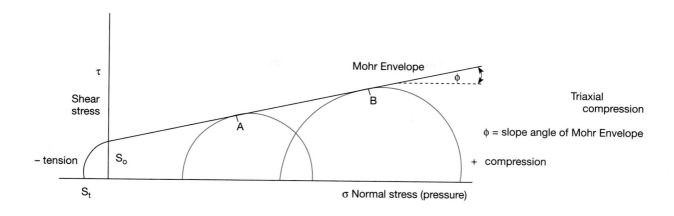

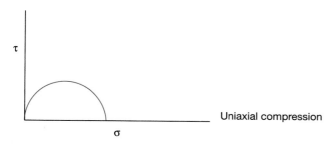

FIGURE **6.3**   Mohr diagrams for rock in compression.

tions of confining pressure and axial pressure. The slope of the Mohr envelope is indicated by ϕ and is a measure of the increase in τ needed to cause failure as the confining pressure increases. The shear strength at zero normal load, S_0, is sometimes referred to as the cohesion c or "shear strength" for the intact rock. On the basis of this relationship, the shear strength τ for any confining pressure σ can be determined from the following equation:

$$\tau = S_0 + \sigma \tan\phi$$

It is similar to the equation written $\tau = c + \sigma \tan\phi$ used for representing shear and normal stresses for soil testing. Another example problem is provided.

EXAMPLE PROBLEM: TRIAXIAL COMPRESSION TESTING _____

Two triaxial compression tests were performed on two samples prepared from the same sandstone. The samples are 2 in. in diameter and 4 in. long

(the solution is shown in Figure 6.4). For the first sample, a confining pressure of 1000 psi was used. The sample failed at an axial pressure of 11,000 psi. For the second sample, a confining pressure of 3000 psi was used. The sample failed at an axial pressure of 17,500 psi. Plot the Mohr circle for each test. Determine S_0 and ϕ.

ANSWER

For the first test, $\sigma_1 = 11,000$ psi and $\sigma_3 = 1000$ psi. These points are located on the σ axis. Using a diameter of $11,000 - 1000 = 10,000$, the Mohr circle is drawn. For the second test, $\sigma_1 = 17,500$ psi and $\sigma_3 = 3,000$ psi. These points are located on the σ axis. Using a diameter of $17,500 - 3000 = 14,500$, the Mohr circle is drawn.

A straight line is drawn connecting the points of tangency of the two circles. This is extended downward to intersect the τ axis. The intercept value is S_0, which equals 2100 psi. The slope

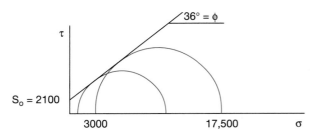

FIGURE 6.4   Mohr diagram for triaxial test example.

angle of the straight line (Mohr failure envelop) is 36°. Based on this information the equation is

$$\tau = S_0 + \sigma \tan\phi$$

$$= 2100 + \sigma \tan 36° \text{ in psi units}$$

The shear stress on the failure plane at failure is indicated by the point of tangency for each test. For the first test τ_{ff} = 4500 psi; for the second, test, τ_{ff} = 7000 psi.

The tensile strengths of rock are considerably less than their compressive strengths—on the order of only 10% as great. Tensile strengths range from 500 to 3400 psi for sandstone, about 700 to 3400 psi for limestone, 970 to 2800 psi for marble, and 970 to 3400 psi for granite (3500 to 24,000 kPa for sandstone, 4400 to 24,000 kPa for limestone, 7000 to 20,000 kPa for marble, and 7000 to 24,000 kPa for granite). Tensile strength governs behavior when a rock is under bending stresses. Rock is sufficiently weak in bending that it is seldom used to span any appreciable distance between supports. Concrete is also weak in tension and it is only through the embedment of reinforcing steel within the concrete that sufficient tensile strength is obtained for engineering structures.

Elasticity of Rocks

Compressibility is a measure of the amount a solid will reduce in length under load. Some of this deformation will be recovered when the load is removed but a portion is not regained. The recoverable deformation is termed elastic and the nonrecoverable part is termed plastic deformation.

Commonly, the elastic deformation of rock is directly proportional to the applied stress. When this is

the case, a proportionality coefficient E can be determined of the form

$$E = \frac{\text{Stress}}{\text{Strain}} = \frac{\sigma}{\epsilon} = \frac{P/A}{\Delta L/L}$$

where

σ = compressive stress
ϵ = axial strain
P = applied load in pounds (or newtons)
A = the loaded cross section in square inches (or square meters)
ΔL = the decrease in length in inches (or meters) = axial deflection
L = the original length in inches (or meters).

The proportionality coefficient E is known as the modulus of elasticity or Young's modulus. In a practical sense it is simply the slope of the stress-strain diagram.

The modulus of elasticity is used to determine the deformation that will occur in a foundation when the load of the structure is placed on it. Using the known load and measured or estimated E value, the strain or deformation can be determined. As an example, deformation of the concrete lining in a water tunnel under internal water pressure, such as in penstocks for dams or in some irrigation projects, can be a problem. If the rock deforms much more than the concrete, the lining will lose contact with the rock and become unsupported, thus causing the lining to rupture under the load.

Returning to the statement that E is the slope of the stress-strain diagram, we quickly realize that this is not a unique value for a loading diagram, as illustrated in Figure 6.5. Three possible E values are shown in Figure 6.5. The proper modulus value to use is that which best relates to the behavior of the rock under load. It is apparent that a secant and tangent modulus can be determined at any point and hence the possibilities become infinite. The accepted practice, however, is to use the tangent modulus at 50% maximum load as the representative value for the curve. This is appropriate because rock is loaded to much less than 50% of the failure load in a rock foundation so that one-half the failure stress represents a high upper limit of loading. That stress level would provide only a factor of safety of 2 relative to failure in compression; a factor of 10 or more is typical for rock foundations.

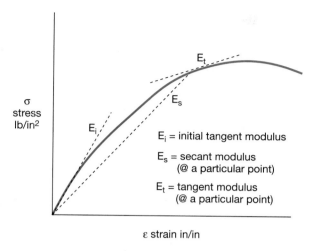

FIGURE 6.5 Typical stress–strain diagram for rock.

Rock is not an isotropic material, that is, its properties are not the same in all directions. Instead it is anisotropic, particularly those rocks with prominent bedding or foliation. Therefore, the elastic properties of rocks will vary depending on the direction of measurement. A lower E value is obtained for rocks measured perpendicular to bedding or foliation than for those measured parallel to these directional features. This results in a larger amount of strain or deflection when loading is perpendicular to bedding than when parallel. For this reason, rocks should be loaded during testing in the same direction as will occur during loading of the engineering structure. For horizontally bedded rocks, samples should be loaded perpendicular to bedding to depict gravity effects. If vertical rock cores are taken, the loading effects of the engineering structure on the rock foundation are determined by testing these cores along the vertical axis.

Also of interest regarding compression testing of rock is the extension that occurs perpendicular to the direction of loading. This lateral extension is typically expressed as a fraction of the vertical strain. The term is known as *Poisson's ratio* (μ) and is expressed in the following way:

$$\mu = \frac{\text{Lateral strain}}{\text{Axial strain}} = \frac{\Delta B/B}{\Delta L/L}$$

where B is the lateral dimension and L is the length or axial dimension. Poisson's ratio for a perfectly elastic material equals 1/3. Rocks, however, are not perfectly elastic so Poisson's ratio for different rocks ranges from 0.10 to 0.50. As with the modulus of elasticity, Poisson's ratio varies depending on the direction of loading.

Values for unconfined compressive strength (q_u), angle of shearing resistance (ϕ), modulus of elasticity (E), and Poisson's ratio (μ) are given in Table 6.1. Also included are unit weights and porosities for some of the common rocks.

EXAMPLE PROBLEM: ROCK DEFORMATION

An N_x core of fine-grained syenite, 2.125 in. in diameter and 4.25 in. long, is tested in unconfined compression. The modulus of elasticity for the rock is 9.2×10^6 psi, Poisson's ratio is 0.21, and its unconfined compressive strength is 24,500 psi. If the core is loaded to one-quarter its unconfined compressive strength, give the answer to the following questions:

1. What is the load on the sample in pounds?
2. What axial strain occurs at this load, what deflection?
3. What lateral strain occurs at this load, what lateral deflection?

ANSWERS

1. $\sigma = \dfrac{P}{A}$; $P = \sigma A = \dfrac{24{,}500}{4} \times \dfrac{\pi\,(2.125)^2}{4} = 21{,}723$ lb

2. $\epsilon = \dfrac{\sigma}{E}$ as $E = \dfrac{\sigma}{\epsilon}$

$$= \frac{24{,}500 \text{ psi}}{9.2 \times 10^6 \text{ psi}} \times \frac{1}{4} = 0.6657 \times 10^{-3}$$

$$= \frac{\Delta L}{L}, \text{ so } \Delta L = \epsilon L = 0.6657 \times 10^{-3}\,(4.25 \text{ in.})$$

$$= 2.83 \times 10^{-3} \text{ in.}$$

3. $\mu = \dfrac{\Delta B/B}{\Delta L/L}$, $\Delta B/B = \text{Lateral strain} = \mu\dfrac{\Delta L}{L}$

$$= 0.21(0.6657 \times 10^{-3})$$

$$= 1.398 \times 10^{-4}$$

$$\frac{\Delta B}{B} = 1.398 \times 10^{-4}, \quad \Delta B = \text{Lateral deflection}$$

$$= B\,(1.398 \times 10^{-4})$$

$$= 2.125(1.398 \times 10^{-4})$$

$$= 2.97 \times 10^{-4} \text{ in.}$$

TABLE **6.2** Engineering Classification of Intact Rock Based on Ultimate Strength (Adapted from Deere and Miller, 1966).

Class	Level of Strength	Uniaxial Compressive Strength	
		psi	kPA
A	Very high	32,000	220,000
B	High	16,000–32,000	110,000–222,000
C	Medium	8,000–16,000	55,000–110,000
D	Low	4,000–8,000	27,500–55,000
E	Very low	4,000	27,500

ENGINEERING CLASSIFICATION OF INTACT ROCK

Deere and Miller (1966) proposed an engineering classification of intact rock based on the uniaxial compressive strength and the modulus of elasticity of the rock. Intact rock is that which can be tested in the laboratory and is free from large-scale weakness planes such as jointing, bedding, and shears. The compressive strength is determined on specimens with a length-to-diameter ratio of at least 2 (in accordance with ASTM D2938), and the modulus of elasticity is the tangent modulus at one-half the ultimate strength of the rock. The engineering classifications of rock have also been related to the stability of rock slopes (West, 1979).

Rocks are subdivided into five strength categories as shown in Table 6.2. Strength categories follow a geometric progression. The 32,000-psi (220,000-kPa) line marks the upper limit of most common rocks with only quartzite, diabase, and dense basalt occurring above it. The B category (between 16,000 and 32,000 psi or 110,000 to 220,000 kPa) includes most igneous rocks, the stronger metamorphic rocks, most limestones and dolomites, plus the well-cemented sandstones and shales. The C category, medium-strength rocks (between 8000 and 16,000 psi or 55,000 to 110,000 kPa), includes most shales, the porous sandstones and limestones, and the prominent schistose varieties of metamorphic rocks such as chlorite, mica, and talc schists. The D and E categories contain low and very low strength rocks including friable sandstone, porous tuff, clay-shale, rock salt, and weathered or altered rocks of various lithologies.

The second property involved in Deere's classification is the modulus of elasticity. However, rather than using the modulus alone for the comparison, the modulus ratio is used, that is, the ratio of the modulus of elasticity to the unconfined compressive strength. These three subdivisions are shown in Table 6.3.

The summary plot for igneous rocks is shown in Figure 6.6. The middle, diagonal zone in this figure is designated as that of M, medium or average modulus ratio. Rocks with interlocking textures and little anisotropy fall into this category and it contains the majority of the igneous rocks. Rocks are classified by both strength and modulus ratio with such examples as BH, CM, and BM.

Figure 6.7 is the summary plot for sedimentary rocks. Limestones and dolomites fall primarily within strength categories B and C although some plot as high as A and as low as D; they have from high to medium modulus ratios. Both sandstones and shales extend into the zones of low modulus ratios.

Figure 6.8 presents the summary plot for metamorphic rocks. The scatter in results is due to the

TABLE **6.3** Engineering Classification of Intact Rock Based on Modulus Ratio (Adapted from Deere and Miller, 1966).

Class	Level of Modulus Ratio	Modulus Ratio Value
H	High	500
M	Average or medium	200–500
L	Low	200

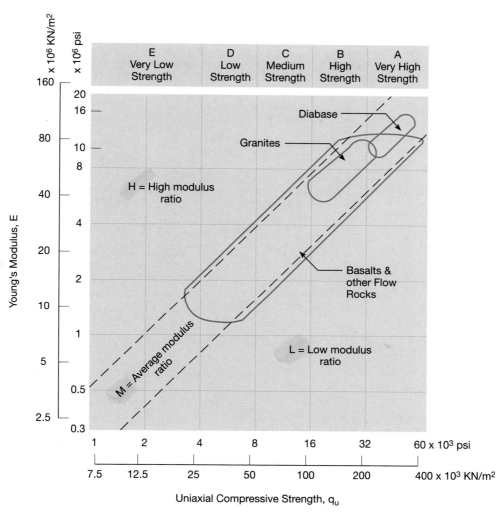

FIGURE 6.6 Engineering classification for intact rock-summary plot, igneous rocks.

great range in mineralogy and fabric (and anisotropy) for metamorphic rocks as a whole. Many quartzites plot in similar fashion with other dense, equigranular, interlocking fabrics, i.e., diabase and dense basalt. Gneisses plot mostly in the BM location in a similar fashion to granite except that gneisses show a somewhat lower average strength and greater scatter in the modulus ratio. This is in keeping with the considerable variation in foliation (gneissic structure and some schistosity) that gneisses exhibit as a group.

Schists are represented by two zones in Figure 6.8, 4A is for schists with steeply dipping foliation or a high angle (45° or more) between a horizontal plane and the direction of foliation. As specimens were tested along a vertical load axis, foliation nearly paralleled the load axis and thus strengths were reduced. This also yields a high modulus ratio by virtue of the low strength, q_u, in the denominator (modulus ratio = E/q_u).

By contrast, for the 4B zone, the foliation is at a low angle from the horizontal and therefore at a high angle to the vertical load axis. The strength is increased somewhat but the modulus of elasticity is reduced because of the closure of microcracks that par-

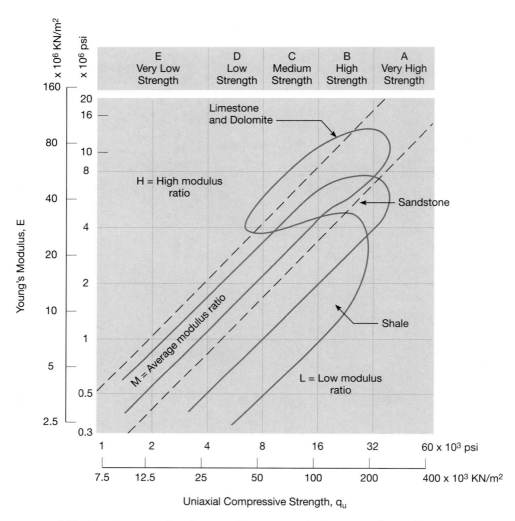

FIGURE **6.7** Engineering classification for intact rock-summary plot, sedimentary rocks.

allel the foliation. This yields a lower modulus ratio for the 4B group. For marble (zone 3) a high modulus ratio is consistent along with a medium strength.

Deere suggests that this engineering classification be included with the lithologic description of the rocks. He also suggests that the rock quality designation (RQD) and fracture frequency (fractures per foot) be included in the core log description as well. RQD is related to the length of pieces of core obtained in the rock-coring process of subsurface sampling. It equals the summary of lengths for all pieces of core 4 in. long or greater (100 mm) divided

by the length of the hole drilled, usually 5 or 10 ft (1.5 or 3 m) in length. More information is provided on this subject in Chapter 19 on subsurface investigation and site selection.

TESTING CONSTRUCTION MATERIALS

As mentioned previously, the primary concerns regarding rock construction materials (in addition to economics) are strength and durability. Strength of these materials is not determined by compression

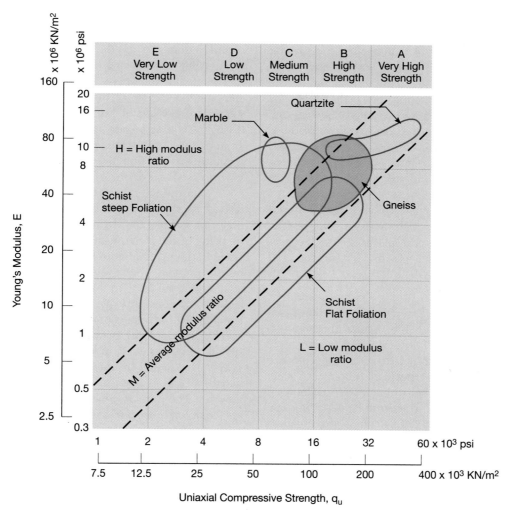

FIGURE 6.8 Engineering classification for intact rock-summary plot, metamorphic rocks.

tests as it is for rock foundations, instead an abrasion test is used. Abrasion and durability tests are discussed after considering several definitions.

Definitions

An aggregate or mineral aggregate is an important construction material. It is defined as an aggregation of sand, gravel, crushed stone, slag, or other material of mineral composition used in combination with a blending medium to form bituminous and Portland cement concrete, macadam, mastic, mortar, and plas-

ter, or alone as in railroad ballast, filter beds, base courses, and various manufacturing processes such as fluxing. Another material, riprap, consists of larger pieces and is defined as broken stone or boulders used as a protective layer on the upstream face of earth dams or on river banks, lake shores, and harbor structures for protection from wave action, currents, and general erosion by water. Concrete aggregates consist of a specified size gradation of high-quality rocks, which along with cement and water are used to make Portland cement concrete. Coarse aggregate consists of pieces greater than 3/16 of an inch and fine aggregate is less than 3/16 in. in size.

The distinction between a quarry and a pit is also significant. Quarries are surface excavations where rock is removed from its natural location in the bedrock as stone production proceeds. A pit by contrast refers to a surface excavation where gravel, sand, or other loose or unconsolidated material is removed from its source. Crushed stone is obtained by crushing rock material from a quarry and dimension stone is rock cut into specific sizes for use as facing stone for buildings or as stone monuments.

Gravel and crushed stone are used both for concrete aggregates and for highway base courses. The base course is the rock base placed below the concrete or asphalt pavement to provide support and drainage for the highway.

Abrasion Resistance

The test commonly used to measure abrasion resistance of aggregates is the Los Angeles abrasion test. The test, designated ASTM-CI31[2], involves a sample weight or charge of 5 kg (11 lb) with a specific size gradation along with a specified number of steel spheres loaded into a steel drum, which contains an interior, projecting shelf. The drum is rotated for 500 revolutions and the resultant sample is sized using a #12 sieve (0.141 mm). A ratio equal to the weight of the fine material (finer than #12) divided by the original weight yields the percent loss. The maximum allowable abrasion loss for concrete aggregates and base courses is established by highway construction specifications for each state in the United States and by federal agencies involved in construction. This value is in the range of 35% to 50% for various highway departments for concrete aggregates and typically a somewhat higher value is allowed for base courses.

Durability

The durability of an aggregate is a measure of its ability to withstand deterioration due to wetting and drying, heating and cooling, and freezing and thawing during the period of performance. Several tests are used to determine durability, with each state highway department and federal agency designating the test to be used to determine acceptance or rejection of materials. The two most common aggregate tests with regard to durability are the sulfate soundness test and the freezing and thawing test.

The sulfate soundness test entails soaking a specific, graded sample of the aggregate in a saturated solution of Na_2SO_4 or $MgSO_4$ followed by complete drying of the aggregate in an oven (ASTM C-88). Temperatures and minimum times are given in the specifications. In this procedure the sulfate solution forms crystals on drying; these growing crystals exert a force on the internal pores of the rock, tending to disrupt its structure and break pieces from it. Five cycles of soaking and drying are performed followed by sizing of the pieces using a specified sieve. A ratio equal to the material finer than this sieve size divided by the original weight yields the sulfate soundness loss. A loss of 12% to 15% is a typical maximum allowed for concrete aggregates, with a 15% to 18% loss allowed for base course materials.

The freezing and thawing test is also run on a specific, graded sample of loose aggregate material (T 103[3]). Depending on the specified test, the sample may be frozen in air or in water and thawed in air or in water. The primary concern is that the rock pieces are completely frozen (including the centers of all aggregates) and completely thawed during each cycle of the test. Typically the number of freeze-thaw cycles is 25. Following completion of the test, the sample is sized using a specified sieve. A ratio equal to the amount finer than the specified size divided by the original weight of the sample yields the percent loss. Typically the same amount of loss is allowed for this test as for the sodium sulfate soundness test.

To determine the performance of concrete in freezing and thawing, concrete beams rather than unconfined aggregate pieces can be tested. At the beginning, and after every 25 cycles of the test, the fundamental transverse frequency N of the beam is determined using an appropriate laboratory apparatus. The relative dynamic modulus E_r of the beam is determined based on the following equation, and when E_r has been reduced by 60%, failure of the beam is indicated. Three hundred cycles of the test

---

[2] American Society of Testing and Material's standard test designation.

[3] T 103 is the designation for the AASHTO (American Association, State Highway Officials) standard test for the freeze-thaw of aggregates.

are run (unless the beam reaches the reduced value sooner) and if the relative dynamic modulus remains greater than 60%, the concrete beam passes the test.

$$\text{Relative dynamic modulus of elasticity} = E_r$$
$$= (N_c^2/N_0^2) \times 100$$

where N_0 is the frequency before testing and N_c prevails after c number of cycles of freezing and thawing occurs.

Freeze-thaw testing of concrete beams examines not only the performance of the aggregate under these conditions, but also that of the hardened paste that binds the aggregates. This is a more complete procedure than testing aggregates alone, but it is quite time consuming and may cause delays in the decision-making process for aggregate selection. A related procedure involves the measurement of length change for the concrete specimens after a specified number of cycles.

Several concerns have been raised regarding the sulfate soundness test. Reproducibility of results within a single testing lab and between testing labs is difficult to achieve if the specified ASTM procedure is not carefully followed. Temperature is important regarding saturation of the solution and the solution should be replaced periodically as prescribed by the specifications. Another concern is the relationship between specifics of the test and what actually occurs in nature. The question is one of how well the formation of sulfate crystals simulates the effects of ice crystal growth during freezing and thawing, or expansion during wetting and drying and heating and cooling. On this basis, freezing and thawing or wetting and drying tests would seem more applicable for testing purposes, but another consideration is involved. To predict the long-term performance of rock by means of a short-duration procedure, the test must be more severe than the natural weathering effects, so that results can be obtained in a reasonable period of time. The sulfate soundness test is severe and certainly qualifies as an accelerated test.

The test can also be used in those states that receive few if any freeze-thaw cycles in a winter season. In states such as Florida it is difficult to convince aggregate producers that the freeze-thaw test should be used to evaluate their materials. But other weathering effects such as heating and cooling and wetting and drying do occur and these are also simulated by the sulfate soundness test. In areas with less severe weather, aggregates with a higher allowable soundness loss can be used. The sulfate soundness test is particularly destructive to argillaceous (clayey) rocks and to those with extremely coarse-grained textures. Experience has shown that clay-rich carbonate rocks yield concrete with low durability and poor long-term quality.

The amount of water absorption of aggregates is another specified test used for aggregate selection. This test is based on the equation provided previously in this chapter in the discussion on specific gravity and repeated below:

$$\text{Absorption} = \frac{\text{Saturated weight} - \text{Dry weight}}{\text{Dry weight}} \times 100$$

A common, maximum allowable absorption value is 5% for concrete aggregates and for those used in bituminous pavements. The absorption value is a general indication of the freeze-thaw resistance of aggregates and, although there are exceptions, high absorptions tend to indicate nonresistant materials.

A final criteria for judging the suitability of aggregates for concrete involves the amount of weak constituents present. Weak materials include shale, siltstone, weathered argillaceous carbonates, iron concretions, friable sandstones, deeply weathered rocks, coal, wood, and low density cherts (SpG < 2.40). These materials will pop out from the concrete as weak aggregates fracture when exposed at the concrete surface during freeze-thaw cycles.

ASTM Standard C-33 indicates a maximum allowable percentage of these deleterious materials relative to the intended use of the concrete. For pavements and driveways that undergo severe weathering conditions, a maximum of 3% clay lumps and friable particles are allowed, 5% low-density chert, 0.5% coal, with the sum of clay lumps, friable particles, and low-density chert set at 5%. For exposed architectural concrete under severe weathering conditions these values are reduced to 2% clay lumps and friable particles, 3% low-density chert, and 0.5% coal, with a total of 3% allowed for the first two categories. Some state highway departments stipulate a 3% maximum for low-density chert for aggregates used in concrete pavements. All areas of the eastern U.S. except Florida and portions of the adjacent southernmost states are included in the region where severe weathering conditions occur.

ROCK MASS PROPERTIES

Rock mass properties dictate the overall characteristics of a large mass of rock as related to engineering construction. Examples include the abutments (side slopes) and foundation (base) of large dams, rock slope stability of a highway cut, and the mass of rock through which a tunnel is to be driven.

The strength and behavior of a rock mass are determined by the nature of its discontinuities or weakness planes. This illustrates a difference between soil and rock and therefore a distinction between the procedures used in soil mechanics versus rock mechanics. Soil and soil mass properties are commonly assumed to be similar and adjustments can be made to account for mass properties. However, for rock, or those materials with an unconfined compressive strength greater than about 100 psi (14,000 psf or 700 kPa), weakness planes completely control the strength. This involves all truly rock-like materials.

For rock weakness planes, the important concerns are the attitude of the planes (position in space) relative to the exposed face, the extent of continuity of the planes, the spacing of weakness planes, the nature of the surface (irregularity, smoothness) of the weakness planes, the degree of infilling of material along these planes, and the ground-water conditions. These features are explored in detail in the section on slope stability found in Chapter 14.

There are several natural weakness planes in rock, including bedding, foliation (slaty cleavage, phyllitic cleavage, schistosity, and gneissic structure), flow banding in lava rocks, joints, faults, and shear zones. These structural geology features are considered in Chapter 10.

Measurement of rock mass properties is not easily accomplished because of the large volume of rock involved. Field testing can be performed to determine rock mass properties. A detailed discussion of this subject, however, is beyond the scope of this basic textbook.

ROCK STRENGTH AND PETROGRAPHY

Petrography involves the detailed description of rocks, as related to both the mineralogy and texture of specimens. Rock strength, measured by unconfined compression, Los Angeles abrasion, or other various means, is also a consequence of these petrographic details.

Textures and Strength

Rock textures, without considering origin directly, are limited to a few basic varieties: 1) interlocking crystals, 2) clastic pieces cemented by various amounts of crystalline material, 3) clastic pieces with infilling of matrix, and 4) glassy materials or noncrystalline varieties.

The interlocking crystals can range in size from microscopic (as small as 5 μm or 5×10^{-4} cm) to those more than 3 cm (about 1 in.) in size, and can occur in igneous, sedimentary, or metamorphic rocks. In some respects glassy rocks act like those with extremely small crystals because they are essentially homogeneous and form conchoidal fractures on breaking.

Fractures in Rock

For both clastic grains and interlocking crystals, fractures occur either through the grains and crystals or around them. Rocks composed of clastic grains surrounded by finer noncemented matrix typically fail through the weaker matrix material. These rocks are held together by compaction forces originating from overburden stresses developed during rock burial, so they are also easily weakened by wetting, drying, or freezing and hence have poor weathering resistance. As long as the matrix is considerably weaker than the grains, failure surfaces will occur primarily through the matrix. Such rocks are seldom if ever used for aggregates.

Clastic rocks with the minor amounts of cementing agent, such as a friable sandstone, tend to fail through the cement, sometimes after only minor agitation. Many quartz sandstones have a high pore volume much of which is interconnected, which provides the high permeability typical of these rocks. Sandstone outcrops, contrary to their friable nature, are extremely resistant to weathering, simply because water seeping from the porous rocks deposits minerals, calcite primarily, on evaporation. This increases rock strength at the surface. The effect, known as case hardening, provides quartz sandstone outcrops with a high level of resistance to both mechanical and chemical weathering.

Clastic rocks with a major portion of their interstices filled with crystallized cement can develop fractures either through the grains or through the cement. Commonly the cement is calcite, which is weaker than most grains and hence fracturing occurs mostly through the cement. These fractures occur both through the center of the cement filling and at the intersection between grain and cement. As explained later, grain size (or crystal size) and mineralogy determine whether failure will occur at the boundaries or through the crystallized cement.

For interlocking crystals or for clastic rocks with strong, well-crystallized cement, the fractures can occur either around the crystals (or grains) or through them. Rocks composed of small particles have a much greater total boundary surface between particles than do coarse-grained rocks. This is a fundamental feature indicated by increased specific surface values (the surface-to-volume ratio) that occur with decreasing grain size. This is why crushing and grinding are used to increase the surface reaction of materials by providing greater surface area as the particles are made smaller.

Coarse-grained rocks composed of minerals with good cleavage are much weaker than their fine-grained equivalents, which consist of much smaller crystals. This is because cleavage yields a planar break directly across the entire crystal, whether large or small. A group of smaller, randomly oriented grains will yield a more irregular surface composed of many small breaks with changing orientation. By contrast the larger grains provide a single straight-line fracture yielding a shorter total distance with less resistance to failure.

For both clastic rocks with crystallized cement and rocks with interlocking crystals, larger grained rocks yield lower strengths than fine-grained rock. This is because the surface of fracture is smaller for coarser rocks. This develops for two reasons: 1) fracture around grain boundaries (coarse-grained rocks are weaker because the boundary surface area is smaller) and 2) fractures through the grains (coarse-grained rocks are weaker because the fracture takes a shorter, less circuitous path through the larger crystals).

The mineralogy of the rock has an effect on strength because some minerals are harder and stronger than others and some minerals cleave more readily than others. This explains why marble with an average grain size of 1 mm would be weaker than a granite of the same size, assuming both are unweathered.

The final feature of texture, related to both origin and mineralogy, is the interlocking of the crystals. Two considerations are involved: the shape or configuration of grain interlock and the mutual orientation between elongated grains. Granites because of the minerals involved contain rounded crystals of quartz and orthoclase that do not interlock as well as the angular grains of plagioclase and augite that occur in diorites. Therefore, a diorite with the same grain size as a granite would have a greater strength. The other feature is the interlocking texture caused by randomly oriented, elongated rectangular grains yielding the so called "tepee structure" of a diabase. These features have the general repeat pattern of an Indian tepee or an A-frame building. Fractures around the grains must take a circuitous route following the random orientation of the elongated plagioclase. This yields increased strength because of the increased distance, making diabases one of the strongest rocks, with unconfined compressive strengths of 40,000 psi (275,000 kPa) or more.

PORTLAND CEMENT CONCRETE

Portland cement concrete or simply concrete is an engineering material in wide use throughout the world today. It can be precast into structural members or cast in place during the construction process. It is both strong and durable and its widespread use has made possible many of the construction achievements of the twentieth century. Dams, bridges, highways, buildings, tunnel linings, retaining walls, and sidewalks are constructed from concrete. In addition, concrete is in many ways similar to rock, both in physical characteristics and in the methods used for analysis. Because of this similarity, it is appropriate to include a brief discussion on Portland cement concrete in this chapter.

Concrete Composition

Concrete is a composite material consisting essentially of a binding medium, the cement paste, which surrounds pieces of relatively inert mineral filler. This binder is formed as a reaction product from Portland cement and water. The mineral filler is the aggregate and ranges in size from fine sand to pebbles or larger gravel-fragments or crushed stone.

In hardened concrete the aggregate comprises about 75% of the volume. Typically, particles less than 3/16 of an inch (4.75 mm) are considered fine aggregate or sand, and those larger than that size are coarse aggregate. The purpose of the aggregate is threefold: 1) Provide an inexpensive filler for the cementing medium, 2) provide a considerable volume of the concrete, which is resistant to applied loads, abrasion, moisture penetration, and weathering action, and 3) reduce the volume change effects that occur in the binding medium during hardening and from moisture changes thereafter.

Cement, water, and aggregate are combined in overall proportions to accomplish the following requirements: 1) When mixed the fresh concrete is workable such that it can be placed in the forms, which will hold it until hardening has occurred; 2) when hardened the concrete will be sufficiently strong and durable to accomplish the purpose intended; and 3) the cost of the total product is as low as possible in keeping with the necessary quality required.

The space between the aggregate, the remaining 25% or so, is filled with cement paste and air voids. Some entrapped air remains in the concrete after placement even though it has been well compacted. This volume is usually between 1% and 2% of the total volume. In recent years, special air-entraining agents have been incorporated when proportioning the concrete. These agents provide small air voids throughout the paste and their primary purpose is to provide freeze-thaw resistance and salt scaling resistance to the concrete during its service life. Concrete durability is discussed in more detail later in this section. For freeze-thaw resistance the desired amount for air content is from 5% to 6% of the total volume of the concrete. This would include both the entrapped air and the entrained air.

The presence of air voids, however, has an effect on the strength of the hardened concrete. This must be considered when proportioning the concrete. Typically more cement must be included in air-entrained concrete to counteract the strength reduction brought about by the addition of the entrained air. If consid-erably more air is included in the paste, a further reduction in strength occurs. Hence, a detrimental effect of entraining too much air in the paste (above the 5–6%) is a reduced compressive strength of the hardened concrete, perhaps below that required.

The solid materials in the hardened concrete consist of mineral aggregate and hardened cement paste. The latter may include some unreacted cement particles plus a major constituent, hydrated cement (CSH, calcium silicate hydrate), which is a product of reaction between the cement and water. Proper curing of the concrete is needed to ensure that the cement has full opportunity to react with water to yield the cement hydrate.

The cement paste has two major functions: 1) To fill the space between the aggregates thereby providing lubrication in the fresh, plastic concrete during placement, and water tightness after the concrete sets, and 2) to provide strength to the hardened concrete. For the hardened paste, the properties are dependent on the following aspects: 1) the characteristics of the Portland cement itself, 2) the relative proportions of cement and water, usually represented as the water-cement ratio by weight, and 3) the extent of chemical reaction between the cement and water.

Cement hydration requires time, favorable temperatures, and the availability of moisture. Curing takes place as the concrete is subjected to suitable moisture and temperature conditions. The curing period for specimens in the laboratory is 28 days under special moist room conditions. The 28-day compressive strength of standard concrete cylinders is commonly used to determine the acceptability of concrete placement (3000, 3500, 4000, and 5000 psi are commonly required values). In construction work the curing period typically varies from 3 to 14 days.

As a final consideration of concrete composition, the contribution of the aggregate to the properties of concrete can be enumerated: 1) The aggregate particles contribute to strength, elasticity, and durability in a direct way, 2) the nature of the surface of the particles is important (for example, roughness decreases workability of fresh concrete and increases bond with the cement paste after hardening), 3) a dense gradation of the aggregate reduces workability and increases the density of the mix, and 4) the higher the percentage of aggregate, the lower the cost and volume changes that occur on drying.

Petrographic Examination of Concrete

As previously indicated, concrete and concrete-making materials can be studied in much the same way as minerals and rocks are evaluated. This involves petrography, the detailed description of these materials based on an evaluation of their texture and mineralogy. Extensive work on concrete petrography has been accomplished by Mielenz (1961) and subsequent workers. A list of the ways in which petrography can be used is provided in Table 6.4.

Determination of the air-void system in concrete can be accomplished according to ASTM C457, "Microscopic Determination of Air Void Content and Parameters of the Air-Void System in Concrete." In this procedure the percentage of air voids is determined along with the average chord length (diameter) of the voids, the specific surface of the void system, and the spacing factor (the average distance water must travel to reach an air void). The purpose of the analysis is to determine whether or not the concrete is resistant to freeze-thaw attack and to salt attack. Well-distributed small bubbles in the paste, totaling about 6% of the total concrete volume, will provide resistance to such attack. This determination is made on a cut surface of the concrete, which has been polished to a smooth planar surface.

Petrographic examination of the polished concrete surface can also provide information on the percentage of constituents. Paste, coarse aggregate, and fine aggregate content in addition to the air content can be determined. This is valuable information when working out details of the original mix design. Cement-aggregate reaction products can also be viewed using petrographic analysis. Alkali-silica and alkali-carbonate reactivity are described in Chapters 3 and 4.

Petrography of concrete can be accomplished in a number of ways. These include 1) the examination of concrete structures in the field to obtain an overall view of the performance, 2) thin sections of the concrete, and 3) polished sections. These are the same procedures used to examine rocks and minerals in the traditional field of geology. In Table 6.5, a list of features that can be studied in a thin section of concrete is provided. A thin section is a thin slice of rock (or concrete) mounted on a glass slide with optical cement. The section transmits light by virtue of it being ground to a thickness of 0.03 mm or less. A polished section by contrast is a smooth, planar surface that has been highly polished to provide a plane on which the constituents can be observed and the voids between them as well. These are viewed in reflected light. Both the constituents of cement clinker and concrete can be studied using polished sections.

TABLE 6.4   Uses for Petrographic Examination of Concrete and Concrete-Making Materials

Description of the Concrete
1. Mix proportions
2. Internal structure
3. Cement-aggregate relationships
4. Deterioration

Description of the Cement
1. Composition especially presence of free CaO or MgO in undesirable amounts
2. Relative fineness
3. Identification of certain additives

Description of the Aggregate
1. Composition, grading, and quality
2. Identification of the type, kind, and source of the aggregate
3. Presence of coatings
4. Detection of contamination

Identification of Certain Admixtures
1. Mineral admixtures
2. Pozzolans
3. Siliceous aids to workability

Evaluation of the Microscopic and Megascopic Void System
1. Air content of the concrete
2. Size and spacing of the voids in the cement paste

Determination of the Cause of Inferior Quality or Failure of Concrete
1. Cement-aggregate reaction
2. Attack by aggressive waters
3. Freezing and thawing
4. Unsound aggregate
5. Unsound cement, especially excessive free CaO and MgO
6. Inadequate proportioning, mixing, placing, curing, or protection
7. Structural failure, abrasion, or cavitation

Determination of the Cause of Superior Quality and Performance of the Concrete

TABLE 6.5   Features Viewed in a Thin Section of Concrete

Fine aggregate	Hydration products
Composition	Texture and microstructure
Cement-aggregate reaction	Admixtures and additions
Alteration in place	Calcium hydroxide
Coarse aggregate	Alteration of the hydration products
Same as fine aggregate	Microfractures
Cement paste	Extent, width, continuity
Unhydrated clinker particles	Origin and relationships
Maximum and average size	Secondary deposits
Frequency	Voids
Composition	Frequency
Evidence of associated cracking	Size
	Special relationships
	Secondary deposits

EXERCISES ON ENGINEERING PROPERTIES OF ROCKS

1. Specific gravity information on an argillaceous dolomite is desired. It weighs 250 g in the dry state, 259 g when saturated (surface dried), and 144 g when saturated and weighed in water.
   (a) What is the bulk specific gravity; bulk specific gravity, saturated surface dried; apparent specific gravity; and absorption of the rock?
   (b) What is the volume of the rock sample?
   (c) What would be its weight per cubic foot in the dry state?
   (d) Would the rock make a suitable concrete aggregate? Explain.

2. Two triaxial compression tests were performed on two prepared samples of the same limestone. At a confining pressure of 2000 psi, the first sample failed at 9000 psi. Then at a confining pressure of 5000 psi, the other sample failed at 21,000 psi.
   (a) Plot the two test results on a shear stress-normal stress diagram and draw in the Mohr envelope.
   (b) Determine the ϕ value for the limestone.
   (c) What is the S_0 value?
   (d) What was the shear stress on the failure plane at failure for both tests?

3. Limestone similar to that tested in Exercise 2 exists in a mountainous region at a depth of 600 ft.
   (a) If the vertical stress is the confining stress of the rock at that depth, what is this value in psf and psi? Assume a unit weight of 160 lb/ft^3 for the rock.

   (b) At this confining pressure, what horizontal stress would be required to cause a shear failure in the rock? Which is σ_1 and which σ_3? Use the plot constructed in Exercise 2 to obtain the answer. What is the shear stress on the failure plane?

4. (a) Under what conditions would the tensile strength of rock be significant? (Consider bending conditions for a rock mass.)
   (b) A sample of clastic limestone yielded a compressive strength of 6000 psi and a shear strength of 1100 psi. What tensile strength value would you select for this rock if testing was not possible? Explain how you would arrive at this value.

5. A core sample of basalt, 4 in. long and 2 in. in diameter, was tested in unconfined compression. The basalt is known to have a modulus of elasticity = 9.1×10^6 psi.
   (a) At a pressure of 10,000 psi, what would be the reduction in length of the sample?
   (b) At this pressure what would be the vertical load on the sample?
   (c) At 10,000 psi stress the core showed an increase in diameter of 5.5×10^{-4} in. What is the Poisson's ratio for the sample?

6. A concrete gravity dam with a trapezoidal shape as shown below will be founded on sedimentary rock. The concrete has a unit weight of 150 lb/ft^3.

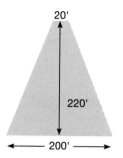

(a) What is the average pressure exerted on the rock foundation in lb/ft²? What is the maximum pressure? Explain.
(b) Refer to the different sedimentary rocks listed in Table 6.1. Using average compressive strength and shear strength values for these sedimentary rocks compare this to the average pressure calculated in part (a). Are any so low to be of concern? Explain.
(c) Assume that the pressure from the weight of the dam is dissipated within 100 ft below the rock surface and the full pressure acts over this distance. How much settlement will occur in the center of the dam because of the gravity force for the different sedimentary rocks listed in Table 6.1, using average values?

7. Excavation for a gravity dam foundation revealed a unit of poorly cemented sandstone 100 ft wide within a massive dolomite sequence. The dolomite has an E value of 8×10^6 psi and the sandstone has an E value of 1×10^6 psi.
(a) What would be the concern if both rocks were loaded equally by the gravity dam? What would be the effect on the dam?
(b) How could this problem be alleviated? Consider both replacement and strengthening possibilities.

8. An 1100-ft-long, horseshoe-shaped tunnel is to be driven through massive granite with a unit weight of 170 lb/ft³. The dimensions of the tunnel cross section are shown. The maximum rock cover above the tunnel in the middle of the mountain is 1200 ft.

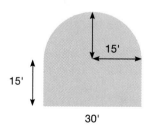

(a) What is the total overburden stress at the midway point in the tunnel in psi and psf?

(b) Because of arching effects that transfer load around the tunnel opening, the roof or crown of the tunnel does not commonly support the full overburden stress but instead is typically much less. If the roof stress was found to be 2500 psf, how many feet of rock above the tunnel does this represent? How do you suppose this load would be supported in the tunnel?
(c) If the tunnel excavation was completed in 80 working days with two shifts per day, how many lineal feet of tunnel were excavated per day? How many feet per shift? What was the average volume of rock removed per day, per shift? (*Hint:* Calculate the area of the cross section.)
(d) The tunnel project was bid at $1,878,000. What would be the cost per lineal foot of the tunnel?

9. An arch dam transfers much of the water load onto the abutments (the rock mass on the sides of the dam) (see drawing for details). The E value for the concrete is 6×10^6 psi.

PLAN VIEW

(a) What would occur if the abutments were shale with an E of 3×10^6 psi?
(b) What if the abutments were diabase with an E of 14×10^6 psi? Which of the two rocks would be preferred, the shale or the diabase? Explain why.
(c) What would be the effect if the thrust (force direction from the dam) were parallel to the slaty cleavage direction in the abutments? What could be a conclusion of this arrangement?
(d) What type of force is preferred within the arch dam: compression or tension? Explain why. Which type of force would develop if the abutments had a high modulus of elasticity?

10. For room and pillar underground mines, the pillars must carry the total weight of the overburden.
(a) Why does the arching effect that acts in tunnels not contribute any support for these mines?
(b) If a coal mine is 500 ft deep, can 40% of the coal be removed and still yield a factor of safety (FS) of 3 relative to compressive strength of the coal in the pillars? Assume an unconfined compressive strength of 3000 psi for the coal and use the equation

$$FS = \frac{\text{Rock strength } (1 - \text{fraction removed})}{\text{Rock load}}$$

(*Hint:* Use a unit weight of rock = 160 lb/ft^3. What is the FS?)

(c) What percent of the coal can be removed at a depth of 800 ft using an FS = 3 and the same compressive strength as in part (b)?

11. A massive hillside of rhyolite contains joint planes that dip at an angle of 40° toward a stream valley. A rectangular-shaped rock bounded by other joint planes rests on the dipping joint plane. The block measures 5 ft high, 15 ft wide, and 10 ft long. The unit weight of the rhyolite is 172 lb/ft^3. Assume a dry slope.
    (a) What is the weight of the rock block?
    (b) Assuming that only friction holds the block in place (no cohesion along the joint plane), what is the minimum frictional force needed to resist sliding?
    (c) What coefficient of friction is needed to produce this minimum force?
    (d) What would be an actual coefficient of friction for this rock block, that is, rhyolite sliding against rhyolite? A friction angle is commonly used, the coefficient of friction being equal to the tangent of that angle. Determine that angle of friction.

12. A stream gravel is to be used as a concrete aggregate for a concrete road in central Ohio. What tests should be run on the aggregates to determine if they are suitable materials? What allowable maximum values would be specified for the individual tests?

13. From the following rocks prepare a list in order of their decreasing unconfined compressive strength. Opposite each rock, give the reason for your selection.
    (a) fine-grained limestone
    (b) fine-grained basalt
    (c) medium-grained diabase
    (d) porous tuff
    (e) friable quartz sandstone
    (f) coarse marble
    (g) medium-grained granite

14. When shale is excavated from a deep foundation for a heavy structure such as a power plant or concrete dam, layers of shale may loosen from the bedrock after a few hours of exposure. To prevent this, a concrete slab is placed over the shale soon after excavation and rock bolts may also be placed through the concrete into the shale.
    (a) Why does shale have this problem whereas limestone, for example, does not?
    (b) How is this related to the petrography of the rocks, including texture composition and origin?

15. (a) Why is tuff an extremely weak rock whereas welded tuff is so massive that it is commonly mistaken for rhyolite or andesite?

(b) How does the porosity of a rock generally relate to its strength?
(c) The strength of sandstones as a group ranges from about 2800 to 23,000 psi. Describe the types of sandstones that would provide the two extremes of this range.

16. Recent volcanic rocks are some of the most hazardous rocks through which to tunnel. List some of the possible problems that could develop when tunneling through such materials.

17. Refer to Figure 6.4. For the granite family, what is the uniaxial compressive strength and modulus of elasticity? What is the modulus ratio? Does this turn out to be average or medium as indicated in Table 6.3? Discuss.

18. Refer to the entry for diabase on Table 6.1. What is the highest compressive strength given for diabase? How does this compare with designations in Table 6.2? What is the highest E value given for this rock in Table 6.1? What modulus ratio value is obtained using these two values? What level of modulus ratio does this yield relative to Table 6.3? Check also the highest value for basalt in this respect. What two-letter designation would these examples yield?

19. A certain sandstone rates a CL designation by the Deere classification. What does this mean specifically according to Tables 6.2 and 6.3? Where would it plot on Figure 6.7?

20. Coal is listed in Table 6.1. Using the lowest values for unconfined compressive strength and for the modulus of elasticity, determine where this would plot on Figure 6.7. Calculate the modulus ratio in the process.

21. In hardened concrete three major materials are present: coarse aggregate, fine aggregate, and cement paste. Describe the nature of these three constituents.

22. What are the two types of voids that are present in the paste? How did they get into the paste and what is their function if any?

23. What is meant by hydration of the cement and by the term curing? How long are standard cylinders cured?

24. Why does the aggregate tend to reduce the volume change of concrete during heating-cooling and wetting-drying cycles? Why does a higher water-cement ratio in the paste yield a higher volume change effect?

25. What are the two primary functions of the paste in concrete? On what primarily do the properties of the hardened paste depend?

26. Petrography can be used to examine hardened concrete just as it is used to examine various types of rocks.

Why does this seem to be a logical extension from rocks to concrete? What similarities are involved?

27. Megascopic examination and thin section and polished section viewing of concrete are useful petrographic procedures. What is involved in each of these? What is a thin section? What is a polished section? How are they examined?

28. Why is the air-void system of hardened concrete commonly of interest in an investigation? Why is an air content that is too high or one that is sometimes too low a serious concern? Explain.

29. Petrography can be used to examine both the raw materials for concrete construction and the hardened concrete that results. What materials does this include?

30. Review the discussion on alkali-silica reaction in concrete (Chapters 3 and 4). How can petrography be used to evaluate this problem? Provide a thorough answer.

Additional Readings

American Society for Testing and Materials, 1992, Concrete and Aggregates, Vol. 04.02, ASTM, Philadelphia.

Attwell, P.B., and Farmer, I.W., 1976, *Principles of Engineering Geology,* Chapman and Hall, London.

Deere, D.U., and Miller, R.A., 1966, *Engineering Classification and Index Properties for Intact Rock,* Technical Report AFWL-TR-65-116, Air Force Weapons Laboratory, Kirtland Air Force Base, New Mexico.

Farmer, I.W., 1968, *Engineering Properties of Rocks,* E & FN Span, London.

Krynine, D.P., and Judd, W.R., 1957, *Principles of Engineering Geology and Geotechnics,* McGraw-Hill, New York.

Mielenz, R.C., 1962, "Petrography Applied to Portland-Cement Concrete," *Reviews in Engineering Geology,* Vol. 1, p. 138, Geological Society of America.

Stagg, K.G., and Zienkiewicz, O.C., Eds., 1968, *Rock Mechanics in Engineering Practice,* Chaps. 1 and 2, John Wiley & Sons, New York.

West, T.R., 1979, *Rock Properties, Rock Mass Properties and Stability of Rock Slopes,* in Selected Geotechnical Design Principles for Practicing Engineering Geologists, Short Course, Association of Engineering Geologists, Chicago, Illinois, Annual Meeting.

Elements of
Soil Mechanics

Soil mechanics is the branch of mechanics concerned with the action of forces on soil masses. Stated another way, it involves the scientific approach to understanding and predicting the behavior of soil for engineering purposes, as in foundations for buildings, soil embankments, or use as construction materials. Soils engineering is the engineering specialty that involves the principles of soil mechanics.

A term more recently established in this field of study is *geotechnical engineering*. This term refers to the knowledge of soil mechanics and other engineering disciplines combined with a basic knowledge of geology. This combined knowledge is applied with judgment to develop an engineering design. Geotechnical engineering consists primarily of soils and foundation engineering, plus some aspects of highway engineering.

In this discussion on the elements of soil mechanics, only the basic considerations of the subject are included. The purpose is to provide an overview of the subject with an emphasis on those concepts related to geologic investigation and analysis. The following topics are covered in this overview of the subject: index properties of soils, soil structure, compaction, effective

103

stress, shear strength, stress distribution, and consolidation. Some additional subjects related to soil mechanics are presented in other chapters of the text; for example, soil moisture and permeability (or hydraulic conductivity—the ability of soil or rock to transport water) are included in Chapter 15 on ground-water geology, and soil classifications are considered in Chapter 8 on rock weathering and soil development.

INDEX PROPERTIES

Index properties are those common properties of soil that serve as indexes of behavior. These are the observable physical characteristics that have a significant influence on a soil's behavior. They include the description of soil particles, soil density, phase relationships, relative density, moisture content, and soil consistency.

Description of Soil Particles

Soil Texture

Based on grain size distribution, soils can be subdivided into different classes according to their texture. The term *texture* involves the appearance or feel of a soil as determined by its particle size, shape, and gradation. The major textural classes are gravel, sand, silt, and clay. Sand and gravel have coarse textures, whereas fine-textured soils consist primarily of grains too small to discern individually with the naked eye. Silts and clays are fine-textured soils.

In addition, for engineering purposes soils are subdivided into two types, cohesive and noncohesive (or cohesionless) soils. Cohesive soils are those in which the particles stick together even without confinement. They contain clay minerals and possess a property known as plasticity, the ability to be rolled into a thin thread before breaking into small pieces. Mud sticking to one's shoes is an example of soil cohesion. Sands are nonplastic and cohesionless, whereas clays are both plastic and cohesive. Silts have properties intermediate between the extremes of sand and clay; they are fine grained but typically nonplastic and cohesionless.

Cohesionless soils have no shear strength unless confined; that is, they lack the ability to hold together without confinement. Instead they obtain their strength from grain-to-grain contact, which provides frictional resistance between the grains. This requires that an external force be applied to provide the contact and the strength. Granular particles, those consisting of gravel, sand, and silt, are cohesionless.

Grain Size Distribution

Particle size has an important effect on engineering behavior especially for granular soils. For classification purposes, the gradation or grain size distribution of the sample is particularly significant. The range of particle size in soils is enormous; soil particles can range from boulder size (10^3 mm) down to extremely fine colloidal materials (10^{-5} mm) or across a range of 10^8. Because of this extreme range in grain size, it is convenient to use four- or five-cycle logarithmic paper to plot grain size distribution curves.

The different textural classifications of soil used by various organizations and agencies are presented in Chapter 8, Rock Weathering and Soils. They include the Wentworth scale used in geology; the AASHTO, ASTM, FHWA, and Unified Soil classifications, which are engineering related; and the USDA or agricultural classification.

Particle sizes in the United States are designated by either British or metric units. Sizes are measured in millimeters for soil less than 0.25 in., but for larger sizes, inches and fractions of inches are commonly used. This system applies to grain size distributions of soil as well as for crushed stone and gravel.

Grain sizes between 5 and 0.074 mm are classified based on the U.S. standard sieve numbers. The relationship between the common sieves and their size openings is shown in Table 7.1. In the past,

TABLE 7.1 Common U.S. Standard Sieve Sizes and Their Corresponding Dimension of Openings

U.S. Standard Sieve Number	Sieve Opening (mm)
4	4.75
10	2.00
20	0.85
40	0.425
60	0.25
100	0.15
200	0.074

microns were the units of convenience used for designating sizes finer than the No. 200 sieve (where 1 micron = 1 μ = 10^{-3} mm = 10^{-6} m). For example, the No. 200 size is 0.074 mm or 74 μ. Under the SI designation for metric units, the term micrometer, abbreviated μm, is preferred rather than micron, but both equal 10^{-6} m.

Particle size distribution is determined for soils using a mechanical analysis test. For coarse-grained soils this involves a sieve analysis performed on the dry sample shaken mechanically over a stack of successively finer sieves. For sizes smaller than the No. 200 standard sieve, sieve analysis is impractical because of the extremely small openings needed in the wire mesh for these sieves. Consequently, for fine-grained soils a hydrometer analysis is used instead. The test is based on Stokes' law, which pertains to the settling velocity of spheres in a liquid.

The grain size distribution curve is obtained by plotting a cumulative frequency diagram. Particle size in millimeters is plotted against the percentage by weight (or mass) finer than that size. The ordinate is

an arithmetic scale along which the percent finer is plotted, whereas the abscissa is the logarithmic scale used for plotting grain size. Typically, five-cycle semi-log paper is required for grain size distribution curves. By convention these curves are plotted with the grain size decreasing from left to right. Figure 7.1 illustrates several grain size curves.

In Figure 7.1 the well-graded curve has an even distribution of particle sizes over a wide range. By contrast a poorly graded soil composed mainly of one size of particles is a curve in which the grain size is concentrated in a rather narrow zone, yielding a very steep plot. A gap-graded or skip-graded soil lacks certain sizes in the complete gradation. This yields the flat portions shown on the curve where gaps in grading occur.

Several measures are used to describe gradation curves or to select information from them. A common convention is to employ a D with a subscript number to indicate the diameter for which that percent passing or percent finer will occur. For example, D_{10} represents a grain diameter for which 10% of the

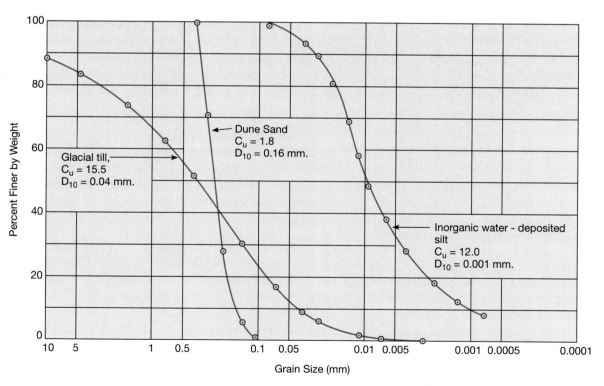

FIGURE 7.1 Grain size distribution curves.

sample will be finer than it. More simply, 10% of the sample by weight is smaller than diameter D_{10}. The D_{10} is sometimes called the effective size and it is used as an estimate of permeability (hydraulic conductivity) for coarse-grained soils. This is Hazen's approximation, which states:

$$k \text{ (cm/sec)} = 100D_{10}^2$$

with D_{10} in centimeters. This obviously is an empirical relationship because the units are not consistent from one side of the equation to the other.

A rough measurement of the shape of gradation curves is provided by the uniformity coefficient C_u, which is defined by

$$C_u = \frac{D_{60}}{D_{10}}$$

where D_{60} = diameter (in millimeters) for which 60% of the sample is finer than D_{10} diameter (also in millimeters) for which 10% of the sample is finer.

The smaller the value for C_u the more uniform the gradation (or the steeper the curve); $C_u = 1$ is the minimum possible value and indicates a material of only one size. For example, Standard Ottawa sand has a C_u of 1.1; ordinary beach sand, which has undergone some winnowing action by the waves, would have a C_u from 2 to 6. A well-graded soil would have a C_u from 15 to 25 and, typically, glacial tills of the Midwest would have a $C_u = 30$. Values of C_u as high as 1000 have been measured.

A second shape parameter sometimes used in conjunction with the uniformity coefficient is the coefficient of curvature. It is defined by

$$C_c = \frac{(D_{30})^2}{(D_{10})(D_{60})}$$

This is a measure of the second moment of grain distributions and it indicates more than the central tendency of the curve shown by C_u. A soil with a coefficient of curvature between 1 and 3 is thought to be well graded as long as C_u is also greater than 4 for gravels and 6 for sands.

Particle Shape and Roundness
Particle shape is a measure of the extent to which particles are equidimensional or the extent to which they deviate from that shape. On this basis particles can be equant, rod-like, flat, elongated, or combinations of these. Sphericity, the degree to which a particle approximates a sphere's equant dimensions, is another way of describing particle shape.

Roundness is an indication of the angularity of the edges and corners of a particle and is not related to shape. A sphere is both equant in shape and highly rounded, a cube is equant but very angular. More details on shape and roundness of natural materials are available in the literature on sedimentary petrology.

In soil mechanics the equant-shaped particles are described as bulky, whereas others are designated as needle-like (rod-like) or flakey (flat). Mica flakes are flakey and Ottawa sand particles are bulky. Particle shape influences the maximum density, compressibility, and shear strength of soils. In fact, in some instances, shape has a greater influence on these properties than does particle size.

Soils containing platey particles, regardless of the particle size, are typically more compressible than those composed of bulky particles, either rounded or angular. Only a moderate amount of mica in sand will increase the compressibility markedly. Coarse-grained soils with angular particles have both greater strength and bearing capacity than those with rounded particles. In the case of clays, the addition of interlocking rod-shaped particles such as those comprising the mineral halloysite will provide added strength.

Phase Relationships
A mass of soil consists of solid grains of minerals with voids between them. The voids can be filled with air or water or filled in part by both of these constituents. (The symbols used in this discussion are shown in Table 7.2.)

The total volume V_t is equal to the volume of the solid V_s plus the volume of voids V_v. In the same fashion, V_v is equal to the volume of the water V_w plus the volume of air V_a. In equation form,

$$V_t = V_s + V_v = V_s + V_w + V_a$$

The three phases of the soil—solid, air, and water—can be represented schematically as though the soil were compartmentalized into these three components. Figure 7.2 is a phase diagram, or solid–water–air diagram, for the soil. Volumes are indicated on the

TABLE 7.2   Symbols and Terminology for Soils

Symbol	Term	Preferred Units
e	Void ratio	Decimal
D_r	Relative density	Decimal
G_s	Specific gravity of solid	Decimal
M_s	Mass of solid	kg
M_t	Total mass	kg
M_t'	Submerged mass	kg
M_w	Mass of water	kg
n	Porosity	%
S	Degree of saturation	%
V_a	Volume of air	m³
V_s	Volume of solids	m³
V_t	Total volume	m³
V_v	Volume of voids	m³
w	Water content	%
ρ	Total or wet density	kg/m³
ρ'	Buoyant density	kg/m³
ρ_d	Dry density	kg/m³
ρ_s	Density of solids	kg/m³
ρ_{sat}	Saturated density	kg/m³
ρ_w	Density of water	kg/m³
γ	Unit weight (used in British engineering units in same capacity as is ρ in the preferred SI units)	lb/ft³

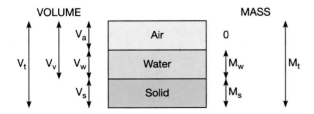

FIGURE 7.2   Phase diagram, or solid–water–air diagram, for soil.

left side, and on the right side the corresponding masses are given. Volumes are in cubic meters or cubic centimeters on the metric system and in cubic feet for the British system. In geotechnical engineering practice, the total volume V_t, the mass of water M_w, and the mass of the dry solid M_s are typically measured. The other terms or parameters can be calculated from these measurements using the phase diagram relationships. Most of the terms are not dependent on sample size and some are dimensionless. In solving problems, it is helpful if the phase diagram relationship is sketched first.

Three volumetric ratios commonly used to describe soils can be developed directly from the phase diagram. These are void ratio e, porosity n, and degree of saturation S. Void ratio e is defined by

$$e = \frac{V_v}{V_s}$$

Void ratio is expressed as a decimal. The maximum mathematical value for e is infinity and the minimum is zero. Practically speaking, however, values for granular soils range from about 0.3 to about 1.0, and for cohesive soils (clays) they range from 0.4 to about 1.5 with values as high as 5.0 for organic clays.

Porosity n is defined by

$$n = \frac{V_v}{V_t} \times 100(\%)$$

Porosity by custom is expressed as a percentage. The maximum, mathematically, is 100% and the minimum zero, but common values range from about 10% to 90% for soils, with clays and organic clays typically having higher porosities than granular soils. Although n is expressed as a percentage, in equations and calculations in geotechnical engineering, its decimal form is commonly used.

By substituting equivalent terms into the preceding equations for e and n, the following can be shown

$$n = \frac{e}{1+e} \quad \text{and} \quad e = \frac{n}{1-n}$$

The degree of saturation S is defined by

$$S = \frac{V_w}{V_v} \times 100(\%)$$

The degree of saturation indicates the percentage of the total volume of voids that is filled with water.

TABLE 7.3 Conversion Factors for SI Units to British Engineering Units

Linear Measurements
  1 inch, in. = 25.4 mm = 0.0254 m
  1 foot, ft = 0.3048 m; 1 meter = 3.28 ft
  1 yard, yd = 0.9144 m
  1 mile (statute) = 1.609×10^3m = 1.609 km
  1 mile (nautical) = 1.852×10^3m = 1.852 km
  1 angstrom, Å = 1×10^{-10} m

Area
  1 square inch = 6.452 sq. cm.
  1 square foot = 0.0929 m²; 1 square meter = 10.764 ft²
  1 acre = 43,560 ft² = 4046.8 m² = 0.405 hectares
  1 hectare = 2.47 acres = 10,000 m²
  1 square mile = 2.590 km²; 1 km² = 0.3861 mi²

Weights and Masses
  1 pound mass, lb_m = 0.4536 kg
  1 short ton = 2000 lb_m = 907.2 kg
  1 gram, g = 10^{-3} kg
  1 metric ton = 1 tonne, t = 10^3 kg = 10^6 g = 1 Mg
  1 slug (1 lb_f/ft/sec²) = 14.59 kg

Force
The SI unit of force is obtained from F = Ma and is known as the newton or N; 1000
  newtons = 1 kilonewton or 1 kN.
  1 lb_f = 4.448N
  1 short ton = 8.896×10^3 N = 8.896 kN
  1 kip = 1000 lb_f = 4.448×10^3 N = 4.448 kN
  1 metric ton_f = 1000 kg_f = 9.807×10^3 N = 9.807 kN
  1 dyne (gram-cm/sec²) = 10^{-5} N

Pressure
The SI unit for pressure or stress is the pascal or Pa and is defined as 1 newton per square
  meter (N/m²).
  1 psi (lb-force/in.²) = 6.895×10^3 Pa = 6.895 kPa = 6.895×10^{-3} MPa
  1 psf (lb-force/ft²) = 1/144 psi = 47.88 Pa
  1 kg-force/cm² = 9.807×10^4 Pa = 98.07 kPa
  1 metric ton-force/m² = 9.807×10^3 Pa = 9.807 kPa
  1 short ton-force/ft² = 9.576×10^4 Pa = 95.76 kPa
  1 ksi (kip/in.²) = 1000 psi = 6.895×10^6 Pa = 6.895×10^3 kPa
  1 atm at STP° = 1.013×10^5 Pa = 101.3 kPa
  1 bar = 1×10^5 Pa = 100 kPa
For calculating overburden pressure, $\sigma_v = \rho g h$ in SI units (N/m²). In terms of British
  engineering units, $\sigma_v = \gamma h$(lb$_f$/ft²), or for previous metric units, $\sigma_v = \gamma h$(kg$_f$/cm²).

Density and Unit Weight
Density or mass per unit volume in the SI system is in kg/m³. The density of water ρ_w, at 4°C
  is exactly 1 g/cm³. Hence, ρ_w = 1 gm/cm³ = 1000 kg/m³ = 1.000 Mg/m³ is used in these
  calculations. γ is unit weight in lb$_f$/ft³.
  62.4 lb$_f$/ft³ = 1,000 kg/m³
  or 1 lb$_f$/ft³ = 16.018 kg/m³

Velocity
  1 mi/hr = 17.60 in/sec = 1.6093 km/hr = 44.71 cm/sec
  10^{-6} cm/sec = 1.035 ft/yr

°STP = standard temperature and pressure.

Completely dry soil has $S = 0\%$ and for fully saturated soil, $S = 100\%$. Most cohesive soils, even those exposed to the Earth's surface under drying conditions, contain some moisture and therefore have a value of S greater than zero.

Moving to the right side of the phase diagram, the mass relationships are examined next. The first term of interest is the ratio known as the water content w. One of the most significant soil parameters, it is defined as

$$w = \frac{M_w}{M_s} \times 100(\%)$$

where M_w = mass of water and M_s = mass of solids. Water content is typically expressed as a percentage and values range from zero (completely dry soil) to more than 200%. Because M_s is used in the denominator rather than M_t, values greater than 100% are possible. Values greater than 300% have been measured for organic clays.

The one remaining determination of major significance that can be calculated based on the phase diagram relationship is density ρ, or the mass per unit volume. The unit used for density in the SI system is kilograms per cubic meter. For those unfamiliar with the magnitude of these units, 1.000 g/cm^3, the density of water = 1000 kg/m^3, thus equating the cgs units with the SI units. In like manner 1000 kg/m^3 = 1.000 Mg/m^3 where Mg stands for a megagram or 1 million grams. Therefore 1.000 g/cm^3 = 1.000 Mg/m^3 = 1000 kg/m^3.

In the British system, unit weight is used despite a persistent confusion between weight and mass. The equivalence in this respect is 1.000 g/cm^3 = 62.4 lb/ft^3 = 1000 kg/m^3. Therefore, soil with a unit weight of 100 lb/ft^3 = (100/62.4) (1000) = 1603 kg/m^3 = 1.063 g/cm^3 = 1.063 Mg/m^3. Table 7.3 lists the factors used to convert SI units to British engineering units. End-papers located on the inside cover at the end of this textbook also show conversion factors used in geotechnical engineering and engineering geology.

Density equals mass divided by volume and therefore density is used to relate mass to volume, or the right side to the left of the phase diagram in Figure 7.2. This allows one to calculate density if the mass and volume are known or, generally speaking, for the three parameters, to calculate the third one if the other two are known.

Several density measurements are used in geotechnical engineering. Two of these are 1) total or wet density ρ and 2) density of the solids, ρ_s. These are defined in terms of masses and volumes as shown in Figure 7.2.

$$\rho = \frac{M_t}{V_t} = \frac{M_s + M_w}{V_t} \quad \text{and} \quad \rho_s = \frac{M_s}{V_s}$$

where ρ can range from just above 1000 kg/m^3 to as high as 2400 kg/m^3 or so. The density of the solids ρ_s, equals G_s, the specific gravity of the mineral, times 1000 kg/m^3, which is the density of water. Hence,

$$\rho_s = G_s \rho_w$$

where ρ_w = 1000 kg/m^3 (in SI units) and G_s ranges from about 2.0 to 3.0 for common rock-forming minerals, but for most soil minerals it lies between 2.65 (quartz) and 2.80.

Three other densities are used in geotechnical engineering: the dry density ρ_d, the saturated density ρ_{sat}, and the submerged density ρ'. They are defined as follows:

$$\rho_d = \frac{M_s}{V_t}$$

$$\rho_{sat} = \frac{M_s + M_w}{V_t} \quad \text{where} \quad V_a = 0 \text{ and } S = 100\%$$

$$\rho' = \rho_{sat} - \rho_w$$

The details for calculations involving the terms just defined, according to the phase diagram, can best be shown through a series of sample problem solutions.

A useful relationship for soil density involves w, ρ, and ρ_d:

$$\rho_d = \frac{\rho}{1 + w.}$$

For saturated soils the phase diagram simplifies to two phases when V_a goes to zero. The sample problems will begin with a two-phase example.

EXAMPLE PROBLEM 7.1

Given ρ_w, e, and G_s for a saturated soil, find w in terms of the given parameters.

ANSWER

First draw the phase diagram of Figure 7.3. For purposes of this solution we can assume a term equal to 1 and solve the problem relative to that assumption. This is possible because any volume of soil can be assumed for the problem, this

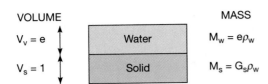

FIGURE 7.3   Phase diagram, two-phase, *e* given.

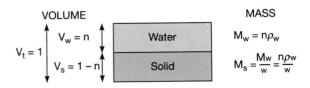

FIGURE 7.4   Phase diagram, two-phase, *n* given.

simply selects a volume that simplifies the calculation.

Because *e* is given and $e = V_v/V_s$, assume $V_s = 1$ because 1 in the denominator eliminates the fraction. Hence,

$$V_s = 1, \quad e = \frac{V_v}{1} = V_v.$$

Because the soil is saturated, $V_v = V_w = e$. Now $M_w = V_w \rho_w$ because $\rho_w = M_w/V_w$:

$$M_w = e\rho_w$$

$$M_s = V_s G_s \rho_w$$

because $G_s = M_s/V_s\rho_w$

$$M_s = 1G_s\rho_w = G_s\rho_w$$

Enter these values on the phase diagram you drew. Realizing that $w = M_w/M_s$, we have
$w = e\rho_w/G_s\rho_w = e/G_s$.

EXAMPLE PROBLEM 7.2

Given ρ_w, *n*, and *w* for a saturated soil (hence another two phase problem), find G_s in terms of the given parameters.

ANSWER

First draw the phase diagram as presented in Figure 7.4. Because *n* is given and $n = V_v/V_t$, let $V_t = 1$ to simplify the calculation. Therefore,

$$n = \frac{V_v}{1} = V_v = V_w$$

for the saturated case. By inspection, $V_s = V_t - V_w = 1 - n$. Also, $M_w = V_w\rho_w = n\rho_w$. These can be added to the diagram. Next we note that *w* is given and by definition

$$w = \frac{M_w}{M_s} \quad \text{or} \quad M_s = \frac{M_w}{w}, \quad \text{but} \quad M_w = n\rho_w$$

Therefore,

$$M_s = \frac{n\rho_w}{w}$$

To find G_s we note that $M_s = V_s G_s \rho_w = (1 - n)G_s\rho_w$. But $M_s = n\rho_w/w$ so

$$\frac{n\rho_w}{w} = (1 - n)(G_s)\rho_w$$

$$G_s = \frac{n\rho_w}{w(1 - n)\rho_w} = \frac{n}{w(1 - n)}$$

EXAMPLE PROBLEM 7.3

This is a three-phase problem. Given ρ_w, *e*, G_s, and *S*, find *w* in terms of the given parameters.

ANSWER

First draw the phase diagram of Figure 7.5. With *e* given, $(e = V_v/V_s)$, let $V_s = 1$, then $e = V_v$. Because $S = V_w/V_v$,

$$V_w = SV_v = eS$$

The mass side of the equation can now be completed:

$$M_s = V_s G_s \rho_w = G_s\rho_w$$

$$M_w = V_w\rho_w = eS\rho_w$$

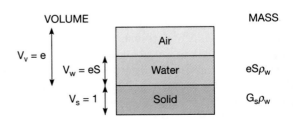

FIGURE 7.5   Phase diagram, three-phase, *e* given.

Then

$$w = \frac{M_w}{M_s} = \frac{eS\rho_w}{G_s\rho_w} = \frac{eS}{G_s}.$$

EXAMPLE PROBLEM 7.4

Given ρ_w, n, w, and S, find G_s in these given terms.

ANSWER

First draw the phase diagram presented in Figure 7.6. With n given, we observe $n = V_v/V_t$. Let $V_t = 1$ and $V_v = nV_t = n$.

$$S = \frac{V_w}{V_v} \quad \text{so} \quad V_w = SV_v = Sn$$

By subtraction $V_s = V_t - V_v = 1 - n$

$$M_w = \rho_w V_w = \rho_w Sn$$

Observing that $w = M_w/M_s$ and solving for M_s, we have

$$M_s = M_w/w = \rho_w nS/w$$

Also, $M_s = \rho_w G_s V_s$ or $\rho_w nS/w = \rho_w G_s (1 - n)$. Cancelling ρ_w and solving for G_s,

$$G_s = \frac{nS}{w(1 - n)}$$

EXAMPLE PROBLEM 7.5

Here we determine w from soil measurements. A soil sample taken from the field had a mass of 396 g. After drying, the sample had a mass of 327 g. What was its natural moisture content?

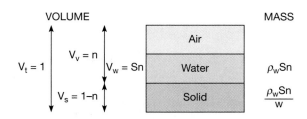

FIGURE 7.6   Phase diagram, three-phase, *n* given.

ANSWER

$$w = M_w/M_s \times 100\%$$
$$M_s = \text{mass of the dry soil} = 327 \text{ g}$$
$$M_w = \text{mass of the water} = M_t - M_s = 396 - 327 = 69 \text{ g}$$
$$\therefore w = 69/327 \times 100 = 21.1\%$$

These example problems serve to show how void ratio, porosity, water content, and the specific gravity of solids are related. The various densities of the soil have not been considered. They are now evaluated in the following section.

Density

The densities are best considered relative to void ratio, degree of saturation, and the specific gravity of solids. We begin with a two-phase system, that is, a saturated condition ($S = 100\%$). This is illustrated in Figure 7.7.

With e given, as before we let $V_s = 1$. Hence, $V_w = V_v = e$. Considering the not submerged situation $M_w = \rho_w V_w = e\rho_w$ and $M_s = V_s G_s \rho_w = G_s \rho_w$ as $V_s = 1$ is assumed. Therefore,

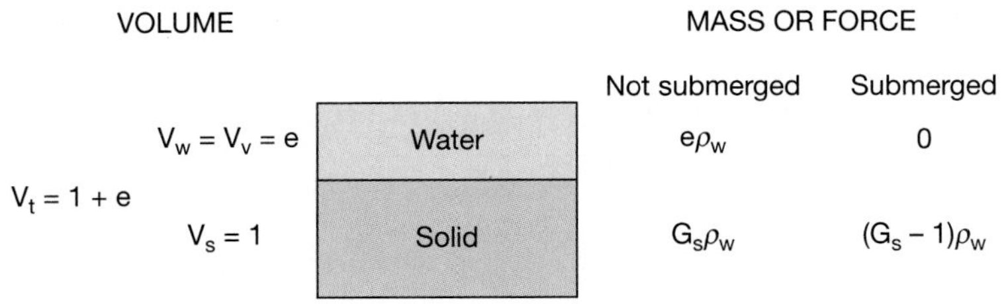

FIGURE 7.7   Phase diagram, two-phase, submerged versus not submerged.

$$\rho_{\text{sat}} = \frac{M_w + M_s}{V_t} = \frac{e\rho_w + G_s\rho_w}{1 + e} = \frac{\rho_w(e + G_s)}{1 + e}$$

$$\rho_d = \frac{M_s}{V_t} = \frac{G_s\rho_w}{1 + e}$$

Next we consider the submerged mass. This equals the mass in air minus the buoyancy effect. The buoyancy effect is the volume times ρ_w, the density of water. For the water portion on the mass side we have

$$M_w - \text{buoyancy} = e\rho_w - e\rho_w = 0$$

For the solid portion on the mass side we have

$$M_s - \text{buoyancy} = G_s\rho_w - 1\rho_w = (G_s - 1)\rho_w$$

$$\rho_{\text{submerged}} \text{ or } \rho' = \frac{(G_s - 1)\rho_w}{1 + e} = \frac{M'_t}{V_t}$$

Another way of expressing this is

$$\rho' = \rho - \rho_w = \frac{e\rho_w + G_s\rho_w}{1 + e} - \rho_w$$

$$= \frac{e\rho_w + G_s\rho_w - \rho_w(1 + e)}{1 + e} = \frac{G_s\rho_w - \rho_w}{1 + e}$$

$$= \frac{(G_s - 1)\rho_w}{1 + e}$$

The final example involves the three-phase system for submerged densities, as illustrated in Figure 7.8. Working in terms of e, S, and G_s we start with e and set $V_s = 1$, $V_v = e$, $V_w = eS$, and $V_a = e(1 - S)$. For the mass side, not submerged, this yields

$$M_s = V_sG_s\rho_w = G_s\rho_w, \quad M_w = V_w\rho_w = eS\rho_w,$$
$$M_a = 0$$

Hence

$$\rho = \frac{M_w + M_s}{V_t} = \frac{eS\rho_w + G_s\rho_w}{1 + e} = \frac{\rho_w(eS + G_s)}{1 + e}$$

$$\rho_d = \frac{M_s}{V_t} = \frac{G_s\rho_w}{1 + e}$$

The ρ_{sat} situation would occur when the air voids are filled with water so that $V_w = e$ and $M_w = e\rho_w$ and

$$\rho_{\text{sat}} = \frac{G_s\rho_w + e\rho_w}{1 + e} = \frac{\rho_w(e + G_s)}{1 + e}$$

Now we consider the submerged mass. For the water portion,

$$M_w - \text{buoyancy} = eS\rho_w - eS\rho_w = 0$$

For the solid portion,

$$M_s - \text{buoyancy} = G_s\rho_w - 1\rho_w = (G_s - 1)\rho_w$$

For the air portion,

$$M_a - \text{buoyancy} = 0 - e(1-S)\rho_w = -e(1-S)\rho_w$$

$$\rho_{\text{submerged}} \text{ or } \rho' = \frac{(G_s - 1)\rho_w + 0 - e(1-S)\rho_w}{1 + e}$$

$$= \frac{\rho_w[(G_s - 1) + e(S - 1)]}{1 + e}$$

Note that $\rho_s = G_s\rho_w$ and ρ_w is known to be 1000 kg/m³ or 1 g/cm³; therefore, ρ_s or G_s can be determined directly when the other is provided.

Two additional, useful relationships are:

$$\rho_d = \frac{\rho_s}{1 + e} \quad \text{and} \quad \rho_d = \frac{\rho}{1 + w}$$

and from these, by substitution,

$$\rho = \rho_s \frac{(1 + w)}{(1 + e)}$$

Another sample problem is presented to illustrate density calculations.

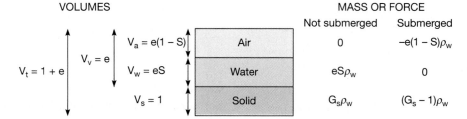

FIGURE 7.8 Phase diagram, three-phase, submerged versus not submerged.

EXAMPLE PROBLEM 7.6

Given $e = 0.75$, $w = 18\%$, and $\rho_s = 2.65$ Mg/m³, find ρ, ρ_d, S, ρ_{sat}, ρ', and w for $S = 100\%$.

ANSWER

First draw the phase diagram presented in Figure 7.9. Let

$V_s = 1$, $V_v = e$, $V_t = 1.75$

$M_s = V_s G_s \rho_w = V_s \rho_s = 1 \times 2.65 = 2.65$

$w = M_w / M_s$ or $M_w = w M_s = (0.18)(2.65) = 0.477$

$M_w = \rho_w V_w$

$V_w = M_w / \rho_w = 0.477$

Then

$$\rho = \frac{M_w + M_s}{V_t} = \frac{0.477 + 2.65}{1.75} = 1.79 \text{ Mg/m}^3$$

$$\rho_d = \frac{M_s}{V_t} = \frac{2.65}{1.75} = 1.51 \text{ Mg/m}^3$$

$$S = \frac{V_w}{V_v} \times 100\% = \frac{0.477}{0.75} \times 100 = 63.6\%$$

$$M_{w(sat)} = \rho_w V_v$$

$$= 1.0(0.75) = 0.75$$

$$\rho_{sat} = \frac{M_s + M_w(sat)}{V_t} = \frac{2.65 + 0.75}{1.75} = 1.94 \text{ Mg/m}^3$$

$$\rho' = \rho_{sat} - \rho_w = 1.94 - 1.00 = 0.94 \text{ Mg/m}^3$$

$$w \text{ for } S@ 100\% = \frac{0.75}{2.65} \times 100(\%) = 28.3\%$$

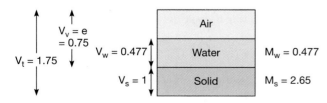

FIGURE 7.9 Phase diagram, three-phase, density calculation.

Relative Density

A final consideration before leaving the subject of density is the concept of relative density. It is extremely important for granular soils.

Experience indicates that it is not the absolute or numerical value of density, expressed as mass per unit volume, that is important for granular soils. Instead, the *degree* of density relative to its loosest and densest possible conditions controls the behavior of these cohesionless materials. This relationship is known as relative density D_r and is defined as follows

$$D_r = \frac{e_{max} - e}{e_{max} - e_{min}}$$

where e is the void ratio of the sample, or that for which D_r is being calculated; e_{max} is the maximum value of the void ratio (the loosest state) for a dry or submerged condition; and e_{min} is the minimum value of the void ratio (the densest state). Numerically D_r ranges from zero for the loosest relative density to one for the greatest density. A summary of the equations used to calculate soil properties is provided as an end-paper on the front inside cover of this textbook.

Soil Consistency

Consistency denotes the relative ease with which a soil can be deformed. It is described by such terms as soft, firm, or hard. Consistency depends greatly on the nature of the soil minerals present and on the water content. From a practical point of view, only fine-grained soils or the fine fraction of coarser soils are of significance relative to consistency. These are the soils whose deformability is most subject to change with changing water content.

Natural water contents of soils are used to describe and compare subsurface conditions at a site and to aid in correlation of materials. Moreover they are used to anticipate engineering behavior. To accomplish this, some engineering standard of performance is needed for comparison purposes. The Atterberg limits accomplish this need because they designate critical water contents, which mark the boundaries or fields of behavior for soils at critical stages.

Atterberg Limits

Atterberg, a Swedish soil scientist, proposed in 1911 that certain limits of moisture content be used to define soil behavior. The consistency of soils according to the Atterberg limits is shown in Figure 7.10.

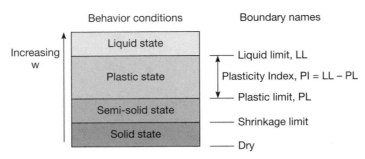

FIGURE 7.10   Consistency of soils.

The liquid limit (LL) is the lowest water content at which a soil will behave as a viscous liquid, and the plastic limit (PL) is the lowest water content at which it will behave as a plastic material. The range in water content over which soils behave plastically is the plasticity index (PI) and is equal to the LL minus the PL. The behavior of a soil depends greatly on its present water content as compared to the LL or PL. If a soil's moisture content is near its liquid limit, it will be more compressible and probably less permeable than if it is near the plastic limit, where it is stronger and less compressible.

Another limit term suggested by Atterberg is the sticky limit, the lowest water content at which soil adheres to metal tools. This limit is related to agriculture and to plowing procedures. The shrinkage limit is the moisture content below which soils do not decrease further in volume as they continue to dry out, and this limit is related to brick manufacture and ceramics.

The liquid limit and plastic limit must be defined through a specific procedural test because otherwise the definitions become too arbitrary. Liquid limit is defined as the moisture content at which a small standard groove cut by a standard grooving tool will close for a distance of 1 cm when the standard liquid limit device is dropped 25 times for a distance of 1 cm onto a hard rubber base.[1] It is virtually impossible to moisten soil to just the right moisture content such that exactly 25 blows will close the groove the standard distance. Fortunately, a straight-line relationship holds between the water content and the log of the number of blows to close the groove. In past years,

three points were used to determine this curve; more recently two points have been used. If the slope of the plot is known in advance, only one point is needed to define the line. These slopes are related to the geologic origin of the material, and if that is known with some certainty, the liquid limit can be approximated reasonably well with one point.

The plastic limit test requires some experience before reproducible results can be obtained. The plastic limit is the moisture content at which a thread of soil crumbles when rolled into a diameter of 3 mm (1/8 in.). It should break into segments 3 to 10 mm (1/8 to 3/8 in.) long.

The range of liquid limits can vary from zero to over 600, but most soils have values below 100. Plastic limit values range from zero to about 125, but the PL for most soils is 40 or less.

The PI or plasticity index as previously stated equals the LL minus the PL. The PI is useful in engineering classifications of fine-textured soils and has been correlated with a number of soil properties and soil types. The plasticity chart shown in Figure 7.11 is one of the most useful associations available for soils descriptions.

Liquidity Index

The Atterberg limits also provide a rough measure of sensitivity for a soil. This is indicated by the liquidity index, which is defined by

$$LI = \frac{w_{nat} - PL}{PI}$$

where w_{nat} is the natural water content for the soil in question. When the liquidity index is 1.0, the soil is at the liquid limit and it possesses little strength. This suggests a high sensitivity. Values greater than 1.0

[1]ASTM D4318, standard test method for liquid limit, plastic limit, and plasticity index of soils.

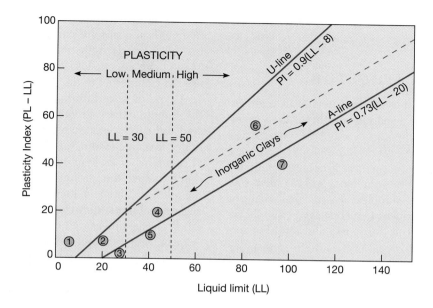

Key:
1) Cohesionless soils
2) Inorganic clays, low plasticity
3) Inorganic silts, low compressibility
4) Inorganic clays, medium plasticity
5) Inorganic silts and organic clays, medium compressibility
6) Inorganic clays, high plasticity
7) Inorganic silts and organic clays, high compressibility

FIGURE 7.11 Plasticity chart.

indicate ultrasensitive or quick clays (values as great as 1.9 have been measured but most lie between 1 and 1.2). The St. Lawrence River Valley in eastern Canada and clays in Scandanavia are well-known examples of these ultrasensitive soils. These values indicate soils that are extremely sensitive to a sudden collapse in structure when sheared. If disturbed in any way they can flow like a viscous liquid.

A liquidity index of zero indicates that the soil is at its plastic limit and is probably not sensitive. If a negative value for LI is obtained, it means the soil is below its plastic limit and will fail as a brittle material when sheared. The advantage of the liquidity index is that a good indication of soil sensitivity can be obtained from the index properties of the soil, w_{nat}, LL, and PL.

Activity
The final consistency index for consideration is an expression known as the activity. This is a measure comparing the clay-like behavior of a soil to the portion of the soil that contributes that behavior. Therefore, the higher the activity the more clay-like the soil must be.

The activity A of a clay is defined as

$$A = \frac{PI}{\text{clay fraction}} = \frac{PI}{\% < 2 \ \mu m}$$

The 2-μm (micrometer or micron = 10^{-6} m) level is the upper boundary for colloidal clay. Clays with activities less than 0.75 are inactive clays; those with activities between 0.75 and 1.25 are active clays. A fair to good correlation exists between activity and clay mineral types and it develops that smectite (montmorillonite) generally has the highest activity and kaolinite and halloysite have the lowest for the common clays. Other factors such as particle size, cation type, and concentration influence activity as well. However, the correlation between clay minerals and Atterberg limits is so strong that the activity adds little if any additional information not revealed by the clay type present.

SOIL STRUCTURE

According to modern usage in geotechnical engineering, the structure of a soil includes both the geometric arrangement of the particles and the interparticle forces acting between them. Soil fabric, by contrast, pertains only to the geometric arrangement of the particles or grains. For granular or cohesionless soils the forces between particles are so small that soil structure, in fact, reduces to a concern for soil fabric. This is particularly true for gravel and sand but to a lesser extent for silt.

In cohesive soils, the interparticle forces are much greater so that both the fabric and the attractive forces must be considered in the analysis of structure. We know that structure in both cohesive and cohesionless soils has much to do with their behavior, so a detailed consideration of the subject is warranted. Because of basic differences in interparticle forces it is convenient to discuss soil structure for cohesionless and cohesive soils separately.

Fabric of Cohesionless Soils

Individual soil particles that settle from a fluid medium as independent grains yield an open packing known as a single-grained structure. These grains are generally coarse silt size or larger than 0.02 mm and typically consist of gravel, sand, or mixtures of sand and silt. These particles are controlled strongly by gravity in the settling process rather than by surface forces.

The medium of deposition can either be water or air. Materials formed in water include stream deposits and alluvial fans, beaches, and deltas; those settling from air include loess, sand dunes, and pyroclastic deposits (mostly volcanic ash).

Single-grain structures are described according to their relative density, a relationship considered in the previous section. These soils can be loose (high void ratio, low density, low relative density) or dense (low void ratio, high density, high relative density) or somewhere between these extremes. The relative density of a soil has a strong influence on its engineering behavior. Loose soils tend to densify under sudden vibrational loading, yielding a sudden decrease in volume and momentary loss of strength. Saturated fine sands and silts are particularly prone to a sudden loss of shear strength during earthquake vibrations. This phenomenon, known as liquifaction is considered in Chapter 14 in the discussion of slope stability.

A type of open fabric that forms in some cases is the honeycombed structure, which has a very high void ratio. It is metastable and sensitive to collapse under dynamic loading or vibrations.

Sampling of loose granular soils at any appreciable depth below the ground surface is very difficult. Samples of these materials will densify with the slightest vibration during the sampling process so that density values measured on such samples are always suspect even if careful techniques are employed. The *N* value, that is the values from the standard penetration test (discussed in Chapter 19), can be roughly correlated with density. The *N* values are usually more reliable than density measurements on loose samples performed some time later in the laboratory. If the deposit contains sizable gravel pieces, this may lead to erroneous results in the standard penetration test because the sampler may bear partially on a gravel piece when driven into the ground, yielding a higher resistance in the process.

Note, however, that relative density alone may not fully describe the structure of a granular soil. This can be illustrated, for example, by two sand samples that have identical gradations and identical void ratios but very different soil fabrics. An open structure versus an evenly dense one is an example. Therefore, their engineering behavior will be quite different although the standard tests of gradation and void ratio determination would suggest closely similar soils. A measure of the grain contacts per unit area in cross section is needed to show a numerical difference between these two materials.

Structure of Cohesive Soils

As mentioned previously, the structure of cohesive soils is dependent on both the interparticle forces between the grains and on the fabric or geometric arrangement. In this discussion on cohesive soils these two factors are considered.

Interparticle Forces of Clays

In many discussions involving the interparticle forces of clays, the analysis begins with a review of clay mineralogy. This pertains to the two-layer and three-layer clays and the properties that develop because of these individual clay structures. In this text, clay minerals are considered in Chapter 2 as part of the discussion on minerals. A review of that section will provide details regarding clay mineralogy.

The properties of clay minerals are easily related to the plasticity chart discussed previously in this chapter. The individual clay groups plot in specific locations on the plasticity chart as shown in Figure 7.12. Glacial clays from the Great Lakes region consist predominantly of illite and they plot just above the A line on this figure.

SPECIFIC SURFACE AND PARTICLE SIZE   Specific surface is the ratio of the surface area of a solid to its volume, that is,

$$\text{Specific Surface} = \frac{\text{Surface}}{\text{Volume}}$$

For a sphere it equals

$$\frac{\pi d^2}{\pi d^3 / 6} = \frac{6}{d}$$

This parameter is discussed in Chapter 8. The units are in reciprocal millimeters or reciprocal inches.

Specific surface is inversely proportional to grain size. Since water molecules are attracted to the surface of soil grains, the amount of moisture per unit volume of soil will also increase with a decrease in grain size. This is why cohesive soils tend to have a higher water content than do granular or cohesionless soils with their much larger grain diameters.

Individual particles comprising the different clay minerals are by no means the same size. The more active clays tend to be much smaller in size than those with low activity. This is shown diagrammatically in Figure 7.13. Montmorillonite (smectite) is considerably smaller than illite or chlorite (about one-tenth the thickness) and considerably smaller than kaolinite. It is no wonder that the swelling characteristics of montmorillonite are much greater than those of kaolinite. The relative sizes of montmorillonite and kaolinite with regard to a layer of water is shown in Figure 7.13. Clay particles are almost always hydrated by a layer of water surrounding each crystal. This is known as adsorbed water, that is, water held as an extremely thin layer on the surface of the clay.

TYPES OF INTERPARTICLE BONDS IN CLAYS   Several types of forces or bonds are known to act on small particles such as cohesive clays and considerable differences

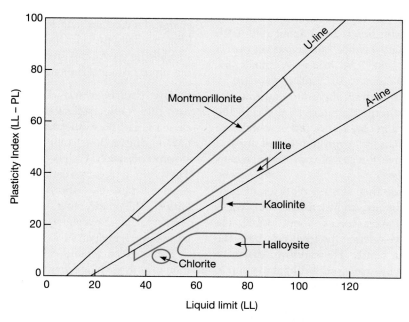

FIGURE **7.12**   Location of clay minerals on plasticity chart.

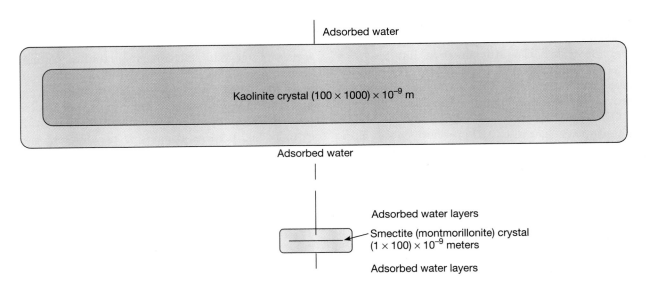

FIGURE 7.13   Relative sizes of different clay mineral particles.

are evident in the strength of these forces. The strongest are primary valence bonds, which bind atoms together. It is not likely that any of these are broken in engineering soils applications but they serve as a point of comparison.

Hydrogen bonds occur because of the dipole nature of water, and they cause the positive end of the water dipole to attach to oxygens or hydroxls on the clay surface. Hydrogen bonds have about one-tenth the strength of primary valence bonds and act over a distance of about 2 to 3 Å, where Å stands for Angstrom unit $= 10^{-8}$ cm $= 10^{-10}$ m.

Another force that attracts a water layer is the cation bond. Cations in the water are attracted by the negative charges on the clay layers. These bonds are somewhat weaker than hydrogen bonds and are less important as attractive forces between the clay surface and the water layer.

An additional force that acts on particles of molecule size and larger are known as van der Waals forces. These are weak forces about 1/100 the strength of primary valence bonds and 1/10 the strength of hydrogen bonds. They act over distances of about 5 Å, and are inversely proportional to r^3 up to r^7 depending on the specific molecules involved (where r = radius of the molecule).

The electrostatic bond is the last and the weakest

of the forces involved in the attraction of water layers. They act over distances much greater than 5 Å.

CAUSES FOR NEGATIVE CHARGE ON CLAYS   The source of the negative charges on the clay surfaces, which are neutralized by the water layers, is related to two primary causes. The substitution of cations in the silica and the octahedral sheets of the clays yields charge deficiencies, and terminated crystal lattices or broken edges provide the other. In this regard, the smaller the crystal the greater the specific surface and the more crystal terminations.

CATION EXCHANGE CAPACITY   Various types of clay have different crystal sizes and different charge deficiencies, and thus different capacities to accept these neutralizing cations. One cation can be exchanged stoichiometrically (valence for valence) for another. Therefore, a plus-two cation (Ca^{++}, for example) can replace 2 plus-one cations (Na^+, for example) on the lattice. The ability to take on cations by a clay is known as cation exchange capacity. The values for this parameter are given in Table 7.4.

Some exchangeable cations are particularly common in soils such as calcium and magnesium, whereas potassium and sodium are less abundant. Aluminum and hydrogen are the common cations of acidic soils. Marine clays contain primarily sodium and magne-

TABLE **7.4** Cation Exchange Capacity of Clay
Minerals in Milliequivalents per 100 g of Clay

Clay Minerals	Exchange Capacity
Kaolinite	3–15
Halloysite	5–40
Illite	10–40
Montmorillonite (smectite)	80–150

sium, which are the most common cations in the oceans. The presence of organic matter also complicates cation exchange in soils.

In attenuation-type, sanitary landfills, where soils are used to enclose solid wastes, the clays in the containment soils remove heavy cations from the leachate by the base exchange procedure. Cohesionless soils have a limitation in this regard because their higher permeability and low clay content allow leachate to be more quickly released through the soil and not be cleansed of the heavy cations.

Some cations have a greater ability to replace existing ions on the clay than do others. Higher valence cations can replace those of lower valence. When the ions are the same valence, the larger the ion the greater the replacement capability. The approximate order of replacement ability is displayed here. The exact order depends on the type of clay involved, the ions to be replaced and the concentration of the ions in the water. The order of replacement in increasing order is essentially

$$Li^+ < Na^+ < H^+ < K^+ < NH4^+ < Mg^{++}$$
$$< Ca^{++} < Al^{+++}$$

Interaction of Forces
Individual clay particles interact with each other through their layers of adsorbed water. When two particles approach each other, they are attracted by their van der Waals forces, cation bonds, and electrostatic bonds. When the particles become so closely associated that their adsorbed water layers overlap, repulsion occurs. Hence the particles arrange themselves in response to these attractive and repulsive components.

The presence of various ions and organics and their relative concentrations influence the nature of these forces. The net effect is that individual particles can flocculate together, yielding a flocculated arrangement, or be repelled from each other, providing a dispersed arrangement. A schematic of these structures is shown in Figure 7.14. The flocculation that occurs most commonly in water with a high salt concentration such as seawater, is shown in Figure 7.14(a). Under these conditions the particles stick together in the fashion in which they first make contact. This is an example of face-to-face contact.

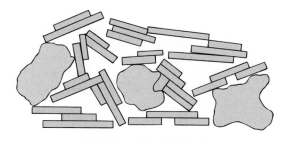

(a) high-salt flocculation

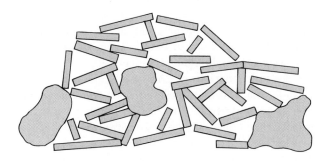

(b) Low-salt (acid) flocculation

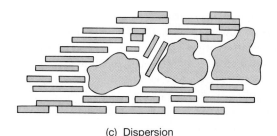

(c) Dispersion

FIGURE **7.14** Schematic diagram of clay structure.

The edge-to-edge attraction tends to occur in acidic water where the hydrogen ions attach themselves to the edge of the particles. This is known as nonsalt flocculation and is illustrated in Figure 7.14(b). It is an example of an edge-to-face structure. The dispersed structure occurs when the adsorbed water layer is increased, yielding an overlap of the water layers between clay particles, as illustrated in Figure 7.14(c). The dispersed structure commonly predominates in clays carried in suspension by freshwater. When this water flows into the ocean, the salt solution causes flocculation to occur because of the high salt concentration. This converts the clay structure to that shown in Figure 7.14(b).

In summary, the three possible configurations of flocculated particles are edge-to-face, edge-to-edge, and face-to-face. Of these, edge-to-face is the most common. The tendency toward flocculation is enhanced by an increase in 1) concentration of electrolyte, 2) valence of the ion, and 3) temperature; or by a decrease in 1) dielectric constant of the fluid, 2) size of the hydrated ion, 3) pH, and 4) anion adsorption (Lambe, 1958).

Two properties of clay structures are of major interest in this discussion: sensitivity and thixotropy. Sensitivity is the ratio of the undisturbed strength to the remolded strength for a soil sample tested at the same water content. It occurs because of disruption of the structure when the clay is remolded. Sensitivity for clay soils ranges from 1.5 to more than 8. Ordinary clays have a sensitivity of less than 4, whereas sensitive clays range from 4 to 8. Clays with sensitivities greater than 8 are called extrasensitive or quick clays.

The reduction in strength occurs when edge-to-face bonding is changed to face-to-face bonding on remolding. This pushes the attractive forces farther apart, yielding a lower shear strength in the remolded state. Quick clays are also formed by the leaching of marine clays by freshwater. The addition of rainwater lowers the cation concentration, thus reducing the attractive forces in the clays. Cation bonds are reduced in this way, giving the clay a reduced shear strength.

Thixotropy is the gain in shear strength with time following remolding. It is a recovery of some of the strength lost by sensitive soils on remolding and it develops with time at a constant water content. Remolding of the clay initially causes a change in structure toward the dispersed arrangement. With time the particles begin to reorient toward a more stable flocculent structure to result in a stronger material. Thixotropic behavior is more pronounced for montmorillonite and other expansive clays than it is for kaolinite and similar minerals.

Cohesive Soil Fabrics

Cohesive soil fabrics are not simple collections of grains or particles that have become intergrown. Instead, they are thought to be aggregates of flocculated particles of differing sizes and assemblages with pore spaces between them. The structure is most easily understood by considering this feature on two scales, the macrostructure and the microstructure of the clays.

Macrostructure versus Microstructure

Macrostructure has an important influence on the behavior of soils with regard to their engineering properties. The behavior of a soil mass is dependent on the weakness zones or defects within it and these include, for example, the effects of bedding. This may be manifested as sand or silt layers, varves, zones of increased organic content, or layers in the soil profile. Other defects usually lie at a high angle to the bedding and they include joints, fissures, root holes, and root fillings. Great care must be taken to observe macrostructure effects in soils during site investigation because shear strength, permeability, and compressibility may have critical values at these locations.

Details of clay microstructure have been discerned by scanning electron microscope (SEM) studies. Results show that large collections of aggregated or flocculated particles known as peds can be observed with the naked eye. Peds consist of smaller assemblages of clay particles known as clusters. These clusters are observable with a petrographic microscope using transmitted, visible light, and the particles lying in the range of 1 to 5 μm can be resolved.

The smallest aggregates of clay particles are congregated into submicroscopic units known as domains. Domains are aggregates of the clay crystal platelets that collect because of surface forces. Therefore, in order of increasing size the units are domains, clusters, and peds. These are illustrated in Figure 7.15.

The microstructure of clays is more important from a fundamental viewpoint, and the macrostructure is more significant where engineering behavior is concerned. The microstructure records the geologic

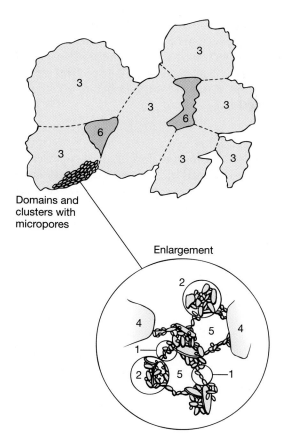

Domains and
clusters with
micropores

Enlargement

FIGURE 7.15 Schematic representation of soil microstructure and macrostructure: (1) domain, (2) cluster, (3) ped, (4) silt grain, (5) micropore, and (6) macropore. (Adapted from Holtz and Kovacs, 1980.)

history of the deposit, including details of deposition, environment, and weathering history.

COMPACTION

Compaction of soils, as considered in soil mechanics, means densification of soil through the application of mechanical energy. It may involve the modification of water content but the primary densification is obtained by rearrangement of particles without any outflow of water required. By contrast, consolidation in soil mechanics terms is the process of densification that occurs in clayey soils as water is squeezed from its pores under load. This time-dependent phenomenon is considered in a later section of this chapter. Both terms, compaction and consolidation, have different mean-

ings than these when used in the geological sense where they are applied to the process by which sediments change into sedimentary rocks. These geologic definitions are discussed in Chapter 14.

Achieving Compaction

Cohesionless soils are most easily compacted by vibration. On construction projects compaction of sands and gravels may be accomplished with hand-operated vibration plates and motor-driven, large vibratory rollers. Sand can also be compacted using rubber-tired equipment and more dramatically by large weights (demolition or "headache" balls) dropped from heights by construction cranes to compact loose, cohesionless soils.

Cohesive soils are compacted in the field with sheepsfoot rollers, rubber-tired rollers, and, for jobs with little clearance, by hand-operated tampers. A common practice for accomplishing compaction of soil fills is to route the soil-hauling equipment over the entire area during construction rather than allowing localized traffic in a restricted location.

Construction Objectives

Several construction objectives are accomplished by compaction of a soil, including the following:

1. Postconstruction settlement can be prevented or reduced by accomplishing the densification through compaction.
2. The strength of soil is increased though increased density, which improves—among other characteristics—the slope stability of the fill.
3. The bearing capacities of pavement subgrades and base courses can be improved by this densification.
4. The tendency for volume changes such as that caused by frost action or by expansive soils can be reduced or controlled.

Proctor Compaction Curves

Details on the compaction of soils were set forth in the early 1930s by Proctor. Based on this early work, the Proctor test for laboratory compaction was developed.

Controlling Variables

For the compaction of soils, Proctor described four variables: 1) dry density ρ_d, 2) water content w, 3) compactive effort, and 4) soil type and gradation (including clay minerals, types, and percentages). Compactive effort is the amount of mechanical energy applied to the soil during compaction.

Standard and Modified Proctor Tests

The standard Proctor test, designated as ASTM D698 or as AASHTO T991, is a common laboratory test used to evaluate soil compaction. Equipment for the test includes a 2.495-kg (5.5-lb) hammer and a cylinder or mold 944 cm^3 or nearly 1 liter (1/30 ft^3) in volume. Three layers of soil are added to the cylinder with each layer compacted by dropping the hammer a distance of 0.3048 m (1 ft) 25 times on each of the layers. The layers are scarified with a knife to encourage better contact between them. This procedure is repeated on four or five samples of the same soil each at different water contents. A plot of water content versus dry density yields the standard Proctor compaction curve shown in Figure 7.16. It should be recalled that $\rho_d = \dfrac{\rho}{1 + w}$ The mass density ρ is determined by dividing the mass of the soil in the cylinder by the volume of the cylinder or M_t/V_t.

Observe in Figure 7.16 that a maximum dry density is reached and then a decrease in density occurs with increasing water content. This maximum density is known as the optimum density and the water content at which it occurs is the optimum water content. Another standard compactive effort used is the modified Proctor compactive test (ASTM D1557 and AASHTO T180). This test utilizes a heavier hammer dropped a greater distance with more soil layers than the standard Proctor test. In the modified Proctor test the hammer is 4.536 kg (10 lb) dropped a distance of 457 mm (1.5 ft) using five layers with 25 blows each. The same mold is used (944 cm^3 or 1/30 ft^3). The surface is scarified before addition of the next soil layer.

Compactive Effort

The modified Proctor test illustrates the significance of the compactive effort. The optimum dry density will be higher and the optimum moisture content lower for the same soil when comparing the results of these two types of Proctor tests. Compactive effort should be determined numerically for the two Proctor tests.

In British units, energy input per unit volume is given in ft-lb/ft^3 and in SI units this is in joules per cubic meter or J/m^3. It develops from conversion factors that 1 ft-lb/ft^3 = 47.88 J/m^3. Therefore, the calculation for the standard Proctor test is as follows:

$$\text{Compactive effort} = \frac{5.5 \text{ lb}_f \times 1 \text{ ft} \times 3 \times 25}{1/30 \text{ ft}^3}$$

$$= 12375 \frac{\text{ft-lb}_f}{\text{ft}^3}$$

or

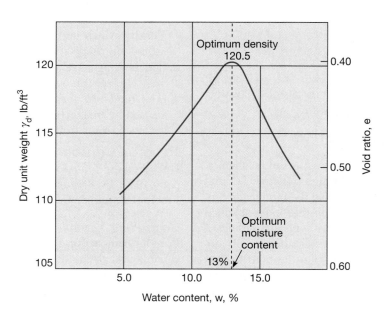

FIGURE **7.16**   Standard Proctor compaction Curve.

Compactive effort =

$$\frac{2.495 \text{ kg} \times 9.81 \text{ m/sec}^2 \times 0.3048 \text{ m} \times 3 \times 25}{0.944 \times 10^{-3} \text{ m}^3}$$

$$= 592.7 \text{ kJ/m}^3$$

or $12375(0.04788 \text{ kJ/m}^3) \times 1/\text{ft-lb}_f/\text{ft}^3 = 592.5 \text{ kJ/m}^3$, yielding the same answer when round-off effects are considered. For the modified Proctor test:

$$\text{Compactive effort} = \frac{10 \text{ lb}_f \times 1.5 \times 5 \times 25}{1/30 \text{ ft}^3}$$

$$= 56,250 \frac{\text{ft/lb}_f}{\text{ft}^3}$$

or

Compactive effort

$$= \frac{4.536 \times 9.81 \text{ m/sec}^2 \times 0.457 \text{ m} \times 5 \times 25}{0.944 \times 10^{-3} \text{ m}^3}$$

$$= 2692.8 \text{ kJ/m}^3$$

or $56250(0.04788 \text{ kJ/m}^3) \times 1/\text{ft-lb}_f/\text{ft}^3 = 2693.2 \text{ kJ/m}^3$

which is the same answer when round-off effects are considered. We see that the modified Proctor compaction test has $56,250/12,375 = 4.55$ times the compactive effort as the standard Proctor test.

Compactive Effort and Soil Type

Two other points of interest concerning soil compaction are the variation in the compaction curves with compactive effort and the variation for these curves with changing soil type. In Figure 7.17 the effects of compactive effort are shown. With a change from standard to modified Proctor compactive effort, the optimum point on the moisture-density curve is shown to increase in dry density and decrease in water content. The soil type is kept constant; a lean clay or CL as indicated. This shows that the line of optimums slopes downward to the right.

The zero air voids line is also shown on Figure 7.17. This represents saturated conditions (or zero unfilled voids) for different water contents. The equation for this condition can be easily derived in terms

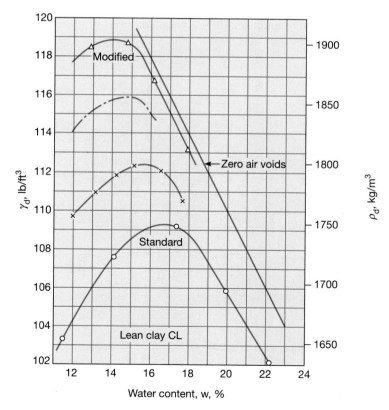

FIGURE 7.17 Modified and standard Proctor compaction Curves.

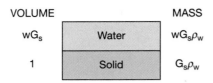

FIGURE 7.18 Derivation of dry density under zero air voids condition.

of ρ_w, G_s, and w using a two-phase soil relationship, as shown in Figure 7.18. Let

$$V_s = 1, \quad M_s = V_s G_s \rho_w, \quad w = M_w/M_s \quad or$$

$$M_w = w M_s = w G_s \rho_w$$

$$M_w = V_w \rho_w \quad or \quad V_w = M_w/\rho_w = w G_s \rho_w/\rho_w = w G_s$$

Finally

$$\rho_{(ZAV)} = \frac{G_s \rho_w}{1 + w G_s}$$

The change in the compaction curves with different soil types is shown in Figure 7.19, which illustrates two points. The first is that clayey soils have a wider range of moisture content with which to influence the optimum, or as the soil becomes more silty and then

sandy the optimum moisture content is more steeply peaked. This signifies that moisture control is very critical for silty and sandy soil if the optimum dry density is to be obtained. The second feature shown is that densities increase from clayey soils to silty soils to sandy silty soils and the optimum water content decreases. This again yields a line of optimums that slopes to the right.

Properties of Compacted Cohesive Soils

Compacted soils can be described by comparing them to the Proctor compaction curve, yielding such conditions as *dry of optimum, near optimum,* or *wet of optimum.* If compactive effort is held constant, with increasing water content the soil fabric becomes increasingly more oriented.

Dry of optimum, soils have a flocculated fabric, but wet of optimum they are more oriented or dispersed in nature. Permeability at a constant compactive effort decreases as the water content increases, reaching a minimum at about the optimum water content. The coefficient of permeability is about an order of magnitude greater (10 times more) for soils compacted dry of optimum compared to

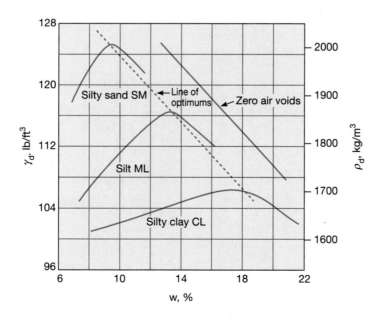

FIGURE 7.19 Compaction curves for different soils, but same compactive effort.

those compacted wet of optimum. If the compactive effort is increased, permeability decreases because of the reduction in void ratio.

Compressibility of soil is dependent, along with several other factors, on the level of stress imposed on it. At low stress levels, clays are more compressible if compacted wet of optimum. For high stress levels the opposite is true, because clays compacted wet of optimum are greatly compressed under such loads, making only a small amount of further settlement possible.

Clays compacted dry of optimum have a greater tendency to swell than those compacted wet of optimum. On the dry side, water is more deficient and hence soils can absorb more water and swell more. Therefore, soils dry of optimum are more subject to change with changes in environment, water content included. For shrinkage, the opposite situation exists: Soils compacted wet of optimum undergo the greatest shrinkage.

With regard to the strengths of compacted clays, those compacted dry of optimum have greater strengths than those compacted wet of optimum. However, if the samples are soaked, because of swelling effects, the wet side soils can be somewhat stronger. Relative to compactive effort the strength is increased on the dry side of optimum with increasing compactive effort but on the wet side less strength is obtained with higher and higher compactive efforts.

This fact is most significant in the design and construction of earth fills for engineering works. A summary of the properties of soil relative to the optimum water condition is given in Table 7.5.

Compaction and Excavation Equipment

Soil or common borrow for placement in a compacted fill is obtained from a borrow pit or borrow area. The equipment needed includes draglines, power shovels, and scrapers to excavate the borrow material. Self-propelled scrapers, known also as pans (Figure 7.20 (a)), remove the soil from the borrow area and transport it to the fill where it is placed in thin layers or lifts. Dozers (Figure 7.20(b)) may be used to help the scrapers during loading by pushing them through the borrow area.

If considerable distances of haulage are involved, trucks may be used to transport fill material (Figure 7.21 (a, b)). These would be loaded by shovels and draglines, or possibly front-end loaders if the soil is in a loose condition. The soil is spread over the fill area where drying can be accomplished to lower the water content if needed, or water trucks can add water to increase the water content if so desired (Figure 7.21(c)).

TABLE 7.5 Properties of Compacted Soils Relative to Optimum Water Content

Property	Comparison
Structure	
Particle arrangement	Dry side more random or flocculated. Wet side more oriented or dispersed.
Swelling	Dry side imbibes more water, yielding greater swelling.
Effect of environmental change	Dry side more sensitive to change.
Permeability	Dry side more permeable.
Compressibility	Wet side more compressible at low pressure levels, dry side more compressible at high pressure levels.
Strength	
As molded	Dry side higher.
After saturation	Dry side somewhat higher if swelling prevented, wet side likely higher if swelling permitted.
Pore water pressure at failure	Wet side higher.
Stress strain modulus	Dry side much greater.
Sensitivity	Dry side more likely to be sensitive.

(a)

(b)

(a)

(b)

FIGURE **7.20**   Excavation and compaction equipment. (a) Scraper or pan. (b) Dozer— used to push scrapers in the borrow area or pull rollers across the fill. (Caterpillar, Inc.)

In the fill area, dozers and motor patrols or motor graders, known informally as blades (Figure 7.22(a)), are used to spread the soil in the desired thickness or lifts. Lift thickness ranges from 0.15 to 0.5 m (6 to 18 in.) depending on the compaction equipment, gradation and size of the soil particles, and design of the embankment. Much of the compaction of the fill is usually accomplished by compacting equipment with a roller of some sort. As previously mentioned, the heavy equipment is routed over the entire fill to provide additional compaction. The rollers can induce compaction in several ways: by pressure, impact, vibration, or kneading.

(c)

FIGURE **7.21**   (a) Borrow area where trucks are loaded by track excavator. (b) Truck climbing up slope from borrow area. (c) Water truck adding water to fill area.

Types of Rollers

A smooth wheel or drum roller has 100% contact with the ground (or 100% coverage) and provides contact pressures up to 380 kPa (55 psi). It can be used on all types of soils except those containing large rocks (Figure 7.22(b)). These rollers are commonly used in the final rolling of subgrades (proof rolling) and in compacting asphalt pavements. They are also effective in micaceous soils where the mica flakes are forced into a flat-lying position with greater density than would occur with most other rollers. When additional drying of the soil is required, discing equipment is pulled through the fill to expose more surface area for drying (Figure 7.22(c)).

The pneumatic or rubber-tired roller provides about 80% coverage (from closely spaced tires) and has tire pressures up to 700 kPa (100 psi). As in the case of the smooth wheel roller, the rubber-tired roller is used in both cohesive and noncohesive soils for both highways and dam construction.

Perhaps the best known piece of compaction equipment is the sheepsfoot roller (Figure 7.22(b)). This roller has many round, rectangular, or foot-shaped protrusions known as feet that extend from a central steel drum. In the past, other names were used based on the shape of the protrusion, but sheepsfoot is the overall term commonly applied today.

Sheepsfoot rollers provide about 8% to 12% coverage and very high contact pressures, from 1400 to 7000 kPa (200 to 1000 psi) depending on drum size and whether the drum is water-filled or not. Four-foot- and five-foot-diameter rollers are the most common on heavy construction projects. On a project where a high degree of compaction is required, the 4-ft rollers are preferred because the feet do not shear or knead the soil as much as the 5-ft roller, thereby providing a denser soil structure.

The feet extend about 0.15 to 0.25 m (6 to 10 in.) from the drum and they compact the soil immediately below this depth. The roller compacts the soil higher and higher up in the lift with successive passes until it eventually "walks out" as the upper portion is compacted. Sheepsfoot rollers are best suited for cohesive soils.

Another roller with protrusions is the tamping foot roller, which can be either towed or self-propelled. These provide about 40% coverage and develop

(a)

(b)

(c)

FIGURE 7.22 (a) Motor patrol or motor grader commonly used to smooth haul roads for earth-moving equipment. (Caterpillar, Inc.) (b) Self-propelled smooth wheel or drum roller on left, sheepsfoot roller on right. (c) Discing equipment used to promote drying in fill area.

high contact pressures: from 1400 to 8400 kPa (200 to 1200 psi) depending on roller size and whether the drum is water-filled. These rollers have special hinged feet, which rotate in the soil under load to provide a kneading action. They are best suited for fine-grained soils and will "walk out" when proper compaction occurs.

Mesh or grid pattern rollers provide about 50% coverage and pressures from 1400 to 6200 kPa (200 to 900 psi). With their open mesh these are ideally suited for rocky soils, gravels, and sands. A vibrational motion is also provided when the roller is pulled at a high towing speed across the fill.

Vibratory compaction is particularly effective for granular soils because of the particle rearrangement that vibrations produce. Compaction equipment with vertical vibrators are available on both smooth wheel and tamping foot rollers to provide efficient compaction of granular soils. Small pieces of equipment that compact granular soils in restricted areas are also available. Vibrating plates and rammers ranging in size from 58 to 290 cm^2 (9 to 45 in.^{2}) that weigh from 50 to 3000 kg (100 to 6000 lb) are available. They have limited depths of penetration for compaction, extending downward only 1 m (3 ft) or less. This requires that granular soils be placed in lifts of 0.5 m (1.5 ft) or less between vibrational tamping. The subject of vibratory compactors is an extensive one and cannot be presented here in detail.

Field Compaction Control

The common concern on earth-moving projects is that the soil fill or embankment becomes adequately compacted to yield the density and water content desired. A more complete view of the situation involves other important engineering properties of the soil, such as permeability, compressibility, strength, and sensitivity, which should come as a consequence of proper density and water content. Despite these details, density and water content are the two parameters usually controlled.

Density and Water Content

Only two decades ago density alone was specified for compaction control without the full realization that very high or very low water contents, although accomplishing the desired dry density, would not provide the desired engineering properties. Modern specifications indicate the percent relative density required (relative to either the standard Proctor or the modified Proctor compaction) and a range of water contents between which the compacted soil can vary and still be acceptable. This provides the contractor with the information needed to manipulate the soil in the fill area in order to achieve the acceptable compaction.

Developing Specifications

The procedure for developing compaction control specifications is accomplished in a series of steps. First, samples are obtained from the borrow area for laboratory testing. About 14 kg (31 lb) of sample are needed to perform a Proctor compaction curve. Based on this and other information, the earth structure is designed and the compaction specifications are written in keeping with the design. Field compaction control is specified, such as 95% standard Proctor density with a 3% moisture range on either side of optimum, or a 90% modified Proctor density within 2% water content on the dry side and 3% on the wet side of optimum, or other similar designations. The inspectors for construction control will conduct tests to ensure that these specifications are met.

The compaction control procedure described previously is for end-product specifications. This procedure is followed on most highways and building foundation projects. The equipment or method for achieving the specified density or water content is not dictated; the contractor can use whatever means deemed best to achieve these end-product requirements. Competitive bids and the economics of the marketplace are assumed to provide the incentive for use of the most efficient equipment available.

Another procedure sometimes used involves a different approach, known as method specifications. In this case, the type of roller, number of passes, and lift thickness (the methods) are specified. Material types may also be specified based on different borrow areas designated for the project. If compaction control testing indicates that the embankment soil does not meet the necessary density or moisture content, then the contractor is paid extra for additional rolling. Method specifications require that the engineer/owner know in advance the detailed behavior of the soil during compaction in order to predict its behavior accurately. Such knowledge requires detailed testing, which typically can be justified only on very large earth-moving projects such as major dam construction.

Testing Procedures

Test procedures for field control of density and moisture content, like many kinds of testing, fall into two groups: destructive and nondestructive. Nondestructive testing, accomplished by nuclear methods, is described following a discussion of destructive methods.

DESTRUCTIVE TEST METHODS For destructive testing the sample is excavated from the fill area. The mass of the sample in its wet condition is determined, and after oven drying its dry mass is obtained. These measurements yield the water content. The volume of the hole left by the excavated sample (and of the soil sample itself) is obtained by means of 1) the sand cone method, 2) the balloon method, or 3) by pouring water or oil of a known density into the hole, if the soil has a low permeability. Another method is accomplished by driving a steel tube of known volume into the ground (the drive-cylinder method, ASTM D2937).

In the sand cone method, dry sand of a known density is allowed to flow into the hole (ASTM D1556). The volume is determined based on the weight of the sand used. For the balloon method, air pressure is used to push water into a balloon and against the side of the hole (ASTM D2167). Changes in water volume within the apparatus are noted as this occurs. In the third method listed, the soil sample, which extends from a steel tube, is cut level on both ends, yielding a full tube of soil. The volume and weight of the tube are known in advance. The soil, following extraction, is weighted before and after drying. For all methods, the dry density is obtained from the dry mass and volume measured.

NUCLEAR DENSITY MEASUREMENTS Nondestructive testing is accomplished by nuclear density and water content meters (ASTM D2292). These tests have several advantages over the destructive tests: 1) They can be performed quickly with the results available in a few minutes, rather than in hours and 2) with more tests possible a greater distribution of tests over the fill area can be accomplished providing a better statistical evaluation.

Nuclear density meters are not without disadvantages. They have a relatively high initial cost and there is a potential danger of exposure to radioactivity. Strict safety standards must be followed when using these devices.

Two different types of radiation are used in nuclear nondestructive testing. Gamma rays are used to determine soil density because their amount of scatter is proportional to total mass. Neutrons are scattered by hydrogen atoms and this provides the means for water content measurements. Thus, two emitters, one for gamma rays and another for neutrons, are included in the nuclear density devices.

EFFECTIVE STRESS

Definitions

The intergranular or effective stress is defined by the equation

$$\sigma = \sigma' + u$$

where

σ = the total normal stress or total stress

σ' = the intergranular stress, grain-to-grain pressure, or effective stress

u = pore water pressure, pore pressure, or neutral stress.

The total stress can be calculated using the densities and thicknesses of the layers involved, and the neutral stress is found by calculating the hydrostatic pressure that develops below the ground water table. Effective stress must be found by subtraction because it cannot be measured directly ($\sigma' = \sigma - u$).

For the British engineering system, total pressure is obtained by the equation $\sigma = h\gamma_{soil}$. For example a 10-ft-thick section of soil weighing 120 lb/ft^3 has σ = 10 ft (120 lb/ft^3) = 1200 lb/ft^2. Similarly $u = h\gamma_{water}$ so the neutral stress 5 ft below the ground water table is u = 5 ft (62.4 lb/ft^3) = 312 lb/ft^2.

In SI units, pressure is given by $\sigma = \rho g h$. In the case of a 5-m-thick section of soil with a density of 1800 kg/m^3,

$$\sigma = 1800 \text{ kg/m}^3 \times 9.81 \text{m/sec}^2 \times 5 \text{ m}$$
$$= 88{,}290 \text{ kg/m/sec}^2$$

but one newton or 1 N = 1 kg m/sec^2. Therefore,

$$\sigma = 88{,}290 \text{ N/m}^2 = 88.29 \text{ kN/m}^2 = 88.29 \text{ kPa}$$

(1 N/m^2 = 1 Pa or one pascal). The neutral stress 3 m below the ground water table is

$$u = \rho_{water}\, gh = 1000 \text{ kg/m}^3 \times 9.81 \text{ m/sec}^2 \times 3 \text{ m}$$
$$= 29{,}430 \text{ N/m}^2 = 29.43 \text{ kN/m}^2 = 29.43 \text{ kPa}$$

Conversions from pressure in English units (in lb_f/ft^2) to SI units (in kN/m^2) can be made directly using 1 $lb_f/ft^2 = 47.88$ N/m^2. For example,

$$\sigma = 1200 \ lb_f/ft^2 = 1200 \ (0.04788) = 57.46 \ kN/m^2$$
$$u = 312 \ lb_f/ft^2 = 312 \ (0.04788) = 14.93 \ kN/m^2$$

and conversely

$$\sigma = 88.29 \ kN/m^2 \times 1/0.04788 = 1844 \ lb_f/ft^2$$
$$u = 29.43 \ kN/m^2 \times 1/0.04788 = 614.7 \ lb_f/ft^2$$

Basic Calculations

The calculation of total, neutral, and effective stress is best illustrated by two sample problems. Based on the following cross section, the total stress, neutral stress, and effective stress are calculated at the depths shown and the data presented in a profile diagram in Figure 7.23.

Total Stress Calculation

$$\text{At 4 m deep, } \sigma = \rho g h = \frac{1900 \times 9.81 \times 4}{1000}$$
$$= 74.5 \ kN/m^2$$

$$\text{At 10 m deep, } \sigma = 74.5 + \frac{2180 \times 9.81 \times 6}{1000}$$
$$= 74.5 + 128.3 = 202.8 \ kN/m^2$$

$$\text{At 18 m deep, } \sigma = 202.8 + \frac{1925 \times 9.81 \times 8}{1000}$$
$$= 202.8 + 151.1 = 353.9 \ kN/m^2$$

$$\text{At 30 m deep, } \sigma = 353.9 + \frac{1990 \times 9.81 \times 12}{1000}$$
$$= 353.9 + 234.0 = 587.9 \ kN/m^2$$

For the Neutral Stress

$$\text{At 4 m, } u = 0$$
$$\text{At 10 m, } u = 0 + \frac{6(1000) \times 9.81}{1000} = 58.9 \ kN/m^2$$
$$\text{At 18 m, } u = 14 \times 9.81 = 137.3 \ kN/m^2$$
$$\text{At 30 m, } u = 26 \times 9.81 = 255.1 \ kN/m^2$$

Therefore, the *effective stress* is $\sigma' = \sigma - u$ or

$$\text{At 4 m, } \sigma' = 74.5 \ kN/m^2$$
$$\text{At 10 m, } \sigma' = 202.8 - 58.9 = 143.9 \ kN/m^2$$
$$\text{At 18 m, } \sigma' = 353.9 - 137.3 = 216.6 \ kN/m^2$$
$$\text{At 30 m, } \sigma' = 587.9 - 255.1 = 332.8 \ kN/m^2$$

A similar problem can be calculated using British engineering units, as shown in Figure 7.24.

Total Pressure

$$\text{At 6 ft deep, } \sigma = h\gamma_d = 6(127) = 762 \ lb/ft^3$$
$$\text{At 10 ft deep, } \sigma = 762 + 4(136) = 1306 \ lb/ft^2$$
$$\text{At 21 ft deep, } \sigma = 1306 + 11(120) = 2626 \ lb/ft^2$$
$$\text{At 32 ft deep, } \sigma = 2626 + 11(125) = 4001 \ lb/ft^2$$

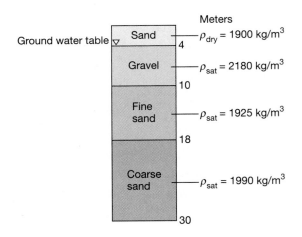
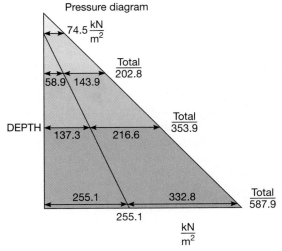

FIGURE 7.23 Variations in pressure with depth, SI units.

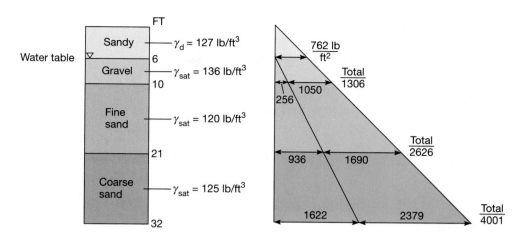

FIGURE 7.24 Variations in pressure with depth, British engineering units.

For the Neutral Stress

At 6 ft, $u = 0$

At 10 ft, $u = 4(62.4) = 256$ lb/ft²

At 21 ft, $u = 15(62.4) = 936$ lb/ft²

At 32 ft, $u = 26(62.4) = 1622$ lb/ft²

Therefore, the *effective stress* $\sigma' = \sigma - u$ is

At 6 ft, $\sigma' = 762 - 0 = 762$ lb/ft²

At 10 ft, $\sigma' = 1306 - 256 = 1050$ lb/ft²

At 21 ft, $\sigma' = 2626 - 936 = 1690$ lb/ft²

At 32 ft, $\sigma' = 4001 - 1622 = 2379$ lb/ft²

Effects of Water Flow

The effects of water flow can now be considered by looking at three cases: 1, no flow conditions; 2, downward flow through the soil mass; and 3, upward flow through the soil mass. Case 1 is shown in Figure 7.25. At the bottom of the soil (or Point A)

$$\sigma = (H_1\rho_w + H\rho_{sat})g$$
$$= [(H_1+H)\rho_w + H\rho']g$$

Also at Point A:

$$u = (H_1+H)\rho_w g$$
$$\sigma' = [(H_1+H)\rho_w+H\rho' - (H_1+H)\rho_w]g = H\rho'g$$

Note that the effective stress σ' is independent of the height of water (H_1) above the saturated soil column. Therefore, effective stress in the sediment at the bottom of the ocean basins is related only to the thickness of sediment and not to the water depth. Deep water does not make the sediment compress because the effective stress (intergranular stress) controls the settlement of the sedimentary column and it is independent of water depth.

Case 2, downward flow through the soil mass, is shown in Figure 7.26. At Point A

$$\sigma = (H_1\rho_w + H\rho_{sat})g$$
$$= [(H_1+H)\rho_w + H\rho']g$$

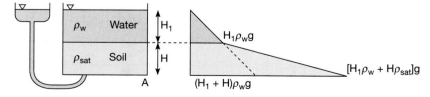

FIGURE 7.25 Case 1, no flow conditions.

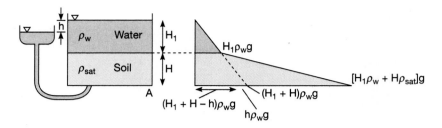

FIGURE **7.26**   Case 2, downward flow through the soil mass.

Also at Point A

$$u = (H_1+H-h)\rho_w g$$

In this case h is negative because the water would flow downward from the water-soil column toward the standpipe. Downward flow is the opposite of the buoyancy or uplift force (u) and hence it is subtracted from that value. Therefore, at Point A

$$\sigma' = \sigma -u = [(H_1+H)\rho_w+H\rho'- (H_1+H-h)\rho_w]g$$
$$= (h\rho_w + H\rho')g$$

Hence, the effective stress is increased by a downward flow of water through the soil system.

Case 3, upward flow through the soil mass, is shown in Figure 7.27. At Point A

$$\sigma =(H_1\rho_w + H\rho_{sat})g$$
$$= [(H_1+H)\rho_w +/ H\rho']g$$

Also at Point A

$$u = (H_1H+h)\rho_w g$$

where h is positive in this case because the water flows upward from the standpipe through the water-soil column. This is in the same direction as the buoyancy force (upward) and hence is added to that value. Therefore, at Point A

$$\sigma' = \sigma - u = [(H_1+H)\rho_w+H\rho']g - (H_1+H+h)\rho_w g$$
$$= (H\rho'-h\rho_w)g$$

This indicates that the effective stress is reduced by an upward flow of water through the soil column.

Critical Hydraulic Gradient

Another significant feature concerning effective stress occurs when the upward flow is sufficient to reduce σ' to zero. These details are shown next.

If $\sigma'= 0$, $\sigma = u$, or $0 = \sigma - u$, then substituting,

$$0 =[(H_1+H)\rho_w+H\rho']g - (H_1+H+h)\rho_w g$$
$$= (H\rho' - h\rho_w)g$$

Cancelling g and transposing, $H\rho' = h\rho_w$ or $\rho'/\rho_w = h/H = i_{cr}$, where i_{cr} = the critical hydraulic gradient and h/H is a measure of head loss (h) over a distance of flow (H), which is a hydraulic gradient. The i_{cr} value is the hydraulic gradient for a zero effective stress level where the soil has a zero intergranular strength or zero shear strength. This denotes a quick condition, which is a critical case for stability.

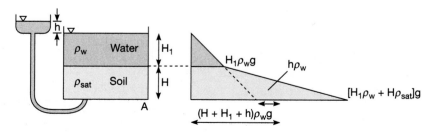

FIGURE **7.27**   Case 3, upward flow through the soil mass.

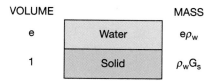

VOLUME   **MASS**

e   Water $e\rho_w$

1   Solid $\rho_w G_s$

FIGURE 7.28 Phase diagram for calculating submerged density.

The value of i_{cr} turns out to be about equal to one ($i_{cr} \sim 1$) for cohesionless soils. This is shown in the next calculation, which is based on Figure 7.28 for a two-phase soil condition. The soil is saturated and considered in terms of e, G_s, and ρ_w.

$$\rho' = \frac{\rho_w G_s - \rho_w(1)}{1+e} = \frac{\rho_w(G_s - 1)}{1+e}$$

In this situation, $G_s \sim 2.7$ and e for cohesionless soil is about 0.7. Therefore,

$$\rho' \simeq \rho_w \frac{(1.7)}{(1.7)} \quad \text{or} \quad \rho' \simeq \rho_w \text{ and } \frac{\rho'}{\rho w} \sim 1$$

But

$$\frac{\rho'}{\rho_w} = \frac{h}{H} \quad \text{or} \quad \frac{h}{H} = i_{cr} \simeq 1$$

An example of a critical hydraulic gradient is shown for a concrete dam in Figure 7.29.

$$i = \frac{h}{H} = \frac{50 \text{ ft}}{50 \text{ ft}} = 1,$$

thus signaling that a quick condition exists at the toe of the dam.

SHEAR STRENGTH OF SOILS

If the load or stress on a soil is increased until deformation is excessive or if the soil mass suddenly

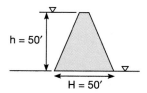

$h = 50'$

$H = 50'$

FIGURE 7.29 Configuration for dam with a critical hydraulic gradient problem.

yields a major displacement, the soil is said to have failed. This indicates the strength of the soil, or the maximum stress that the soil can sustain. In soil mechanics the shear strength is the main concern because most failures result from applications of excessive shear stresses.

Stress at a Point

The usual approach is to consider the state of stress at a point. This is accomplished by viewing an infinitesimally small cube of material acted on by normal stresses and shear stresses as shown in Figure 7.30. The three stresses acting perpendicular and inward on the cube faces are normal compressive stresses. Those acting parallel to the faces are shear stresses. In addition to the nine stresses shown, another nine are not provided. These nine additional stresses are distributed over the three remaining faces, the back and bottom of the cube, to yield equal and opposite stresses to those shown. Because the cube is infinitesimal (a point), these stresses would all be in equilibrium.

Principal Stresses

A fundamental principle of mechanics states that through any point there exist three mutually perpendicular planes that have only normal stresses acting on them (no shear stresses). These comprise the three principal planes, which have the three principal stresses acting perpendicular to them. They are shown in Figure 7.31.

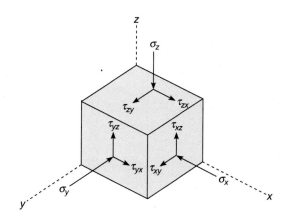

FIGURE 7.30 Shear and normal stresses acting through a point.

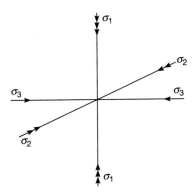

FIGURE 7.31 Three principal stresses acting through a point.

The greatest stress is called the major principal stress or σ_1, the least is the minor principal stress or σ_3, and the third is the intermediate principal stress or σ_2. In many soil mechanics problems the behavior is considered to be independent of σ_2, which yields a two-dimensional problem involving only σ_1 and σ_3. Based on this we can now show the relationship in two dimensions in Figure 7.32. Planes parallel to σ_1 or σ_3 will have only normal stresses acting on them but all other planes have both a normal stress σ and shear stress τ that act collectively.

Equations for σ and τ inclined at the angle θ from the maximum principal plane are:

$$\sigma = \sigma_1\cos^2\theta + \sigma_3\sin^2\theta = \sigma_3(\sigma_1-\sigma_3)\cos^2\theta$$

$$\tau = (\sigma_1-\sigma_3)\sin\theta\cos\theta$$

Checking at the boundaries to show that the equations apply there,

for $\theta = 0$, $\sin\theta = 0$ and $\cos\theta = 1$:

$$\sigma = \sigma_1 + 0 = \sigma_1; \quad \tau = (\sigma_1 - \sigma_3)(1)(0) = 0$$

for $\theta = 90°$, $\sin\theta = 1$ and $\cos\theta = 0$:

$$\sigma = 0 + \sigma_3(1) = \sigma_3$$

$$\tau = (\sigma_1 - \sigma_3)(1)(0) = 0$$

and in another form,

$$\sigma = \frac{\sigma_1+\sigma_3}{2} + \frac{\sigma_1-\sigma_3}{2}\cos2\theta$$

$$\tau = \frac{\sigma_1+\sigma_3}{2}\sin2\theta$$

Checking again at the boundaries

for $\theta = 0$, $\cos2\theta = 1$ and $\sin2\theta = 0$:

$$\sigma = \frac{\sigma_1+\sigma_3}{2} + \frac{\sigma_1-\sigma_3}{2}(1) = \sigma_1$$

$$\tau = \frac{\sigma_1+\sigma_3}{2}(0) = 0$$

for $\theta = 90°$, $\cos2\theta = 1$ and $\sin2\theta = 0$:

$$\sigma = \frac{\sigma_1+\sigma_3}{2} + \frac{\sigma_1-\sigma_3}{2}(-1) = \sigma_3$$

$$\tau = \frac{\sigma_1-\sigma_3}{2}\sin2\theta = \frac{\sigma_1-\sigma_3}{2}(0) = 0$$

Mohr Circle

If all values of σ and τ are plotted on a σ-τ diagram as they vary with θ from 0 to 360°, a circle is obtained. The circle represents all the possible combinations of

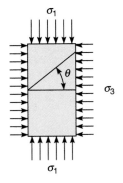

FIGURE 7.32 Two-dimensional stress field.

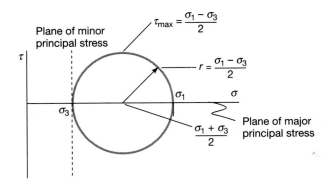

FIGURE 7.33 Mohr circle diagram for stress at a point.

σ and τ acting in different directions through that point, as is shown in Figure 7.33. As indicated before, σ_1 and σ_3 are the values of the major and minor principal stress respectively and the locations where $\tau=0$.

The center of the circle is located at $(\sigma_1+\sigma_3)/2$ and the radius of the circle is $(\sigma_1-\sigma_3)/2$. The dotted vertical line through σ_3 is the plane of minor principal stress and the σ-axis is the plane of maximum principal stress. The maximum shear stress is equal to the radius of the circle of $(\sigma_1-\sigma_3)/2$. The resultant stress at the point R is equal to $\sqrt{\sigma^2 + \tau^2}$. In addition, $\sigma_1-\sigma_3$ is referred to as the stress difference or as the deviator stress.

EXAMPLE PROBLEM 7.7

Figure 7.34 shows the details for this problem.
Given the major and minor principal stresses at a point of 70 and 25 psi, respectively, solve for the following: a) Find the value of the maximum shearing stress at the point. b) What is the value of the normal stress on the plane of maximum shearing stress? c) What is the value of σ, τ, and the resultant on a plane at an angle of 22° to the plane of major principal stress? d) What is the value of σ, τ, and the resultant on a plane at an angle 38° to the plane of minor principal stress?

ANSWERS

a) Maximum shear stress occurs at point (A) on Figure 7.34. It equals the value of the radius or

$$\frac{\sigma_1-\sigma_3}{2} = \frac{70-25}{2} = \frac{45}{2} = 22.5 \text{ psi}$$

b) Normal stress at point (A). It equals the same as the value for the center of the circle, or

$$\frac{\sigma_1+\sigma_3}{2} = \frac{70+25}{2} = \frac{95}{2} = 47.5 \text{ psi}$$

c) A plane 22° from the plane of maximum principal stress is measured from the horizontal,

$$\sigma = \sigma_1 \cos^2\theta + \sigma_3 \sin^2\theta; \quad \theta = 22°$$

$$= 70(0.9272) + 25(0.3746)$$

$$60.17 + 3.508 = 63.7 \text{ psi}$$

$$\tau = (\sigma_1-\sigma_3) \sin\theta \cos\theta$$

$$= (70-25)(0.3746)(0.9272) = 15.6 \text{ psi}$$

$$R = \sqrt{\sigma^2+\tau^2} = (63.7^2 + 15.6^2)^{1/2} = 65.5 \text{ psi}$$

d) A plane 38° to the plane of minor stress is $90 - 38 = 52°$ from the plane of major stress.

$$\sigma = \sigma_1 \cos^2\theta + \sigma_3 \sin^2\theta; \quad \theta = 52°$$

$$= 70(0.6156)^2 + 25(0.7880)^2$$

$$= 26.53 + 15.52 = 42.0 \text{ psi}$$

$$\tau = (\sigma_1-\sigma_3) \sin\theta \cos\theta$$

$$= (70-25)(0.6156)(0.7880) = 21.8 \text{ psi}$$

$$R = \sqrt{\sigma^2 + \tau^2} = (42.0^2 + 21.8^2)^{1/2}$$

$$= (1764 + 475.2)^{1/2} = 47.3 \text{ psi}$$

The obliquity angle is the angle between the origin (or the origin of stresses) and the point of interest on

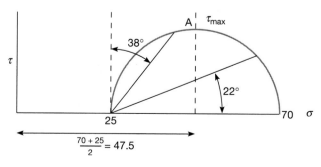

FIGURE 7.34 Diagram for Example Problem 7.7, Mohr circle.

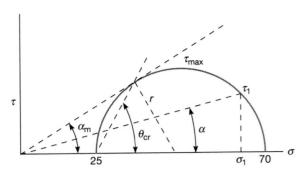

FIGURE 7.35　Obliquity angle (α) on Mohr circle diagram.

the Mohr circle diagram. This is related in the following way:

$$\alpha = \tan^{-1}(\tau/\sigma)$$

and is illustrated in Figure 7.35. The maximum obliquity angle occurs at the point of tangency to the circle. By recalling that the radius equals $1/2(\sigma_1 - \sigma_3)$ and the distance to the center of the circle is $1/2(\sigma_1 + \sigma_3)$ we have

$$\sin\alpha_m = \frac{\sigma_1 - \sigma_3}{\sigma_1 + \sigma_3} \quad \text{and} \quad \theta_{cr} = 45 + 1/2\alpha_m$$

Note that the point of maximum shear stress or $(\sigma_1 - \sigma_3)/2$ is not the point of maximum obliquity, and conversely the point of maximum obliquity has a shear stress less than the maximum. This will become more significant later in this discussion.

Another example problem is now appropriate for consideration.

EXAMPLE PROBLEM 7.8 _____

Given the same data as in Example Problem 7.7, find the angle of maximum obliquity and determine the τ and σ values for that orientation.

ANSWER

$$\sin\alpha_m = \frac{\sigma_1 - \sigma_3}{\sigma_1 + \sigma_3} = \frac{70 - 25}{70 + 25} = \frac{45}{95} = 0.4736$$

$$\alpha_m = 28.27°$$

$$\theta_{cr} = 45 + 1/2\alpha_m = 45 + 1/2(28.27) = 59.13°$$

$$\sigma = \sigma_1 \cos^2\theta + \sigma_3 \sin^2\theta$$

$$= 70(0.5131)^2 + 25(0.8583)^2$$

$$= 18.42 + 18.42 = 36.84 \text{ psi}$$

$$\tau = (\sigma_1 - \sigma_3)\sin\theta\cos\theta$$

$$= (70 - 25)(0.8583)(0.5131) = 19.82 \text{ psi}$$

Check $\tau = \sigma \tan\alpha_m = 36.85 \tan 28.27 = 19.82$ psi.

_____

Mohr-Coulomb Failure Criterion

The Coulomb equation for the strength of soil is written as

$$\tau = c + \sigma \tan\phi$$

and is illustrated in Figure 7.36.

The angle ϕ is known as the angle of internal friction, which provides for an increase in ϕ (shear strength) with an increase in σ (normal stress). The other term, c, is the cohesion of the soil and is the inherent strength of a soil that persists even when unconfined, that is, at zero normal stress. ϕ and c are collectively known as the strength parameters.

For certain types of soil, either the c or ϕ parameter can equal zero. When $\phi=0$, $\tau=c$ and when $c=0$, $\tau=\tan\phi$. These are illustrated in Figure 7.37.

In 1900 Mohr stated that the shear stress on the failure plane at failure has a value that is a function of the normal stress alone on that plane. This can be written $\tau_{ff} = f(\sigma_{ff})$ where τ and σ are the usual shear stress and normal stress, respectively. The first f in the subscript stands for the failure plane and the second f stands for at failure. So the equation is read shear stress on the failure plane at failure is a function of the normal stress on the failure plane at failure.

If a series of soil tests is run with different σ_1 and σ_3 values and loaded to failure, a line enclosing all the failure circles can be drawn. This is known as the

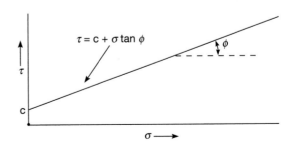

FIGURE 7.36　Mohr-Coulomb failure criterion.

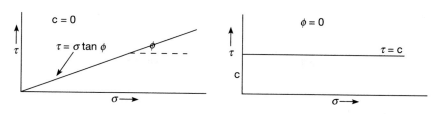

FIGURE 7.37   Plots for τ versus σ when *c* or φ = 0.

Mohr failure envelope. No Mohr circle can exist for soil that extends beyond the envelope because failure would occur before extending beyond that limit. In like manner, the soil cannot fail if its Mohr circle lies within the envelope without touching it. These details are illustrated in Figure 7.38.

When the Coulomb strength equation is combined with the Mohr failure criterion, the Mohr-Coulomb strength criterion is obtained. We do not know who first combined these relationships, but today this is one of the most widely used concepts applied to soil studies. It is given by

$$\tau_{ff} = c + \sigma_{ff}\tan\phi$$

This can now be shown graphically as in Figure 7.39 including an example of a Mohr circle for failure conditions. From Figure 7.40 we can observe that

$$\sin\phi = \frac{R}{D} = \frac{(\sigma_1 - \sigma_3)/2}{(\sigma_1 + \sigma_3)/2 + c\,\cot\phi}$$

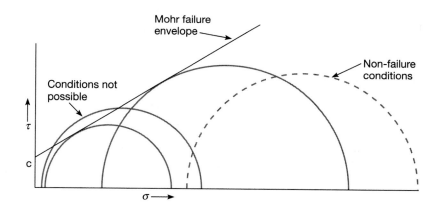

FIGURE 7.38   Mohr envelope and failure conditions.

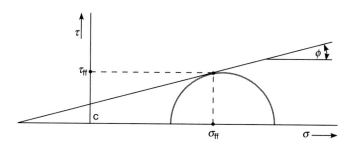

FIGURE 7.39   Mohr-Coulomb failure, including cohesion.

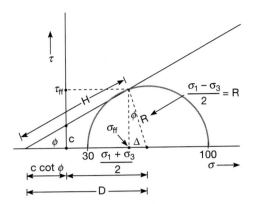

FIGURE 7.40 Trigonometric relationships, including cohesion.

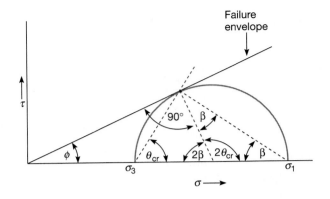

FIGURE 7.41 Trigonometric relationships, cohesionless case.

If $c = 0$ then this becomes

$$\sin\phi = \frac{\sigma_1 - \sigma_3}{\sigma_1 + \sigma_3}$$

which is the same as the obliquity relationship developed previously, and ϕ can be substituted for α_m.

Several other relationships can now be developed regarding cohesionless soils ($c = 0$). Figure 7.41 is used to illustrate this. In Figure 7.41 θ_{cr} is the angle of inclination from the maximum principal plane to the failure plane, β is the opposite angle measured from the maximum principal plane, and 2β is the corresponding central angle.

$$2\beta + \phi + 90 = 180°$$
$$2\beta = 90 - \phi$$
$$\beta = 45 - \phi/2$$
$$2\beta + 2\theta_{cr} = 180°$$
$$\beta = 90 - \theta_{cr}$$

Equating relationships for β,

$$45 - \phi/2 = 90 - \theta_{cr}$$
$$\theta_{cr} = 45 + \phi/2$$

In addition it can be shown by trigonometry that

$$\frac{\sigma_1}{\sigma_3} = \frac{1 + \sin\phi}{1 - \sin\phi}$$

$$\frac{\sigma_1}{\sigma_3} = \tan^2(45 + \phi/2)$$

$$\frac{\sigma_3}{\sigma_1} = \tan^2(45 - \phi/2)$$

$$\sigma_{ff} = \sigma_3(1 + \sin\phi) = \sigma_1(1 - \sin\phi) = (\sigma_1 - \sigma_3)\frac{\cos^2\phi}{2\sin\phi}$$

$$\tau_{ff} = \sigma_{ff}\tan\phi = \sigma_3\tan\phi(1 + \sin\phi) = \sigma_1\tan\phi(1 - \sin\phi)$$
$$= (\sigma_1 - \sigma_3)1/2\cos\phi$$

EXAMPLE PROBLEM 7.9

Given a Mohr circle for failure conditions for a cohesionless sand as follows:

$$\sigma_1 = 100 \text{ psi and } \sigma_3 = 30 \text{ psi. } c = 0$$

a) Find ϕ, θ_{cr}, σ_{ff}, and τ_{ff}. If $c = 0$,

$$\sin\phi = \frac{\sigma_1 - \sigma_3}{\sigma_1 + \sigma_3} = \frac{70}{130}$$

$$\phi = 32.57°$$

$$\theta_{cr} = 45 + \phi/2 = 45 + 16.28 = 61.3°$$

$$\sigma_{ff} = \sigma_3(1 + \sin\phi) = 30(1 + 35/65)$$
$$= 46.15 \text{ psi}$$

$$\tau_{ff} = \sigma_{ff}\tan\phi = 23.5 \text{ psi}$$

b) If $c = 10$ psi, find ϕ, σ_{ff} and τ_{ff}. Refer to Figure 7.40.

ANSWERS

$$D = \frac{\sigma_1 + \sigma_3}{2} + c\cot\phi$$

$$\sin\phi = \frac{R}{D} = \frac{(\sigma_1 - \sigma_3)/2}{(\sigma_1 + \sigma_3)/2 + c\cot\phi}$$

$$R = \frac{100 - 30}{2} = \frac{70}{2} = 35$$

$$\frac{\sigma_1 + \sigma_3}{2} = \frac{100 + 30}{2} = 65$$

$$\sin\phi = \frac{35}{65 + 10\cot\phi}; \quad \sin\phi(65 + 10\cot\phi) = 35$$

$$\text{or} \quad 65\sin\phi + 10\cos\phi = 35$$

Solving by an interactive method:

Try $\phi = ?$ $65\sin\phi + \cos\phi = 35$

30°	41.1
25	36.5
23	34.5
23.5	35.08
23.4	34.992

Therefore,

$$\phi = 23.4°; \quad \theta_{cr} = 45 + \phi/2 = 45 + 11.7 = 56.7°$$

$$\frac{R}{H} = \tan\phi, H = \frac{R}{\tan\phi} = \frac{35}{\tan 23.4} = 80.88$$

$$D = 65 + 10\cot\phi = 88.10, R = 35$$

Check

$$D^2 = R^2 + H^2 = 35^2 + 80.88^2$$

$$D = 88.1$$

$$\frac{\tau_{ff}}{H} = \sin\phi; \quad \tau_{ff} = H\sin\phi = 80.88\sin 23.4 = 32.12$$

$$\frac{\tau_{ff}}{\sigma_{ff} + c\cot\phi} = \tan\phi$$

$$(\sigma_{ff} + c\cot\phi)\tan\phi = \tau_{ff}$$

$$\sigma_{ff}\tan\phi + c = \tau_{ff}$$

$$\sigma_{ff} = \frac{\tau_{ff} - c}{\tan\phi} = \frac{32.10 - 10}{\tan 23.4} = 51.12$$

$$\frac{\Delta}{R} = \sin\phi; \quad \Delta = R\sin\phi$$

$$\Delta = 35\sin 23.4 = 13.9$$

$$\frac{\sigma_1 + \sigma_3}{2} - \Delta = \sigma_{ff} = 65 - 13.9 = 51.1$$

For clay-rich soils (CH by Unified Soil classification) the ϕ angle is equal to zero. In this case the Coulomb equation for shear strength reduces to $\tau = \sigma$ and the Mohr-Coulomb strength criterion is $\tau_{ff} = c = 1/2q_u$ where q_u is the unconfined compression strength. This is illustrated in Figure 7.42.

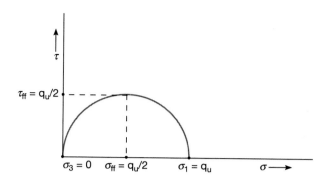

FIGURE 7.42   Shear strength relationship, $\phi = 0$.

EXAMPLE PROBLEM 7.10

Given that the unconfined compression strength for a highly plastic clay is 3000 lb/ft² and its unit weight is 125 lb/ft³ (143.6 kPa and 2003 kg/m³, respectively), find the following:
a) What is its shear strength, cohesion, or cohesive strength and is it affected by depth?
b) At what depth is the vertical overburden stress σ_v equal to the shear strength?
c) At a depth of 30 ft, what is the ratio of vertical overburden stress to the shear stress? How would this affect the ability to tunnel through this plastic clay at that depth?

ANSWERS

$$\text{a) } \tau_{ff} = q_u/2 = \frac{3000}{2} = 1500 \text{ lb/ft}^2$$

This is not affected (or increased) with depth as $\phi = 0$.

$$\text{b) } Z = \frac{\tau_{ff}}{\gamma} = \frac{1500 \text{ lb/ft}^2}{125 \text{ lb/ft}^3} = 12 \text{ ft}$$

$$\text{c) } P_z = \gamma Z; \quad P = (125)(30) = 3750 \text{ lb/ft}^2$$

$$\text{Ratio} = \frac{\text{Overburden stress}}{\text{Shear strength}} = \frac{3750}{1500} = 2.5$$

At this high stress ratio, the soil will yield dramatically when excavated. Loss of ground will occur above the tunnel and surface subsidence will result.

STRESS DISTRIBUTION

When a load or stress is applied at a boundary (the Earth's surface for example) the stress is distributed

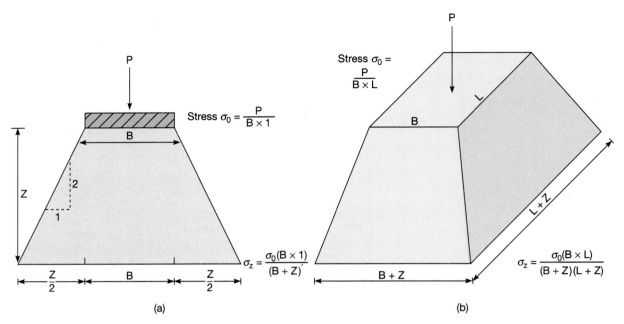

FIGURE 7.43 The 2 : 1 approximation for distribution of vertical stress with depth: (a) strip footing and (b) rectangular footing.

or transferred to the different levels below. Depending on the size of the loaded area, the added stress will reduce with greater depth, that is, with increasing distance from the boundary stress. The manner in which this pressure dissipates with depth is the stress distribution.

Two-to-One Method

A simple way to estimate this distribution of stress with depth is to use the two-to-one (or 2 to 1 or 2:1) method. As the same vertical force is spread over a larger and larger area, the unit stress must decrease with depth, as illustrated in Figure 7.43.

The 2:1 method provides data on the *average* stress at some depth; however, the *maximum* stress at that level (vertically below the center of the footing) is somewhat greater. A value of 50% is typically added to yield the maximum value.

Boussinesq Formula

Another method used to estimate the vertical stress at a point below a concentrated applied load at the boundary is by use of a Boussinesq formula:

$$\sigma_z = N_B \frac{P}{Z^2}$$

It allows the calculation of stress directly below P and at horizontal distances away from the vertical location. This is illustrated in Figure 7.44.

Westergaard Formula

Both the Boussinesq and Westergaard factors are shown in Figure 7.44 as they vary with r/z. For values of $r/z < 1.5$, Boussinesq indicates higher stresses than does Westergaard. When $r/z \geq 1.5$ both provide about the same answer. In practice the 2 : 1 ratio method, though fairly crude, is commonly used because it involves a contact area for the load rather than a point load, and it is easy to apply. An example problem provides a comparison of the three methods and is illustrated by Figure 7.45.

The Westergaard theory was derived assuming thin horizontal sheets and likely is more appropriate for bedded sedimentary rocks and soils. It also typically provides the smallest stress level of the three shown in Example Problem 7.11. This may be of concern to design engineers who typically would not select the lowest value of stress using alternative calculation methods.

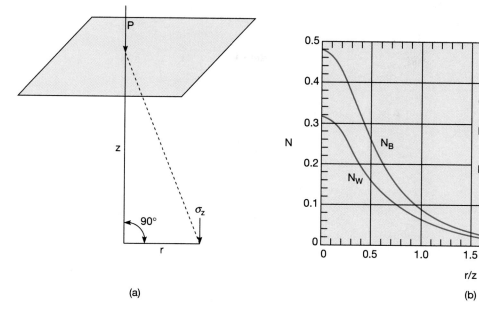

(a)

(b)

FIGURE 7.44 (a) Boussinesq relationship for stress distribution. (b) Boussinesq (N_B) and Westergaarel (N_W) factors for stress reduction with depth and lateral distance from point load.

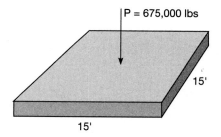

P = 675,000 lbs

15'

15'

FIGURE 7.45 Example Problem 7.11 on stress distribution.

EXAMPLE PROBLEM 7.11

Given a footing 15 × 15 ft with a column load of 675 kips, find the stress level at 15 and 30 ft directly below the center of the footing using a) 2 : 1, b) Boussinesq, and c) Westergaard methods.

ANSWERS

a) By the 2 : 1 method:

$$\sigma_z = \frac{P}{(B+Z)(L+Z)} = \frac{P}{(B+Z)^2} \text{ for square footing}$$

$$\sigma_0 = \frac{675,000}{15^2} = 3000 \text{ psf}$$

$$\sigma_{15_{avg}} = \frac{675,000}{(30)^2} = 750 \text{ psf}; \quad 1.5\sigma_{15_{avg}} = 1125 \text{ psf} = \sigma_{max}$$

$$\sigma_{30_{avg}} = \frac{675,000}{(45)^2} = 333 \text{ psf}; \quad 1.5\sigma_{30_{avg}} = 499.5 \text{ psf} = \sigma_{max}$$

b) Boussinesq:

$$\sigma_{15} = N_B \frac{P}{Z^2} = 0.48 \frac{675,000}{15^2} = 1440 \text{ psf}$$

$$\sigma_{30} = \frac{0.48(675,000)}{30^2} = 360 \text{ psf}$$

c) Westergaard:

$$\sigma_{15} = N_w \frac{P}{Z^2} = 0.32 \frac{675,000}{15^2} = 960 \text{ psf}$$

$$\sigma_{30} = N_w \frac{P}{Z^2} = 0.32 \frac{675,000}{30^2} = 238 \text{ psf}$$

Newmark Influence Chart

A final method for computing vertical pressure σ_z that can accommodate a uniformly loaded area of any shape is provided by the Newmark influence chart. The point in the subsurface for which the stress level is desired may be inside or outside the loaded area.

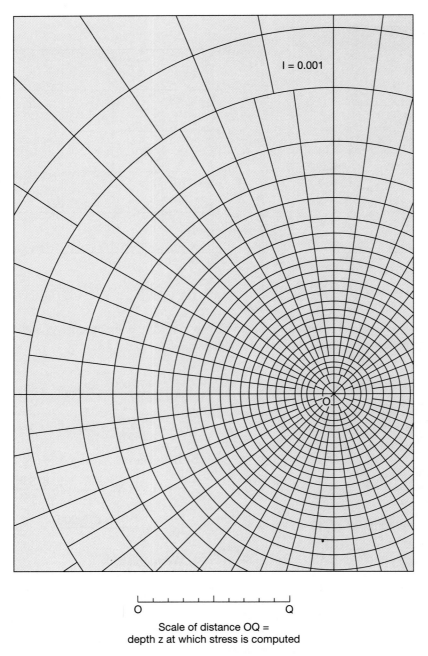

I = 0.001

O   Q

Scale of distance OQ =
depth z at which stress is computed

FIGURE **7.46** Newmark chart for vertical stress on horizontal planes at depth. (After Newmark, 1942.)

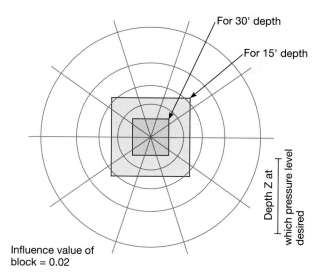

FIGURE 7.47 Example problem using the Newmark influence chart, rectangular loaded area.

This is accomplished by placing the point of interest over the center of the chart.

A Newmark influence chart is shown in Figure 7.46. The scale of the chart is set with respect to the depth of interest for which the stress level is being measured. On the chart the yardstick on the scale of distance is shown. This is set equal to the depth z, the distance below the surface for which the vertical stress σ_z is desired. This distance is used as a scale to draw the outline of the loaded area. The vertical stress is found according to the following equation:

$$\sigma_z = q_0 I \times \text{Number of blocks covered}$$

where q_0 is the contact pressure of the loaded area, I is the influence value specified on the chart, and the number of blocks is the individual small area segments that occur within the boundary of the contact area. As indicated above, the point in the subsurface for which the stress level is desired is set at the center of the chart.

EXAMPLE PROBLEM 7.12

You are given two different loaded areas, the first 15 × 15 ft as shown in Figure 7.47 and the second one an irregular shape as shown in Figure 7.48. The pressure on the first is 3000 psf and on the second is 5000 psf. Find the stress at 15 and 30 ft for both loaded areas for the first in the center and for the second at the inside corner.

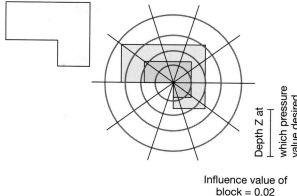

FIGURE 7.48 Example problem using the Newmark influence chart, L-shaped loaded area.

ANSWERS

For the rectangular-loaded area:

For a 15-ft depth, about 14.5 blocks are involved:

$$\sigma_z = q_0 I \times \text{Number of blocks}$$
$$= 3000 \text{ lb/ft}^2 (0.02) \times 14.5 = 870 \text{ lb/ft}^2$$

For a 30-ft depth, about 5 blocks are involved:

$$\sigma_z = q_0 I \times \text{Number of blocks}$$
$$= 3000(0.02)(5) = 300 \text{ lb/ft}^2$$

For the L-shaped loaded area:

For a 15-ft depth, about 19 blocks are involved:

$$\sigma_z = q_0 I \times \text{Number of blocks}$$
$$= 5000 \text{ lb/ft}^2 (0.02)19 = 1900 \text{ lb/ft}^2$$

For a 30-ft depth, about 9 blocks are involved:

$$\sigma_z = q_0 I \times \text{Number of blocks}$$
$$= 5000 \text{ lb/ft}^2 (0.02)9 = 900 \text{ lb/ft}^2$$

CONSOLIDATION

Consolidation is the reduction in volume of clays under external loading as water drains from the pores. This reduction of excess pore water pressure or

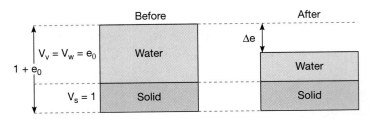

FIGURE 7.49   Derivation of consolidation equation.

dissipation of pore pressure depends on the permeability of the soil and, therefore, is a time-dependent phenomenon. Settlement takes place as the drainage occurs.

Consolidation can be illustrated using the phase diagram in Figure 7.49. The saturated case is involved. Considered in terms of e, for one-dimensional consolidation, the change in height as compared to the original height can be related as follows.

$$\frac{\Delta H}{H_0} = \frac{\Delta e}{1+e_0}$$

and

$$\Delta H = H_0 \frac{\Delta e}{1+e_0}$$

EXAMPLE PROBLEM 7.13

Given a very compressible soil layer 8 ft thick with an original void ratio of 1.1, but following consolidation it has a void ratio of 0.95, estimate the settlement of the soil layer.

ANSWER

$$\Delta H = \frac{\Delta e}{1+e_0} H_0 = \frac{1.1 - 0.95}{1 + 1.1}(8) = 0.57 \text{ ft} = 6.9 \text{ in.}$$

Laboratory Testing

A consolidation test can be performed in the laboratory to determine the compressibility of the clay. This is accomplished by compressing the soil under load in a device known as a consolidometer (Figure 7.50).

An undisturbed soil specimen is trimmed and placed in the confining ring. Porous stones are located above and below the sample. An applied load is added to the loading plate and the deformation of the

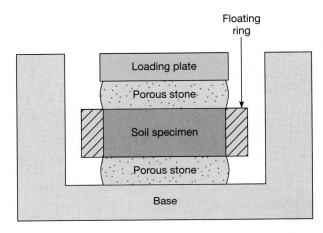

FIGURE 7.50   Schematic drawing of a floating ring consolidometer.

sample is carefully measured. The stress is determined by dividing the load by the area of the specimen.

Loads are added in increments, but after each increment the sample is allowed to consolidate until little or no further reduction occurs. This signals that very little if any excess pore pressure remains and the final stress is an effective stress. Load increments are added until sufficient data points are obtained to define the consolidation curve properly.

Typically, the void ratio is plotted against the logarithm of the effective consolidation pressure or effective stress (Figure 7.51). The slope of the virgin compression curve, as shown in Figure 7.51, is designated C_c or the compression index. It can be seen that

$$C_c = \frac{e_1 - e_2}{\log \sigma'_2 - \log \sigma'_1} = \frac{e_1 - e_2}{\log \sigma'_2/\sigma'_1}$$

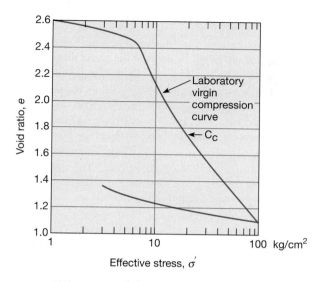

FIGURE 7.51 Consolidation test data, void ratio versus log effective stress.

The void ratio is dimensionless and the units cancel in σ'_2/σ'_1, so C_c is a dimensionless value.

If you wish to find C_c graphically, note that if you select $\log \sigma'_2/\sigma'_1 = \log 100/10 = \log 10 = 1$, then C_c is equal to Δe for that one log cycle.

EXAMPLE PROBLEM 7.14 _____

Find C_c for the Figure 7.51.

ANSWER

For the portion from 10, 2.1 to 43, 1.45

$$C_c = \frac{2.1 - 1.45}{\log 43 - \log 10} = \frac{0.65}{\log 43/10} = \frac{0.65}{\log 4.3} = 1.02$$

Alternately, using one log cycle for the portion 10, 2.1 to 100, 1.1

$$\Delta_e = 2.1 - 1.1 = 1.0 = C_c \text{ for 1 cycle}$$

If we next examine the two equations presented previously,

$$\Delta H = H \frac{\Delta e}{1 + e_0} \quad \text{and} \quad C_c = \frac{\Delta e}{\log(\sigma'_2 - \sigma'_1)}$$

solving the second equation for Δe yields

$$\Delta e = C_c \log(\sigma'_2 - \sigma'_1) = C_c \Delta \log \sigma'$$

but

$$\Delta \log \sigma' = \log(\sigma' + \Delta \sigma') - \log \sigma'$$
$$= \log \frac{\sigma' + \Delta \sigma'}{\sigma'} = \log(1 + \frac{\Delta \sigma'}{\sigma'})$$

and

$$\Delta e = C_c \log(1 + \frac{\Delta \sigma'}{\sigma'})$$

Finally

$$\Delta H = \frac{H}{1 + e_0} C_c \log(1 + \frac{\Delta \sigma'}{\sigma'})$$

EXAMPLE PROBLEM 7.15 _____

Given the following information, calculate the settlement of the clay: In a 10-ft-thick clay layer with a compression index of 0.95, the initial void ratio is 2.1 and the effective stress is 1550 psf. The pressure increase for the clay is 500 psf.

ANSWER

$$\Delta_H = \frac{H}{1 + e_0} C_c \log(1 + \frac{\Delta \sigma'}{\sigma'})$$
$$= \frac{10}{1 + 2.1} (0.95) \log(1 + \frac{500}{1500}) = 0.372 \text{ ft}$$
$$= 4.4 \text{ inches}$$

Many details of consolidation have not been included in this discussion because of the considerable extent of the subject. Overconsolidated clays and details of their settlement relationships have not been included. The time rate of settlement related to consolidation (how long it takes for settlement to occur) is another aspect of interest in soils engineering, which is not considered in this discussion. The reader is referred to textbooks on soil mechanics or geotechnical engineering for a detailed study of this subject matter.

EXERCISES ON ELEMENTS OF SOIL MECHANICS

1. *Grain Size Distribution Curves:* The following sieves were used to determine a grain size distribution; #4, #10, #20, #40, #60, #100, and #200. The weight of material retained on each sieve is provided in the following table:

U.S. Std. Sieve No.	Sieve Opening (mm)	Weight retained (g)	% Retained on Sieve	% Passing Sieve
4	4.75	12.7	6.2	93.8
10	2.00	24.8	12.1	81.7
20	0.85	21.5		
40	0.425	38.8		
60	0.25	23.8		
100	0.15	19.5		
200	0.074	46.7		
Minus No. 200		17.3		
		205.1 Total		

(a) Complete the table by calculating the values based on the information supplied.
(b) Using five-cycle semilog paper, plot the grain size distribution for this sieve analysis.
(c) Give the D_{60} and the D_{10} values for this sieve analysis. Calculate the k value for this analysis using Hazen's approximation (k in centimeters per second = $100 D_{10}^2$, D_{10} in millimeters).
(d) Calculate the coefficient of uniformity and the coefficient of curvature for this analysis. Is the sample well graded? Explain why or why not.

2. What is the difference between shape and roundness? Describe a cube and a sphere relative to these terms. What is an equant particle called in soil mechanics terms?

3. Given that $n = V_v/V_t$ and $e = V_v/V_s$ show that $n = e/(1+e)$.

4. The total volume V_t, the total mass M_t of a fully saturated soil, the dry mass M_s, and the specific gravity of the solids G_s are known. Draw the phase diagram for this situation and, using the given parameters, derive the equation $e = wG_s$.

5. The void ratio e, the specific gravity of solids G_s, and the degree of saturation S of a soil are known. In terms of these parameters derive an equation for ρ, the wet density. Begin by drawing the phase diagram.

6. The porosity n and specific gravity of solids G_s of a saturated soil are known. Derive the equation for ρ, the wet density, and for w, the water content, based on this information. Begin by drawing the phase diagram.

7. A moist sand sample has a volume of 40.5 cm^3 in its natural state and a mass of 50.2 g. When dry its mass is 48.3 g; the specific gravity of solids is 2.68. Determine the void ratio, the porosity, water content, and degree of saturation. Begin by drawing the phase diagram.

8. A sample of saturated clay has a mass of 1526 g in its natural state and 1053 g after drying. The value of G_s was found to be 2.70. Calculate w, ρ_d (the dry density), and ρ (the wet density) of the soil.

9. For saturated conditions of a soil, derive the equation for ρ', the submerged density in terms of n, G_s, and ρ_w. Derive the formula in two ways, the first based on the fact that $\rho' = \rho - \rho_w$ and the second using

$$\rho' = \frac{M_t - V_s\rho_w - V_w\rho_w}{V_t}$$

that is, the mass minus the buoyancy effect divided by the volume.

10. A sample of mica flakes has a dry mass of 145 g. The G_s for the sample is 2.82. In the loose state, the sample had a volume of 1000 cm^3. After vibration, the volume of the sample was reduced to 400 cm^3, and finally after vibrating the sample with a load placed on the mica flakes, the volume was reduced to 200 cm^3.
(a) For the three conditions described (loose, vibrated, vibrated/loaded) calculate the void ratio and porosity for each. Record the answers in a table of the following form.

Loose		Vibrated		Vibrated and Loaded	
e	n	e	n	e	n

(b) For the three conditions, calculate the dry density, saturated density, and submerged density for each. Record the answers in a table of the following form. Next show the unit weight for each in lb/ft^3 for the dry density, saturated density, and submerged density. Include these answers in the same table. Use kg/m^3 and lb/ft^3.

Loose			Vibrated			Vibrated and Loaded		
Dry	Sat.	Sub.	Dry	Sat.	Sub.	Dry	Sat.	Sub.

11. *Given:* $\rho_s = 2680$ kg/m^3, $\rho = 2050$ kg/m^3, $\rho_d = 1765$ kg/m^3. *Find: e, w, S, n,* and ρ'. Assume $V_t = 1$ kg/m^3 and use the appropriate phase diagram to illustrate the problem.

12. For a certain sandy soil, $e_{max} = 0.72$ and $e_{min} = 0.45$. In its natural state this soil was found to have a density of 2.01 Mg/m^3, a water content of 17%, and a G_s of 2.65. What is the relative density and the degree of saturation? Use the appropriate phase diagram to illustrate the problem.

13. *Given:* An unsaturated soil has a density of 1850 kg/m^3, a water content of 8.6%, and a G_s of 2.65. *Find:* The relative density D_r if $e_{max} = 0.642$ and $e_{min} = 0.462$.

14. The liquid limit of a soil sample is 35 and the plastic limit is 22. What is the plasticity index? What is meant by liquid limit, plastic limit, and plasticity index? Where would this soil plot on the plasticity chart shown in Figure 7.11? If the natural water content of the soil was 30%, what is the liquidity index? What does this value signify about the soil?

15. An inorganic clay with a liquid limit of 80 is estimated to plot directly on the A line. What would be the PI if it did plot on the A line? Referring to Figure 7.12, what clay mineral would likely make up the sample?

16. Distinguish between the terms *soil fabric* and *soil structure*. Why is only soil fabric important for cohesionless soils, whereas both soil fabric and soil structure are of interest to cohesive soils? What is a single-grain structure? Why are loose grain structures subject to sudden densification under vibratory loads?

17. The specific surface of a sphere equals $6/d$ where d is the diameter of the sphere. What would be the specific surface of a fine sand grain 0.05 mm in diameter?

18. List the types of interparticle bonds in clays. What is their strength relative to primary valence bonds? Over what distance do they act? An answer in the form of a table should prove useful.

19. What is meant by the term *cation exchange capacity?* Explain why the mineral kaolinite might have a value of 10 millequivalents per 100 g of sample whereas smectite could equal 140. In a water softener, Mg^{++} and Ca^{++} ions are replaced by the cation from the salt added. What is that cation?

20. What are the three possible configurations for clay structures? Under what circumstances do each of these occur? Which configuration would be most compact? Explain.

21. A clay sample was found to have a maximum strength of 800 lb/ft^2 and a residual strength of 150 lb/ft^2. Is this a sensitive clay? Explain.

22. Macrostructure and microstructure of clays are of importance. Bedding is one of the types of macrostructure; name several others. How are the effects of bedding indicated or manifested? Concerning microstructures, what are the different aggregates that have been designated for clays? List them from the smallest to the largest.

23. Distinguish between compaction and consolidation in terms of soil mechanics definitions. What are the construction objectives of compaction? What four variables control the extent of compaction?

24. Refer to Figure 7.16. What would 90% of the optimum density equal for this curve? If the specifications require 95% of standard Proctor density and within ±2% of the optimum water content, what range of values would be acceptable for this test?

25. Refer to Figure 7.17. Which is a greater density: 100% standard Proctor density or 90% modified Proctor density? What is the optimum moisture content for the standard and the modified Proctor tests? How much of a decrease does this indicate from the lesser compactive effort to the greater one?

26. For a dry density of 1795 kg/m^3 and a G_s of 2.65, what water content is required to yield zero air voids? What would be the wet density and void ratio for this situation?

27. Refer to Table 7.5 on the properties of compacted soils. Why are soils compacted dry of optimum more permeable than those compacted wet of optimum, assuming the same compactive effort? Refer also to the text. If a compacted soil dry of optimum had a permeability of about 10^{-6} cm/sec what would it be wet of optimum? On the dry side of optimum the compacted soil is also more sensitive to cracking. Why is this detrimental in earth dam construction? Explain.

28. A specific sheepsfoot roller has a coverage of 11% and a contact pressure of 600 psi. How many passes of the roller are needed before 100% coverage of the fill is obtained? Why is it important that the specified lift thickness in the fill not be exceeded? What would be a likely lift thickness allowed for this roller? Explain.

29. Distinguish between end-product and method specifications. Which of these requires more study prior to letting the bids? Explain.

30. What are the various methods used for field control of density and moisture content? Which of these is a nondestructive test? How does it work?

31. Given the following cross section, calculate the values related to effective stress. Use SI units exclusively:

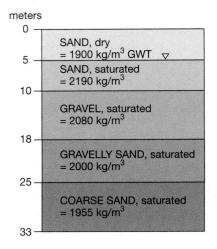

(a) Calculate σ, u, and σ' for depths of 5, 10, 18, 25, and 33 m.
(b) Complete the following table and prepare a drawing of the soil profile, showing σ, u, and σ' values, labeled and drawn to scale. (See Figures 7.23 and 7.24 as examples.)

Depth (m)	Total Pressure σ (kPa)	Neutral Pressure u (kPa)	Effective Pressure σ' (kPa)
5			
10			
18			
25			
33			

32. Given the following cross section calculate values for effective stress. Use British engineering units.

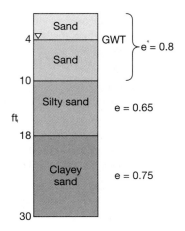

In the soil profile above, the ground water table occurs at a depth of 4 ft. Assume that $G_s = 2.65$ for all materials.
(a) Calculate the unit weight for the four different zones in the cross section. Assume the soil above the water table is completely dry and that below it is saturated.
(b) Calculate the total stress, neutral stress, and effective stress for depths of 4, 10, 18, and 30 ft. Supply the answers in table form similar to that in Exercise 31.
(c) Draw a pressure diagram of the soil profile showing total stress, neutral stress, and effective stress. Draw to scale.

33.

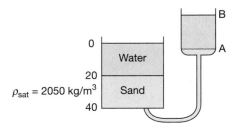

A standpipe consists of 20 cm of water and 20 cm of saturated sand below. A column of water held at a constant head is attached to the base of the standpipe as shown.
(a) Consider that the water level is maintained at location A, 8 cm below the water level in the standpipe. Calculate the total, neutral, and effective stresses at 20 and 40 cm for the standpipe. Draw a pressure diagram to the right of the standpipe showing this relationship.
(b) Consider that the water level is maintained at location B, 10 cm above the water level in the standpipe. Calculate the total, neutral, and effective stresses at 20 and 40 cm for the standpipe. Draw a pressure diagram showing this relationship.
(c) Calculate what is needed to yield a quick condition at the base of the sand column. What is required of the neutral stress and of the effective stress for this to occur?

34. Give the equation for the critical hydraulic gradient i_{cr} in terms of submerged density of the soil mass and density of water. What would be the critical hydraulic gradient for a 5-ft-thick layer of barite ($BaSO_4$) placed to reduce uplift effects? Barite has a specific gravity of 4.5; assume the barite layer, consisting of a gradation of gravel-sized pieces, has a porosity of 30%.

35. A gravelly sand channel slopes away from a reservoir, through the reservoir rim, to an adjacent drainage area.

The channel slopes at a gradient of 75 ft vertically to 100 ft horizontally. In cross section the channel is 6 ft thick and 15 ft wide. The channel intersects the adjacent drainage area at a horizontal distance of 500 ft from the reservoir.

(a) At minimum pool elevation in the reservoir the water is 5 ft above the sandy gravel channel. What is the total pressure head in the gravelly sand seam at the upstream end (or at the reservoir)? What is the total pressure head at the downstream end of the channel where it daylights into the adjacent drainage area 500 ft away? What is the hydraulic gradient? What is the average velocity of flow if $k = 10^{-2}$ cm/sec? What is the flow loss in cubic feet per second and in gallons per minute for the channel (use Darcy's Law, $v = ki$ relationship, see also Chapter 15, Ground-Water Geology)? How long does it take for the water to flow from the reservoir to the adjacent drainage area?

(b) At maximum pool the water stands at 50 ft higher than minimum pool. What is the flow loss in cubic feet per second and gallons per minute for this case? Does a quick condition occur when the pool reaches this maximum level? If not, what is required to cause a quick condition? Explain.

36. Explain why the sediment on the ocean bottom can be just as loosely compacted when it lies at a water depth of 1000 ft as it can be at a water depth of only 10 ft. What is required to compact the sediment? Explain.

37. This question pertains to the Mohr circle procedure for shear strength of soils. The major and minor principal stresses at failure for a point in a cohesionless sand are 65 and 20 psi, respectively. Begin by drawing the Mohr diagram.

(a) What is the value of the maximum shearing stress at that point?

(b) What is the value of the normal stress on that plane of maximum shearing stress?

(c) What are the normal and shearing stresses on a plane 27° from the plane of major principal stress? What is the value of the resultant stress on this plane?

(d) What is the value of σ and τ on a plane 30° from the plane of minor principal stress? What is the resultant stress on this plane?

(e) What is the value of the maximum obliquity angle? This angle is α_m or ϕ.

(f) Using the relationships

$$\sigma_{failure} = \sigma_3(1+\sin\phi)$$

$$\tau_{failure} = (\sigma_1-\sigma_3)\,1/2\,\cos\phi$$

find σ_f and τ_f.

(g) If $\theta_{failure} = 45 + 1/2\phi$, find θ_f. Show all relationships on the Mohr diagram.

38. Triaxial test results on a clayey silt indicate a c value of 900 lb/ft^2 and a ϕ value of 28°.

(a) Draw a Mohr envelope for the soil.

(b) If the effective stress σ' on this material was 30 psi, what would the maximum shear stress be under failure conditions?

(c) Assume that the water table occurs at the ground surface of the clayey silt deposit under study. The saturated unit weight was found to be 130 lb/ft^3. At what depth would a vertical normal effective stress of 30 psi occur? Show calculations.

39. The unconfined compression strength q_u for a highly plastic clay is 2000 lb/ft^2.

(a) What is its q_u in psi, kg/cm^2, and N/m^2?

(b) What is its cohesive strength in these units?

(c) What is its ϕ value? Why?

(d) If the unit weight of the clay is 125 lb/ft^3, at what depth is the overburden stress σ_v (vertical stress) equal to the cohesive strength?

(e) At a depth of 32 ft, what is the ratio of overburden stress to compressive strength?

(f) What effect would this ratio have on the ability to tunnel through this clay at a depth of 32 ft?

40. A sandy deposit ranges from a clean sand to a slightly clayey sand. For the clean sand, failure conditions were found to be $\sigma_1 = 90$ psi, $\sigma_3 = 25$ psi, and $c = 0$. For the clayey sand, failure conditions were $\sigma_1 = 92$ psi, $\sigma_3 = 25$ psi, and $c = 8$ psi.

(a) Find ϕ, θ_{cr}, σ_{ff}, and τ_{ff} for the clean sand. Use $c = 0$ in this calculation.

(b) Find ϕ, θ_{cr}, σ_{ff}, and τ_{ff} for the clayey sand. Use $c = 8$ in this calculation.

41. A fine-grained granite was tested in a series of triaxial tests. A q_u value of 30,000 psi was also determined. For the triaxial tests, a ϕ value of 55° and an S_0 value (τ intercept) of 8000 psi was obtained. Draw the Mohr circle for the q_u test. Why does the q_u diagram not fall in line with the Mohr envelope for the triaxial test? Explain.

42. A footing 10 × 10 ft is designed to carry a 220-ton load. Find the stress level at 0, 5, 10, 15, and 20 ft using the 2 : 1, Boussinesq, and Westergaard methods. Compare the answers obtained. Which would likely be used for checking the adequacy of the soil? Explain.

43. If a footing 8 × 12 ft is used instead of the footing in Exercise 42, with the same 220-ton load, find the stress level at a depth of 10 and 20 ft using the Newmark influence chart from Figure 7.46.

44. Convert Exercise 42 to a problem with SI units. Use kilograms, meters, and N/m^2 as units. Calculate an-

swers for the same depths but use only the Boussinesq method.

45. *Consolidation of Clay:* Given the following drawing and the accompanying information, what is the ultimate settlement in the clay under the applied load?

P, 125 Tons

Sand, γ_d = 105 lb/ft^3 ——— GWT

5'

Silty sand, γ_{sat} = 127 lb/ft^3

16'

Clay, γ_{sat} = 121 lb/ft^3, n = 0.629, C_c = 0.82

22'

Assuming the sand will not settle, determine the ultimate settlement in the clay using:

$$\Delta H = \frac{H}{1+e_0}C_c \log(1+\frac{\Delta\sigma'}{\sigma'})$$

Proceeding in steps:

(a) Determine the effective stress for the cross section prior to loading. Show the stresses at depth using an appropriate diagram.

(b) What is $\Delta\sigma'$ on the clay layer caused by the imposed load? (*Hint:* Calculate this for the center of the clay layer at 19 ft deep.)

(c) What is the value of e_0 based on the original porosity of the clay?

(d) How much settlement will occur in the clay?

Additional Readings

DAS, B.M., 1979, *Introduction to Soil Mechanics,* Iowa State University Press, Ames.

DUNN, L.S., ANDERSON, L.R., and KIEFER, F.W., 1980, *Fundamentals of Geotechnical Analysis,* John Wiley & Sons, New York.

HOLTZ, R.D. and KOVACS, W.D., 1981, *An Introduction to Geotechnical Engineering,* Prentice Hall, Englewood Cliffs, New Jersey.

HOUGH, B.K., 1957, *Basic Soils Engineering,* Ronald Press Co., New York.

LAMB, T.W., 1951, *Soil Testing for Engineers,* John Wiley & Sons, New York.

LAMB, T.W., 1958, The Structure of Compacted Clay, *Soil Mechanics and Foundation,* Amer. Soc. Civil Engrs., Vol. 84, No. 2.

LEONARDS, G.A., 1962, Ed., *Foundation Engineering,* McGraw-Hill, New York.

NEWMARK, N.M., 1942, *Influence Charts for Computation of Stresses in Elastic Formations,* University of Illinois Engineering Experiment Station Bulletin, Series No. 338, Vol. 61, No. 92, Urbana.

PERLOFF, W.H. and BARON, W., 1976, *Soil Mechanics, Principles and Applications,* Ronald Press Co., New York.

SCHROEDER, W.L., 1980, *Soils in Construction,* 2nd ed., John Wiley & Sons, New York.

TAYLOR, D.W., 1948, *Fundamentals of Soil Mechanics,* John Wiley & Sons, New York.

TERZAGHI, K. and PECK, R.B., 1948, *Soil Mechanics in Engineering Practice,* John Wiley & Sons, New York.

8

Rock Weathering and Soils

CHAPTER OUTLINE

Most rocks are formed under conditions drastically different from those at the Earth's surface. Minerals in igneous and metamorphic rocks reach equilibrium at elevated temperatures and generally must also adjust to high confining pressures. Sedimentary rocks are compressed by overburden stresses at temperatures that increase with depth. This compression yields adjustments in the clay constituents, overgrowths on certain crystals, and deposition of cementing minerals between grains. By contrast, in the near-surface environment, low temperatures prevail, there is little or no confining pressure, and water and oxygen are abundant. Consequently, rocks at the surface will undergo changes. This mechanical and chemical breakdown of solid rock (bedrock) to form soil or unconsolidated material is known as weathering.

Mechanical weathering involves the physical disintegration or degradation of rock pieces without a change in composition. The size of particles is simply reduced. This may occur by the disaggregation of separate crystals in a rock or by cross-fracturing of massive, fine-grained rocks.

Chemical weathering is accomplished by decomposition whereby one mineral species is changed into another through various chemical processes. Water commonly plays a major role in chemical weathering.

FACTORS CONTROLLING WEATHERING

Role of Water

The amount of water available to the rock is the primary ingredient controlling the type of weathering. To be sure, the rock mineralogy is important but, by and large, if abundant water (not ice) is available, chemical weathering will prevail. In the absence of water, mechanical weathering takes control.

Water availability can be related to climate (humid versus arid), elevation (high altitudes yield ice formation), and latitude (high latitudes also yield water in the frozen state). This is why mechanical weathering prevails in the deserts and rapid chemical weathering in the tropics. In mountainous regions, blocks of broken rock line the slopes and the headward areas of alpine glaciers persistently supply rocks through mechanical breakage. The high latitudes yield colder weather and a lower occurrence of liquid water. Again, mechanical weathering prevails.

Water's role in chemical weathering is that of universal solvent. Most chemical reactions on the Earth's surface take place in water because it provides oxygen for reaction and mobility for the ions. With water present, an increased temperature typically yields an increased rate of reaction. This explains the high chemical weathering rate of the tropics, but without water, elevated temperatures do not have this effect. Therefore, mechanical weathering prevails in desert areas.

Topographic Expression

When mechanical weathering prevails, as in arid regions, sharp angular topography develops. In sedimentary terrains sandstones and limestones (including dolomites) form cliffs. Because water is not abundant, the carbonate rocks are not subject to solution but persist as resistant rocks. The Grand Canyon in Arizona illustrates dramatically the angular forms that develop in an arid region. The angular topography of the Grand Canyon is illustrated in Figure 8.1.

By contrast, when chemical weathering prevails, rounded, softened contours develop. Sandstone still forms cliffs in wet regions, but limestones (and dolomite) and shales form slopes. The Appalachian Plateau is a region of horizontal sedimentary rocks that have weathered chemically. The outcrops are more rounded and soil accumulation is deeper than in the arid regions, such as Arizona. Also thick vegetation, another consequence of humid conditions, blankets the slopes. In these areas the valleys are typically bottomed by shale and limestone, and the ridges are capped with sandstone.

PROCESSES OF MECHANICAL WEATHERING

Several processes act to reduce the size of massive rock. The primary factors include release of confining stress at the surface, differential expansion and con-

FIGURE 8.1 Angular topography of an arid region in the Grand Canyon.

traction of rock constituents, and the wedging action of water freezing in rock fractures. This last condition, known as frost wedging, will propagate cracks through rock because of the expansion that occurs when ice crystals form. Each freezing and thawing cycle can extend the fracture until a rock piece is broken from the mass. Most areas of the world experience periods during the year when frost wedging can operate to disintegrate rock.

Frost wedging is caused by a 9% increase in volume that occurs when water freezes to form ice. The force generated by each freezing cycle is about 1550 psi (225 kPa), which represents from about 5% to 50% of the unconfined compressive strength of most rocks. This stress is exerted repeatedly with each freeze-thaw cycle until the rock is broken apart.

The conditions required to cause extensive frost wedging effects are 1) a supply of moisture; 2) fractures, bedding planes, or other openings in the rock for water to enter; and 3) temperature fluctuations across the freezing point of water. Frost wedging is particularly effective in alpine areas where snowmelt forms during the day and refreezes at night after seeping into rock fractures. It is less prevalent where water is permanently frozen or in desert areas where water is scarce.

Piles of angular rock pieces, known as talus, accumulate at the base of steep cliffs as a product of frost wedging. Finer-sized materials downslope from the talus slopes yield half-circle-shaped features called alluvial fans. They are formed by surface runoff, which carries the finer sediment from the steep cliffs. These deposits are discussed in more detail in Chapter 14 on landslides and related phenomena.

Differential Expansion and Contraction

Differential expansion and contraction develop between the individual minerals in a coarse-grained rock when heating and cooling occurs. The coefficient of thermal expansion is not the same for all minerals so when rocks are heated and cooled the mineral grains are subjected to differential stresses. In some cases this will induce and propagate fractures in rocks whereas in others, individual mineral grains are disaggregated from the mass. Known as granular disintegration, this process is particularly prevalent in coarse-grained, quartz-rich rocks. Quartz has a larger coefficient of expansion than other common minerals

and it is highly directional, with the greatest expansion occurring along the axis of elongation of mineral crystals (the *c* axis). Consequently, intrusive igneous bodies in desert environments tend to weather by granular disintegration, yielding a volume of loose, coarse-sand size crystals of quartz, feldspar, and hornblende. These deposits are known as grus.

Sheeting and Exfoliation

Release of confining stresses in a rock mass exposed at the surface allows the rock to expand elastically. This occurs in all rock masses and it is the reason why joints in rocks open when the rock is exposed, or conversely why joints tend to tighten at depth. Exfoliation is the term used for the action that takes place when rocks break loose from the surface along a roughly parallel fracture. Massive igneous and metamorphic rocks with high residual stresses are particularly prone to this occurrence. For dimension-stone quarries in granite this is known as sheeting and special care is taken to prevent its occurrence if possible. In such quarries, large blocks of rock are processed for the facing of buildings or for gravestones, and sheeting can ruin valuable stone. Quarries may be developed in preferred directions designed to minimize the sudden stress release, or the rock may be kept in compression by long rock anchors that prevent stress release.

Exfoliation is common in massive igneous rock bodies such as stocks and batholiths. It can be observed in Yosemite National Park, California, which is located within the Sierra Nevada Batholith. Half Dome is an exfoliation form that occurs on El Capitan, the prominent wall of rock exposed in Yosemite by alpine glaciation. Exfoliation preserves the vertical face of rock first carved by glacial ice moving down the valley. Exfoliation in Yosemite National Park is shown in Figure 8.2.

Spheroidal Weathering

Spheroidally weathered boulders are rounded, large rocks of equal dimensions that form in place on a bedrock surface by mechanical weathering. It is likely that they result from a combination of pressure relief, frost wedging, and expansion of feldspar crystals when weathered to kaolinite. Joints and exfoliation planes form the initial boundaries in the rock mass, then pressure relief and chemical weathering along the surface cause an inward-developing skin of weathered rock to form.

FIGURE 8.2   Exfoliation of massive rock in Yosemite National Park.

Other Mechanical Weathering Processes

Several other mechanical weathering processes act to a lesser extent to disintegrate rocks. Erosion by wind and running water mechanically abrades rock masses, but these agents work primarily to move soil or sediments that have already been disaggregated. Root wedging by plants can contribute to mechanical weathering. This activity is observed in some of the most massive rock types such as granite, quartzite, and sandstone where roots of pines and other trees penetrate what seems to be solid rock to obtain support and sustenance.

Forest fires can cause mechanical breakdown of rocks because of the great expansion that occurs at elevated temperatures. From the time of early humans up until the sixteenth century, fire and water-quenching were used as a mining procedure to break rocks loose and expose valuable minerals. Only after the advent of explosives was this procedure discontinued.

PROCESSES OF CHEMICAL WEATHERING

Mechanical weathering or disintegration, in a sense, aids chemical weathering because it exposes additional surfaces to air and water. This additional exposure is only effective, of course, if water is available (in liquid form) to foster the chemical reaction.

Surface Area Effects

Chemical weathering is basically a surface phenomenon; therefore, the greater the surface area exposed, the more intense the reaction. Increased surface area results when large blocks are subdivided into smaller ones. The total volume remains the same but the surface area increases logarithmically. A measure of this feature is the specific surface or area to volume ratio, and it also increases logarithmically as the blocks are subdivided.

The relationship between length of sides for cubes and their surface area is shown in Figure 8.3. From the graph it can be determined that a 1-cm cube, when divided into eight smaller cubes, which are in turn subdivided in like fashion for a total of 13 times, will yield cubes 1 μm on a side (10^{-6} m). This provides about one trillion (10^{12}) cubes with a total surface area of 5 m^2 or about ten thousand times (10^4) that of the original. It is no surprise, then, based on this example that the subdivision of rock pieces by mechanical means greatly accelerates chemical weathering.

Solution, Oxidation, Hydrolysis, and Effects of Plants

Water is a polar liquid that has the ability to dissolve both ionic and organic substances to a certain degree. The water molecule is illustrated in Figure 8.4. Because of its dipolar nature water is able to dissolve many chemical compounds. In addition to the solution effect, water aids decomposition through acid action, oxidation, and hydrolysis. Acids in the soil are rather weak. Carbonic acid forms when carbon dioxide in the air reacts with rainwater in the following manner:

$$H_2O + CO_2 \rightarrow H_2CO_3 \leftrightarrows H^+ + HCO_3^-$$

As water percolates downward through the soil, concentrations of the acid are greatly increased when carbon dioxide is provided by decaying organic mat-

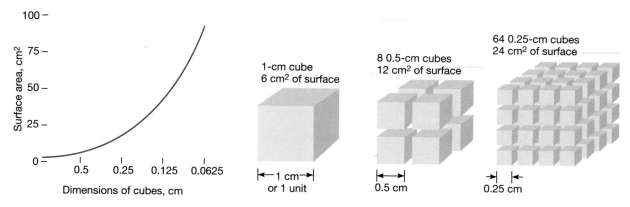

FIGURE **8.3** Cube dimensions versus surface area.

ter. Hydrogen ions (H^+) are extremely effective in decomposing minerals. They begin by neutralizing the broken bonds at the edge of the mineral. This is illustrated by the decomposition of potassium feldspar (orthoclase):

$$2KAlS_3O_8 + 2H^+ + H_2O \rightarrow 2K^+ + Al_2Si_2O_5(OH)_4 + 4SiO_2$$

(Potassium   (Hydrogen ion) (Water)   (Potassium   (Kaolinite)   (Silica)
feldspar)   ion)

In this chemical equation the reaction between a mineral and water to yield a secondary mineral is known as hydrolysis.

Oxidation of sulfides, native metals, and ferrous oxides occurs when rocks weather. Oxygen is dissolved in the water, which supplies the needed mobility for the reactants.

Soluble materials including some dissolved silica are carried away by subsurface water. These include potassium, sodium, magnesium, and calcium cations plus chloride, carbonate, and sulfate anions. Iron and aluminum oxides, clay minerals, and some silica are left behind. This process, known as leaching, involves the removal of soluble materials by water from the bedrock or weathered soil.

The contribution of plants to rock weathering bears further consideration. Roots contribute to mechanical weathering by wedging through fractures in the rock. Chemical weathering is promoted by plants through the presence of hydrogen ions around their roots when they are alive. Following this, on decay of the organic matter, more hydrogen ions are supplied for chemical weathering.

Resistance to Weathering

Some minerals are more susceptible to chemical weathering than others. This feature is related to their temperature of formation, which has a direct influence on the mineral structures that develop on cooling. Silicate minerals, ranging from a high temperature of formation to one of low temperature, portray an increased sharing of oxygens with decreasing temperature of formation. This greater oxygen sharing provides those silicates formed at lower temperatures with a greater resistance to chemical weathering. In Figure 8.5 Bowen's reaction series, showing the order of formation of silicate minerals from a silicate melt, is compared to Goldrich's mineral-stability series. The similarities are obvious and illustrate the greater resistance of orthoclase, muscovite, quartz, and clay minerals to chemical attack.

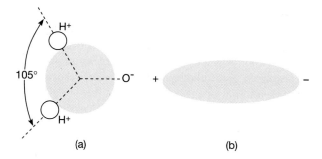

FIGURE **8.4** (a) Diagram of a water molecule and (b) shown as a dipole.

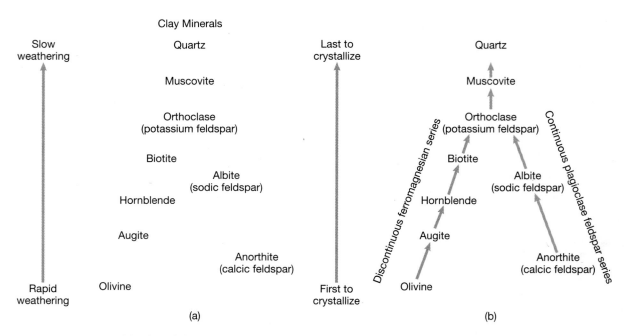

FIGURE 8.5   (a) Goldrich stability series compared to (b) Bowen's reaction series.

In addition to quartz, other less common minerals are also resistant to weathering. Some precious metals (such as gold and platinum) plus ilmenite (a titanium ore), cassiterite (a tin ore), zircon, and diamond persist after rock weathering. The metallic ores also concentrate by differences in particle settling velocity during stream flow because of their higher density to yield what are known as placer deposits. Subsequently, these can be mined with great success. In fact, many of the gold strikes of Colorado, Montana, California, and Alaska occurred as placers.

The effects of weathering on different rocks and the rates involved are of particular interest. The Goldrich mineral-stability series provides insight into the relative rates of chemical weathering. For all practical purposes only minor amounts of quartz are decomposed in humid climates, but most other common silicates are ultimately destroyed under humid conditions. In like manner limestone and dolomite (plus the more soluble rock gypsum and rocksalt) are dissolved by the action of ground water. Only impurities are left behind, consisting of clay minerals, some silica, and iron oxide. This oxide in the ferric state yields the orange to brick-red color of terra rosa, so common in soils derived from limestone.

Effects of Climate

As stated previously, in arid climates, weathering results from mechanical processes, which dominate in the absence of water. Rocks simply become smaller in size or disaggregate into their constituent minerals. This occurs in deserts and also in high altitudes and high latitudes where liquid water is not generally available.

In humid tropical areas chemical weathering is even more pronounced. One consequence is the increased pH of the soil and of soil water. Under these conditions silica becomes much more soluble than in other humid areas and consequently quartz also will be removed from the soil profile during weathering. The residual soil, laterite, is red in color and rich in iron and aluminum oxides. It is soft and earthy when formed but becomes extremely hard on exposure. A crust-like layer, 15 ft (5 m) or more thick, may develop in places.

Bauxite is another form of deeply weathered soil. It is common to the tropics and subtropics. Bauxite contains mostly hydrous aluminum oxide with little iron oxide or silica. Apparently bauxite is derived from a parent rock with a low iron content because of the absence of iron in the residuum. In view of the

TABLE 8.1 Weathering Details for Rocks Under Various Climatic Conditions

Rock Type	Minerals Present	Weathering Conditions	Products
Granite	Quartz, orthoclase	Humid	Quartz pieces, clay, oxides
Gabbro	Calcic plagioclase, olivine	Humid	Iron oxide and clay
Quartz sandstone	Quartz	Humid	Quartz pieces
Granite	Quartz, orthoclase	Arid	Pieces of quartz and orthoclase
Cherty limestone	Calcite, chert, impurities	Humid	Chert, iron oxide, clay
Granite	Quartz, orthoclase	Humid, tropical	Clay, oxides
Muscovite schist	Muscovite, hornblende	Humid	Muscovite pieces, clay, oxides

low mobility of iron oxide when present, it would persist in the soil if contained in the parent rock. A feldspar-rich granite is probably the original rock from which the silica is removed under tropical weathering conditions to yield bauxite. Bauxite, $Al_2O_3 \cdot nH_2O$ is the primary ore of aluminum.

With this background information on chemical weathering, predictions can be made concerning the end products of weathering for specific rock types under different climatic conditions. The mineral constituents of the rock provide the needed information for this analysis. Examples that depict the various possibilities are shown in Table 8.1.

SOIL PROFILES

One consequence of weathering is the formation of the soil profile, the material so vital to human existence. It is certainly one of the primary resources of any country and has much to do with the agricultural base that can be established.

Definitions for Soil
Engineering Definitions
The term *soil* is used in several technical fields, each with a different definition. In civil engineering, soil is the earth material that can be disaggregated in water by gentle agitation, whereas from a construction viewpoint it is material that can be removed by conventional means, that is, without blasting. Both definitions have a practical orientation because the behavior of the material under a specific condition determines how it is classified. These definitions compare favorably to the geologic term regolith which includes all the loose material lying above bedrock.

Agronomy Definition
In soil science, agronomy and agriculture, soil consists of the thin upper layers of the Earth's crust formed by surface weathering that are able to support plant life. Many geologists adopt this definition as well. However, in the engineering geology specialty where soil and bedrock are commonly considered as construction or foundation materials, the practical definitions for soil as provided in the previous subsection are used.

Bedrock
In an engineering sense, material lying below the soil is bedrock. It can be defined more completely as the solid, continuous mass of rock lying below the soil or exposed at the Earth's surface that must be removed by blasting. Other terms for soil and bedrock in the engineering sense are unconsolidated material (or overburden) and consolidated material, respectively. Again the meanings contrast loose material versus a lithified rock mass because unconsolidated in this sense signifies nonlithified material.

Soil Versus Soil Profile
Differences in the meaning for soil can be resolved for the engineer or engineering geologist by referring to the thin upper, weathered layers as the soil profile and the total nonlithified material as soil. The soil profile consists of three major horizons A, B, and C. These horizons are sometimes referred to as zones, which more accurately indicates their nature. The A horizon is the top soil or the zone of leaching from which downward percolating water has removed some clays and soluble ions. It also is commonly rich in organic matter. The B horizon is the subsoil or

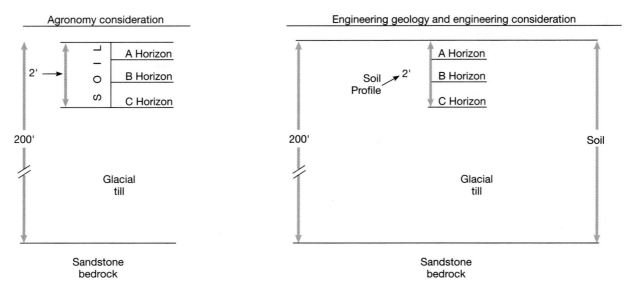

FIGURE 8.6   Illustrations of the term soil.

zone of accumulation. Clays are more prevalent here and organic matter is less abundant. The C horizon marks the transition from the soil profile to the unweathered parent material below.

Distinctions between meanings for soil become more apparent when an example is considered. Figure 8.6 shows an idealized cross section of a soil profile developed on glacial till. (Glacial till is the unsorted debris deposited directly by a melting glacier without being reworked by water.) The left section demonstrates the agronomy aspects of soil, and the right shows the engineering considerations. Soil in the engineering sense extends for 200 ft, the entire depth to the sandstone bedrock for the drawing on the right side. By contrast, soil only pertains to the upper 2 ft or so for the agronomy cross section.

Consideration of the unconsolidated material as strictly soil material is not without its problems. Take, for example, the glacial till in Figure 8.6. When drilling and sampling the till in a subsurface investigation for a building foundation, the origin or geologic nature of the material (such as "most likely Wisconsin till") should be noted on the drilling log as well as its soil description ("silty clay with traces of pebbles"). This geologic information should prove useful in making correlations across the site and decisions considering the nature of the soil mass.

Figure 8.12, near the end of this chapter, is a drilling log for a soil boring.

FACTORS CONTROLLING SOIL PROFILE DEVELOPMENT

The soil profile is an end result of surface weathering brought about primarily by downward percolation of water. As previously stated, the A horizon is the zone of leaching where organic material may be abundant and the B horizon is the zone of accumulation, mostly of clays left behind after water migration. Figure 8.7 shows typical soil profiles developed on glacial till and on granite bedrock. Saprolite consists of a combination of the granite (parent rock) and residual soil.

The most significant factors controlling soil profile development are parent material, climate, time, topography, and vegetation. Parent material is the original substance from which the soil zones develop. It controls, at least in the early stages of formation, much of the specific details involved with the soil layers such as permeability, acidity, and chemical composition. The two primary subdivisions for parent material are 1) bedrock weathered in place and 2) transported regolith or unconsolidated material, in other words, transported soil in the engineering sense.

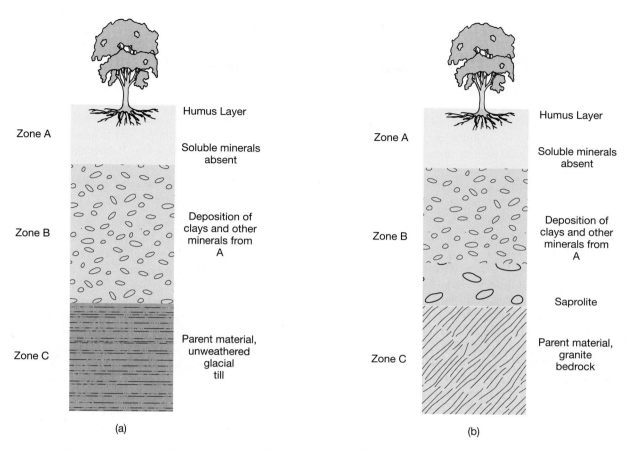

Zone A — Humus Layer

Zone A — Soluble minerals absent

Zone B — Deposition of clays and other minerals from A

Zone C — Parent material, unweathered glacial till

(a)

Zone A — Humus Layer

Zone A — Soluble minerals absent

Zone B — Deposition of clays and other minerals from A

Saprolite

Zone C — Parent material, granite bedrock

(b)

FIGURE 8.7 Typical soil profiles: (a) glacial till and (b) granite bedrock.

Parent Material

Weathered Bedrock

The bedrock parent material can consist of a variety of rocks in sedimentary, igneous, and metamorphic terrains. Shale, limestone, and sandstone dominate in the sedimentary areas, whereas many igneous rock domains are granite to granodiorite in composition, representing intrusive bodies, or basalts and andesites for the extrusive situations. In metamorphic rock areas, gneiss, schist, and phyllite are common parent materials. Each rock type weathers by chemical and/or mechanical means, depending on climatic conditions, to yield the soil. In most cases the overall mineralogy of the rock is sufficient to provide a range of chemical constituents in the soil profile. In some, however, the range is quite limited, for example, in quartz-rich rocks such as quartz sandstone, quartzite, and some siltstones. These residual soils may be generally lacking in chemical diversity, which reduces their ability to support vigorous plant growth.

Clay Minerals

Parent material seems to be the primary factor controlling the type of clay prevailing in a soil profile. The clay mineral prevalent in the parent rock is also dominant in the upper soil horizons. Illite appears to dominate in marine shales, whereas kaolinite and illite are prominent in freshwater sediments. Montmorillonite (smectite) is abundant only in the relatively recent sediments, typically Cenozoic in age, and is particularly common in rocks containing pyroclastic rocks, that is, volcanic ejecta.

Some influence of climate on clay mineral stability does seem to exist, however. Although illite is widely distributed, kaolinite seems restricted to areas of high rainfall. To a lesser degree, montmorillonite is associated with soils of a dry climate. Kaolinite development seems to be favored in acid or low pH soils where good drainage and active leaching prevail. The K, Na, Mg, and particularly Ca ions must be removed quickly from the soil as soon as they weather from the parent material if kaolinite is to prevail. These are mobil ions that are easily removed by water, so high rainfall is conducive to kaolinite formation. Formation of illite seems favored by the presence of potassium and to a lesser degree sodium. Montmorillonite develops in the presence of sodium and potassium but magnesium seems to be an essential ingredient. Clay minerals are also discussed in Chapter 2.

Transported Soils

Transported soil is soil that has been carried from one area and deposited in another. It consists of a variety of particle sizes, ranging from clay to gravel size and in some cases even boulders. The transporting agents are those that carve the land surface in the erosive processes. Running water, wind, glacial ice, and gravity are the most important agents. Littoral drift along seacoasts or major lakeshores is another agent. Glacial ice provided the thick blanket of deposits by continental glaciation in the north central section of the United States as well as the smaller scale deposits in mountainous regions of the western states. Gravity supplies the landslide and colluvial deposits that lie adjacent to most slopes, and the wind yields mostly sand dunes and loess deposits. Water, of course, provides the vast deposits in stream beds and floodplains but also the slope wash that tops most sloping areas.

Climate

Climate is extremely important in soil profile development. The relative abundance of water in liquid form is dictated by the climate and it is this water, moving downward from the surface, that develops the anisotropic or nonisotropic layers of the soil profile. An isotropic substance is one that has the same properties in all directions, which is in direct contrast to conditions in a soil profile with its obviously layered nature. The annual rainfall and annual temperature range determine whether an area is termed hot or cold, humid or arid. Evaporation rate is also determined by these climatic conditions, which also determine the local vegetation.

Types of Soil Profiles

The prevailing climate in a region has a great influence on the physical, chemical, and biological processes that act on the soil. In warm, wet climates organic activity is high and frost action is absent. In dry regions the dissolving and removal of carbonates, normally quite soluble materials when water is present, are greatly inhibited and instead carbonate precipitation occurs within the B horizon. Different major soil groups are formed under contrasting climatic conditions. The three most important groups are pedalfers, pedocals, and laterites.

PEDALFERS   A pedalfer is a soil in which much clay and iron have been added to the B horizon. They occur in about the eastern half of the United States where the annual precipitation is 25 in. (63 cm) or greater. This is a humid region with a moderate to cold climate. In pedalfers the soluble carbonates have been removed and aluminum, iron, and colloidal clay have been carried downward into the B horizon. Podsols are pedalfer soils that have been intensely leached by humic acid. The upper part of the B horizon has an ash gray color in podsols, in contrast to their dark humus-rich A horizon.

There are several other varieties of the pedalfer group in addition to podsols. Included are the red and yellow soils of the southeastern United States and the gray brown podsolic soils of the northeastern quarter of the United States and southeastern Canada. This latter group was formed in relatively cold regions under heavy forest cover where humus accumulates in the soil. Prairie soils are transitional between the pedalfers in the east and the pedocals in the west.

PEDOCALS   Pedocals are soils that contain an accumulation of calcium carbonate. These soils are found where the temperature is relatively high, the rainfall is slight, and the vegetation is mostly grass or brush. The pedocals typically occur in the western half of the United States where the annual precipitation is less than 25 in. (63 cm). Pedocals contain an accumulation of calcium carbonate in the B horizon. This may range from a few scattered concretions to a solid layer of $CaCO_3$ known as caliche. The depth below the ground surface to the caliche layer is a function of precipitation: Caliche lies deeper in high rainfall areas and nearer the surface in drier ones.

Soils containing abundant calcium carbonate are alkaline in nature, which makes them conducive to the growth of grass instead of trees. Moist, flat areas are typical sites for grassland prairies. A mat of accumulated humus in the A horizon develops as a consequence of the continued grass cover yielding a dark brown upper zone and a lighter B zone. This feature typical of prairie soils or chernozems is in contrast to forest soils, which have a lighter colored, A horizon.

Desert areas lack the needed moisture to maintain a grass cover, and humus is essentially absent in the A horizon. Therefore, the soil cover is determined by the color of the minerals that persist. Ferric iron oxide, stable where the humic acids of vegetation are absent, yields a red color but the other mineral colors are light in color including caliche. Typically, desert soils are gray or tan but the red or brown influence of iron oxide is sometimes evident.

LATERITES Laterites, the third group of soil types, were considered briefly in the discussion on the effects of climate. These are the deep red soils of the tropics in which all silicates have been completely weathered away—even silica—leaving mostly aluminum and iron oxides. If little iron was present in the parent rock, for example, a high-silica igneous rock such as granite, then bauxite is formed. Tropical soils are not very fertile materials for crop production because the small amounts of humus in these soils are soon used up by crop growth. Elevated temperatures in the humid climate encourage higher chemical and biological activity, which quickly oxidizes the organic matter. For this reason, many lateritic soils can be used for only a few years before they become barren and must be abandoned. The procedure of clearing by burning and then farming for one or two years before abandoning the land is commonly practiced in some of the tropical, developing nations.

Time

Time is also a factor in soil development and is most significant for those soils developed within the past few hundred thousand years. For bedrock terrains exposed throughout the Pleistocene period (2 million years or so), the time factor plays a lesser role. In the glacial terrain of the north central United States, whether the deposit was formed during the Wisconsin age (ice retreated 8000 to 14,000 years ago) or during the Illinoian or pre-Illinoian age (about 100,000 to 1 million years old) is significant. Soils of central Indiana (Wisconsin age) can be contrasted with those of central Illinois (Illinoian age). For instance, the Wisconsin age material has less accumulation of clays in the B horizon and the soil profile in general is not as thick as that in central Illinois. This smaller amount of clay makes the younger soil more permeable and in general helps to promote better soil drainage and more fertility. The nature of the parent material (glacial till) is virtually the same for the two areas so it is the time factor that sets apart the soil profile characteristics. In older glacial deposits, Kansan, for example, which have long been exposed at the surface (northern Missouri area), erosion has prevailed to a major extent and the flat, youthful topography of younger till plains and their highly productive farm land no longer prevail. This erosion yields rolling hills cut by headward eroding tributary streams.

Topography and Vegetation

The concern about erosion provides a link to the next factor in our study of soil profile development, topography. Steep slopes encourage greater runoff and erosion. Consequently, the soil profile thins near the hilltops and thickens in the low lying areas between them. Gravity also moves the loose soil to the lower areas, mainly by soil creep, so organic accumulation is greater in these depressions as well. An example of this is provided by the relatively flat, youthful terrain that is common to Wisconsin-age till plains of the north central United States. They appear lighter in the high areas because of less moisture and organics, and darker in the low portions.

Vegetation controls the extent of runoff from the land surface and hence also the amount of erosion. When vegetation is present, soil remains in place, held by roots of trees and grasses. Farm land, particularly that planted in row crops, is subject to surface erosion following plowing and planting and prior to substantial growth of the crop. To a large extent, natural vegetation is dependent on climate and to a lesser degree on soil parent material. In modern agricultural practice, natural vegetation is replaced by crops, pasture, or other vegetation, which in many cases makes the land more subject to erosion.

CLASSIFICATION OF SOIL PARTICLE SIZE AND TEXTURE

Particle size and size distribution are important characteristics that affect soil behavior. Various agencies and professional organizations have devised classifications that provide subdivisions which are the most useful for specific applications in their specialty. Several of the most common classifications are given in Figure 8.8.

Wentworth Scale

The Wentworth scale of 1922, used widely by geologists, employs 2 mm as the primary unit with subdivisions based on a ratio of 2. Consequently, the sand-silt boundary is $(1/2)^4$ or 1/16 mm and the silt-clay boundary is $(1/2)^8$ or 1/256 mm. It is presented as a comparison to the engineering and agriculture classifications. Interestingly enough, the Wentworth scale allows the use of logarithms to the base 2 for designation of particle sizes rather than using the diameter directly. Proposed by Krumbein (1934) this is known as the phi scale. It is used widely by sedimentary petrologists but to a much lesser extent by engineering geologists.

One incentive for the engineer to learn and understand geologic literature is that a wealth of information is available on many subjects in geology that can be related to engineering studies if some of the terminology is known. This is certainly true for grain size distributions and studies of sediments by geologists. Knowledge of the phi scale should prove useful in supplying valuable information.

Phi Scale

The phi scale is the negative logarithm to the base 2 of the millimeter size of a particle. It is an extension of the Wentworth scale using logarithms rather than fractions. The result is an arithmetic scale beginning with a value of 0ϕ equal to 1 mm and progressing as $+1\phi$, $+2\phi$, $+3\phi$, and $+4\phi$ for values of 1/2, 1/4, 1/8, and 1/16 mm, respectively. Negative phi values of -1ϕ, -2ϕ, -3ϕ, and -4ϕ are equal to 2, 4, 8, and 16 mm, respectively. The negative logarithm is used because in grain size analysis studies, commonly much more of the sample is less than 1 mm in size yielding $+\phi$ units than is the portion larger than 1

mm. This results in predominantly positive ϕ units for most analyses.

Plotting size-frequency data using the phi transformation permits easy computation of the arithmetic mean size, a measure of central tendency of the size distribution. The standard deviation can also be calculated for the distribution, using the usual formulation for standard deviation to provide a measure of the spread of the grain size distribution. Figure 8.9 provides an example of a frequency distribution of sand relative to the phi scale. The mean and standard deviation are shown on the figure.

AASHTO, Bureau of Public Roads, and ASTM Classifications

The AASHTO (American Association of State Highway and Transportation Officials) classification is used primarily for highway construction materials. Shown in Figure 8.8 along with several other soil classifications, its primary textural classes are gravel > 2.0 mm; sand, 2.0 to 0.075 mm (or No. 10 sieve to No. 200 sieve); silt, 0.075 to 0.005 mm; and clay < 0.005 mm.

A highway classification that predates the AASHTO version is the U.S. Bureau of Public Roads classification. It employs a triangular texture diagram of sand, silt, and clay. Grain size designations used are gravel > 2.0 mm; sand, 2.0 to 0.05 mm; silt, 0.05 to 0.005 mm; and clay < 0.005 mm. This is illustrated in Figure 8.10.

The ASTM (American Society for Testing and Materials) classification is a general-purpose soil classification referenced in specifications for numerous construction projects, as are countless other ASTM materials specifications. This classification differs from AASHTO regarding the gravel-sand boundary. Its primary textural classes are gravel > 4.75 mm (No. 4 sieve); sand, 4.75 to 0.075 mm; silt, 0.075 to 0.005 mm; and clay < 0.005 mm.

USDA Classification

The USDA classification is widely used for agricultural studies by agronomists and soil scientists. The boundary for silt and clay is 0.002 mm, considerably lower than that used in the Wentworth (0.0039) and the ASTM (0.005) classifications. It was set low intentionally to ensure that particles in the clay fraction would very likely be restricted only to clay minerals. This brings up a significant concern deserving further discussion, which is presented in the following section.

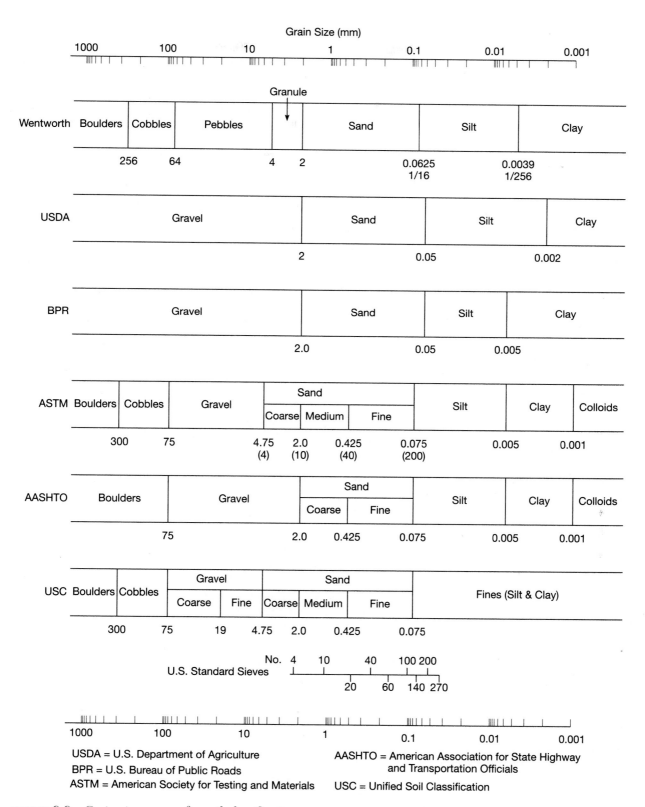

FIGURE 8.8 Grain size ranges for soil classification systems.

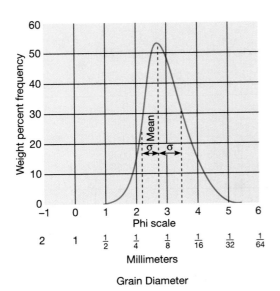

FIGURE **8.9** Frequency distribution curve based on the phi scale.

Clay Size Versus Clay Mineral

The concept of clay size versus clay mineral has plagued soils workers for many years. Clay minerals are fine-sized platy silicates (phyllosilicates) that have the property of plasticity, that is, they can be rolled into a thin thread which adheres together at low moisture levels. Clay size, by contrast, is a small particle size designation that does not always ensure that its constituents will be plastic. In actual fact, some fine-sized quartz and feldspar grains can occur in the range of 0.005 mm and it is these nonplastic materials that complicate the problem. However, very few feldspar and quartz grains occur at 0.002 mm and below, so by placing the clay boundary at that lower value, the clay size and clay mineral designations coincide quite well.

Unified Soil Classification

In the Unified Soils classification the problem is faced in a different way. Because the silt-clay size boundary for the ASTM classification (0.005 mm) does not

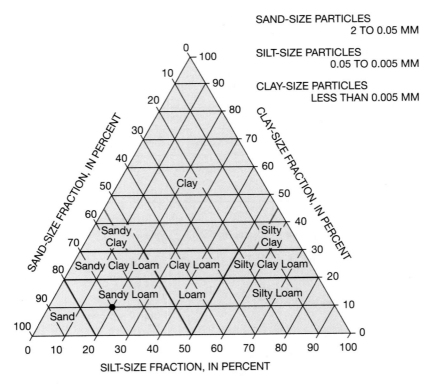

FIGURE **8.10** Bureau of Public Roads textural classification chart.

ensure that finer material will consist of clay minerals (and be plastic) and coarser materials will be nonplastic, the entire size group below 0.074 is collectively referred to as fines. If the fines are plastic, then the material is considered to be clay and, in like manner, if the fines are not plastic then they are silt. This leads to the distinction of plastic fines and nonplastic fines, replacing the decision based on particle size.

The Unified Soil classification is used in most construction projects for soil except for highways, airports, and a few other specialized applications. For foundation design and support for buildings and in dam construction and similar major earth-moving projects, the Unified Soil classification prevails. This classification also supplies a shorthand notation for designating the soil constituents in the following way:

G—gravel	W—well graded	H—high liquid limit
S—sand	P—poorly graded	Pt—peat, humous
M—silt	U—uniformly graded	O—organic
C—clay	L—low liquid limit	

Well graded means that the various size gradations are fairly well represented and poorly graded signifies that not all the sizes are present or that one size tends to dominate. Uniformly graded means all sizes are equally represented, which contrasts directly with poorly graded. A poorly graded gravel, for example, would indicate a gradation in which one size dominates with few other sizes available to occupy the voids or interstices between the particles. This gravel would have a higher permeability than would a well-graded gravel with its more complete gradation and representative sizes.

In sedimentary geology the term *sorted* is used to describe grain size distributions. Ocean waves along the coast are able to separate grain sizes by removing the finer sizes, which leaves larger grains behind. Therefore, a well-sorted deposit has a narrow range of grain size and a poorly sorted deposit has a wide range of sizes present. We can now compare the terms *graded* and *sorted*. Well graded, indicating a

wide distribution of grain sizes, is also poorly sorted. Similarly, poorly graded, or essentially consisting of one size, corresponds with well sorted.

L and H stand for low liquid and high liquid limit, respectively. A low liquid limit would be a less plastic material and a high liquid limit would indicate a highly plastic material. Therefore, pure silts would likely be designated with an L, whereas highly plastic clays (also known as heavy or fat clays) would have an H.

As suggested by this last discussion, combinations of letters are used to indicate various soils. A GW is a well-graded gravel and a CH is a highly compressible (high liquid limit and high plasticity) clay. In addition, coarse-grained soils with appreciable fines have a two-letter designation, both indicating a textural term. The most prevalent material is listed first with the less prevalent one second. Examples are GM, GC, SM, and SC, which stand for silty gravel, clayey gravel, silty sand, and clayey sand, respectively.

Some fine-grained soils, which classify near the boundary between soil types, are given double hyphenated designations such as ML-MH. This indicates a silty soil that has some plasticity but is intermediate between the two designations. Other possible double hyphenated designations are CL-CH, OL-OH, CL-OL, CH-MH, and CH-OH.

Soil Textural Diagrams

Soil texture is the appearance and feel of the material as determined primarily by its particle size. The major classes are gravel, sand, silt, and clay. In addition the term *loam* is used to depict about equal amounts of sand and silt. Triangular charts have been prepared by several agencies to aid in describing soil textures. A grain size analysis of the sample is needed prior to classifying soil textures by these charts.

Bureau of Public Roads Classification Diagram

Two triangular, textural classification charts are shown in Figures 8.10 and 8.11. Figure 8.10 is the U.S. Bureau of Public Roads classification. Only sand, silt, and clay are represented in the triangular diagram. A sample point is designated by locating two of the constituent percentages with the third determined automatically because the three must total 100%. Note that the size fractions are sand, 2 to 0.05; silt,

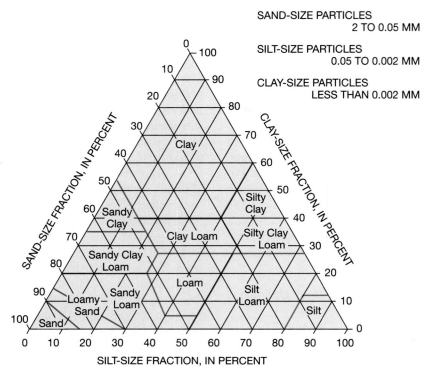

SAND-SIZE PARTICLES
2 TO 0.05 MM

SILT-SIZE PARTICLES
0.05 TO 0.002 MM

CLAY-SIZE PARTICLES
LESS THAN 0.002 MM

FIGURE 8.11   USDA textural classification chart.

0.05 to 0.005; and clay < 0.005; all in millimeters. The apex or points are either 100% sand, silt, or clay as labeled and the sides opposite are 0% of that same constituent. Lines parallel to the opposite side show increments of 10% of the constituent. A point of 70% sand, 20% silt, and 10% clay is shown on the diagram. This falls in the category of sandy loam.

USDA Classification Diagram

Figure 8.11 is the textural chart for the USDA. As indicated, the size fractions are sand, 2 to 0.05; silt, 0.05 to 0.002; and clay < 0.002; all in millimeters. This is not the same breakdown as that used in Figure 8.10. Consequently, the same size gradation (or sample) would likely plot at two different locations relative to Figures 8.10 and 8.11. Note also that the subdivision boundaries are drawn at different locations for the two classifications and the labeled areas even have different names in some cases.

When gravel is present in a sample in significant amounts, the size fractions of interest must be recalculated, thereby excluding the gravel percentages from consideration. For example, if 25% each of gravel, sand, silt, and clay were present, if we exclude the gravel, there is 25/75 of each constituent in the sample. Readjusting, this yields 33.3% for sand, silt, and clay or by direct calculation it becomes 25/100 × 100/75 = 33.3%.

Field Descriptions of Soils

Soils are described and classified during the site investigation for engineering construction projects. Boring logs are prepared by an engineering geologist, a geotechnical engineer, or the drilling foreman from samples obtained in the subsurface investigation for the sites. Samples are collected from borings advanced by drilling equipment such as truck-mounted drilling rigs. This process is discussed in detail in Chapter 19, Subsurface Investigation and Site Selection.

The samples are described visually in the field after retrieving them from the boring. Color, moisture conditions, and textures are included in the description. Textural classification, based on a visual inspection, can be refined following laboratory testing of the sample. Typically, this would include a grain size analysis, hydrometer analysis for the clay fraction, and Atterberg limits testing.

Figure 8.12 is a boring log for a boring drilled in the glacial terrain of northern Indiana. The samples were taken with a split spoon sampler, a device that is driven into the ground ahead of the drill augers. The samples were taken at 5-ft intervals. The *N* values, or the blows per foot required to drive the split spoon sampler, are also included. An explanation of the sampling data and drilling techniques is provided in Chapter 19.

In the sample description, a Unified Soil classification designation is provided in parentheses after the verbal description (see Figure 8.12). These designations were previously discussed in the section on the Unified Soil classifications.

AGRICULTURAL SOILS MAPS

Agricultural soils information is supplied on large-scale maps (1:4800), which depict different soil series. These have been delineated by an agronomist based on field study aided by interpretation of aerial photography. In many cases, modern soils maps are printed on an airphoto base. This yields quite a large map, sections of which are bound in a volume comprising the agricultural soils report. A county 20 miles square, for example, would yield a single map 22 × 22 ft at the 1:4800 scale if presented on one large sheet.

A soil series is made up of a location name and a texture name, for example, Russel silt loam, and it is subdivided according to the slope of the land surface. The soil textural terms are those discussed previously and include such terms as silt, clay, sand, loam, and combinations thereof.

Agricultural soils maps are commonly prepared on a county-by-county basis. They are based on the upper 3 or 4 ft of the Earth's surface. Each soil horizon or zone is described on the basis of color, texture, consistency, depth, porosity, stoniness, moisture aspects, and perhaps other features. The top soil (A horizon), subsoil (B horizon), and parent material (C horizon) comprise the subdivisions described in the report. Soils are subdivided into slope subclasses with the following typical ranges for each: 0% to 2%, level to nearly level; 2% to 6%, gently sloping (undulating); 6% to 12%, sloping (rolling); 12% to 20%, strongly sloping (hilly); 20% to 35%, steep; 35+%, very steep.

Modern agricultural soils maps (those produced in the last 15 years) also contain information relating engineering soils data and engineering uses to the agricultural soils information. The individual soils subclass, example: Russel silt loam, 2% to 6% slope, has an accompanying rating relative to engineering use. Such uses may relate to building foundations, ponds, homesites, roadways, septic tank absorption fields, cemeteries, basements, sanitary landfills, and excavations. The various subclasses are rated good, moderate, and severe relative to these various proposed uses.

Agricultural maps of larger areas, an entire state, for example, or those that depict a county on a single text-sized sheet (about 1:150,000 scale) cannot provide enough detail to show information on an individual soil series. In this case, soil associations are used as the mapping unit. A soil association is a collection of soil series that form a characteristic landscape wherever they occur. The associations are named for soils within the group (example Miami-Russel-Fincastle) that have similar soil type, physiographic location, and parent materials. Counties in northern Indiana (about 400 mi^2 in area) typically contain six or seven soil associations.

Agricultural soils maps are used along with other sources of information for site selection for engineering construction. These and other maps are used for reconnaissance purposes prior to detailed subsurface exploration. The contributions of these soils maps are discussed further in Chapters 15 and 19.

ENGINEERING PROPERTIES, WATER CONTENT, AND CONSISTENCY OF SOILS

The following material was discussed in detail in Chapter 7, but some of the same subjects are included here because of their direct application to soil classification.

Water Content

All natural soils contain at least some moisture, even in the driest of seasons. The terms *water content* and

Project Name _____ Location _____ Date _____

Weather Conditions _____ Elevation _____

Boring Number _____ Logged By _____

Depth		Blows/Ft.	Description
0			Top soil, black organic clayey silt
5	SS - 1	7	Brown, moist silty clay (CL) with a trace of rock pebbles
10	SS - 2	12	Gray, moist clayey silt (ML) with a trace of rock fragments
15	SS - 3	6	Brown, wet, medium, silty sand (SM)
20	SS - 4	13	Gray, moist, clayey silt (ML) with a trace of rock fragments
25	SS - 5	16	Same as previous sample
			Bottom of hole, 25 feet

(handwritten annotations in Description column: "Clay low liq lom", "Silt Low LL", "SAND SILT", "SILT LO", "5")

Water in hole on completion, 12.5 feet

SS = Split Spoon Sample

FIGURE **8.12** Drilling log of a soil boring.

moisture content are used interchangeably to refer to this degree of moisture in engineering studies. The most common way to express water content is on the mass basis, which is the mass of water divided by the mass of dry solids. This is illustrated by the following relationship:

$$w = \frac{M_w}{M_s}(100) = \frac{M_w}{M_t - M_w}(100)$$

where

w = water content in %
M_w = mass of water
M_t = total mass of sample of wet mass of soil
M_s = mass of solids.

Soil Consistency

Soil passes through several states of consistency as it dries from a wet, soft condition to a dry one. Consistency, which denotes the degree of firmness, is designated by such terms as soft, firm, or hard. Actually, only soils with some plasticity show this change with decrease in moisture content.

As a soil changes consistency, there is an attendant change in its engineering properties. Thus shear strength, compressibility, and bearing capacity vary with consistency. A textural classification alone will not provide this information but its consistency, natural moisture content and Atterberg limits are needed also. Atterberg limits were discussed in more detail in Chapter 7.

Soil at a very high moisture content will behave as a liquid. This is the first state of consistency. As the soil drys, it changes to three other states: plastic, semisolid, and solid, as was shown diagrammatically in Figure 7.10. When soil goes from the liquid state to the plastic state it passes through the liquid limit. This is the lowest moisture content at which soil behaves as a liquid. Similarly, the plastic limit is the lowest moisture content for plastic behavior of the soil.

The shrinkage limit is the lower boundary of the semisolid state. This is the point beyond which the soil will no longer reduce in volume despite further drying. The PI or plasticity index is equal to the liquid limit (LL) minus the plastic limit (PL). It signifies the range in moisture content over which the soil behaves plastically. Soils with a large plasticity index tend to be more compressible than those with a low PI.

The plasticity chart can be used to designate the nature of soils based on the Unified Soil classification. This is shown in Figure 7.11. It is a plot of LL versus PI. A liquid limit of 50% is the boundary between high compressibility and low compressibility soils. LL varies in a general way with clay and/or organic content. The A line separates inorganic soils (above the line) from organic ones. At a specific liquid limit, a reduction in plasticity index yields an increase in permeability and a reduction in compressibility.

The field moisture content of soil can supply valuable information about a site. If the moisture content is near the liquid limit, the soil is of fairly low strength and subject to significant strength reduction on remolding. If instead the moisture content is closer to the plastic limit, the soil will be relatively firm and show greater strength. Natural moisture contents can also be used to estimate the extent of inhomogeneity at a site. Wide ranges in moisture content suggest the presence of different soils or at least the same soil under very different conditions. The liquid limit, plastic limit, and natural moisture content are used along with the soil texture to determine the extent of variation at the site.

CONSTRUCTION PROBLEMS ASSOCIATED WITH VARIOUS SOILS

A multitude of problems can develop when different soils are used to support engineering structures. They may vary considerably from one area to another in the United States and only a general list can be compiled. It is presented as follows:

1. *Compressible soils:* Occur in highly organic soils including some glacial deposits and certain floodplain areas. Highly plastic clays in some glacial deposits and in coastal plains and offshore areas can yield compressible soils. Problems involved are excessive settlement, low bearing capacity, and low shear strength.
2. *Collapsing soils:* Settlement in loose sands and silts primarily. Densification occurs by movement of grains to reduce the volume. Typically includes shallow subsidence. May occur in sandy coastal plain areas, sandy glacial deposits, and alluvial deposits of intermountain regions in the western United States.

3. *Expansive soils:* Soils containing swelling clays, primarily smectite (montmorillonite), which increase in volume when absorbing water and shrink when losing it. Climate is closely related to the severity of the problem. Semiarid to semihumid areas with swelling clays suffer most because the soil moisture active zone has the greatest thickness under such conditions. Foundation supports should be placed below this active soil zone. Expansive soils are most prevalent on the Atlantic and Gulf coastal plain and in some areas of the central and western United States.

4. *Corrosive soils:* Soils of high acidity cause the corrosion of underground metal pipes. Predictions can be made based on the pH of the soil or on its electrical resistivity. Values of 700 ohm cm or less typically indicate a high corrosion potential for soils, 750 to 1000 is moderate, greater than 1000 is low to moderate, and greater than 1750 is very low.

5. *Fine-textured soils:* Can yield problems related to highways performance. Plastic subgrades introduce a problem known as rigid pavement pumping for concrete highways. Rigid pavement pumping is the removal of fines in the subgrade and base course by water when it is forced out from pavement joints with the passing of heavy wheel loads, typically those from trucks. This yields a void under the concrete slab and eventually the concrete fails by flexure adjacent to the joint. A rough pavement develops and concrete adjacent to the joint must be replaced.

Silt-sized soils are frost susceptible. Water moves through this material by capillary action to provide a source of water in a freezing soil. Ice lenses form, which on melting provide insufficient support for the highway. Asphalt highways are particularly prone to spring breakup problems when wheels punch through the flexible pavement causing potholes to form. Gravel secondary roads are also subject to intense, spring breakup problems, making some impassable until gravel fill and road grading can remedy the situation. Base courses for flexible pavements must contain 2% or less of minus No.200 size material (fines) in their gradation to prevent spring breakup from freezing and thawing.

6. *Saturated fine sands and silts:* Are subject to liquefaction failures during earthquake shaking. Alluvial deposits are particularly susceptible to this problem in earthquake-prone areas.

7. *Permeability of soils:* A problem for many different types of construction. Values for permeability vary over nine orders of magnitude from coarse gravel to colloidal clay—this is an enormous difference. For water supply, high permeability is desirable; in underseepage for dams and containment of contaminated liquids, low permeability is desirable. Problems of dewatering and grouting are related to soil permeability. Slope stability, excavations, road construction, and many other construction specifics are related to soil permeability.

EXERCISES ON ROCK WEATHERING AND SOILS

1. This question is similar to Table 8.1. Fill in the blank spaces in the table at the bottom of this page.

2. What is a placer deposit? How would they be mined? What is meant by the phrase "mother lode" and how does it relate to placers?

Rock Type	Minerals Present	Weathering Conditions	Products
Basalt	_____	Humid	_____
Syenite	_____	Arid	_____
Gabbro	_____	Arid	_____
Gneiss	_____	Humid, tropical	_____
_____	Muscovite, quartz (with schistosity)	Arid	_____
Dolomite	_____	_____	Clay minerals and iron oxide
Arkose	_____	Arid	_____

3. What type of weathering would prevail in the following locations? (a) Mojave Desert, California; (b) Houston, Texas; (c) Southern Appalachians; (d) Grand Junction, Colorado; (e) Butte, Montana; (f) Portland, Oregon; (g) Honolulu, Hawaii.

4. Examine Figure 8.5 and Table 8.1. Why does muscovite occur as a common mineral in soils of the Piedmont Plateau, a worn-down mountain area? Explain based on rock type and weathering.

5. Rocks such as graywackes contain pieces of basalt, gabbro, and peridotite as grains in the sedimentary rock. How do you account for this relative to weathering? If the source area had a humid climate, how could this occur?

6. The thickness of residual soils in tropical and subtropical areas can exceed 50 ft (15 m). Why would this be the case? If the parent material was a basalt, what would the soil probably consist of?

7. In an area in the Ozarks of southern Missouri an accumulation of chert pieces can be observed along the top of the soil-bedrock interface. What is the origin of this assortment of irregular loose chert pieces? (*Hint:* The Ozark Plateau consists of a very thick sequence of cherty, carbonate rocks, i.e., limestones and dolomites.)

   If a 2-ft-thick zone of unweathered chert pieces weighed 100 lb/ft³ in place and the chert content in the limestone below was 1.3%, what thickness of rock would be indicated by the residual chert layer? Give all assumptions.

8. This question pertains to particle size classifications of soil and is related to Figure 8.8. On the log scale shown below put in the boundaries for sand, silt, and clay in keeping with the different classifications of Figure 8.8.

9. *Classifications of soil:* Give the names for each of the following classifications described.

Material	Names 1	2	3	4
Gravel	>2 mm	>4.76	>2	>2
Sand	2–0.05	4.76–0.074	2–1/16	2–0.05
Silt	0.05–0.002	<0.74 nonplastic fines	1/16–1/256	0.05–0.005
Clay	<0.002	Plastic fines	<1/256	<0.005

10. What is the percent of sand, silt, and clay for each of the four classifications in Exercise 9 (indicate by number) in the following soil sample. Record this information in the table on the following page.

Size (mm)	Nature of Material	Percent
2–0.074	Gritty, nonplastic	20
0.074–0.0625	" "	30
0.0625–0.05	" "	25
0.05–0.005	Nonplastic	15
0.005–0.0039	"	2
0.0039–0.003	Plastic	3
0.003–0.002	"	3
0.002–Pan	"	2
		100

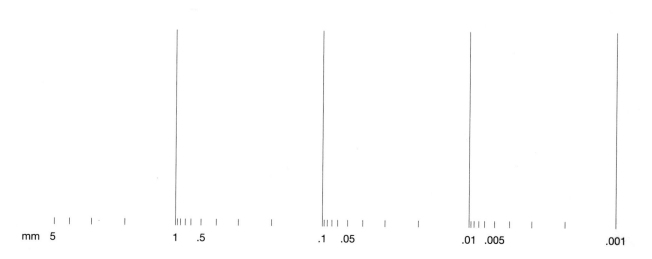

mm 5   1 .5   .1 .05   .01 .005   .001

	Classification Number			
	1	2	3	4
Sand	_____	_____	_____	_____
Silt	_____	_____	_____	_____
Clay	_____	_____	_____	_____

11. For classification Numbers 1 and 4 in Exercise 10, plot these on the appropriate triangular diagram for the two classifications depicted in Figures 8.10 and 8.11. Give the name for each soil based on this diagram.

12. A soil sample has the following size distribution: 4 to 2 mm = 17%; 2 to 0.05 mm = 12%; 0.05 to 0.005 mm = 22%; 0.005 to 0.002 mm = 10%; <0.002 mm = 39%. How would this be plotted on the triangular soils diagram using the USDA classification? Give values.

13. A certain soil was found to have a LL of 28 and a PL of 17. What is its PI? Another soil has a PI of 48 and a PL of 32. What is its LL? Is it a low compressibility soil? Explain. Plot both soils on the plasticity chart (Figure 7.11). In which area do they fall? Supply a

descriptive name if possible. Give a typical PI and LL for an ML soil.

14. What engineering soils problems are likely to occur in the following landform areas:
    (a) Till plain in north central United States
    (b) Gulf coastal plain
    (c) Alluvial sands of the San Joaquin Valley of California
    (d) Floodplain of Mississippi River at St. Louis, MO

15. An agricultural soils map is being used to find information concerning the location of a fossil fuel power plant along a large river. What limitations of this map are likely to develop during this application? Explain.

Additional Readings

CARROLL, D., 1970, *Rock Weathering*, Plenum Press, New York.

GOLDICH, S.S., 1938, "A Study in Rock Weathering," *Journal of Geology*, Vol. 46, pp. 17–58.

KELLER, W.D., 1957, *Principles of Chemical Weathering*, Lucas Brothers Publishing Company, Columbia, MO.

KRUMBEIN, W.C., 1934, "Size Frequency Distribution of Sediments," *Journal of Sedimentary Petrography*, Vol. 4, pp. 65–77.

REICHE, P., 1950, "A Survey of Weathering Processes and Products," University of New Mexico, Publications in Geology, No. 3, pp. 1–95, Albuquerque.

9

Stratigraphy and Geologic Time

CHAPTER OUTLINE

A detailed study of the Earth's history is not typically included in a discussion on physical geology or in a geology text for engineers. Instead, it is the subject of a semester course, Historical Geology, which pertains to the chronological occurrences of Earth history, including the development of life. Some knowledge of this history, however, is required for engineers who work with earth materials in order to understand relationships between geologic units and their impact on engineering construction. Background information is also needed to understand the details depicted on geologic maps. This chapter provides the basic information needed to understand and appreciate the concepts of geologic time and the geologic past with regard to engineering design and construction.

The term *stratigraphy*, which refers to the study or description of rock strata or layers, also includes the chronological sequence of rock origins and the field relationships among them. All rock types are involved, not only sedimentaries, but igneous and metamorphic rocks as well. Because the description and origin of rock units are of importance to stratigraphy, a carryover of knowledge from rock identification is required. The texture and composition of rocks are a consequence of their origin and they determine the name assigned to a rock. A foliated metamorphic rock, a clastic sedimentary rock, or a glassy igneous rock illustrates the significance of this relationship. Stratigraphic concepts are also applied to soil or unconsolidated materials.

The chronological sequence of rock units and their field relationships are a product of the geologic history of an area. Many geologic features, including planes of weakness and conditions of the rock mass, are related to the historical development of rocks. Therefore, a full appreciation of the impact of geology on construction is not possible without proper consideration of geologic history.

For example, jointing in a basalt may occur at different times following solidification of the lava. Joint planes formed early in the rock's history may later become filled with secondary minerals such as calcite. This tends to knit the rock mass together, greatly increasing its strength. If it can be established that the joints are rehealed in this fashion, based on details of geologic history, a lack of through-going fractures may be confirmed. As a result a rock excavation in the basalt can be made at a steeper angle or be stabilized with considerably less support.

Ground-water flow is another condition of concern. If an erosion surface is recognized at a certain elevation, increased water flow in rock cuts or tunnels can be anticipated at that level. Development of the geologic history may establish the presence of erosion surfaces or other permeable zones, which can then be properly considered during design and construction.

The shape and orientation of geologic units are also related to geologic history. Whether a body of sand extends as a two-dimensional layer (a sheet) or as an elongated tubular form is related to both its origin and geologic history. Drainage, ground-water inflow, variation in soil strength, and numerous other factors would differ greatly for these two examples.

STRATIGRAPHY

Rock units are subdivisions of the geologic column, that total sequence of rocks formed since the cooling of the Earth. The basic rock unit or division is known as a formation. It is defined as a distinct lithologic

(rock type) unit that is recognizable in the field based on its physical characteristics. It also must be sufficiently thick (tens to thousands of feet or several to hundreds of meters thick) and have sufficient lateral extent for proper inclusion and easy detection on a geologic map. A formation name consists of two parts, the location name indicating where the rock unit was first described, and a lithologic name indicating the type of rock that prevails in the unit. Examples are St. Louis Limestone, Manhattan Schist, Pierre Shale, Monterey Chert, Salem Limestone, and Navajo Sandstone. If no dominant rock type prevails throughout the unit, the word formation is used instead, such as the Chinle Formation, which is found in the Painted Desert of Arizona.

Subdivisions of a formation are called members and a collection of formations is known as a group. The Borden Group in Indiana, a collection of siltstone units plus a few thin limestone beds, is such an example.

BASIC GEOLOGIC PRINCIPLES FOR RELATIVE AGE DATING

In a discussion of the chronology of rock units several basic principles must be established at the outset. These principles are used to develop the relative ages of geologic units and are listed as follows: 1) original horizontality, 2) superposition, 3) faunal assemblage, 4) cross-cutting relationships, and 5) uniformitarianism.

Original Horizontality

The principle of original horizontality pertains to most sedimentary rocks, because they are deposited in parallel layers to form horizontal beds. This involves marine sedimentary rocks, which comprise the major portion of the sedimentary column. They are deposited on the continental shelf, which has a gentle slope toward the ocean basins. This principle suggests that those sedimentary rocks which deviate significantly from the horizontal position have been folded by earth stresses to yield that inclined position.

There are exceptions, of course. Some sedimentary sequences are formed in an inclined position. Steeply dipping limestone or dolomite reefs are formed at the sloping edges of islands in the ocean (see Figure 4.2). These features are fairly common in Silurian aged rocks (about 400 million years old), which are prevalent in areas surrounding the Great Lakes. Other

[handwritten: Correlation; Piecing together the column over an area]

rocks such as lava flows are typically deposited on gently sloping terrain, also yielding a nonhorizontal orientation when formed. All in all, however, the vast majority of bedded sedimentary rocks do display original horizontality.

Superposition

The principle of superposition states that for a sequence of undisturbed sedimentary rocks (when viewed in cross section), the lowest layer is the oldest and the layers become consecutively younger as you proceed toward the top. This assumes the rocks have not been overturned by folding. This is a simple concept, but when rocks are inclined it is a valuable tool for working out the sequential details. Superposition of sedimentary rocks above an old erosion surface, developed on either igneous, sedimentary, or metamorphic rocks, is a naturally occurring phenomenon.

Faunal Assemblage

The principle of faunal assemblage is related to age dating of rocks by means of the fossils they contain. Properly stated: Like assemblages of fossil organisms indicate like geologic ages for the rocks that contain them. This is related to the evolutionary process of organisms through geologic time.

Cross-Cutting Relationships

The principle of cross-cutting relationships states that faults, dikes, folds, unconformities, and other cross-cutting features are younger than the rock units they cut across. Quite simply, a rock must predate a feature that formed by cutting across it. This is true for microscopic mineral assemblages on up to entire mountain ranges. A simple example of this is a tree stump with saw marks on the cut surface. The tree had to exist before the saw was used to cut through it.

Uniformitarianism

The principle of uniformitarianism establishes the background for determining occurrences in the geologic past. It proposes that the natural laws operating on Earth today are the same as those that prevailed in the geologic past. The past can be studied by viewing these processes at work on Earth today, allowing for reasonable variations in the magnitude of these effects. Thus, details concerning descriptive phenomena of an ancient beach deposit can be resolved by analysis of modern-day deposition in a similar environment. Uniformitarianism is sometimes expressed by the phrase "the present is a key to the past."

UNCONFORMITIES

During the geologic history of an area, rock masses are commonly uplifted from below sea level where they were deposited. In the uplifted position the rocks are subjected to erosion, followed in some cases by lowering below sea level again and a renewal of sedimentation. This gives rise to a buried erosion surface, which separates the two distinct intervals of deposition. Following another cycle of uplift, these surfaces can be observed in profile view near the Earth's surface. A surface of nondeposition or of erosion separating older rocks below from younger rocks above is called an unconformity.

The interval of time represented by an unconformity is an important consideration in the geologic history of a region. It indicates not only that lowering and uplift of the Earth's surface occurred but that a sequence of sedimentary rocks is missing compared to a region that received continuous sedimentation. Some unconformities represent breaks in the rock record of a few thousand years, whereas others may designate several hundred million years. Regional studies may be needed to determine how much of the rock record probably eroded or simply was not deposited during the interval of uplift.

Types of Unconformities *[handwritten: Buried erosion surface / Gap in time]*

The four primary types of unconformities observed in nature are angular unconformity, disconformity, paraconformity, and nonconformity.

Angular Unconformities

In an angular unconformity the older strata (below the unconformity as suggested by superposition) dip or slope at a different angle from that of the younger strata (Figure 9.1). A significant amount of historical detail is indicated by this unconformity. After deposition, the older rocks were folded and uplifted above sea level. Erosion occurred on the dipping beds to form the erosion surface. Following this episode, the rocks sank below the ocean and additional sedimentary rocks were deposited. Later, both units were uplifted and exposed to view at the Earth's surface.

Disconformities

A disconformity is an unconformity in which sedimentary rock beds parallel each other on opposite sides of an irregular unconformity (Figure 9.2). Only the minor relief of the erosion surface caused by

Locality #1e (handwritten, left margin)

FIGURE 9.1 An angular unconformity viewed in cross section. The rocks on angle below the erosion surface underwent folding prior to erosion.

gullies or slope movement provides evidence of the interruption of sedimentation. These are formed when layered rocks are uplifted, undergo erosion to yield an irregular surface, and then lowered below sea level to receive more sedimentation. No folding occurs during this entire episode.

Paraconformities

A paraconformity is similar to a disconformity except that it lacks any relief on the erosion surface and is not discernable from other bedding plane surfaces. Detailed regional study and considerable experience are required to locate the precise unconformable surface within the rock sequence. Fossil evidence and lithologic detail are used to establish the location of the paraconformity.

Nonconformities

A nonconformity is an unconformity that develops when igneous or metamorphic rocks are exposed to erosion, and sedimentary rocks are subsequently deposited above the erosion surface (Figure 9.3). A

Crystalline (handwritten, left margin)

typical example consists of granite lying below a sandstone or carbonate rock with an erosion surface between them. The geologic history depicted in this situation is first the intrusion of granite into the country or host rock followed by extensive erosion as thousands of feet of rock must be removed to intercept the igneous intrusion. Then the ocean advances over the granite mass and deposition of sediments occurs. Following lithification of the sedimentary rock, the area is uplifted to expose the nonconformity.

Unconformity Versus Intrusive Contact

The discussion on nonconformities brings into focus a challenging interpretation involving geologic field studies. Another seemingly similar exposure consists of a contact between granite and sedimentary rock, but it marks the boundary where the igneous intrusion came to rest within the sedimentary rock mass. In this igneous intrusion, the granite is younger than the rock it intrudes (cross-cutting relationships). In the case of the nonconformity, the granite did not

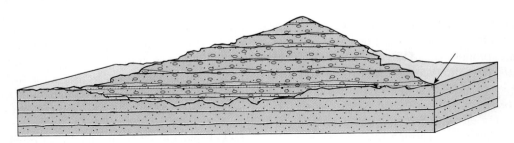

FIGURE 9.2 A disconformity in a sedimentary rock outcrop. An irregular erosion surface separates horizontally bedded rocks into two distinct units above and below the surface.

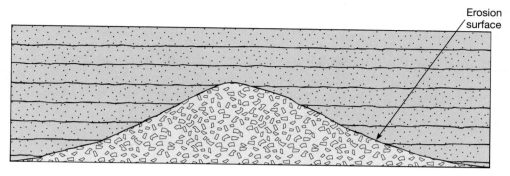

Erosion surface

FIGURE 9.3 A nonconformity. An irregular erosion surface separates the granite below from the sandstone above.

intrude the sedimentary rock above it, and is the older rock unit of the two (superposition). The sedimentary rock was deposited on the erosion surface.

To resolve the problem of igneous contact versus nonconformity, the surface between the two rock units must be examined in detail. For the nonconformity, the erosion surface is reasonably smooth without irregularities of a small dimension. Pieces of granite may occur as conglomerate pebbles in the sedimentary rock, and the bedding tends to parallel the unconformity (Figure 9.4a). Joints in the granite may be filled with sand grains from the rock above and dikes may cut through the granite but stop at the unconformity.

The igneous contact would appear differently. The contact may be quite irregular with small protrusions of the granite into the rock above. The overall trend of the contact may not be parallel to the bedding planes of the sedimentary rock above, and pieces of that rock may be present as partially digested remnants in the granite mass. A zone of contact metamorphism in the sandstone may surround the granite or an outer zone of the granite may be finer grained than the rest (known as a chilled border), indicating that it cooled more rapidly because of its contact with cooler rock above. Cross-cutting dikes may penetrate both the granite and the sandstone (Figure 9.4b).

CORRELATION

Because of missing gaps in the sedimentary sequence, there is no location on the Earth where a continuous rock column from the oldest to the youngest rocks

exists. We must piece the total column together from one place to another by comparing rock units and their local sequences. This method of relating rock units from one locality to another is known as correlation.

When sedimentary rocks show a fairly constant and distinct lithology over a widespread area, the boundaries between units can be connected or extended from one locality to another. An example is illustrated by Figure 9.5. Here two valley walls at the edge of a wide stream valley are shown. Sandstone, exposed at the surface, is the youngest rock in the sequence. These natural exposures of bedrock at the Earth's surface are called outcrops. Below the sandstone is a siltstone, then a coal seam, and below it a shale. A sandstone bed is exposed near the base of the valley wall, and on the west, a limestone bed lies below the sandstone. By extending imaginary lines across the valley, we can conclude that the beds correlate with each other.

On the east side of the valley, correlation strongly suggests that a limestone bed would be found below the lower sandstone unit in the sequence. Because the rock is not exposed, the presence of the limestone must be verified by another means. If this information is needed for engineering or economic purposes, rock cores can be obtained from exploratory drill holes extending downward from the east valley wall to a depth below the base of the stream valley.

Correlation can be accomplished using rock cores when no rock exposures are available for examination. In engineering studies, correlation from one boring to another is used to determine the nature of the

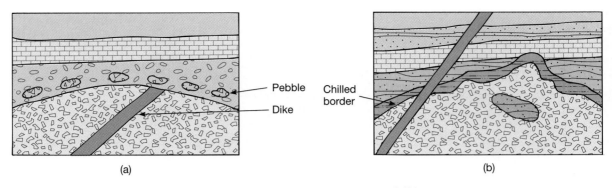

FIGURE 9.4 Contrasting field conditions for (a) a nonconformity and (b) an igneous intrusion, both involving sedimentary rock units.

subsurface conditions. Drilling procedures are discussed in detail in Chapter 19 on subsurface investigation and site selection.

Rocks can also be correlated using fossils rather than lithology alone. In many rock sequences such as the Mississippian System of the midcontinental area of the United States, thick deposits of one major rock type, in this case limestone, occur. Details concerning the differences in color and texture between limestone units are used along with fossil assemblages to determine which formations are present.

Complications in Correlation

Correlation over a large area for an engineering site can be considerably more complicated than the example shown in Figure 9.5. This occurs because sedimentary beds are not always continuous over large areas. Several possibilities exist: 1) The rocks may be cut by an unconformity, 2) the rock units may disappear (pinch out) if the basin of deposition did not extend across the entire area, and 3) the rock unit may grade laterally into a different lithology because of changes in the sediments deposited across the basin. Lateral changes that occur across a rock unit are called facies changes. Rock descriptions relate to lithofacies, and the fossil content to biofacies. Changes in the lithofacies are the specific concern in item 3) just listed.

DATING THE EARTH

By all indications the human race has been interested in the age of the Earth since the earliest recorded time. Before reviewing some early speculations about Earth history, it is worthwhile to examine the reasons why humans seek to know the past.

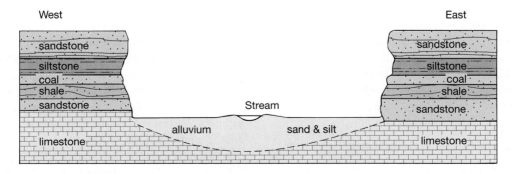

FIGURE 9.5 Correlation of sedimentary rock formations from one valley wall to another across a wide stream valley.

Two apparent reasons for this interest are 1) natural curiosity and 2) economic considerations. Included in curiosity is the seeking out of past events for religious or theological reasons. Contained also would be pure science with its thirst for knowledge for the sake of knowledge alone. Applied science more likely falls under the realm of economic purpose.

The economic aspects of age relationships become more apparent when we realize that certain types of minerals are found in specific kinds of rocks of an appropriate geologic age. Coal in the United States is found mostly in rocks of Pennsylvanian and Permian age (230 to 320 million years ago). The greatest amounts of petroleum are found in the younger marine rocks, those of Cenozoic age (last 65 million years), although older sedimentary rocks also contain oil. The reason why younger rocks contain more oil is because life was more abundant in the oceans during the Cenozoic age, and with more organisms there was more organic material available to produce oil. Also less time has transpired since the petroleum was formed, yielding less chance for loss from erosion to occur.

Metallic ore deposits are commonly associated with certain of the oldest igneous and metamorphic rocks. Gold, silver, and other base metal areas occur generally in isolated zones of the earliest of rocks, those Precambrian in age (more than 600 million years old), although younger igneous and metamorphic rocks can also be ore bearing.

Certain construction problems can be anticipated for certain rock types and geologic age. The expansive shales of the Cretaceous period (65 to 136 million years ago) are well known as is the scarcity of carbonate rocks for construction in the Pennsylvanian of the Midwest. Rocksalt and prevalent solution cavities are also typical in the Permian. Solution cavities and sinkholes in the Mississippian and Ordovician limestones and dolomites are well documented across the United States. The age and history of earth materials play an important role in their nature and engineering properties.

Early Speculations About the Earth's Age

Every society, religion, or cult seems to develop an explanation of how the Earth began and how long ago it happened. Humans obviously have an inherent desire to pursue the details of their beginnings, which is manifested through theological thought and scientific endeavor.

For example, the Brahmins in ancient India thought the Earth was eternal, whereas astrologists of Babylon claimed humans appeared 500,000 years ago. The Persians of 100 B.C. thought the Earth was 12,000 years old and would last about 3000 years more. The priests of Chaldea said the Earth was 2 million years old.

Seventeenth-Century Biblical Scholars

In the seventeenth century, scholars of Western civilization studied the question of the Earth's age. Interpreters of the Old Testament attempted to determine the time of creation by working back through the lineage of people in the Bible. In 1642 John Lightfoot, a scholar at Cambridge University in England, deduced that the moment of creation was 9 A.M., September 17, 3928 B.C. Later in 1658, Archbishop Usher of Ireland claimed that the Earth was created on the evening of October 22, 4004 B.C. This date appeared in some subsequent editions of the Bible. The basis for these two interpretations was the assumption that humans were created soon after the Earth so by working back through human history in the Bible, the time of planetary creation could be ascertained. The Earth was thereby designated to be about 6000 years old.

Ancient Scientific Estimates

Careful observations of natural phenomena formed the basis for unraveling the mysteries concerning the Earth's age. Herodotus, the great Greek historian, noted around 450 B.C. that the Nile River delta had to be the product of many floods because individual floods deposited only thin layers of sediment, yet the river alluvium was quite thick. He reasoned that thousands of years must have occurred to form the Nile delta.

Aristotle and the Greek and Roman naturalists and philosophers that followed continued the scholarly approach of combining observation with deduction. But this method was lost during the Dark Ages (approximately AD 476–1000) and when inquiries about the age of the Earth developed again it was through the biblical interpretations of the seventeenth century. To question such religious pronouncements was heresy, which brought dire consequences to naturalists and scientists of that day. Indeed, it was nearly 200 years, not until after 1830 or so, before the scientific community openly professed the belief that the Earth was considerably older than 6000 years.

Eighteenth- and Nineteenth-Century Scientific Studies

James Hutton

James Hutton, a Scottish geologist, in 1785 was one of the early scientists to cast doubt on the accuracy of the biblical scholars. He proposed the principle of superposition for sedimentary rocks but also made some astute observations about the Earth's features. He noted that geologic processes carved the landscape but at a reasonably slow rate. If the terrain he observed—gorges, mountain passes, plateaus—were carved by these geologic processes, it would take much longer than a few thousand years to accomplish the feat.

William Smith

William "Strata" Smith, a civil engineer and surveyor, noted around 1800 that certain of the flat sedimentary rocks in southern England contained fossils unlike those in any other layer. He could predict the rock strata and the fossils they contained based on the elevation of the rocks as a result of the horizontal nature of the strata in the area. This showed former animal life and a marine origin were involved, suggesting a need for more time to transpire than only a few thousand years. Although Hutton and Smith proposed no specific age for the Earth, they pointed out the apparent conflict between the biblically based time scale and that indicated by natural features.

Georges-Louis Buffon

Physicists also began to labor with the problem of such a short time for the age of the Earth. After Buffon, Kant and Laplace proposed their theories on the origin of the solar system (between 1749 and 1796), there was a basis for estimating the relationship between time and the planetary orbits. The time necessary for the formation and motion of the Sun and planets seemed too great to fit into the brief time span demanded by the biblical studies. Buffon also studied the rates of melting and cooling of iron balls, because he had concluded earlier that the Earth contained an interior not unlike iron because it was so dense. His estimate of 75,000 years for the Earth to cool seemed low to some geologists at the time but nevertheless made the biblical fundamentalists most unhappy.

Lord Kelvin

Different estimates for the age of earth materials came rapidly in the nineteenth century. In 1854, Hermann von Helmholtz, one of the founders of thermodynamics, established that the Sun, based on gravitational contraction, would have burned for 20 to 40 million years. Later in 1897 Lord Kelvin (William Thompson) indicated that the Earth by his estimates had taken 20 to 40 million years to cool. In other papers his estimates had been as high as 75 million years. We know today that Lord Kelvin's calculations were much too low, but several problems beyond his control placed them in error. He had no accurate measure of heat flow from the Earth, values for the thermal conductivity of rock were virtually unknown at the time, and the radioactivity of elements in the Earth's crust, which supply much of the heat, was not considered (because radioactivity had not yet been discovered).

Sediment Accumulation

Estimates on the age of the Earth based on the accumulation of sediments and sedimentary rocks were presented during the latter half of the nineteenth century. In 1854 a statue of Ramses II, the renowned Egyptian pharaoh, was found at Memphis, Egypt, under 9 ft of river laid sediment. The statue was known from historical data to be 3200 years old. The 9 ft of sediment cover yielded an accumulation rate of 3.37 in. per century. Since the total sediment thickness in the river was 40 ft, a total age of 14,200 years for sediment accumulation in the Nile River was obtained. This alone is more than twice the time suggested by the biblical scholars.

Salt Accumulation in the Sea

Two other estimates of the Earth's age were obtained in a similar way. First, a rate of accumulation is determined, followed by a measurement of the total accumulated amount. Elapsed time is found by dividing the accumulated thickness or amount by the calculated rate. Salinity in the sea was used by John Jolly in 1899 for such an estimate. Assumptions included the following: 1) The rate at which Na^+ is added to the sea is constant through time, 2) oceans were fresh (salt free) at the start, 3) only a small amount of Na^+ removal with time is considered, and 4) the rate is determined based on a current estimate of runoff to the oceans and the Na^+ concentration in the water delivered to the sea. Using these assumptions, a value of 90 million years was obtained. Complications are that more Na^+ removal from the oceans has occurred than assumed and today's rate of Na^+ input is probably not a good average of that contributed throughout geologic time.

Limestone Deposition

A final estimate for consideration involves the accumulation of limestones around the world. Through correlation the duplicate sections of rocks deposited simultaneously are excluded and a total thickness of limestone accumulation is obtained. Using the rate of limestone accumulation observed at the present time in the oceans (about 1 in. every 200 years), the total elapsed time is found by

$$\text{Time} = \frac{\text{Total thickness}}{\text{Accumulation rate}}$$

Some 18 determinations were presented between 1860 and 1909 with values ranging from 3 million to 1.5 billion years. Most values were somewhat less than 100 million years.

Common Shortcomings

All of these calculations based on early scientific methods suffer from similar shortcomings. The rates of accumulation are only approximate, the total accumulations cannot be accurately determined, and the assumptions are usually oversimplified. For example, in limestone accumulation it is now known that these rocks did not form during the earliest geologic time so the complete history of the Earth is not represented by the limestone column. By and large, all the age estimates for the Earth turned out to be too low.

Darwin and Evolution

In 1859 Charles Darwin put forth his famous theory of evolution. He knew it would have taken a considerable amount of time for the life he observed as fossils to evolve from those simple forms to mammals and finally to humans. Darwin believed it would require at least 100 million years for this evolution and

he was concerned that Kelvin's 20 to 40 million year figure was much too low.

Relative Dating

William Smith's use of correlation had shown that the ages of rocks could be compared across distances of tens of miles by means of fossil content and physical appearance of the rock units. By use of fossil content alone, it soon became possible to correlate across hundreds and then thousands of miles.

By the middle of the nineteenth century a general geologic column had been fairly well developed. It is a diagram combining in succession from youngest to oldest the sequence of all known strata compiled on the basis of fossil or other evidence of relative age. This geologic time scale and rock column are shown in Table 9.1. Originally, dates were fixed only in relation to other events with the age increasing downward. Values in years before the present were added when absolute dating became available.

The column represents not only a succession of layers but a passage of geologic time. Prior to the 1930s and 1940s no absolute values in years could be assigned to the geologic rock column. Although it could be used only to assign relative dates many of the details of folding, uplift, and erosion had been worked out and placed properly in the sequence. Only the advent of radiometric dating was needed to complete the story.

ABSOLUTE OR RADIOMETRIC DATING

In 1895 and 1896 several events occurred that prepared the way for absolute dating of earth materials. Within a few months of each other, Becquerel discovered radioactivity in uranium salts, William Röntgen discovered x-rays, and Madam Curie isolated radium, a radioactive element. Between 1905 and 1913 the nature of radioactivity and isotopes was clarified and by the 1930s the difficulties concerning various applications had been resolved. Radioactive decay could be used to date rocks that contained these elements.

Emission of Particles

Radioactivity functions in the following way. A few elements, among them uranium and thorium, disintegrate spontaneously into lighter elements when their nuclei give off particles of three different types: alpha, beta, and gamma radiation. Alpha radiation is

TABLE 9.1 Geologic Time Scale and Rock Column

Rocks Time Era	Rocks Time System Period	Derivation of Names	Rocks Time Series Epoch	Time in Years Duration	Time in Years Before the Present	Some Aspects of the Life Record	Some Aspects of Physical Events
Cenozoic	Quaternary	Geologic eras were originally named Primary, Secondary, Tertiary, and Quaternary. The first two names are no longer used; Tertiary and Quaternary have been retained but used as period designations.	Holocene	10,000		Recorded History	
	Quaternary		Pleistocene	2,000,000		Humans	Glaciation
	Tertiary		Pliocene	9,000,000			Formation of Pacific Coast Range
	Tertiary		Miocene	12,000,000		Grass becomes abundant	
	Tertiary		Oligocene	11,000,000			Formation of the Alps and many mountain chains
	Tertiary		Eocene	22,000,000			Volcanic activity in western United States
	Tertiary		Paleocene	9,000,000	65,000,000	Horses appear	
Mesozoic	Cretaceous	Derived from Latin word for chalk (creta) and first applied to extensive deposits that form white cliffs along the English Channel.		71,000,000		Extinction of dinosaurs. Birds appear	Early folding of Rocky Mountains
	Jurassic	Named for the Jura Mountains, located between France and Switzerland, where rocks of this age were first studied.		54,000,000			
	Triassic	Taken from word "trias" in recognition of the threefold character of these rocks in Europe.		40,000,000	230,000,000	Dinosaurs appear	

Era	Period	Description	Years		Events
	Permian	Named after the province of Perm, U.S.S.R., where these rocks were first studied.	50,000,000		Folding of Apppalachian Mountains Widespread glaciation
	Pennsylvanian	Named for the State of Pennsylvania where these rocks have produced much coal.	30,000,000		Coal-forming swamps
	Mississippian	Named for the Mississippi River valley where these rocks are well exposed.	35,000,000		Paleozoic Alps
Paleozoic	Devonian	Named after Devonshire, England, where these rocks were first studied.	60,000,000		
	Silurian	Named after Celtic tribes, the Silures and the Ordovices, that lived in Wales during the Roman conquest.	20,000,000		Vertebrates appear (fish)
	Ordovician	Taken from Roman name for Wales (Cambria), where rocks containing the earliest evidence of complex forms of life were first studied.	75,000,000		First abundant fossil record (marine invertebrates)
	Cambrian		100,000,000	600,000,000	
Precambrian Time		The time between the birth of the planet and the appearance of complex forms of life. More than 80% of the Earth's estimated 4.6 billion years falls within this era.	4,000,000,000		Scanty fossil record Primitive marine plants and invertebrates, one-celled organisms Origin of the Earth
				4.6 billion	

Stratigraphy and Geologic Time

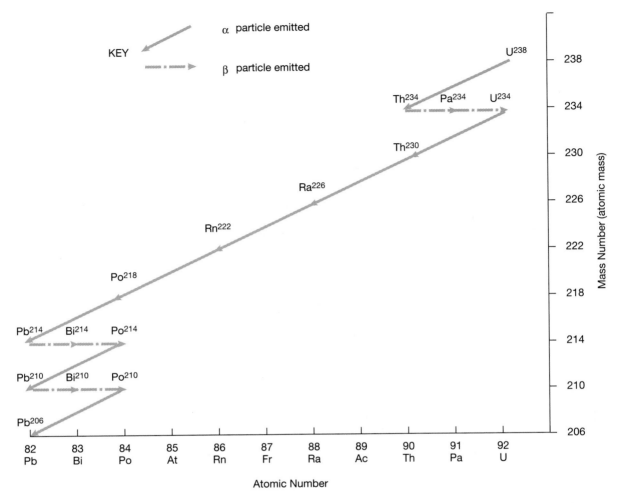

KEY

α particle emitted

β particle emitted

FIGURE 9.6 Radioactive decay of U^{238} to Pb^{206}.

the emission of helium atoms, $_2He^4$, from the nucleus at the speeds of thousands of kilometers per second.[1] This emission converts the original atom into another having an atomic weight of four less and an atomic number of two less. The emitted particles ($_2He^4$) collide with surrounding atoms to generate a considerable amount of heat. This supplies a substantial portion of the heat given off today by the Earth's interior.

Beta particles consist of electrons derived from the disintegration of neutrons in the nucleus, which form protons after the ejection of electrons. Although

[1] The designation $_2He^4$ indicates atomic number = 2, atomic mass = 4.

electrons are expelled at an even higher velocity than are alpha particles, they have so little mass that the heat generated is negligible. Gamma rays are short-wavelength x-rays emitted at the speed of light.

Decay of U^{238}

Emission of either alpha or beta particles from the nucleus of a radioactive atom converts it into a new element (reduces its atomic number). For example, $_{92}U^{238}$ decays through a series of seven emissions of alpha particles and six emissions of beta particles until it reaches a stable nonradioactive isotope $_{82}Pb^{206}$. This is shown in Figure 9.6 in a plot of atomic number versus atomic mass for the $_{92}U^{238} \rightarrow _{82}Pb^{206}$ reaction.

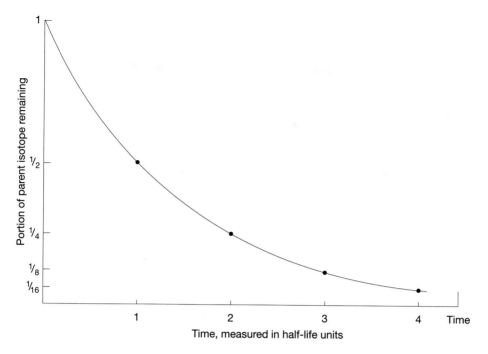

FIGURE 9.7 Graph of decay of an isotope with time.

Uranium has two radioactive isotopes, U^{238} as discussed previously, which provides 99.28% of natural uranium, and U^{235}, comprising only 0.72%. It is U^{235} that is used in nuclear power plants to generate electricity. Several other radioactive elements are also utilized in dating geologic materials and are discussed later in this section.

Decay Constants

The rate of decay is constant for each radioactive isotope, but rates differ considerably from one isotope to another. Disintegration rates are not affected by temperature, pressure, or the chemical environment, which makes it possible to determine the age of geologic materials containing such radioactive elements. Prior to 1950, gravimetric techniques (accurate weight determinations) were used to find the mass of the end member elements of the decay sequence (for example, $_{92}U^{238}$ and $_{82}Pb^{206}$) and, based on the decay constant, the age of the rock was calculated. This typically required large-sized minerals from pegmatites. Today, using the mass spectrometer, only a few milligrams of the two are needed so that small-sized grains can be used for age determinations.

The decay constant λ is used to express the proportion of atoms of an isotope that decays in a unit of time. The decaying or parent isotope continually decreases in amount while the end member isotope, the daughter, continues to increase. The fraction of the total number of parent atoms that decay during a given interval of time is constant but the actual number that decay will decrease because the parent atoms are being continuously depleted. The abundance of the parent isotope decreases exponentially with time as shown in Figure 9.7. Disintegration rates are expressed in terms of the half-life of the radioactive substance, the time required for half of the atoms to disintegrate or decay. Half-life units are shown on the abscissa of Figure 9.7.

Calculations for Decay Constant and Half-Life

The disintegration of radioactive isotopes occurs according to a first-order differential equation, which can be written as:

$$-\frac{dc}{dt} = \lambda c$$

where c = the concentration, t = time, and λ = the decay constant. Separating variables and adding the limits of integration,

$$-\int_{c_0}^{c_1} \frac{dc}{c} = \int_0^t \lambda \, dt$$

or

$$\ln c \Big|_{c_0}^{c_1} = -\lambda t \Big|_0^t$$

$$\ln c_1 - \ln c_0 = -\lambda t$$

$$\frac{c_1}{c_0} = e^{-\lambda t}$$

$$c_1 = c_0 e^{-\lambda t}$$

which is a common form of the equation. Solving for λ,

$$-\lambda t = \ln c_1 - \ln c_0$$

$$\lambda t = \ln c_0 - \ln c_1 = \ln c_0/c_1$$

$$= 2.303 \log c_0/c_1$$

$$\lambda = \frac{2.303}{t} \log c_0/c_1$$

If $c_0 = a$ and $c_1 = a - x$, where a is the initial amount and x is the amount remaining after time t, then

$$\lambda = \frac{2.303}{t} \log \frac{a}{a-x}$$

at the time when one half of the parent isotope is used up, which is the half life or $t_{1/2}$, $a - x = 1/2$ if $a = 1$. Then

$$\lambda = \frac{2.303}{t_{1/2}} \log \frac{1}{1/2} = \frac{0.693}{t_{1/2}} \quad \text{or} \quad t_{1/2} = \frac{0.693}{\lambda}$$

Limitations of Isotope Dating

A list of the radioactive isotopes commonly used in age dating is provided in Table 9.2. The radioactive decay, as stated previously, finally ends in the formation of a stable end product although a number of steps is involved in the process. Some additional information about the isotopes is also provided in Table 9.2.

Stringent conditions must be met in order to provide an accurate age for a rock. In the case of uranium and thorium, different isotopes of lead pro-

vide the end products. The minerals analyzed must be fresh, that is, not altered or weathered, because circulating solutions might leach out the lead and the parent isotopes at different rates. Metamorphism of the mineral would also invalidate the age calculation because lead may be driven out faster than uranium or thorium. Hence the rock or mineral must represent a closed system since the time it was formed. No loss or gain of either parent or daughter isotope can have occurred. Also if the radioactive mineral happened to be closely associated with a lead mineral containing mostly common lead, it would be difficult to sort out the various lead isotopes present.

Age Dating Based on Uranium, Thorium, and Lead

Many minerals meet the rigorous conditions just described. For samples containing uranium isotopes, independent dates can be obtained based on U^{235}/Pb^{207} and U^{238}/Pb^{206}. A date can also be determined on Pb^{207}/Pb^{206} and these are usually quite reliable. If any leaching or other loss occurred to the sample, this ratio should be unaffected although the quantities themselves may change markedly because the two lead isotopes are chemically identical and would be removed in the same fashion. In addition, Th^{232}/Pb^{208} may be used in the same rock for age verification. If the age value obtained by these different ratios agrees within a few percent, the determination is considered to be reliable. Because of the long half-life of these isotopes, rocks must be more than a few million years old to provide accurate dates.

Age Dating Using Rubidium and Strontium

The $_{37}Rb^{87}$ isotope decays to $_{38}Sr^{87}$ by the emission of a β particle. Since the half-life of $_{37}Rb^{87}$ is about 47 billion years, the method is limited to rocks more than a few million years old. As shown in Table 9.2, it is useful for dating minerals in metamorphic rocks.

Common strontium consists of four isotopes, Sr^{84}, Sr^{86}, Sr^{87}, and Sr^{88}, but not all Sr^{87} is derived from the disintegration of Rb^{87}. Hence it is necessary to correct for the Rb^{87} in the mineral not generated by radioactive decay. Because of this needed correction to a mass that is already small, Rb-Sr age calculations are extremely sensitive to errors in the youngest rocks within their dating span, that is, those rocks only a few million years old. They are instead much more

TABLE 9.2   Principal Isotopes Used in Radiometric Age Dating

Isotopes	Half-Life (years)	Effective Dating Range (years)	Earth Materials That Can Be Dated
Uranium-238/ lead-206	4.50×10^9	10^7 to age of Earth°	Zircon Uraninite Pitchblende
Uranium-235/ lead-207	0.71×10^9	10^7 to age of Earth	
Thorium-232/ lead-208	15×10^9	10^7 to age of Earth	Zircon
Potassium-40/ Argon-40	1.30×10^9	10^4 to age of Earth	Muscovite Biotite Hornblende Whole volcanic rock Arkose$^+$ Sandstone$^+$ Siltstone$^+$
Rubidium-87/ Strontium-87	4.7×10^{10}	10^7 to age of Earth	Muscovite Biotite Microcline Whole metamorphic rock
Carbon-14	5730 ± 30	0 to 50,000	Wood Charcoal Peat Grain Tissue Charred bone Cloth Shells Tufa Ground water Ocean water

°Age of the Earth is about 4.6×10^9 years.
$^+$For paleogeographic studies.

reliable for dating rocks Mesozoic in age or older (greater than 65 million years). However, because the Rb-Sr decay path does not contain a gaseous phase as does the K-Ar disintegration (argon is an inert gas), Rb-Sr determinations are affected much less by the alteration that occurs during slight degrees of metamorphism. Hence, many of the older ages of metamorphic rocks are derived from Rb-Sr for increased accuracy.

Experiments show that biotite heated above 350°C no longer gives stable ages for the Rb-Sr determination whereas muscovite must be heated to about 550°C before its lattice permits the loss of strontium. The cooling history of the rock may be dated by such comparisons, or metamorphic occurrences may be so established. Such metamorphism is sometimes said to "reset the clock" for age determination.

Age Dating from Potassium and Argon

The potassium isotope $_{19}K^{40}$ decays simultaneously into two daughters, $_{20}Ca^{40}$ by emitting a β particle and $_{18}Ar^{40}$ by capturing an electron from the innermost shell. Of the decaying nuclei, 89% take the Ca^{40}

route whereas 11% become Ar^{40}. The conversion to Ca^{40} is of no value in geochronology or age dating because Ca is so abundant in nature and the Ca^{40} isotope is not easily separated from the other Ca isotopes.

The Ar^{40} from the sample deterioration must be discerned from the Ar^{40} of the atmosphere, but this can be accomplished with accuracy. In the atmosphere, a constant ratio of Ar^{36} to Ar^{40} persists so that by measuring both Ar^{36} and Ar^{40} in the sample, a correction for the Ar^{40} contribution from the atmosphere can be made. The half life of the K-Ar disintegration process is 1.30 billion years, and rocks approximately 100,000 years or older can be dated. Under extremely favorable conditions, rocks as young as 50,000 years may also be dated.

Argon is an inert gas and therefore does not react to form any chemical compounds. It readily enters and leaves some crystal lattices by diffusion but in certain lattices it seems to be retained indefinitely. The minerals biotite, muscovite, hornblende, and sanidine (the high-temperature form of potassium feldspar, which is common in many volcanic rocks) are particularly good recipients of argon. Any minerals that readily lose argon are not suitable for radiometric dating.

When subjected to high temperatures all rocks and minerals lose argon and hence the K-Ar dates are extremely sensitive to the thermal history of the sample. Hornblende seems to retain argon at higher temperatures than does biotite, so that K-Ar dates on the same rock using these two minerals may yield quite different dates. This can be used to date an occurrence of metamorphism documented by the lower age value obtained. Commonly a U-Pb analysis would be performed as well to corroborate the time of formation of the rock.

Radiometric dates on some sedimentary rocks can also be obtained using the K-Ar method. Although most minerals analyzed are from igneous or meta-igneous rocks, the sedimentary mineral glauconite, a greenish clay-like material, contains potassium and has been successfully dated. It forms on the seafloor prior to cementation of the sediment and is fairly common in shale, siltstone, and sandstone. It is widely accepted that argon diffuses from the glauconite lattice more readily than from the micas and thereby the accuracy of radiometric dates for sedimentary

rocks is less reliable. Typically, sedimentary ages would be bracketed by dates on intrusive rocks, which include those both younger and older than the sedimentary unit.

Radiocarbon Dating

Development of the Method

All of the radiometric methods discussed previously apply to extremely old rocks. In 1947 a method based on C^{14}, or radiocarbon, for determining the ages of younger materials was discovered. Carbon-14 is continuously formed in the upper atmosphere by the bombardment of N^{14} by neutrons from cosmic radiation. Radiocarbon decays by β radiation to N^{14} with a half-life of 5730 years. (Previously a half-life value of 5568 years for C^{14} had been determined so that earlier calculated carbon dates must be adjusted by a factor of 1.03.)

The reaction for the C^{14} relationship is

$$6^{C14} - \beta = {}_7N^{14}$$

It is not necessary, and it would actually be impossible, to measure accurately the amount of daughter N^{14} to obtain an age determination. Instead, the amount of C^{14} is determined relative to the other carbon isotopes (mostly C^{12} with about 1% C^{13}) and the age is found by that means. The details involving the carbon system are described in the following section.

$C^{12}-C^{14}$ Ratio

Radioactive carbon mixes with ordinary carbon, diffusing rapidly through the atmosphere, hydrosphere, and biosphere. The proportion of radiocarbon is essentially constant throughout the system because of the rate of mixing compared to the half-life of C^{14}. As long as the production rate of C^{14} remains constant, the ratio of C^{14} to C^{12} is also constant because the production rate and decay rate are in equilibrium. Introduction of old carbon from the burning of coal since the Industrial Revolution and the variation with latitude on Earth in generation of C^{14} in the atmosphere complicate the interpretations but do not invalidate the method.

As long as an organism is alive, it injests air and water and maintains the equilibrium proportion of C^{14}. After death, the equilibrium is lost as replenishment of air or water ceases and C^{14} is reduced by radioactive decay. Because of the short half-life, C^{14}

dates (time elapsed since death) of as low as 100 years can be measured. After 40,000 to 50,000 years (seven to nine half-lives) so little C^{14} remains that dating is no longer possible. For organic materials less than 50,000 years old, radiocarbon is an invaluable measurement tool. It completely revolutionized the study of archaeology and provides great insight into the Holocene and late Pleistocene time spans.

Sources of Error

Several sources of error are associated with radiocarbon dating. As previously stated, the introduction of "old carbon" by burning coal and the variation in production of C^{14} with latitude location on Earth are part of the problem. A variation of C^{14} concentration in the atmosphere with time is also involved. The flux of cosmic rays varies inversely with the strength of the Earth's magnetic field and directly with solar flares. We know that the strength and polarity of the magnetic field have varied significantly in the geologic past, and as a consequence we know that the C^{14} concentration has varied as well.

Algae in springs whose CO_2 is partly obtained from dissolved limestone or "old carbon" yields tufa with a much lower C^{14} content than the air. It is indeed "born old." Mollusk shells (clams, oysters, etc.) in areas where ground-water springs flow upward into the ocean also obtain old carbon from the water, which typically is several hundred years old. They too would register older than the time of death of the mollusk. By contrast, porous reefs composed of coral and other animal shells, long dead, would be rejuvenated by the present-day water from the sea spray providing a younger age for the corals.

There are problems associated with dating parchment (animal carbon), tree rings, and ancient artifacts. Commonly the radiocarbon dates do not agree with the established historical dates. The problems of flux and mixing account for part of this, but for the tree rings an additional complication occurs. Each tree ring is active for only one year and does not reequilibrate with the next succeeding ring. Therefore, only the last ring to form before the death of the tree would indicate the correct length of time since its demise. These problems notwithstanding, radiocarbon dating has greatly expanded the knowledge of prehistory and of geologic events in the late Pleistocene and Holocene.

Fission Track Dating

Uranium atoms, in addition to disintegration yielding α and β radiation, also break down by fissioning into two essentially equal-sized nuclei. This involves an extremely small number of atoms, at a rate of 1 in 69×10^{16} atoms per year. These fragments fly apart with a great amount of force, striking the crystal lattices of surrounding minerals and leaving extremely small tracks, about 50Å wide (5×10^{-7} cm) and 10 to 20 µm (10 to 20×10^{-4} cm) long. The small imperfections are below the resolving power of most microscopes but they are readily enlarged by etching in sodium hydroxide or hydrofluoric acid.

Such etched surfaces are being used for geologic dating. Minerals known to concentrate uranium are selected from the rock, embedded in plastic, and etched with sodium hydroxide. For glassy rocks, polished surfaces are etched with hydrofluoric acid. The pits and cones of the lattice imperfections are counted on a measured area. Following this determination, the specimens are bombarded by neutrons in a nuclear reactor to produce additional fission tracks from the U^{235} fissioning in the samples. More etching and the new tracks are counted. Since the flux of neutrons is known in the nuclear reactor, the increase in fission tracks is a measure of the uranium atoms present. From this it is possible to compute the time required to cause the tracks first counted.

About ten minerals and natural glasses are known to contain fission tracks suitable for dating. It has also been learned that tracks will heal in samples heated to several hundred degrees Celsius and samples exposed at the Earth's surface will give erroneous results because of the effects of cosmic rays. The best results are likely to be obtained on materials less than 100 million years old.

Fission-track dating is a relatively new method and it has not been tested as thoroughly as the radiometric techniques. The best evaluation will be made by comparing fission track dates and the radiometric methods on the same samples.

Absolute Age of the Earth

The tremendous expanse of geologic time is no longer debatable. It is likely that the Earth is nearly 4.6 billion years old. Rocks approaching this age, from eastern Siberia, have been reported. In Greenland rocks measured at 3.75 billion years old are noted and in South Africa very old granitic rocks 3.50 billion years in age are found. The oldest rocks in North America are gneisses from Minnesota dated at about 3.35 billion years.

The radiometric dating methods best suited for these ancient rocks are the U-Pb methods. K-Ar is too susceptible to loss of argon during subsequent heating of the rocks. Rb-Sr and U-Pb methods are not readily susceptible to "resetting of the atomic clock" by reheating but the U-Pb method applied to zircon is a particularly accurate means for measuring the age of very old rocks.

It is of interest that the 4.6 billion year duration of the Earth is equal to about one half-life of U^{238}-Pb^{206} and more than seven times the half-life of U^{235}-Pb^{208}. Hence, the Earth originally contained 40 times as much U^{235} and twice as much U^{238}. The heat flow from the Earth generated by this radioactive decay must have been many times that occurring today and, consequently, many geologic processes were much more active.

Some meteorites have been dated at 4.6 billion years old, which is thought to be the age of the Earth as a planet. These likely indicate the nature of the planetesimals from which the Earth and other inner planets were formed.

THE AWESOME SPAN OF GEOLOGIC TIME

As mentioned, the age of the Earth is now estimated at 4.6 billion years. The Precambrian Era covers the time span from the Earth's beginning until the start of Cambrian time, a duration of 4 billion years. The earliest organic structures in the form of limey secretions, carbonaceous residues, and other indirect evidences of life first appeared about 3 billion years b.p. (before the present). Simple one-celled animals appeared about 2 billion years b.p. and shelled animals evolved to become abundant 600 million years ago at the start of Cambrian time. The first land plants were established about 500 million years b.p. and dinosaurs appeared 230 million years b.p. and became extinct 65 million years b.p. The early horse appeared at 60 million years and humans at about 2 million years b.p. Recorded history covers only the last 10,000 years.

The 4.6 billion years can be represented by a 24-hour day to emphasize the amount of time between events. The Earth would form at 12 midnight

and the Precambrian era would extend until 8:52 P.M. The first indication of life would occur at 8:35 A.M. with simple one-celled life at 1:34 P.M. and abundant shelled animals at 8:52 P.M. The first land plants evolve at 9:23 P.M. Dinosaurs make their appearance at 10:48 P.M. and became extinct at 11:40 P.M. Humans appear about 38 sec before midnight and recorded history represents about 0.2 sec. The life span of an individual human, three score and ten years, is equivalent to 1.3×10^{-3} sec of this 24-hour day.

HUMAN TIME VERSUS GEOLOGIC TIME

The great span of geologic time provides the backdrop for a difficult comparison for the beginning earth scientist and other students of geology to fully appreciate, that is, the considerable difference between the concept of time in the geological sense and in the engineering or worldly sense. The life of an engineering structure may be 50 or 100 years, the 100-year flood is a long interval of concern, and concrete pavements that last for 40 years without major repair are a resounding success. But erosion continues over millions of years to establish a certain stream gradient, and limestone cave formation may require a similar period of time. Thousands of years are required to develop a fertile soil profile and the retreat of continental glaciers involves much the same span of time.

A point of some interest to the engineer involves limestone bedrock and leakage of a surface reservoir impounded behind a dam. Fissured and cavernous limestones provide conduits to transport leakage away from the reservoir and such limestones are a concern if the regional ground-water gradient encourages such movement. However, limestones free of fissures will not develop conduits to carry sizable quantities of water away in the limits of 50 or 200 years, a normal life span for a dam and reservoir. These underground plumbing systems take many times that number of years to develop.

What is a reasonably long time span in human years may be insignificant from a geological standpoint. Geologically speaking, lakes are a temporary interruption in the erosion cycle of streams and in time their outlets will be destroyed or their basins filled with sediment. This is not to say that the Great

Lakes will soon be lost, for in human time they should enjoy many centuries of existence, but measured in millenniums or in large units of time their demise is to be expected. Of course, humans may be able to stave off the effects of natural erosion to extend this time period considerably.

The downslope movement of soil and rock under the force of gravity is a natural occurrence and part of the erosional process on Earth. Humans attempt to prevent significant slope movements in populated or economically important areas over periods of human years. In some cases they induce greater movement but in others are only innocent bystanders. The purpose, of course, in geology applied to engineering is to minimize the effects of geologic hazards in a variety of ways including restraint, strength improvement, and avoidance. This does not suggest that it is easy to switch from the concept of geologic features in terms of geologic age to a concern for engineering structures designed and built in terms of human years. The two must be considered in turn because they both impact the study of geology applied to engineering.

GEOLOGIC MAPS

Although engineers working in construction would not be expected to prepare a geologic map of a site, they will need to understand engineering geology reports for construction sites or areas of study. By contrast, engineering geologists and geological engineers do have the training skills to prepare geologic maps. Engineering geology reports usually contain, among other drawings, a geologic map, which is the most compact manner for supplying detailed geologic information. A preliminary geologic map may be used as a basis for planning the subsurface exploration program for a project or to investigate possible supplies of construction materials. Following this detailed work, geologic maps and cross sections may be used to summarize the information found in the exploration program.

Geologic maps show the distribution of earth materials located at the Earth's surface or at the bedrock surface. They also show the position in space (attitude) of rock units and indicate their relative age. As previously mentioned, the formation is the smallest geologic unit commonly depicted on geologic maps but the map scale and dimensions of the map will determine this specifically. A geologic map of an entire state of the United States typically can show only groups of rocks or collections of formations.

The line of intersection between geologic units on a map is called a geologic contact or simply a contact. This represents the surface that separates the two units but it appears as a line on the map or in a plan view. Geologic maps produced by the U.S. Geological Survey are drawn on a topographic map base. This makes it possible to determine the elevations of rock contacts and to see the relationship between elevations and rock units.

The legend on a geologic map explains the symbols used for the map. Two letters typically make up the symbol for a formation such as S_W, which would designate the Wabash Formation of the Silurian System.

The two primary types of geologic maps are areal geologic maps and surficial geologic maps. Areal geologic maps are the most common and are sometimes called general-purpose geologic maps or simply geologic maps. They show the distribution of rocks on the Earth's surface as it would appear with all soil or unconsolidated material removed. Map units are depicted by different colors and are described in the legend. Symbols are also used to denote the various units and examples are shown in Figure 9.8. The rock units are also listed in vertical columns in chronological order in the map legend beginning with the youngest at the top. Symbols, showing the attitude of the beds, are added to the map from which the arrangement of the rocks in the subsurface can be determined. This subject requires knowledge of structural geology, which is presented in detail in Chapter 10. A geologic map depicting inclined sedimentary beds is presented in Figure 9.9. A geologic map showing horizontally bedded sedimentary rocks is provided in Appendix B, Topographic Maps.

Surficial geologic maps show the unconsolidated materials that occur at the Earth's surface. These maps are available mostly in areas covered by glacial deposits and hence have few bedrock outcrops. Consequently, they are sometimes called glacial geology maps. Agricultural soils maps also depict surface materials and are included in this general category. They describe and classify the soils based on agricultural concerns but some useful relationships to engineering properties of soil are supplied as well. Surficial maps are useful sources of information in terms of foundation support for engineering structures and

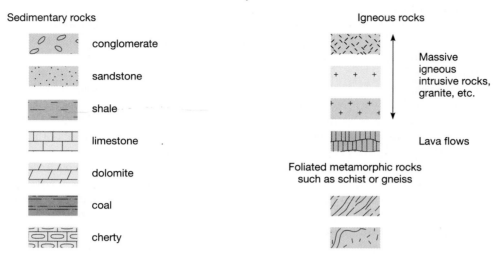

Rock Symbols

Sedimentary rocks

	conglomerate
	sandstone
	shale
	limestone
	dolomite
	coal
	cherty

Igneous rocks

Massive igneous intrusive rocks, granite, etc.

Lava flows

Foliated metamorphic rocks such as schist or gneiss

Letter Symbols

	Symbol
Pleistocene or Quaternary	Q
Pliocene	T_{pl}
Miocene	T_m
Oligocene	T_o
Eocene	T_e
Paleocene	T_p
Cretaceous	K
Jurassic	J
Triassic	T_R
Permian	C_{pm} or P
Pennsylvanian	C_p or P
Mississippian	C_m or M
Devonian	D
Silurian	S
Ordovician	O
Cambrian	$\in$
Precambrian	P$\in$

FIGURE 9.8 Rock symbols and letter symbols used on geologic cross sections and geologic maps.

locations of construction materials including sand, gravel, and common borrow. They are used for planning detailed subsurface investigations, which involve drilling and sampling.

Other maps can be developed from geologic maps by combining or highlighting certain features of the rock units. These may be termed engineering geology maps, environmental geology maps or other similar designations. It should be kept in mind that these are interpretive maps developed through interpretation of geologic map information. They are one more step removed from the specific field data and must be viewed with that understanding.

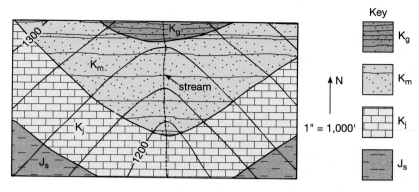

FIGURE **9.9**   Geologic map showing inclined sedimentary rocks.

EXERCISES ON STRATIGRAPHY AND GEOLOGIC TIME

1. Give definitions of the terms *formation, member, group,* and *stratigraphy.*

2. Discuss the three principles of superposition, original horizontality, and cross-cutting relationships.

3. Define unconformity. Supply the names of the different unconformities for the four boxes shown as follows. Draw an appropriate cross section for each.

4. Give the sequence of events that occurred in the four cross sections pictured here. Begin with the oldest unit (the lowest number is always the oldest). Locate and label any unconformities. Indicate how you determined the relative ages of the units (such as superposition, cross-cutting, etc.). You may need to refer to Figure 9.8 for information on lithologic symbols.

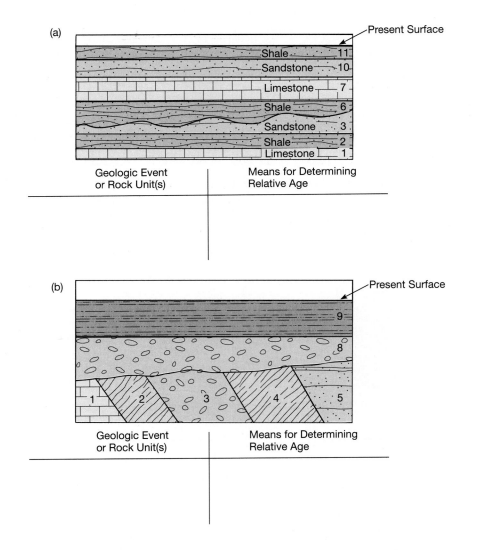

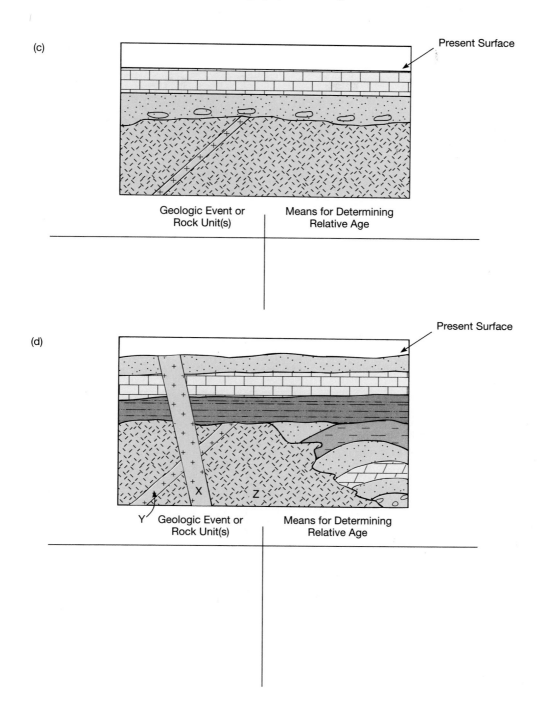

(c)

Present Surface

Geologic Event or
Rock Unit(s) | Means for Determining
Relative Age

(d)

Present Surface

Y   Geologic Event or
Rock Unit(s) | Means for Determining
Relative Age

5. Correlation Studies

For (a) and (b), (on page 196) correlate from one section to another. Connect unconformities, igneous intrusions, and formation contacts as appropriate using dotted lines. For beds that do not carry across, indicate what has occurred (lithofacies change, pinchout, etc.). Diagram (a) depicts two exposures in a plateau region separated by 50 miles; diagram (b) is based on three drill holes spaced a number of miles apart. You may need to refer to Figure 9.8 for information on lithologic symbols.

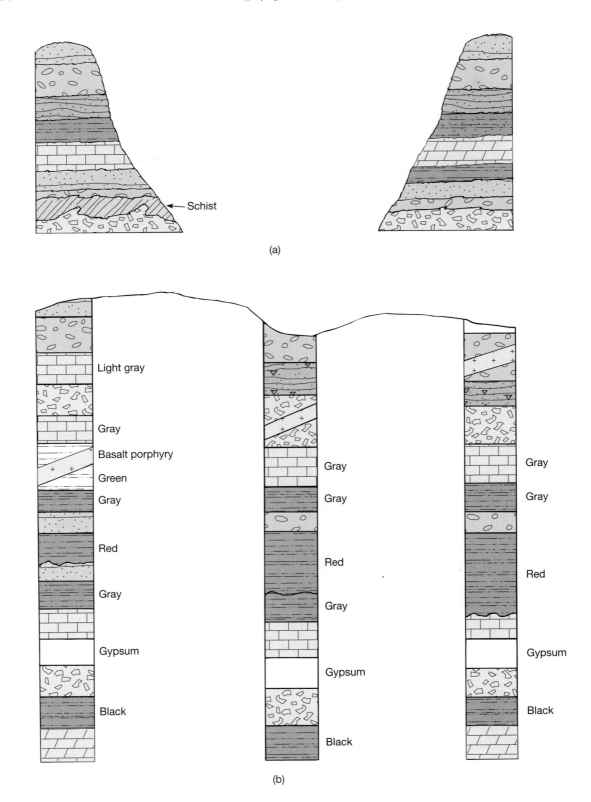

(a)

(b)

6. *Estimating the Age of the Earth:*
   (a) What are the two primary reasons why humans seek to learn the age of earth materials? Into which category would applied science and engineering fall? Why?
   (b) How did the biblical scholars estimate the age of the Earth in the seventeeth century? What age value did they obtain?
   (c) How did James Hutton and William Smith arrive at the conclusion that the Earth's age was greater than that obtained by the biblical scholars? How did the physicists of that time conclude the Earth had to be older? Explain.
   (d) How were estimates on the age of the Earth made using the following considerations? What values were obtained?: cooling of the Earth; sedimentation in the Nile; sodium in the oceans; and limestone formation throughout the world.

7. *Geologic Time:* Supply answers to the following questions (refer to Table 9.1 for assistance).
   (a) How long was the Mesozoic Era? The Paleozoic Era?
   (b) How many years elapsed between the end of the Cambrian and the beginning of the Permian? How long was it between the beginning of the Eocene and the end of the Miocene? How long did it take for the Mississippian System of rocks to form?
   (c) What percentage of the age of the Earth is represented by Precambrian time? By the Cenozoic Era?
   (d) Humans first appeared on Earth about 2 million years ago and recorded history consists of the last 10,000 years. What percentage of the age of the Earth do each of these represent?

8. *Radiometric Age Dating:*
   (a) What are the three different particles given off by radioactive decay of isotopes? What happens to the isotope in the process? Explain.
   (b) Thorium-232 decays to form lead-208 by a ten-step reaction in this order: α, β, β, α, α, α, α, α, β, β, α. In the same form as shown for the $_{92}U^{238}$ to $_{82}Pb^{206}$ reaction (see the text), show this for the $_{90}Th^{232}$ to $_{82}Pb^{208}$ reaction.
   (c) Calculate the value of λ, the decay constant, for U^{238}/Pb^{206}, U^{235}/Pb^{208}, and Th^{232}/Pb^{208}. Include the correct units for λ.
   (d) What is the half-life for carbon-14? How many half-lives occur in 50,000 years, the limit of age dating by this method? What fraction of the original C^{14} remains after 50,000 years? (Calculate as accurately as possible.)

   (e) Why is Rb-Sr dating more accurate in dating some metamorphic rocks if muscovite is used rather than biotite? Explain. What is meant by "resetting the clock" in this respect? If a muscovite sample indicated an age of 138 million years and a biotite sample from the same rock indicated 122 million years, what does this indicate relative to the previous question? Explain.
   (f) Why is the K-A dating particularly subject to errors if considerable metamorphism of the rock has occurred? In a certain metamorphic rock the U^{238}/Pb^{206} date made on a zircon crystal was 275 million years, whereas the K-A date on a hornblende crystal was 204 million years and the K-A date on a biotite crystal was 160 million years. Assuming all of these values are useful, what overall history of this rock can be given?
   (g) How is the clay mineral glauconite used to measure the age of sedimentary rocks? What problem exists with these radiometric dates?
   (h) Relative to C^{14} dating, what is meant by "old carbon" and rocks being "born old"? Why do tree rings and carbon dates not always agree in age? Explain.
   (i) How are fission tracks measured? What age of materials is best measured by this method? Why?
   (j) How do we know that the Earth once contained twice as much U^{238} and 40 times as much U^{235} as it currently does? What effect would that have on volcanic eruption, heat flow, and continental drift if any? Explain.

9. *Great Span of Geologic Time, Human Time Versus Geologic Time:*
   (a) In the text, the divisions of geologic time and geologic events were compared to a 24-hour day. Other comparisons have been made relative to tall buildings. How would such a comparison be made to the Sears Tower in Chicago or the World Trade Center in New York City? Explain. Other than comparisons to tall buildings, name at least one more way this aspect could be dramatized.
   (b) Why is it difficult to keep the comparison between human time and geologic time in perspective? Why must the engineer, working with earth materials, and the engineering geologist deal with this repeatedly? How does the subject of faulting relate to these considerations? Explain.

10. *Geologic Map Reading, Preliminary Aspects:* Answer the following questions on the geologic map depicting inclined sedimentary beds (refer back to Figure 9.9).The four formations shown in Figure 9.9 are the Simpson Shale of the Jurassic System plus the Goodby

Sandstone, the Jacobville Limestone, and the Martinique Formation all of the Cretaceous System.

(a) What is the area of the map and what is its contour
    interval?
(b) In what direction does the stream flow? How is
    this determined? What is its gradient?
(c) Are the sedimentary rocks horizontal? If not, how
    do you know that? If they are, what tells you so?
(d) Which is the correct sequence for the age of the
    formations, older to younger: 1) J_s, K_j, K_m, and K_g
    OR 2) K_g, K_m, K_j, and J_s? Give the correct answer
    and indicate why.

Additional Readings

DOTT, R.H., Jr., and BATTEN, R.L., 1980, *Evolution of the Earth,* 3rd ed., McGraw-Hill, New York.

DUNBAR, C.O., and ROGERS, J.J., 1963, *Principles of Stratigraphy,* John Wiley & Sons, New York.

EICHER, D.L., 1976, *Geologic Time,* 2nd ed., Prentice Hall, Englewood Cliffs, NJ.

LAPORTE, L.F., 1968, *Ancient Environments,* Prentice Hall, Englewood Cliffs, NJ.

STANLEY, S., 1986, *Earth and Life Through Time,* W.H. Freeman & Co., New York.

10

Structural Geology

Structural geology involves the study of rocks that have been deformed by earth stresses and includes a description of their position in space, or attitude, which occurs as a consequence of the deformation. Sedimentary rocks are of particular interest because their layers or bedding express the extent of deformation. Vertical and/or horizontal movements caused by bending and rupture are shown by the relative displacement of the bedding. The spatial relationships of igneous bodies emplaced by magmatic intrusions are also described in terms of structural geology.

Applying the principle of original horizontality, sedimentary rocks that show tilting or folding indicate deformation by compression or tension and uplift of the strata. In areas such as plateaus, large volumes of rock have been uplifted while preserving the original horizontal attitude. For those rocks of marine origin, which comprise much of the sedimentary rock se-

quence, this involves thousands of feet (or meters) of uplift.

Geologic structures range in size from the microscopic features of a fine-grained specimen visible only under great magnification to large geometric forms in a rock mass that may be measured in miles or kilometers. Foliation in metamorphic rocks, flow banding in extrusive igneous rocks, and bedding in sedimentary rocks are examples of structures that can be observed with the unaided eye.

Several related terms are significant in this discussion. Structural geology on a regional scale is called *tectonics* or *geotectonics*. Another related subject is rock mechanics, which involves the mechanics of rock deformation with regard to the design and construction of engineering works. Much of the material discussed in Chapter 6 would fall within the scope of rock mechanics. Petrofabrics is concerned with the

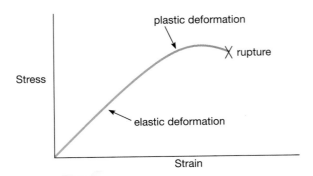

FIGURE 10.1 Schematic diagram of rock deformation in the Earth's crust.

directional aspects of rock texture (for example, rock cleavage) and its relationship to the stress field that deformed the rock.

ROCK DEFORMATION

Rocks are deformed under load in the manner discussed in Chapter 6, Engineering Properties of Rock. In the Earth's crust, as contrasted to testing in the laboratory, rocks deform slowly under high confining pressures and elevated temperature. This causes the rocks to be less brittle so that plastic deformation can occur. The general relationship is shown in Figure 10.1. Stress and strain are the two axes shown in Figure 10.1. Stress is the load per unit area that is applied, whereas strain is the deformation per unit length that is caused by the stress. For deformation in the elastic range, the rock returns to its original shape when the load is removed. An example of this deformation occurs when earthquake waves pass through rock. When the stress is removed the rock returns to its original configuration.

Folding in rocks occurs in the plastic range of the curve. Compressive forces slowly deform the rock but after removal of this load by uplift and erosion, the rock still retains the folded shape. Examples can be observed in highway rock cuts and mountainous terrain. Rocks are typically deformed into a series of troughs and peaks not unlike a sinusoidal wave trace. The geometry of folds is discussed at length in a subsequent portion of this chapter.

Faulting occurs when the rocks rupture. As with folding, the strain is not recovered on release of the

stress. The geometry of faults is also considered in detail in a subsequent section of this discussion on structural geology.

FOLDS IN ROCK

Fold Terminology

Commonly, folds are caused by compressional forces, which buckle rock units. The trough or downwarped portion of the fold is called a *syncline* and the crest portion is an *anticline*. If only one direction of dip prevails in a fold system, it is called a *monocline*. Figure 10.2 shows anticlines and synclines in cross section view and a photograph of these folds in a mountainous region.

The limb of a fold is the sloping portion that connects the crests and troughs. These dip toward the center of the syncline and away from the center of the anticline unless the folds are overturned. More details on overturned folds are presented later in the chapter.

Another essential part of the fold system is the axial plane. This is an imaginary plane used to divide the fold into two equal or nearly equal portions. The intersection of the axial plane with the bedding surface of the rock unit is a line called the axis. This is illustrated in Figure 10.3. In Figure 10.3(a) the axis is horizontal indicating that the fold system trends parallel to a horizontal plane (the Earth's surface). Figure 10.3(b) shows the axis sloping to the rear of the drawing. This indicates that the fold system is not horizontal but plunges (loses elevation) to the rear. This concept is explained later in the discussion.

The axial plane is used to describe the degree of symmetry of the fold system. In Figure 10.3 the axial plane is vertical, which indicates that the fold is symmetrical. If the axial plane is tilted with the limbs dipping in opposite directions but not at the same angle, the fold is asymmetrical [Figure 10.4(a)]. With further inclination of axial plane from the vertical, both limbs dip in the same directions to yield an overturned fold [Figure 10.4(b)]. (The bed on the right side is overturned.) A recumbent fold occurs when the axial plane is essentially horizontal. This is shown in Figure 10.4(c).

Strike and Dip

Before describing the attitude of a fold system in detail we must first visualize a single inclined plane in space,

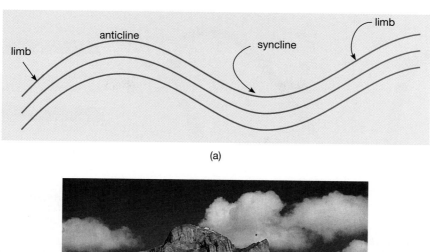

(a)

(b)

FIGURE 10.2 Folded sedimentary rocks. (a) Diagram of a simple fold system. (b) Anticlines and synclines in a mountainous region. (Hamblin, WK, *The Earth's Dynamic System.* 2nd Ed., Macmillan, New York.)

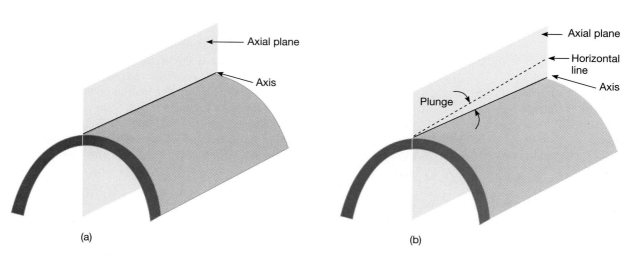

(a) (b)

FIGURE 10.3 Anticline, with axial plane and axis. (a) Nonplunging. (b) Fold plunging to the rear.

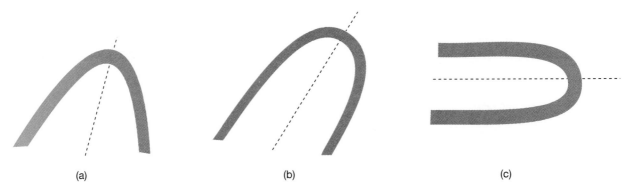

(a)   (b)   (c)

FIGURE **10.4**   Attitudes of asymmetrical anticlines: (a) asymmetrical, (b) overturned, and (c) recumbent.

as shown in Figure 10.5. If a bed is not horizontal, it will form an acute angle when it intersects a horizontal plane. This angle is called the *dip* and it is measured as the maximum inclination between the two planes. The *strike* is the bearing or compass direction (measured from true north) of the line of intersection between the horizontal plane and the inclined plane (Figure 10.5). When a persistent bed intersects the Earth's surface, it will form a ridge running along the strike of the bed. If they are steeply dipping beds (30° dip or more), they are called hogbacks and can be seen, for example, in the Colorado Rockies west of Denver. Gently dipping resistant beds form cuestas, which dip at angles of less than 10°.

Strike and dip directions are mutually perpendicular so, for example, a bed dipping to the east will have a north-south strike, usually designated simply as north. By convention, strike and dip directions are measured from true north so that a bed dipping either to the north or to the south would have an east-west strike, which would be designated as N90°E or N90°W. Again by convention, the strike is designated with respect to north rather than from the south direction.

Several properties of strike and dip can be used as an aid to determine rock attitudes in the field. The strike is simply the direction of a horizontal line on an inclined plane. Consequently, the strike is found by locating such a horizontal line on the dipping plane. Dip is always perpendicular to the strike and is easily determined when the strike is known. Taking another approach, dip is the maximum angle of inclination of the sloping plane. If this maximum slope of an inclined plane is determined by repeated measurement of angles, then the strike occurs at right angles to this maximum slope direction. For a horizontal plane, all lines contained within that plane are horizontal, and there is no unique strike orientation as in a dipping plane. Therefore, strike cannot be defined for a horizontal plane, that is, for a plane that has zero dip.

Anticlines and synclines are extensions of the inclined plane concept. By mental construction, segments of planes inclined at different angles can be assembled to yield the fold system. Therefore, as the inclined beds of the fold system change in dip from one location to another, anticlines and synclines are obtained.

Strike and Dip Symbols

Symbols are used on a geologic map to designate the strike and dip of beds. Figure 10.6 shows these symbols for a system of folds in both cross-section (side) and in plan (map) views. The lowest numbered bed is always the oldest in these diagrams.

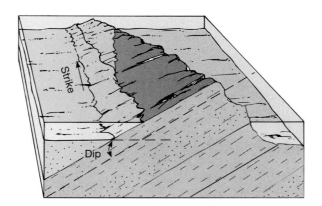

FIGURE **10.5**   Dip and strike of inclined bed.

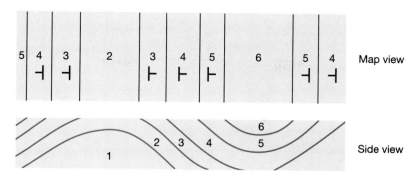

FIGURE 10.6   Anticlines and synclines, with strike and dip symbols.

The strike and dip symbol looks similar to a T, with the stem indicating the dip direction and the crossbar indicating the strike. The amount of dip is sometimes given in degrees opposite the stem portion.

First Rule of Anticlines

Careful examination of the map view portion of Figure 10.6 provides information that leads to the first rule of anticlines. This simply states that for the map view of an anticline, the oldest beds are in the center and the beds become progressively younger in each direction. Note that bed 2 is bounded on both sides by bed 3, which is bounded in turn by bed 4. It can also be noted that the opposite situation is true for synclines, where the youngest bed is in the center and the beds get progressively older in each direction. This rule makes it possible to distinguish between anticlines and synclines in map view, if the relative age of the beds is known, even when strike and dip symbols are lacking.

A review of some of the information presented on folded rocks can be accomplished by analyzing the diagram presented as Figure 10.7. Axial planes are included for reference purposes. Viewing the top portion of the fold system alone in Figure 10.7, the inclination of the axial planes indicates the extent of symmetry of either the anticline or syncline. Also the age of the beds indicates whether an anticline or syncline is present. Strike and dip symbols are shown in a few locations to illustrate how they would be associated with these folds.

Bed Width Relative to Dip

Another feature of folds is illustrated by Figure 10.7, for example, in section C. Note that the width of bed 3 is not the same on opposite sides of the axial plane. On the left side, where the bed is less steep, the width is greater than on the right side where the bed is more steep. This is illustrated in more detail in Figure 10.8.

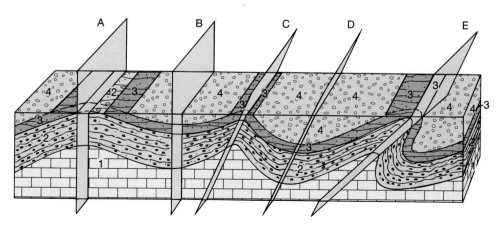

FIGURE 10.7   Diagram of anticlines and synclines.

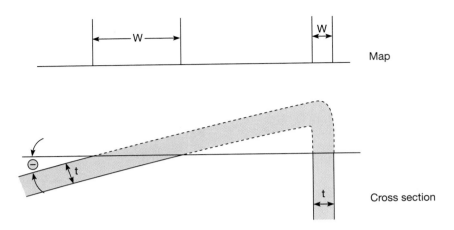

FIGURE 10.8 Outcrop bed width as related to dip.

Observe that $W = t/\sin\theta$ where W is the bed width of the outcrop, t is the thickness of the bed, and θ is the angle of dip (always measured from the horizontal). For vertical beds $W = t$, its minimum width, and for horizontal beds, W becomes infinite. Hence symmetrical folds will have equal bed widths on opposite sides of the axial plane because the opposing dip angles are the same.

Plunging Folds

Having discussed fold types, fold symmetry, and outcrop widths only one major detail remains for consideration concerning fold systems. Figures 10.9(a) and (b) illustrate the plunge of a fold system, a situation we now consider.

Figure 10.9 shows the intersection of a plunging syncline and in turn that of a plunging anticline with

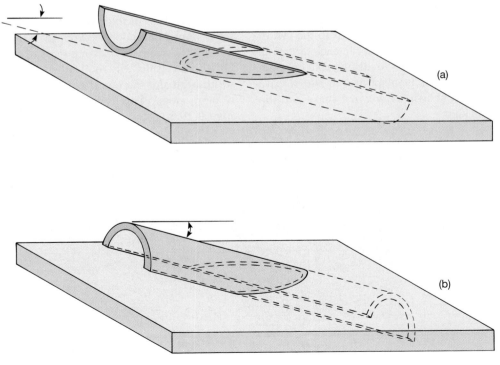

FIGURE 10.9 (a) Plunging synclines and (b) anticlines.

a horizontal plane. The shape and direction of the intersection is of importance. In Figure 10.9(a), the syncline is plunging (losing elevation) to the right and the angle of plunge is shown on the drawing. Note that the nose shape, formed by the intersection with the horizontal plane, points to the left; that is, it points in the direction opposite to the plunge. By contrast, in Figure 10.9(b) for the anticline, the nose points in the same direction as the direction of plunge. The plunge angle is also shown for Figure 10.9(b).

Second Rule of Anticlines
The second rule of anticlines is therefore gained from Figures 10.9(a) and (b). This rule states that for a

plunging anticline, the nose formed by the intersection of the fold system with a horizontal plane points in the same direction as the plunge. As before, synclines are opposite to anticlines; therefore, for them, the nose points in the direction opposite to that of the plunge. The block diagram or three-dimensional view of the fold system (Figure 10.10) can be used to illustrate these details.

Structural Geology Symbols
In a block diagram the upper surface is horizontal and the front and side are vertical sections that are perpendicular to each other and to the upper surface. Consequently, it is a right, rectangular prism with the upper surface horizontal. Included in Figure 10.10

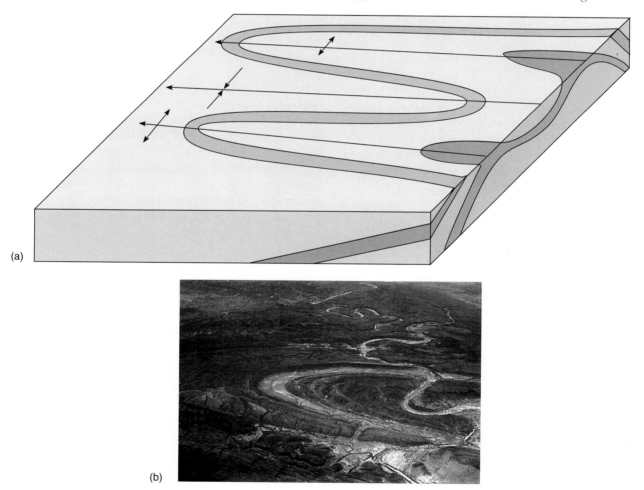

FIGURE 10.10 (a) Block diagram of plunging anticlines and synclines. (b) Aerial view, anticline plunging to the left, the same direction as the nose points. An entrenched stream has cut a water gap through the fold. Northern Territories, Central Australia.

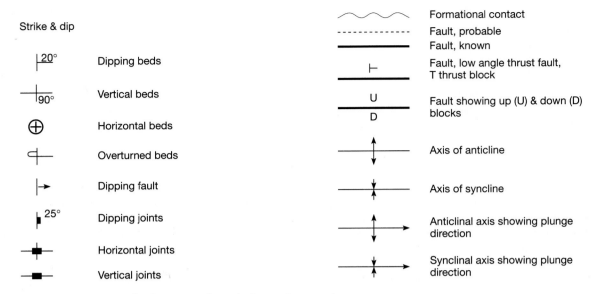

Strike & dip

20° (dipping bed symbol)	Dipping beds
90° (vertical bed symbol)	Vertical beds
⊕	Horizontal beds
(overturned bed symbol)	Overturned beds
→ (dipping fault symbol)	Dipping fault
25° (dipping joint symbol)	Dipping joints
(horizontal joint symbol)	Horizontal joints
(vertical joint symbol)	Vertical joints

∿∿∿	Formational contact
– – – – –	Fault, probable
▬▬▬▬	Fault, known
⊢	Fault, low angle thrust fault, T thrust block
U / D	Fault showing up (U) & down (D) blocks
↕	Axis of anticline
↕	Axis of syncline
↕ →	Anticlinal axis showing plunge direction
↕ →	Synclinal axis showing plunge direction

FIGURE 10.11 Standard structural symbols used on geologic maps.

are standard symbols depicting anticlinal and synclinal axes along with arrows showing the direction of plunge. Note that the first rule of anticlines still holds for plunging folds. The standard structural symbols used on geologic maps to depict folds, faults, and joints are shown in Figure 10.11.

Apparent and True Dip

A final point concerning the dip of beds involves the variation in the dip angle depending on the line of sight of the observer, as illustrated in Figure 10.12. For line of sight A, which is sighting parallel to the strike, the maximum dip or true dip of the bed is observed. For line of sight B, which is sighting perpendicular to the strike, the apparent dip of the bed is zero. As the line of sight increases from 0 to 90° as measured from the strike direction, the apparent dip reduces from the maximum value to zero. Angle α is shown in the diagram. The value of the apparent dip in terms of α is given by $\theta_{apparent} = \theta_{true} \cos\alpha$. As shown in Figure 10.12, $\cos\alpha$ is measured from the strike direction.

Domes and Basins

Three-dimensional fold features also occur in nature. A structural dome is a fold in which the beds slope

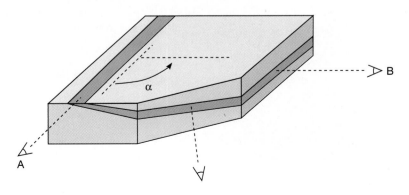

FIGURE 10.12 True dip and apparent dip.

away from the center in all directions. It is in essence a three-dimensional anticline. A structural basin has beds that slope inward in all directions. It is a three-dimensional syncline.

Suggestions Involving Geologic Sections

Several suggestions can be supplied to students constructing geologic cross sections:

1. Keep the thickness of each formation or bed uniform throughout. Variations in the outcrop width are accomplished by varying the dip of the beds.
2. Unless indicated to the contrary by strike and dip symbols, keep dip angles as small as possible in the drawings.
3. Draw the simplest solution (simplest interpretation) to the problem that can be made in keeping with the data supplied.
4. Use straight lines of constant dip if possible. Smooth the intersection points of the straight lines to yield smooth curves as needed.

ROCK FRACTURES

Types of Fractures

Fractures in rock are narrow openings along which the rock mass has lost grain-to-grain contact. Relative movement of the rock blocks on opposite sides of the fracture may or may not have occurred. Based on the extent of movement, three major types of fractures are distinguished: joints, shear zones, and faults.

Joints

Joints are rock fractures along which no movement has occurred parallel to the joint surface. Some displacement perpendicular to the joint may develop because of frost wedging or gravity effects. The attitude of joint planes is described in the same manner as are bedding planes, that is, using the strike and dip of the sloping plane.

There is a tendency for joints to occur in sets of parallel fractures rather than as a single isolated plane. Therefore, the strike and dip of the joint set is supplied along with the joint spacing. Some joints form on release of confining pressures, for example, along the valley wall of a stream or along other steep exposures of rock. Even essentially horizontal sedi-

mentary rocks exhibit joint sets, usually parallel to the slight regional dip and perpendicular to it as well. These joints may be nearly vertical. If a horizontal stress condition has been active in a region, two vertical sets of joints, each about 30° from the direction of maximum principal stress may occur. This is discussed further in a subsequent section of this chapter on the orientation of failure planes in a stress field.

Shear Zones

Shear zones, or simply shears, are fractures in rock along which some movement has occurred but not a great amount. Most movement is due to slippage of the rock to compensate for distortions caused by folding. Typically only a few centimeters of movement are involved but some geologists would allow up to 1 m of movement and still term the feature a shear zone. Shear zones are typically weak, ground-up areas of rock and soil-like material. They are sites of ground-water flow and of low strength and are undesirable areas relative to tunneling, slope excavation, and foundations for engineering works. They do not appear in persistent sets the way joints do, but they tend to be more prominent in certain specific locations within a mountainous region.

Faults

Faults are fractures along which significant movement has occurred, more than that associated with shear zones. Displacement may be measured in meters or kilometers (or in feet and miles), typically with a minimum movement of 1 m to qualify as a fault.

Fault Types and Terminology The attitude of faults is described in a similar fashion to that for bedding planes. The various portions of a fault are labeled in Figure 10.13. The block above the fault plane is called the hanging wall and the portion below it is the foot wall. The fault line is the intersection of the fault plane with the surface of the Earth.

Movement along the fault plane can be either 1) primarily along the dip (as shown in Figure 10.13), 2) primarily along the strike, 3) a combination of movement in both the strike and dip directions, or 4) rotational, in which one block rotates relative to the other. This fourth possibility is not discussed in this introductory presentation on structural geology.

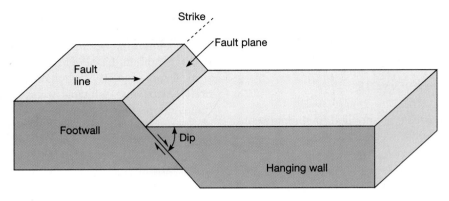

FIGURE **10.13** Fault terminology.

DIP-SLIP FAULTS In Figure 10.13 the hanging wall has moved down relative to the foot wall. This is termed a normal fault and it is a common type of fault. Since the movement is down the dip of the fault plane, it belongs to the group called dip-slip faults. The other type of dip-slip fault occurs when the hanging wall moves up relative to the foot wall. This is a reverse fault and is shown in Figure 10.14. A low angle (dip angle) reverse fault is called a thrust fault. Normal faults involve a lengthening of the Earth's crust whereas reverse faults indicate a shortening of the crust.

A normal fault can be caused by a vertical compressive force or by a horizontal tensile force. By contrast a reverse fault can occur because of a horizontal compressive force or a vertical tensile force.

STRIKE-SLIP FAULTS Strike-slip faults (or wrench faults) occur when the relative movement is essentially all horizontal so that the blocks are displaced along the strike direction (Figure 10.15). The two possible types of these faults are right lateral and left lateral strike-slip faults. Figure 10.15(b) shows a right lateral fault. Standing on the front block of the fault facing the fault plane in Figure 10.15(b), one can observe that the opposite side has had a relative movement to the right. In like manner standing on the back block and facing the fault plane, one sees that the front block has a relative displacement to the right. Figure 10.15(a) shows the reverse situation: The block opposite has been displaced to the left.

TRANSLATION FAULTS The final type of fault movement is a fault that has both a strike-slip and dip-slip component. In Figure 10.16 movement has occurred by translation, which moved the blocks apart from point A to point B. This distance is known as the net slip. No rotation of the blocks is involved because lines that were parallel before movement are still

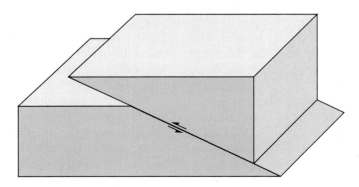

FIGURE **10.14** Reverse fault.

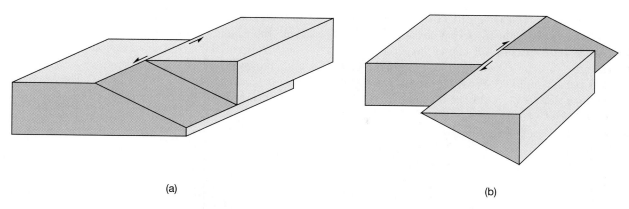

(a)

(b)

FIGURE 10.15 Strike-slip faults: (a) left lateral and (b) right lateral.

parallel afterward. Dip slip and strike slip are as described previously, and it can be seen that they are the component of the net slip in the strike and dip directions respectively. Vertical slip is the vertical component of the net slip (or of the dip slip) whereas horizontal slip is horizontal component of the net slip. It is also known as the heave.

OTHER CONSIDERATIONS   In the exercises at the end of this chapter only strike-slip faults and dip-slip faults are considered, that is, no combined translational faults of the variety illustrated in Figure 10.16 are included. This apparent simplification is justified because most faults in nature show predominantly strike-slip or dip-slip movement.

For strike-slip faults it is rather easy to determine the relative movement involved if a marker bed is present in the fault block. This is shown in Figure 10.17. Strike-slip faults (or wrench faults) in nature

are more likely to be right lateral than left lateral. The famous San Andreas fault of the California coastal area is a right lateral fault.

In dip-slip faults it is more difficult to decipher the relative movements as a result of differential erosion subsequent to the faulting. Typically the uppermost block is eroded much faster than the lower block, yielding a horizontal surface across them as shown in Figure 10.18.

The dip of the marker bed is used to determine the relative movement. Note in Figure 10.18 that the bed dips to the rear of the blocks. If horizontal planes were passed through the block, planes at successively lower elevations would cut the dipping bed further to the rear of the block (because it dips in that direction). Therefore, studying Figure 10.18(b), we see that the left block (the foot wall) has come from a lower elevation than the right block (the hanging wall) because the bed is displaced further to the rear

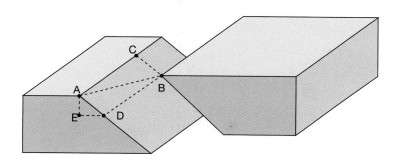

AB – net slip
AC – strike slip
CB or AD – dip slip
AE – vertical slip or throw
ED – horizontal slip or
      heave

FIGURE 10.16 Translation fault.

of the left block. This means that the left block moved up relative to the right block and consequently this is a normal dip-slip fault. Of course, in block diagram problems the arrows showing movement of the blocks would not be supplied, or if the arrows were given, the top or side view would not be completed.

BLOCK FAULTING   Some additional details concerning fault movements require special consideration. In some areas of the world, for example, in the Basin and Range Province of the southwestern United States, *block faulting* is predominant. This involves the movement of three-dimensional blocks of rock

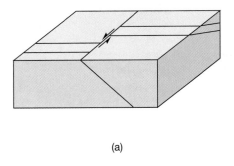

(a)

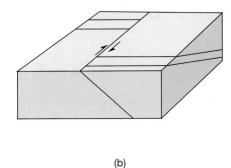

(b)

FIGURE **10.17**   Strike-slip movement and marker beds: (a) left lateral and (b) right lateral.

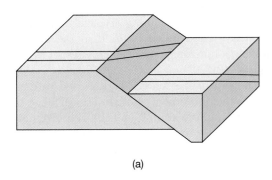

(a)

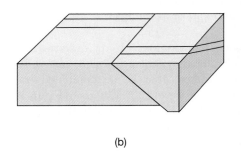

(b)

FIGURE **10.18**   Dip-slip fault related to erosion: (a) before erosion and (b) after erosion.

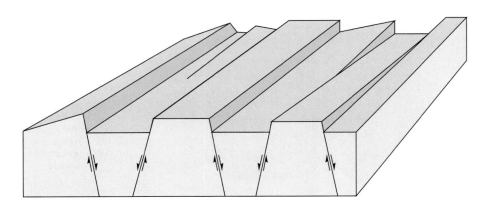

FIGURE **10.19**   Block faulting.

upward and downward along steeply dipping fault planes. The upward displaced blocks, known as horsts, provide the ranges and the down thrown blocks, or grabens, yield basins (geographic basins, not structural basins). This is illustrated in Figure 10.19.

FIELD RECOGNITION OF FAULTING   The recognition of faults in the field is of great importance because faults are of major concern in engineering construction. The existence of faults should be considered early in the site selection investigation. Surface exposures must be interpreted properly to discern the presence of faults.

Several details about faults are used as an aid in their field recognition. In general categories, they are as follows: 1) landform features associated with the fault line, 2) abnormal stratigraphic sequences caused by faulting, and 3) features of the fault plane itself.

One of the landform features depicting the presence of faults is the offset or truncation of geologic structures. This can be the offset of a mountain chain or a valley or the sudden cessation of a mountainous feature. Marked changes in elevation are another clue to faulting. These changes might mark the fault line scarp not yet reduced by erosion. The steep eastern face of the Sierra Nevada along the Nevada–California border is an extensive fault line scarp of major proportions. It significantly retarded the westward migration of early settlers, and except for access along a few, narrow passes through the Sierras, transportation by wagons was prevented.

Other features that develop along a fault scarp are called triangular facets. These occur as a consequence of surface drainage development on the fault scarp, yielding a channel that widens in the downhill direction. The area between the drainages yields a triangular shape when viewed looking toward the mountain (Figure 10.20).

Finally, there is a landform feature of strike-slip faults that disrupts the normal, smooth profile of a stream valley. It occurs where the fault intersects the valley to form a sag pond, a swampy area where the stream profile has been interrupted.

An abnormal stratigraphic sequence related to faulting usually involves the repetition or the omission of the normal beds within that sequence. This is illustrated in the map view presented in Figure 10.21. It shows a series of rocks with a normal sequence of folded sedimentary rocks, which has been disrupted at two locations by faulting (note the F designations).

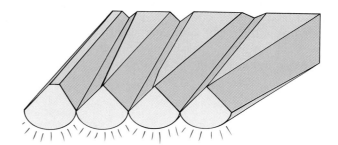

FIGURE 10.20   Triangular facets depicting fault scarps.

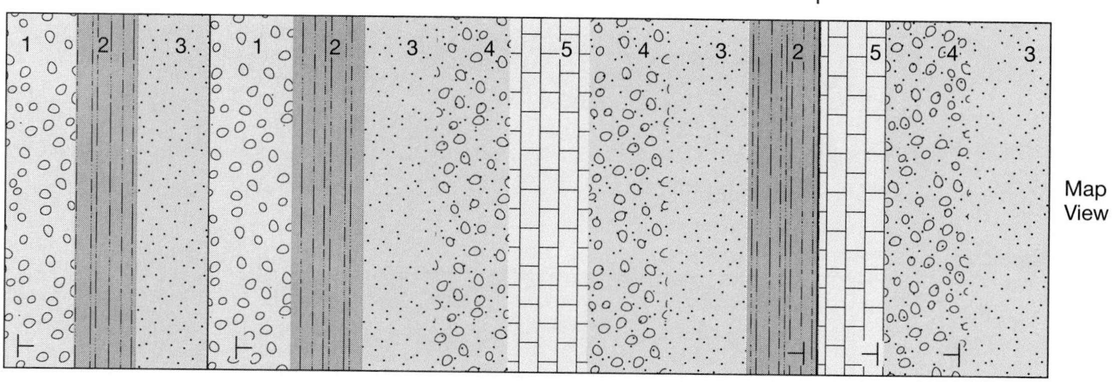

FIGURE 10.21   Faulting indicated by repetition and omission of beds.

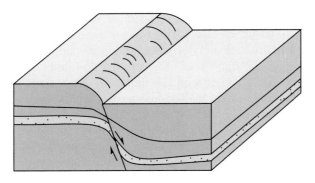

FIGURE **10.22**   Drag of beds along a fault plane.

existence of the fault. *Slickensides* are polished rock surfaces that occur as a consequence of movement. In some rocks, fault displacement causes *gouge* to form. This is fine-ground rock dust, which occurs when hard rock blocks are moved against each other. Fault breccia consists of broken angular pieces of massive brittle rock, which break off from the mass when the blocks slide by each other. Fault breccia and gouge may be tens of feet thick yielding a fault zone. When marker beds are present within the fault blocks, a drag of the beds at the fault plane may occur, which notes fault movement (Figure 10.22).

In Figure 10.21 a repetition of beds is shown on the left side of the diagram and an omission of beds on the right side. The structure is a syncline as indicated by both the age of the beds (note numbering system) and the strike-dip symbols. In this example no change in attitude of the beds occurs at the two fault lines (F designation). In some situations a change in dip might also occur at the fault line.

Features along the fault plane itself, seen in plan view or in cross section, can be used to establish the

FOLDS AND FAULTS COMBINED

One final analysis remains for this discussion on structural geology. It combines plunging folds with dip-slip faults and supplies a final challenge to the student's ability to visualize the relative movements involved. Figure 10.23 shows these two in combination and the caption provides a detailed description. The problem can be unraveled by considering the front view of the fold system relative to horizontal planes passed through it, as shown in Figure 10.24.

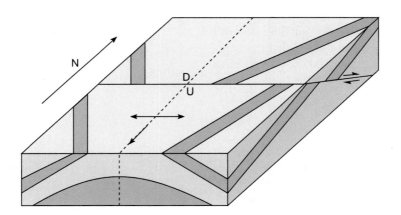

FIGURE **10.23**   South plunging symmetrical anticline cut by an east–west striking normal fault dipping northward.

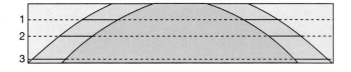

FIGURE **10.24**   Front view of anticline from Figure 10.23.

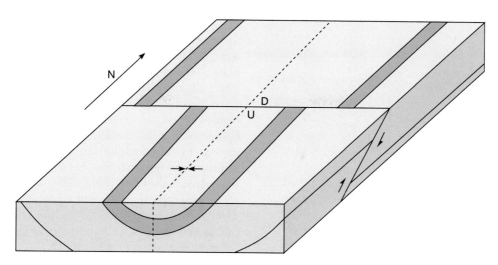

FIGURE 10.25   Symmetrical syncline cut by an east–west striking reverse fault dipping southward.

For the anticline, observe that as you proceed lower in the cross section, the outcrop of the dipping bed will move further apart, that is, it diverges. Therefore, referring to the top view of Figure 10.23, on the back block the bed is closer together than it is on the front block. Therefore, the bed had to have originated at a higher elevation, and consequently the back block has moved downward.

The direction of plunge (toward the front), shown in the side view, is a consequence of the second rule of anticlines (on plunging folds). Note that the nose for the anticline points in the same direction as the plunge. The symbols on the top and side views are examples of information supplied for folds and faults on geologic maps.

Figure 10.25 is an example of a nonplunging syncline and it completes the details for combinations of folds and dip-slip faults. Note that the beds of the syncline converge with depth. Therefore, because the beds are closer together in the front block, that block had to have originated at a lower depth. Consequently, the front block has moved upward relative to the rear block, yielding a reverse fault.

DIRECTION OF STRESS AND FAULT ORIENTATION

The steepness of dip for the fault plane is a consequence of the stress directions that lead to faulting. For this purpose, a triaxial force field can be assumed with orthogonal, principal stress directions for the maximum, minimum, and intermediate values. This can be shown in a two-dimensional view as presented in Figure 10.26. The maximum principal stress σ_1 is in the vertical position; σ_3, the minimum principal stress, is horizontal; and σ_2, the intermediate principal stress, lies perpendicular to the page.

The dotted lines represent conjugate planes of failure, which will occur at approximately 30° from the direction of maximum principal stress (σ_1) for isotropic rocks, that is, those rocks without obvious planar weaknesses. From this, an analysis of normal and reverse faults can be made.

For normal faults the direction of maximum principal stress is commonly close to vertical because excess overburden stresses give rise to this type of faulting. This yields a failure plane 30° from the vertical or 60° from the horizontal. For this reason, most normal faults occur as high-angle faults, that is, faults typically steeper than 45° as measured from the horizontal.

Reverse faults are commonly low-angle faults simply because nearly horizontal compressional forces (σ_1) are typically the cause of such faults. A high-angle reverse fault suggests that an uplift force oriented at a steep angle is the cause of such faulting.

A word of caution is in order here relative to the solutions to exercises on faults at the end of this chapter. It is inappropriate to assume that the type of dip-slip fault (normal versus reverse) can be recognized simply on the basis of the steepness of the dip

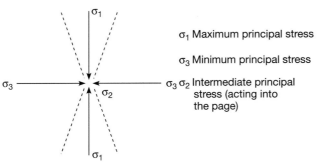

FIGURE 10.26 Directions of principal stress.

angle of the fault plane. This would render it unnecessary to work out the details of the dipping beds. Specific details shown on the block diagrams, in keeping with the discussions on bed displacements in this chapter, should be used to determine block movement.

When rocks are not isotropic (i.e., they are anisotropic), the orientation of the weakness planes can influence the direction of breakage or faulting. These weaknesses are typically known as *s* planes, which consist of prominent bedding in sedimentary rocks, foliation in metamorphic rocks, or flow banding in extrusive igneous rocks.

Laboratory studies have shown that if σ_1 is oriented within 60° of the weakness plane, failure will occur along or be controlled to some extent by the weakness plane. For the 60° portion between a 60° dip to the left and a 60° dip to the right, the *s* planes have little influence on the direction of failure and indeed the failure plane will again occur about 30° from the load axis. This is illustrated in Figure 10.27 (c) and (d). This can be applied to rock faulting. Thinly bedded, horizontal, sedimentary rocks can develop what is known as bedding plane faults when horizontal compressive stresses are imposed on them by earth processes. Low grade metamorphic rocks like argillites may undergo considerable bedding plane slippage during deformation of adjacent mountainous regions.

In addition, this suggests that major forces from engineering constructions should not be transferred to layered rocks in a direction parallel or nearly parallel to these layers. Slippage and failure may occur if this occurs. For this reason the thrust of an arch dam should not be aligned in a direction parallel or subparallel to a weakness zone in the rock. Steeply dipping rock strata are also subject to slippage when gravity loads are placed on them. This subject will be examined in greater detail in a subsequent chapter on landslides and rock slope stability (Chapter 14).

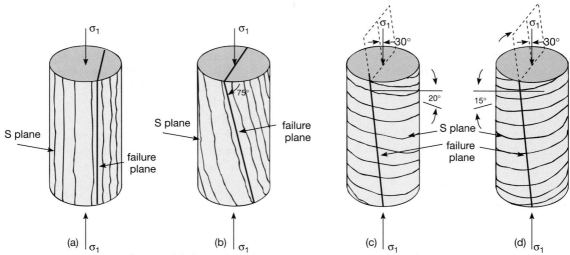

FIGURE 10.27 Dependence of failure plane location on orientation of *s* planes in test specimens.

EXERCISES ON STRUCTURAL GEOLOGY FOLDS AND FAULTS

Reference Views

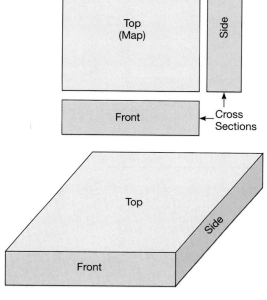

1. *Folds:* Complete (a) through (e) on page 216, locate an axial *plane,* and *describe* the fold. Bed 1 is the oldest. *Be neat.*

2. *Faults*: Complete (a) through (f) on pages 218 and 219, and identify the structures. Assume dip-slip faults unless otherwise noted.

3. *Dipping Beds*: (Drawing with scales, page 219)
   (a) What is the true thickness of bed Q?
   (b) How deep would you have to drill to reach the top of the bed at A?
   (c) What is the apparent thickness of the bed that would be found by drilling at A?

4. A hill slopes 20° to the west. At an elevation of 1000 ft, the top of a sandstone bed, which strikes N-S and dips 30° to the east, is found. Fifty feet down the slope from the 1000-ft elevation, the bottom of the sandstone is found. What is the true thickness of the sandstone? At the 1300-ft elevation, how far would you have to drill to reach the sandstone? Sketch the problem using a scale of 1 in. = 200 ft. Calculate using trigonometry.

5. *Folds and Faults Combined:* (a) through (f) on page 220. Complete the views and describe completely. Assume north for parts (a) through (e) is the same as shown in part (f).

6. Why are sedimentary rock sequences particularly useful in determining the structural geologic history for an area? Relate to the principle of original horizontality.

7. Why would rocks be less brittle when loaded at high temperatures and pressures over long periods of time than when a fast-loading rate is used in the laboratory at room temperatures and pressures? Explain.

8. Distinguish between a hogback and a cuesta.

9. Give the first and second rules of anticlines. Why is it easy to remember how these apply to synclines?

10. How are faults recognized in the field? Give the three general categories and provide examples of each.

11. A coarse gneiss is being tested in uniaxial compression. If the foliation is 10° from the load axis, where is failure likely to occur? If it is 80° from the load axis, where is failure likely to occur? Which of these two tests is likely to provide a greater unconfined compression value? Explain.

12. The Basin and Range Province in the southwestern United States has undergone extensive block faulting. Relate the basins and ranges to the terms *horst* and *graben*. This is a low rainfall area of the United States. What kind of weathering is likely to prevail here? Why?

13. Refer to Figure 10.12 on the true and apparent dip of beds. If a bed strikes N30°E and has a dip of 58°, what would be the apparent dip along a line of sight of N45°E? Give the equation and provide the numerical answer.

14. An anticline contains a bed that is 80 ft thick. The dip angle to the east is 30° and the dip angle to the west is 75°.
   (a) What type of fold is this? Symmetrical, asymmetrical, overturned, or recumbent?
   (b) What is the width of outcrops along a horizontal plane on the east side? On the west side? Show calculations.

15. A coal seam strikes N80°W and dips 5° to the southwest. Depth to the top of the coal is 60 ft at point *x*.
   (a) Assuming a horizontal land surface, how far along the direction perpendicular to the strike will you need to travel before the depth to the coal seam is 100 ft? A sketch should prove helpful.
   (b) What is this direction perpendicular to the strike from point *x*?

Additional Readings

BILLINGS, M.P., 1972, *Structural Geology*, 3rd ed., Prentice Hall, Englewood Cliffs, NJ.

DESITTER, L.U., 1964, *Structural Geology*, 2nd ed., McGraw-Hill, New York.

HILLS, E.S., 1972, *Elements of Structural Geology*, 2nd ed., John Wiley & Sons, New York.

WHITTEN, E.H.T., 1966, *Structural Geology of Folded Rocks*, Rand McNally & Co., Chicago.

(a)
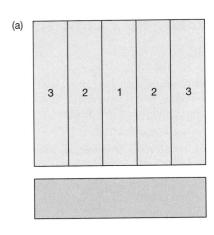

(b)
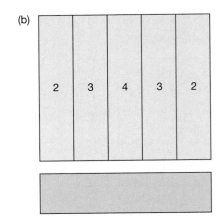

(c)
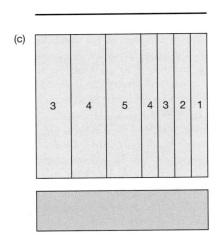

(d)
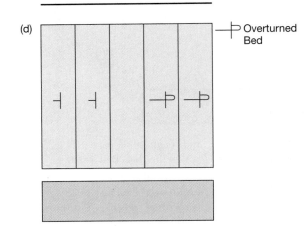

⊢⊅ Overturned
    Bed

(e)
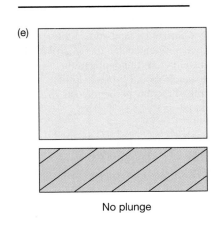

No plunge

(f)
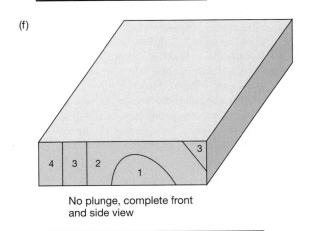

No plunge, complete front
and side view

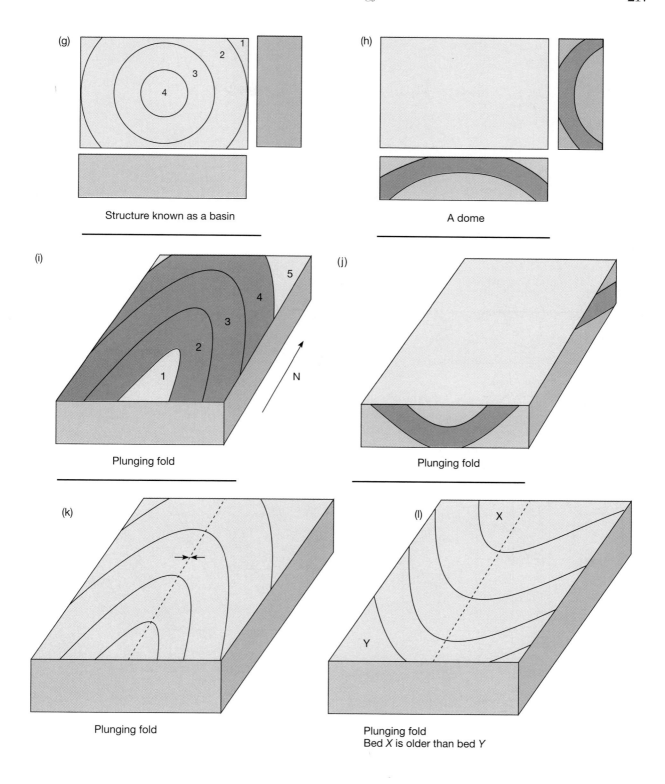

(g) Structure known as a basin

(h) A dome

(i) Plunging fold

(j) Plunging fold

(k) Plunging fold

(l) Plunging fold
Bed *X* is older than bed *Y*

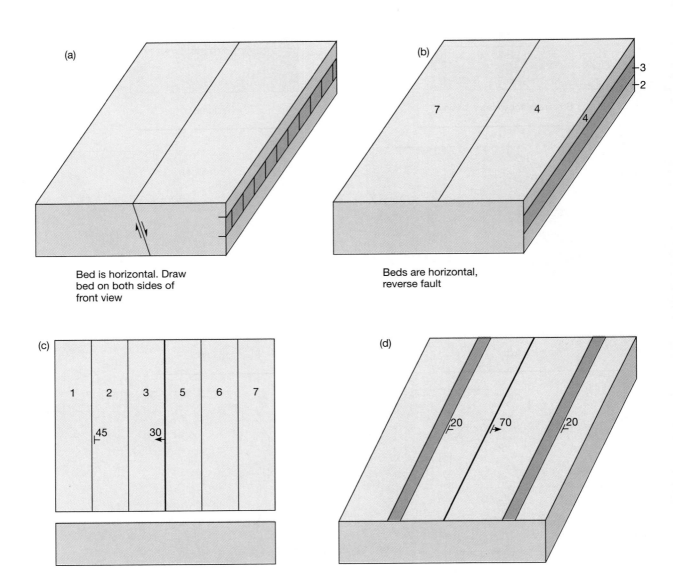

(a) Bed is horizontal. Draw bed on both sides of front view

(b) Beds are horizontal, reverse fault

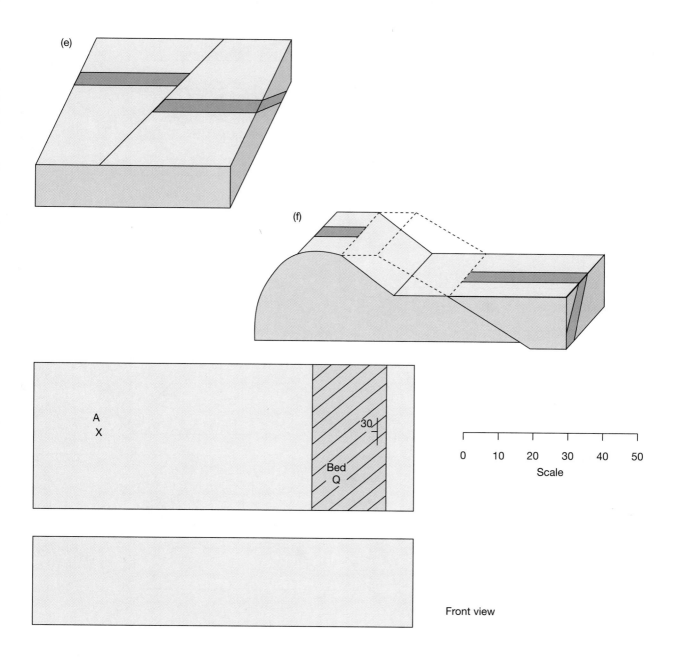

(e)

(f)

A
X

30

Bed
Q

0   10   20   30   40   50

Scale

Front view

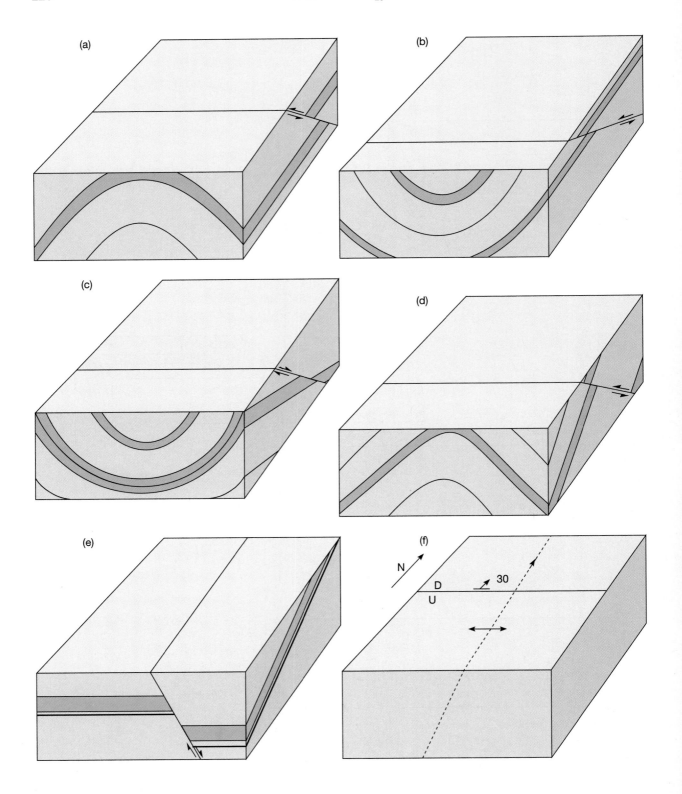

11

Running Water and River Systems

CHAPTER OUTLINE

Hydrologic Cycle and Rainfall Equation
Factors Related to Runoff
Types of Runoff
Stream Flow Terms
Work of Streams
Landform Features of Streams
Engineering Considerations of Streams

HYDROLOGIC CYCLE AND RAINFALL EQUATION

Following rainfall or the melting of snowcover, water runs from the slopes in thin sheets and then collects into rills, gullies, streams, and rivers to arrive finally at the ocean. From the sea it is returned to the atmosphere by evaporation to begin the pattern once more. This process is known as the hydrologic cycle. Illustrated in Figure 11.1, the hydrologic cycle also involves water infiltration into the Earth and groundwater movement, which eventually returns the water to surface streams or the ocean. The rainfall equation relates these factors in the following way:

$$\text{Rainfall} = \text{Evaporation} + \text{Transpiration} + \text{Infiltration} + \text{Runoff}$$

The greatest part of rainfall under natural conditions is returned to the atmosphere directly by evaporation, plus transpiration by plants. A smaller amount flows away on the surface and is known as runoff; the remainder, the smallest portion, soaks into the soil by infiltration.

For humid nonurban regions in the United States from the Mississippi River eastward, about 60% to

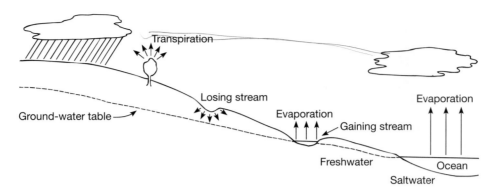

FIGURE **11.1**   The hydrologic cycle.

80% of precipitation is returned to the atmosphere by evapotranspiration. In arid to semiarid regions of the western United States, this may reach an excess of 95%. Some 10% to 25% runs off the land surface into stream systems in the humid areas with as little as 2% to 5% ending up as runoff in arid regions. The remainder, about 10% to 20% in humid regions, but as little as 1% in arid regions, infiltrates into the ground to replenish the ground water system.

FACTORS RELATED TO RUNOFF

Several factors determine the percentage of precipitation that becomes runoff in a local area: 1) relief, 2) surface infiltration, 3) vegetative cover, and 4) surface storage. Relief is the difference in elevation between the highest and lowest points within an area of interest. Areas with higher relief have steeper terrain, which encourages more runoff and less infiltration. In steep terrain the excess runoff can best be controlled by encouraging vegetative cover.

Surface infiltration is related to the permeability of the surface material, whether it is soil or bedrock, and to the pattern of precipitation. Snow versus rainfall is one consideration because snow may melt slowly, allowing a greater opportunity for infiltration. The pattern of rainfall in a storm also has an effect. The infiltration rate is higher in a dry soil so that early in a storm, greater water uptake occurs. As rainfall continues, the pore system becomes temporarily saturated and greater amounts of water must run off. Intense storms will drop more water than can be absorbed

directly and, therefore, more runoff occurs. A low-intensity, long-duration storm will eventually experience a reduced infiltration rate because of soil saturation. However, it tends to yield greater amounts of infiltration than does a high-intensity storm of equal water volume.

The permeability of the soil is determined by the size and nature of its constituent particles. Coarser materials such as sand have a higher permeability than does silt, which in turn yields a considerably higher value than clay. Unweathered bedrock or rock with a thin soil cover has a low permeability and infiltration may be close to zero in such materials. The condition of the soil affects permeability and, therefore, infiltration. Soil saturated by previous rains can absorb only small amounts of additional water. Frozen soil is essentially impermeable and major flooding occurs when snow melts rapidly during a sudden warming trend prior to thawing of the soil.

Vegetative cover promotes infiltration and transpiration and reduces runoff. Soil erosion is a consequence of runoff, so vegetative cover should be maintained to prevent loss of topsoil and the subsequent pollution of streams by eroded sediment. In grading operations on earth-moving projects, only the minimum area that is economically feasible should be denuded of vegetation at any one time. This includes cuts for road construction and site preparation for housing tracts, shopping areas, or industrial development.

Surface storage or ponding allows water to collect for a time rather than running off directly. This reduces the peak level of flow and the erosional effects of runoff. It also allows more time for evaporation and infiltration, both of which reduce the total amount of runoff that occurs.

Before leaving the rainfall equation and its various relationships, the contrast between natural conditions and human-induced situations requires attention. This involves the subject of urban hydrology. Cities contain many hard surfaces including streets, sidewalks, roof tops, and parking lots and as a consequence of their construction, both vegetative cover and infiltration are significantly reduced or eliminated. Runoff increases from a value of about 15% for a humid area east of the Mississippi River to perhaps 60% to 70% in an urban area. Although there may be few soil areas exposed and erosion may not be significant, great quantities of runoff must be accommodated during times of heavy rainfall. Many cities have inherited a combined sewer system, which collects storm water runoff plus sewage. Sewage treatment plants typically cannot accommodate this extreme volume of water during heavy rainfall periods and consequently overflow points in the sewer system are used to reduce the volume of flow. The overflow water mixed with sewage is dumped directly into streams without treatment and this accounts for one of the major contributions to stream pollution of large cities.

The Chicago Deep Tunnel project (known also as TARP or the Tunnel and Reservoir Project) addresses this problem. The first part of the collection systems became operational in 1985 when the underground pumping station was completed. Final work on the 100+ miles of rock tunnels, known as Phase I, is not anticipated until after 1995. This system of collectors and tunnels brings the sewer overflow water from the city and nearby suburban areas into an enlarged sewage treatment facility, which can accumulate the excess runoff during times of peak rainfall. This water is pumped to the sewage treatment plant, which allows for maximum efficiency rather than the normal, daily or hourly fluctuation of sewage flow. To reduce construction and maintenance costs, tunnels ranging from 9 ft to 33 ft in diameter are located primarily in massive rock, 150 ft to 360 ft below the ground surface. Figure 11.2 illustrates TARP. The completed Chicago Deep Tunnel project accommodates increased runoff until it eventually fills the rock tunnels. Any additional runoff from an intense storm

becomes overflow, dumped directly into the rivers. This delayed overflow water is less polluted than that initially filling the combined sewers when the sewage solids are scoured from the invert of sewer. This initial filling, which is now carried to the sewage treatment plant, is called the "first flush."

Current plans call for storage of the additional water (delayed overflow) in surface reservoirs, such as old stone quarries. Design of this project, known as Phase II, which will reduce river flood levels, was begun in 1990 by the U.S. Army Corps of Engineers.

TYPES OF RUNOFF

When water first accumulates on the Earth's surface, it flows in a thin layer across a sizable surface in the form of sheet runoff. The water moves slowly with discrete layers flowing parallel to the base. This type of movement is known as laminar flow. Sheet runoff occurs on cultivated fields where row crops cover only a portion of the soil. Where local irregularities occur in the surface, a surge of flow develops and erosion takes place to remove soil particles. This erosion is called sheet wash and it accounts for the loss of soil over large areas that occurs when bare soil is exposed.

Erosion of topsoil from farm fields has been a problem of major concern in many areas of the United States. A developing farm practice being encouraged by the U.S. Soil Conservation Service is no-till farming. As suggested by the name, plowing of fields no longer occurs; instead, seeds are planted through the stubble of past harvests. First introduced for fields on steeper slopes it is now practiced on an increasing percentage of farm land in the United States.

Within a short distance, sheet runoff becomes concentrated into a confined channel. First it forms into rills then gullies, creeks, streams, and finally rivers. The rate of flow increases in channels and because of this increase and the roughness of the channel boundary, the straight-line flow is replaced by swirls and eddies of movement. This is called turbulent flow, which is effective in eroding and transporting soil.

Extremely high water velocities are obtained in rapids and waterfalls. Called jet or shooting flow, these high velocities are extremely erosive even in bedrock channels. Although not by design, jet flow

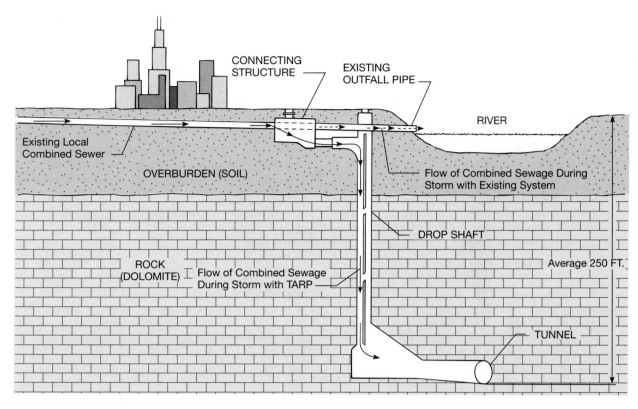

FIGURE **11.2**　Diagram for Chicago Deep Tunnel project for water reclamation from combined sewers.

may occur in spillways for dams when large volumes of water are allowed through the spillway gates. Rapid erosion of concrete in the stilling basin of the dam can be a consequence.

Not all stream channels carry water year round. Many smaller streams carry water only during the wet season or immediately after significant rainfall and are termed *intermittent*. In arid regions many of the stream channels run dry, including in some cases major streams. In humid regions the rivers flow year round as do many of their primary tributaries, the larger creeks. This occurs even if several weeks separate significant storms, because the ground-water table feeds these streams. Streams flowing year round are known as permanent or perennial streams and they are designated on topographic maps as solid blue lines. Intermittent streams are shown instead by dashed blue lines.

Permanent streams are also classified as gaining streams because they gain water from the ground-water table. Intermittent streams provide water to the

water table as they flow along their channels; consequently, they are termed losing streams. This relationship is illustrated in Figure 11.1.

STREAM FLOW TERMS

A list of the most significant stream flow terms would by necessity include the following: velocity, gradient, discharge, and load. Stream velocity is usually measured in feet per second (or meters per second) because of the convenient values that are obtained using these units. A low velocity would be in the range of 0.5 ft/sec and is typical of large rivers flowing under low gradients, whereas 30 ft/sec is a high velocity. This would be typical of a mountain stream flowing under a steep gradient and probably in a bedrock or boulder strewn channel.

Velocity

The velocity of a stream is affected by a number of factors including 1) gradient or slope of the water surface, 2) the nature of the stream bed, that is, its roughness and erodibility, 3) the discharge or volume of flow per unit time, and 4) the load, that is, the material transported by the stream.

Streams do not maintain a constant velocity throughout their cross sections but flow is fastest where the frictional drag is at a minimum. In a straight portion or reach of a stream the water flows fastest in the center of the cross section just below the water surface. The surface velocity is approximately 0.8 times the average velocity of the cross section. Figure 11.3(a) illustrates this point.

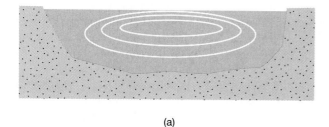

(a)

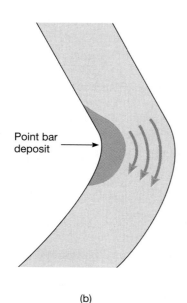

Point bar deposit

(b)

FIGURE 11.3 Variations in velocities for a stream. (a) Cross section showing velocity contours; velocities increase toward the center. (b) Map view of stream velocities at a river bend.

Figure 11.3(b) shows the map view of a river meander. Velocities are greater near the outside of the bend and at a minimum on the inside of the bend. Therefore, active erosion occurs on the outside of the bend and deposition occurs on the inside portion.

Gradient

The stream gradient is the slope of the water surface or, in general, the slope of the stream bed. Measurements show that the gradient of a stream decreases proceeding downstream from the headwaters toward the mouth. This yields a long profile, which is concave up or concave to the sky. The Missouri River, for example, has a gradient of 20 ft/mi near its headwaters but only 1 ft/mi near its confluence with the Mississippi River at St. Louis.

Stream gradients are expressed in feet per mile (meters per kilometer) and may be as high as 200 to 300 ft/mi in the upper reaches in a mountainous terrain to as low as 0.5 to 0.10 ft/mi for large rivers flowing on wide alluvial floodplains. The gradient is determined using a topographic map by locating two points where contour lines cross the stream, finding the vertical difference by subtraction, and measuring the distance of travel between those two points. The drop in elevation divided by the travel distance yields the gradient.

Discharge and Flooding

Discharge is measured in cubic feet per second (cfs) for streams but for water supply and sewage treatment purposes it is typically measured in millions of gallons per day (MGD) (1 ft^3/sec = 0.646 MGD). A small stream may vary from a minimum of several cfs to a maximum of several hundred cfs during periods of flooding. A major river may have an average of 1500 cfs discharge or more. During and after the construction of a dam across a river, discharge must be allowed that is equal to or greater than the established minimum discharge for that stream (determined as an average over a number of years). The minimum discharge is also of interest regarding sewage treatment facilities because it indicates the low-flow conditions and, consequently, the minimum amount of water available for dilution of treated effluent that will occur during the dry season. The maximum discharge, of course, is of interest relative to flooding problems.

For permanent streams the discharge increases downstream. This is the result of additional water being supplied by tributary streams and the input of water from the water table.

Flooding is one of the major geologic hazards that occur in nature, the other major ones are volcanoes, earthquakes, and tsunamis. During periods of intense and repeated rainfall, streams swell to bank full levels and then expand across their floodplains. Recurrence intervals of flood elevations at specific locations along major streams have been determined using recorded information. The 50-, 100-, and 500-year flood levels are established from historical data. They indicate the recurrence interval for a flood elevation with one occurrence in 50 years, one in 100 years, and one in 500 years, respectively. The 100-year flood elevation is commonly used as a basis for minimum elevations for building grades and major highways and are incorporated into building codes. Bridges and railroad grades may be established at the 500-year flood elevation.

The flood of record is another item of recorded information for major rivers at designated locations along the river's course. This involves the date and highest elevation, given as a level above the local flood stage. In many midwestern areas of Ohio and Indiana the flood of record occurred in 1913 or in 1937 when the maximum flood levels were reached. These situations occur during periods of catastrophic precipitation when river levels continue to rise as the extended period of rainfall persists.

Record Flood of 1993

The Mississippi and Missouri River systems at St. Louis, Missouri, provided a record flood in Summer 1993. Intense rainfall on the upper Mississippi River drainage area near Minneapolis was followed by an extended period of rainfall in Iowa and western Missouri on the Missouri River system. Flooding began when the 30.0-ft stage was reached. Flood crests from both the Mississippi and Missouri Rivers arrived in St. Louis at about the same time. The Missouri River has its confluence with the Mississippi River only a few miles north of St. Louis. On August 1, 1993, the Mississippi River reached its highest record level in St. Louis, the 49.6-ft flood stage. The prior record, 43.2 ft, was reached on April 28, 1973.

The flood wall encompassing the major part of the city along the west bank of the river, extends to 52.0 ft

flood stage. Therefore, the river came to within 2.4 ft of topping the flood wall, which protects the major part of the city. Many other cities were not as fortunate; for example, Des Moines, Iowa; Alton, Illinois; and Ste. Genevieve, Missouri suffered major flood damage when levees were overtopped by the flooding rivers.

Floodwater backed up major drainages along the Mississippi River in St. Louis. Many roads including a number of federal highways were flooded. Bridges along both the Mississippi and Missouri Rivers were closed by flooding, which greatly complicated commuter traffic across the rivers. Locks along the Mississippi River were opened, equalizing the water level, and barge traffic was curtailed for an extended period. Fortunately, the interstate highway bridges connecting St. Louis and Illinois stood above the flood level so that these busy arteries continued to carry traffic.

The River Des Peres, a tributary to the Mississippi River in south St. Louis, became bank full then overflowed onto residential neighborhoods nearby. Several major city streets with bridges across the River Des Peres were closed by floodwaters. Extensive sand bagging was accomplished by volunteers, and the National Guard was called out to prevent vandalism or looting. Propane tanks from a flooded industrial plant floated into the floodwaters and the area was evacuated to prevent any casualties should an explosion occur. Utilities were shut off in nearby neighborhoods. Fortunately authorities were able to secure the tanks and no additional problems ensued. As the water receded, clean up of mud and debris provided a major challenge. The river stood above the 30-ft flood stage for 121 days, also breaking the 1973 record of 77 days. Photographs of the flooding conditions in St. Louis, Missouri, during Summer 1993 are provided in Figure 11.4.

Load

Load is the material transported by streams, both solid and dissolved constituents. The ability to erode material and to transport load is increased as velocity increases. This is discussed in detail in a later section of this chapter under the topic of Work of Streams.

Three Views of a Stream

Three possible views can be observed for a single stream channel: the channel pattern, the long profile, and the cross profile. The channel pattern involves

(a)

(b)

(c)

FIGURE 11.4 Photographs of flooding conditions in St. Louis, Missouri, Summer 1993. (a) River Des Peres up to roadway. (b) and (c) St. Louis Arch, Mississippi River, downtown St. Louis.

the map view of the stream. A single stream channel can be described as straight, meandering, or braided. Features of a meandering stream are shown in Figure

11.5, an aerial photograph taken from an altitude of about 19,000 ft (6000 m). These features are discussed later in this chapter. An aerial photograph of a braided stream is shown in Figure 11.6.

The long profile is a cross section in which the slope of the stream is seen in its profile or the upstream to downstream view. The cross profile is a cross section taken at right angles to the direction of flow showing the configuration of the stream at a specific point along the bank. Figure 11.3(a) is a cross profile.

The Economy of Stream Flow

A relationship exists among the parameters that act to establish a stream's cross section. It is not as simple a relationship as that involving flow through an open channel, one constructed, for example, of concrete, which provides the width and gradient for the water course. In a natural stream the banks and bottom can be eroded to widen and deepen the cross section; and the gradient can be changed to some extent by downcutting or by deposition of load to increase or decrease the gradient, respectively. However, a relationship does exist among the parameters involved in stream flow.

First consider the volume of flow for channels, including appropriate dimensions:

$$\text{Discharge, } L^3/T = \text{Width } (L) \times \text{Depth } (L) \times \text{Velocity } (L/T)$$

Therefore, increased discharge can be accommodated by a larger cross section and increased velocity or some combination of width, depth, and velocity that yields the needed increase.

When the cross profile of a stream is viewed without vertical or horizontal exaggeration, the width is shown to be significantly greater than the depth. It is also clear that resistance to water flow for the stream is provided mostly by the frictional drag with the bottom and sides of the stream boundary or, as it is sometimes stated, with the wetted perimeter. When a stream increases in depth, the additional wetted perimeter is that portion added along the upper banks, as shown in Figure 11.7. This increase in percentage of wetted perimeter is considerably less than the increase in percentage of the cross-sectional area.

In Figure 11.7(b) the increase in wetted perimeter is about 16% and the increase in cross-sectional area is about 100%. From this it can be seen that the in-

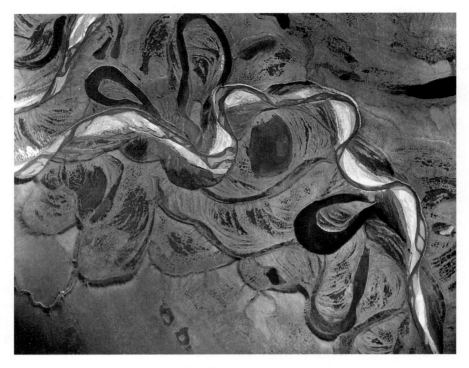

FIGURE 11.5   Aerial photograph of landform features of a meandering stream from an altitude of 19,000 ft (600 m). (Hamblin, WK, *The Earth's Dynamic System.* 2nd Ed., Macmillan, New York.)

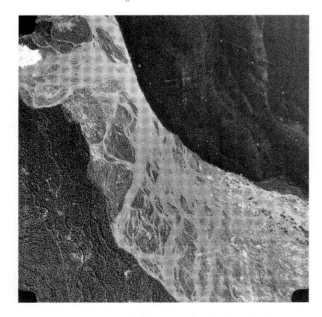

FIGURE 11.6   Aerial photograph of a braided river system. (Hamblin, WK, *The Earth's Dynamic System.* 2nd Ed., Macmillan, New York.)

creased frictional resistance will be more in the range of 16% rather than the 100% increase in the force downhill caused by the weight of the water. This shows that velocity will increase as a consequence of increased depth. This relationship is indicated also by the Manning equation:

$$v = \frac{1.5}{n} d^{2/3} s^{1/2}$$

where

v = velocity
n = coefficient of bed roughness
d = depth
s = slope or gradient.

It is also known that erosive power increases markedly with increasing velocity; actually, it is directly proportional to the square of the velocity, that is, $E \alpha v^2$. From this a sequence of events can be deduced when the stream level rises. When discharge is increased, the depth increases and hence the velocity increases. With an increase in velocity, erosive power is increased so that the cross section is widened and

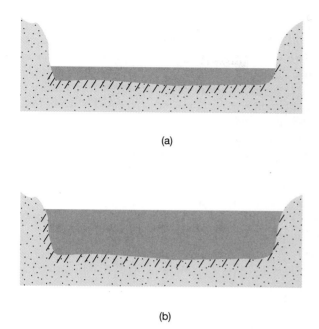

FIGURE 11.7 Cross section of water levels (a) before and (b) after increased discharge showing increased wetted perimeter.

deepened by removal of material. This increases the wetted perimeter until the stream readjusts to the increased discharge. Subsequently, when discharge is decreased, water level drops again, the velocity decreases, and material is deposited in the stream bed. This sequence of events is depicted in Figure 11.8.

A significant problem of river channels is highlighted by Figure 11.8. The scour or removal of bed material at the base of the stream during flood conditions is extensive. In fact, the depth of scour can range from one to four times the amount of increase in height of the water surface. This has an obvious effect on bridge foundations, loading facilities, and other engineering structures located in the stream channel. Undermining of bridge footings is a common consequence of scour when this factor is not properly considered during foundation design.

Through all of this, a relationship between discharge, depth, width, bank erodibility, and gradient develops. An equilibrium condition is reached between these parameters, which holds as long as the discharge remains the same. When that changes the other parameters adjust in an attempt to seek new equilibrium conditions.

Changes in these parameters along the stream's length can be examined next. It is known that proceeding downstream, discharge increases whereas gradient decreases. Measurements at stream gauging stations indicate that the width and depth of streams increase downstream as well. Competency, a measure of the stream's ability to transport material, is also known to decrease downstream. Competency is defined as the largest diameter particle that a stream can transport. This decrease is illustrated by the fact that progressively smaller sized particles occur in the river bed as one proceeds downstream along a major river. As an example, for the Ohio River in its passage along the southern boundary of Indiana, the size of recoverable gravel in the river bed decreases significantly. The effect is that gravel large enough for construction purposes disappears somewhere between Louisville, Kentucky, near the center portion of southern Indiana and Evansville, Indiana, in its southwestern corner.

Considering this decrease in competency, it is surprising to learn that stream velocity actually increases slightly downstream. It would seem that a decrease in competency would suggest a decrease in velocity, because it is the velocity that dictates the size of particles that will be transported by the stream. The matter may best be explained by the concept of

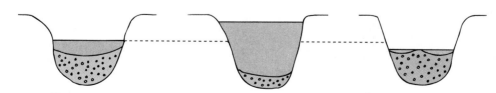

FIGURE 11.8 Cross sections showing changes in channel cross section with changes in stream discharge.

average velocity versus turbulent velocity; the latter being the velocity of turbulence whirls within the stream. Apparently, turbulent velocity decreases as the stream becomes more tranquil, flowing on a lower gradient and, consequently, the ability to transport larger pieces is reduced. This yields a decrease in competency. At the same time, the average velocity increases slightly downstream, contributing to a greater stream discharge.

WORK OF STREAMS

As water flows along in the stream channel it performs three primary functions: 1) erosion of the stream bed, 2) transportation of debris, and 3) deposition of sediment at certain locations. The potential energy of the stream as a result of its elevation is converted to kinetic energy through the mechanism of flow. This energy is expended mostly by frictional resistance with the stream bed with lesser amounts used as internal friction of the water flow itself. The remaining small amount of energy is used for erosion and transportation of material.

Erosion

Running water has the capacity to erode material. In channelized flow it removes material from the stream bed and, as discussed previously, the erosive power is directly proportional to the square of the velocity. A distinction between the two types of stream beds should be made when considering erosion: alluvial channels are those composed of sediment that was previously deposited by the stream (which may range from clay-sized pieces to pebbles, cobbles or boulders), and rocky channels, that is, solid bedrock exposed in the stream bed.

Alluvial Channels

In alluvial channels the particles of alluvium are eroded by direct uplift from the stream bed. They may remain suspended or be rolled or bounced along by the current. The water is in turbulent flow so that the eddies and whirls dislodge particles and lift them into the stream flow. The basic relationship between stream velocity and dislodgment of particles is shown in Figure 11.9. This figure indicates the approximate

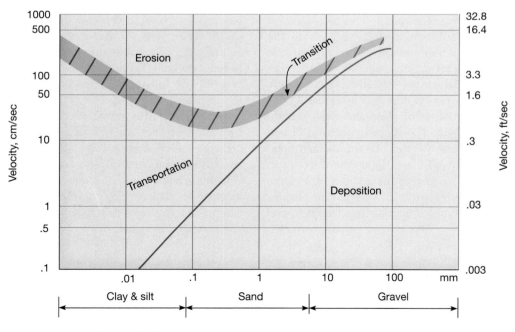

FIGURE 11.9 Relationship between particle size and stream velocity required for erosion, transportation, and deposition of material.

velocity needed to erode clay, silt, sand, and larger particles. It can be seen that medium-sized sand is the easiest grain size to erode because it takes a velocity of only about 1 ft/sec (0.3 m/sec) to dislodge it. Both smaller and larger sized particles take a greater velocity (more energy) for dislodgment. With increasing particle size, mass increases, so that increased velocity is needed. The mechanism is different for finer grained materials. With decreasing size greater cohesion among particles develops so that dislodgment is more difficult. Also the finer materials lie within the laminar flow regime at the base of a stream, which persists at lower velocities. This laminar range is disrupted at higher velocities so that the finer sized particles can then be eroded.

Figure 11.9 also relates the transportation and deposition of particles in regard to stream velocity. Note that for fine particles, much less velocity is needed for transportation (to keep particles in suspension) than that needed for dislodgment. Addi-

tional details on this subject are presented later in this chapter.

Another view of the water velocity needed for direct uplift of alluvial materials is shown in Figure 11.10. In this graph, the relationship between particle size and critical tractive force is shown. This force is the minimum needed to keep particles moving along the base of the stream. The graph is used to determine the maximum safe velocity for flow in a canal which will not cause undesirable erosion of the bed material. For example, a fine gravel particle, 10 mm in diameter (0.4 in.) requires a tractive force of about 0.2 lb/ft^2 to move it along the base of a stream channel. For gravel this size (Figure 11.9) a water velocity of 3.3 ft/sec (100 cm/sec) would be required.

Bedrock Channels
The other forms of erosion occur mostly in bedrock channels or those strewn with cobble- and boulder-sized particles. The suspended solids in the stream

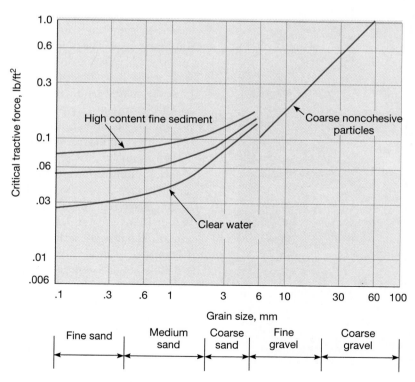

FIGURE 11.10 Critical tractive force for particle size movement versus particle size, canal stability considerations.

abrade the rock. Potholes, or deep circular depressions in the bedrock, may develop from this swirling action of the water charged with abrasive grains. A photograph of potholes formed in quartzite is shown in Figure 11.11.

Smooth, curving surfaces on the rock may form where stream flow is constricted by steep bedrock side walls. Plucking loose blocks of rocks along fractures that were further weakened by water penetration can also occur. This is caused by hydraulic action or "hydraulicing."

Some dissolving action of the water will occur as it flows over bedrock or alluvial materials. In most streams the water carries much smaller amounts of dissolved material than its saturation capacity would allow, simply because insufficient time is provided for greater solutioning to occur. Ground water typically carries a higher concentration of dissolved salts than do streams. When gaining streams receive ground water, this water is typically greatly diluted by the larger volume of surface water. Ground-water volume supplied to a stream becomes its base flow and may constitute the only water it receives during a dry period.

Cavitation occurs at high stream velocities in the range of 25 to 30 ft/sec (7.5 to 9 m/sec). At this high velocity, air bubbles in the stream suddenly collapse, producing an effective loading of up to 2000 psi (45.2 MN/m^2) pressure. This specific erosive process was shown to remove more than a foot of concrete in a few days time from a spillway structure in a major dam. Under natural conditions cavitation is restricted to rapids and waterfalls.

In summary, stream erosion depends on the following variables: 1) water velocity, 2) kind and amount of load, 3) nature of the stream bed, that is, the kind of rock or sediment present, and 4) nature, direction, and spacing of rock structures in rocky channels, for example, jointing, bedding, and foliation.

Transportation

A stream dislodges material from its channel and carries it, sometimes intermittently, downstream toward the ocean. The amount of material that a stream transports is called its load and the maximum amount it can carry is its capacity. The load of most streams, however, is well below its capacity.

The maximum size particle a stream can carry, as previously mentioned, is its competency. This diameter varies approximately with the square of the stream's velocity. For example, a medium-sized sand grain 0.5 mm in diameter is carried by a velocity of about 10 cm/sec (0.3 ft/sec). If the velocity is doubled to 20 cm/sec (0.6 ft/sec) the particle size transported is four times greater, that of a large sand grain of 2 mm. In like manner if the velocity is doubled again to 40 cm/sec (1.2 ft/sec), the particle size transported is a pebble about 8 mm in diameter.

Streams are able to transport material in three ways, by solution, suspension, and traction, the last including rolling, sliding, and skipping of particles along the bottom of the stream. The load is also considered to consist of three portions: solution load, suspended load, and bedload.

FIGURE 11.11   Potholes formed in quartzite. Swirling pebbles in a fast-flowing stream form these features.

Solution

Most natural streams, especially those in humid areas, consist of extremely fresh water, showing much less hardness than the ground water, which typically contains abundant cations of calcium and magnesium and a lesser amount of iron. On the other hand, surface water is not absolutely pure—it does contain some dissolved chemicals. Streams carry measurable amounts of calcium and magnesium cations plus the common anions: chloride, nitrate, sulfate, and silica. Streams deliver these nutrients to the oceans of the world and thereby supply the basic needs for the food chain of the sea.

Suspension

Particles of solid material that are suspended in the stream of water as it flows along are designated the suspended load and are said to be carried in suspension. This is distinct from material chemically dissolved by the water or pushed along at its base. As flow occurs, the particles begin to settle but turbulence whirls lift them back into the main flow before settlement is accomplished. This provides a relationship between stream velocity and the maximum size of suspended particles that can be accommodated, because turbulence typically increases with an increase in velocity.

Settlement of solids in water is similar to that in air, the main differences being that the frictional drag of a particle falling in water is considerably greater than that in air. Therefore, just as a projectile fired horizontally through the air from a cannon falls vertically as it moves horizontally, suspended particles fall through the water as the water moves along. They fall at a velocity that increases with increasing particle diameter, assuming that the specific gravity and shape of all particles is essentially the same. If the particles were not lifted up again by turbulence whirls, they would settle and the water would clarify.

Figure 11.9, discussed previously, also illustrates the manner in which different size materials are transported. Observe that for fine-sized particles much greater stream velocity is needed to dislodge these particles from the alluvial stream base than is needed to keep them suspended in the stream. For coarse particles this difference narrows considerably.

Because the greatest velocity occurs during periods of maximum flow, it should be no surprise to learn that the greatest amount of suspended load is carried during periods of flooding. Because of this, more than three-quarters of the total suspended load

transported in a year's time may occur in the few percent of time represented by periods of high-level flooding. This suggests a somewhat catastrophic effect to stream erosion and transportation, rather than a steady regulated effect year round.

The coarsest material of the suspended load is carried in the lowest part of the stream where the greatest turbulence occurs. This would consist of coarse sand pieces for a velocity of about 1 ft/sec (0.3 m/sec). The finer materials, silt- and clay-sized particles, are well distributed throughout the stream depth. Clay particles may remain suspended even after the water has come to rest in a lake or settling basin. Such suspended material is referred to as turbidity. Flocculating agents, for example, aluminum sulfate, are used in water treatment plants to remove the fine sediment for municipal water supplies.

Bedload

The largest pieces transported by flowing water are moved along the stream bottom as bedload. At high flow rates, as in times of flooding, more than half the total sediment may be moved along the bed.

The mechanism for bedload movement is traction, which may be accomplished by rolling, sliding, or saltation. The largest pieces are moved by rolling or sliding, which may occur in an intermittent fashion depending on turbulence whirls near the bottom. Saltation involves the jumping and skipping of particles along the bottom. The particle is picked up into the stream flow momentarily but settles quickly again to be picked up once more by the more turbulent flow.

Deposition

When stream velocity decreases owing to reduced depth or gradient, deposition of sediment occurs. The coarsest particles carried in the bedload are dropped first, with the coarsest suspended material transferred to the bedload. Further decreases in velocity yield deposition of the next coarsest material so that finer and finer materials are lost as the stream slows down. At a reduction in velocity to 1 ft/sec (0.3 m/sec), pebbles about 4 mm in size are deposited, medium sand is dropped out when the velocity slows to 0.1 ft/sec (0.03 m/sec), and silt will be deposited below about 0.02 ft/sec (0.6 cm/sec).

The ability of a stream to do work is closely related to its base level. This is defined as the lowest elevation to which a stream can cut its channel. There are several base levels involved. The ocean is the ultimate base level to which the entire length of the stream must adjust, because running water is physically unable to erode below sea level. In the geologic past, sea level stood lower than it does today so that rivers on the East Coast of the United States, such as the Chesapeake, Hudson, Potomac, and Delaware Rivers, have channels that extend well below the present sea level and out into the continental shelf.

The local base level for a tributary stream is set by the elevation of the point of confluence with the main stream. Some local base levels may be of a temporary nature, such as a lake that forms because of human activity or a naturally formed dam that crosses the stream. Streams can be dammed because of landslides, earthquakes, volcanic eruptions, or glacial action. Within the framework of geologic time, such dams will be destroyed and the previous gradient reestablished.

When the base level is raised, for example, by the construction of a dam, the reduced gradient causes the stream to deposit sediments until the same slope is established but at a higher elevation. If the base level is suddenly lowered by breaching of the dam, the stream will cut down through its own sediments to reestablish the former gradient.

LANDFORM FEATURES OF STREAMS

Both through erosion and deposition, streams alter the appearance of the land surface. As streams cut down through soil and bedrock they form a branching network known as the drainage pattern. The actual pattern depends on the nature of the underlying materials and on the history of the stream itself.

Drainage Patterns

A stream whose branching habit is like that of the limb of a deciduous tree is called dendritic. Such a pattern develops on horizontally bedded sedimentary rocks or various homogeneous materials, for example, massive igneous and metamorphic rocks or thick soil sequences.

Trellis drainage consists of elongated, parallel channels with short, nearly perpendicular tributaries yielding an orthogonal effect, which is similar to a garden trellis. The elongated channels mark the strike valleys of softer, folded rocks with short perpendicu-

lar tributaries flowing from the higher, resistant ridges. The folded rocks of the Ridge and Valley Province of the Appalachians and similar portions of the Rockies typically express this drainage pattern.

Rectangular drainage consists of perpendicular segments of streams without the dominant elongation of one orientation as typified by trellis drainage. A combination of strong foliation and joint control usually gives rise to this feature. Radial drainage is caused by streams radiating from a high central point such as a volcanic peak. These drainage patterns can be used to deduce information about the geologic structure of an area. Drainage patterns are illustrated in Figure 11.12.

Drainage Basins and Divides

The drainage basin is the total area from which a stream and its tributaries obtain runoff. It may also be referred to as a watershed. The drainage basin of a major river such as the Ohio or the Missouri is quite extensive—both involve considerable portions of the United States. The Mississippi River would include all of this drainage area because both the Ohio and Missouri Rivers are its tributaries. A drainage divide is the imaginary line connecting the high points between drainage basins. The continental divide separates drainage basins that supply water to the Gulf of Mexico and to the Atlantic Ocean from those yielding water to the Pacific Ocean.

Depositional Forms in Alluvial Valleys

For streams with wide valleys the features are depositional in nature except for the valley walls, which mark the sides of the valley. The flat area between these side walls is called the floodplain. Large streams or rivers flowing on low gradients in alluvial channels develop these wide valleys. They consist of the landform features discussed next and are illustrated in Figure 11.13. This figure shows the valley walls, previously discussed, and also the backswamp deposits. These are poorly drained areas away from the stream channel, which accumulate organic materials and are highly compressible when loads are placed on them. Natural levees are the high banks that parallel the river, formed by silt deposition during overbank flow. They form a lip that lies above the general elevation of the floodplain. For this reason some tributary streams must parallel the main course for a

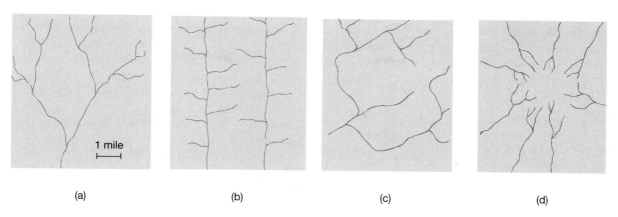

FIGURE **11.12** Drainage patterns that develop on different landscapes: (a) dendritic, (b) trellis, (c) rectangular, and (d) radial.

considerable distance before flowing into it. Named the Yazoo effect, after the Yazoo River of Mississippi, which depicts this feature, this is illustrated in Figure 11.13.

A meander bend occurs when the channel migrates laterally because of local differences in bank erodibility. The outside of a meander loop undergoes erosion, whereas in the inside of the loop, alluvial deposits known as point bars are formed. During periods of high water level a chute cutoff can occur, yielding a short cut across the meander bend. Such features may be successful in taking over the primary flow of the river. Meanders subsequently cut off from the main stream are called oxbow lakes. These features are also shown in Figure 11.13.

Meander scars are high areas adjacent to the channel, marking the location of a former meander loop. They are the remnants of a natural levee held in place by vegetation, particularly those types of trees with a high water demand. Because a meander bend migrates in the direction of the outside loop, the previous depression will be filled with loose, fine

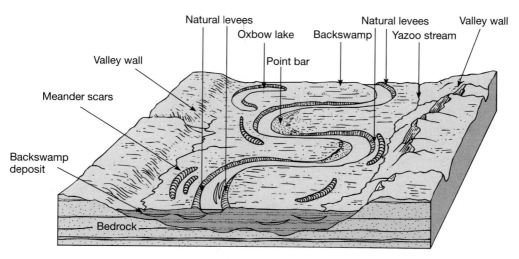

FIGURE **11.13** Schematic diagram showing landform features of wide alluvial valleys along major rivers.

sediments. In some locations a clay core is deposited in the ponded water. This illustrates the diversity of soil materials in an alluvial floodplain. Construction problems associated with floodplains are examined later in the chapter.

Channel Patterns

The channel patterns of streams flowing in alluvial channels were mentioned briefly earlier in this discussion. A channel can be either straight, meandering (Figure 11.13), or braided. If a stream is carrying an excess of sediment in relation to its velocity (and indirectly to its discharge and gradient), material will be deposited in the stream bed. This yields a complicated network of small channels choked by sandbars and islands, which gives an overall appearance somewhat likened to braided hair (Figure 11.6). This braided pattern occurs in streams choked with sediment but also in alluvial fans and glacial outwash deposits. Features related to glaciation are discussed in Chapter 12.

Deltas, Alluvial Fans, and Stream Terraces

Deltas

Other depositional features related to running water include deltas, alluvial fans, and stream terraces. Deltas form where a stream flows into a standing body of water or into one with a much lower velocity and transporting capability. This occurs when the stream reaches the ocean, a lake, or its confluence with a larger stream. The depositional feature that forms takes on a shape similar to the Greek capital letter delta, Δ.

The delta forms as the transported sediment is dropped because of decreased velocity. As the debris collects in one location, the channel shifts to a place with lower elevation until that channel builds up and yet another shift occurs. These are called distributary channels. Overall, it yields a deposit, triangular in shape, with the apex pointing upstream. This deposit eventually extends further and further into the body of standing water. Deltas formed in human-made reservoirs accumulate silt from the upper reaches of the stream and eventually extend downstream into the main body of the lake. Where a tributary stream flowing on a steeper gradient meets a larger stream, the delta may extend far enough into the channel to deflect the direction of flow of the major stream. If

constriction of the channel between the delta and the opposite bank occurs, erosion opposite the delta may develop, thus altering the direction of stream flow.

Some large rivers have extensive deltas where they meet the ocean. The Mississippi River delta is a prime example. However, other rivers form negligible deltas either because they carry no significant amounts of debris or because the ocean currents carry the material away as soon as it is deposited.

Alluvial Fans

An alluvial fan is likened to a delta on dry land or at least where no permanent streams are established. They are commonly found in arid and semiarid regions but may also occur in humid locations if the topography and other conditions are conducive to their formation. Alluvial fans occur where an intermittent stream has a sudden decrease in gradient, for instance, where a mountain stream slopes onto a plain. As with delta formation, the flow channels shift when debris builds the channel base higher than adjacent deposits and eventually a semicircular feature sloping toward the plain is formed. This can be observed in Figure 11.14. As the high area recedes because of erosion over geologic time, a sequence of alluvial fans will be formed at higher and higher elevations and in position close to the mountainous area. Because of the coarse nature of these deposits, alluvial fans are good sources of ground water. More discussion of this subject is included in Chapter 15 on ground-water geology.

Stream Terraces

A stream terrace is a flat or gently sloping surface that runs adjacent and parallel to the valley, with a steep bank separating it from the floodplain below or from a lower lying terrace. It indicates the former floodplain location for the stream when located at a higher elevation. This is illustrated in Figure 11.15. A sequence of events is suggested by this feature. In the past, the stream deposited sediments in the valley to the level of the terrace. Reasons why it aggraded its channel (deposited materials) up to this level include a slow rise in base level, an increase in load, or a decrease in discharge. In any event, the stream aggrades to this level and under equilibrium conditions establishes the level floodplain primarily by meandering of the stream back and forth from valley

FIGURE **11.14**   Alluvial fans in the arid southwestern United States. (Hamblin, WK,
The Earth's Dynamic System. 2nd Ed., Macmillan, New York.)

wall to valley wall. In the next episode the stream begins a role of active downcutting to establish a channel at a lower elevation. A climatic change could cause the rapid downcutting, yielding either a lowering of the base level or an increase in discharge. Again, the stream establishes equilibrium conditions and the floodplain forms. This in turn may be left as another terrace if renewed downcutting occurs or it may remain as the present-day floodplain. Terraces on opposite sides of the valley are called paired terraces. This feature occurs at Lafayette, Indiana, on

opposite sides of the Wabash River. The two terrace levels are depicted schematically in Figure 11.15.

Erosional Features
Water Gaps and Wind Gaps

Several erosional features of stream channels require special attention. Water gaps and wind gaps are of particular interest. When streams are established on a gently sloping surface of uniform resistance to erosion, a channel oriented parallel to the maximum slope will develop. As the stream erodes, the channel

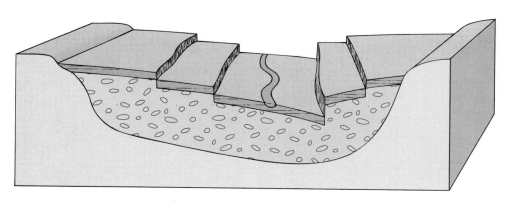

FIGURE **11.15**   Schematic drawing of stream terraces along the Wabash River,
Lafayette, Indiana.

becomes deeper and the overall land surface is slowly reduced. If this well-established stream intercepts a resistant, folded rock mass at depth, it will continue eroding through the rock material although perhaps more slowly. Eventually the softer rocks at the surface erode, leaving the folded, resistant rock standing at a higher elevation than the valley cut by the stream. The depression or pass formed where the stream cuts through the mountain is called a water gap. The gap is cut through the steep, resistant beds. Actually, the stream was superimposed on the structure and through downcutting it exposed the resistant bed. If this gap should become abandoned by the stream after stream piracy or by some other means, the dry gap remaining is termed a wind gap. Stream piracy occurs when a headward eroding stream with greater erosive activity takes over portions of the drainage area of another stream. Figure 11.16 is a photograph of a water gap located in a resistant bedrock area.

Regional Stage

A final detail of erosional features of streams involves the regional dissection obtained in stream development. This is a useful means for describing the appearance of topography and is known as the regional stage. If a flat gently sloping surface several square miles (tens of square kilometers) in size is uplifted from the sea bottom, runoff would subsequently yield a stream channel network in the downslope direction. Early in the development only a few channels would occur but as time passes more tributaries would be added by erosion, as illustrated in Figures 11.17(a) and (b). The areas between the stream channels would remain flat and thus are described as flat, interstream divides. Such an overall condition is known as regional youth. As stream dissection increases, less flat area remains between the stream channels and at some point late youth is reached, as pictured in Figure 11.17(c). When the dissection is sufficiently great that only rounded areas exist between channels, regional maturity has occurred. This is shown in Figure 11.17(d). With further passage of time late maturity occurs when small rounded areas remain between the channels [Figure 11.17(e)] and theoretically, at least, the point when no interstream divides persist is reached and this is called regional old age, which is shown in Figure 11.17(f).

The concept of a regional stage proves most useful in describing areas that began as flat, planar surfaces.

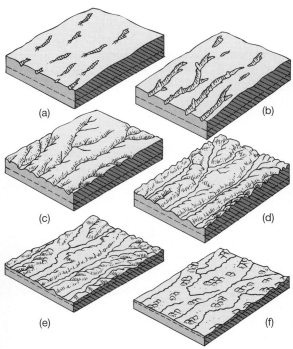

FIGURE 11.17 Regional stage: (a, b) early youthful topography, (c) late youth with flat interstream divides, (d, e) mature topography with rounded stream divides, and (f) old-age topography.

FIGURE 11.16 A water gap established in a resistant bedrock area.

This is true for many continental glacial deposits, for example, and for areas such as the Interior Lowlands of the central United States or the Great Plains that lie to the west of them. A youthful topography is typified by much of Indiana, Illinois, and Kansas. Northern Missouri, which is glaciated, shows a mature topography because of the several hundred thousands of years since glacial deposition occurred. Central Kentucky and Tennessee depict a mature topography as well, because these areas were developed on horizontal sedimentary rocks.

In many texts the concept of stage is applied to individual streams as well as to regions, with the net effect that youthful, mature, and old-age streams are distinguished. Youthful streams are said to have straight channels with steep gradients and few stream deposits, whereas old-age streams have low gradients, extensive deposits, and a meander pattern. Another procedure is to describe the overall area by regional stage and the individual stream by its specific characteristics, which for a mature stream would be a wide floodplain, low gradient, and meandering pattern. By contrast, youthful streams typically flow over rocky channels at high gradients with a straight pattern. This direct description is preferred by some geologists and is also preferred in this text rather than simply calling the stream "youthful."

ENGINEERING CONSIDERATIONS OF STREAMS

Many of the engineering concerns related to streams involve the use of wide floodplain areas. The upstream portions of smaller tributaries have engineering concerns but generally to a lesser extent. Both of these are considered in the following discussion.

Problems of Wide Floodplain Areas

Wide floodplains are composed of a diversity of materials as observed in Figure 11.13. Much of the soils consists of sands, sandy silts, and clayey silts of relatively low strength. Because of the high ground-water table, these materials are generally saturated as well. All the soils are somewhat compressible, but the organic clays of the backswamp deposits are of an extreme nature. A major problem is the difficulty involved with predicting where the backswamps are located in the subsurface because as the river meanders and changes elevation, new backswamp areas are

formed. The organic deposits may be 10 feet or more below the surface and provide no indication of their existence at ground level.

During construction of interstate highways across wide floodplains, the location of compressible organic soils must be considered. Highway embankments are constructed 10 or more feet above the floodplain near the valley wall so as to extend above the flood of record, and even to a greater height near the stream itself. In some low areas of the floodplain and for bridge overpasses, the embankments may be 30 to 50 ft (9 to 15 m) high or more. This leads to settlement of the underlying compressible soils.

Subsurface exploration for organic pockets of soil is difficult. Closely spaced borings only 25, 50, or 100 ft apart across a wide floodplain are not economically feasible. Instead, probing using a steel rod may be used to determine penetration resistance to a depth of 10 to 15 ft. When compressible soil pockets are located, excavation and replacement is a possible solution, or construction of a surcharge on the embankment followed by an extended time interval. A surcharge consists of an extra thickness of soil fill, causing accelerated settlements, which is then removed prior to paving. This encourages all the settlement to occur before the pavement is placed. The use of flexible (bituminous) rather than rigid (concrete) pavement is also a possibility to reduce cracking of the pavement if additional settlement occurs.

A high ground-water table makes deep excavations more difficult to accomplish. Basements and underground parking garages would not be typically included for most building constructions on floodplains. For structures where deep excavation is required, such as dams, locks, bridge foundations, and power-generating stations, the subsurface must be dewatered prior to and during construction.

Because of the availability of ground water in a floodplain, industrial facilities and communities may elect to obtain their water from the subsurface rather than directly from the river. Ground water tends to carry more dissolved ions than does surface water, but typically it has less turbidity and is less likely to be contaminated. Surface water is processed to remove the turbidity and to ensure that pathogens are eliminated, whereas ground water may be softened to remove nuisance cations but further processing, other than chlorination, may not be needed. Floodplains typically supply the greatest volume of ground water compared to other geologic situations. Ground-water pollution is a major concern and this subject along with water supply is discussed in Chapter 15.

The base of the alluvial soils in a major flood plain typically lies a hundred feet or more below the surface. This base may be bedrock or another geologic material such as glacial deposits, which predate stream development. The thick sequence of alluvial soils tends to consist of loose sands and silts and compressible clays of low shear strength. Consequently, they provide poor support for heavy structures. Deep foundations are required, such as steel H piles or possibly timber piles, which can transfer the load by end-bearing to a stronger material below the alluvium. This leads to expensive foundation support and economics may dictate that the structure be located away from the floodplain where foundation conditions are better. For some power-generating facilities, the power plant itself is located on the valley wall with only the water intake structures and cooling water ponds positioned directly on the floodplain.

The migration of meander bends is a concern. We should always assume that erosion will continue on the outside of a meander bend. Permanent buildings and other structures should not be constructed in these locations because it may be impossible to prevent erosion from occurring and undermining these structures.

Flooding is a problem of all floodplain areas. Just a small amount of reflection on the subject makes us realize that the floodplain was formed by the action of the stream both through lateral migration and flooding; and the active floodplain marks the location of previous high-volume flows of water. Under the appropriate weather conditions, serious flooding will occur again and the wide floodplain is needed to transmit this water. The recurrence interval for flooding may be 10 years or less or it may measure in the hundreds of years.

Constricting floodplains by encroachment of buildings or by building flood walls and levees prevents the river from carrying the water volume it possesses during flood time. This transfers the flood effect further upstream by backing up the water and it extends the effect further downstream as well. Floodplain encroachment by new construction is fairly well eliminated today because building permits are no longer issued for the floodplain zone. National flood insurance is also not available for such homes so that home mortgages will not be granted for new construction.

Construction in the floodway of a stream is prohibited by law today. This is the area where the stream transmits water by current flow. Adjacent to the active floodway is the floodplain, where water backs up but where active flow does not occur. Encroachment on the floodplain is allowed for structures not subject to property damage.

The question may be raised concerning why people construct homes in areas prone to flooding. Scenic beauty is one factor and individual rights is another claim sometimes expressed. In some mountainous regions, the only flat areas seem to exist in the valleys that are used for agricultural endeavors, home construction, and transportation routes. These locations are particularly prone to flooding because the runoff rate is high on the steep slopes and the narrow valleys can carry only relatively small volumes of water.

Navigational problems are associated with major waterways of the United States as an aid to shipping. Silt is removed from the channel by dredging. Erosion of the banks is reduced by the placement of riprap where the stream changes direction, and revetments are used to keep the main channel in one specific location rather than allowing it to shift its course. Levee construction and maintenance are necessary for both navigation and flood control purposes.

Streams are used as the primary means for removing liquid wastes from cities and industry. Sewage treatment plants, after removing much of the BOD (biological oxygen demand) from the sewage, dump their effluent into streams. Flowing water has the ability through its dissolved oxygen (DO) to eliminate the remainder of this BOD and then rejuvenate the oxygen level as it flows along. With greater volumes of sewage have come more advanced sewage treatment plants, so that secondary treatment is mandatory and tertiary treatment is becoming more prevalent. However, in periods of low flows (summer mostly) the dilution capability of the stream is obviously reduced and the DO is also lower at this time because of higher water temperature. If the DO drops to a critical level, fish kills can occur and an anaerobic condition can develop, yielding a foul-smelling, polluted stream condition. This situation must be avoided; further sewage treatment is needed to prevent its occurrence.

As mentioned previously in this chapter, urban runoff is a major consideration today. The hard surfaces prevalent in cities yield increased volumes of runoff, which must be accommodated by the storm-sewer systems. Because sanitary and storm water were commingled in many sewer systems at one time, the consequence is a greatly increased volume of polluted water. Overflow points occur in the system and at these locations the combined water is dumped directly into the surface stream. As explained previously, in Chicago a solution to the problem is being implemented. Other cities in the United States, Milwaukee, Wisconsin, for example, are also constructing deep tunnels to collect the urban runoff.

Problems of Upland Stream Locations

The major problems in upland stream areas are erosion, landslides, and flash floods. It is also true that flash floods occur along major streams in arid regions such as in west Texas, southwestern United States, and parts of the western Great Plains and the Rockies. Vegetation may be sparse and the terrain steep. This yields rapid amounts of runoff and deep erosion as a consequence.

The upland portion of tributary streams is the steepest and headward erosion is active. That, of course, is how youthful topography develops into mature topography. Areas are prone to landslides because of the steep slopes and undercutting by the stream channel. These areas, however, are quite scenic and because of the steep nature of the terrain may remain forested because clearing for agricultural purposes has not been advantageous. These locations make attractive homesites and may require special attention to prevent loss of topsoil—and possibly of buildings—by erosion and landslides.

EXERCISES ON RUNNING WATER AND RIVER SYSTEMS

Map Reading

Soda Canyon, Colorado; 15-min quadrangle map

This map is located in the Colorado Plateau province. The basic structure is a plateau of horizontal, sedimentary rocks, mostly sandstone.

1. What is the fractional scale, size, and contour interval?

2. What type of drainage is shown? What does it suggest about the underlying bedrock?

3. What is the regional stage of the quadrangle? Describe the nature of the stream channels; gradients (in general), bed material, valley type, etc.

4. Is the rainfall heavy or light in this area? Explain.

5. What is the origin of the small ponds at the head of Grass and Greasewood Canyon? Their shape should prove helpful in your determination.

6. Sketch on a separate sheet a cross section from the northcentral part of Sec. 30, T34N, R14W, southward to Ute Trail. Label the stream locations.

7. What do the springs (i.e., Mancos Spring) tell about the character of the underlying bedrock?

8. What is the gradient of the lower half of the Mancos River? What is the gradient of Johnson Canyon? Why should it be greater than the gradient of Mancos River?

9. Why does the Mancos River meander? Relate this to its gradient determined in Exercise 8.

10. What engineering problems might you expect in the area?

Kimmswick, Missouri–Illinois; 15-min quadrangle map

11. What is the fractional scale, size, and contour interval?

12. What is the regional stage of the area (exclusive of the Mississippi River floodplain)?

13. Sketch on a separate sheet an E-W cross section across the Mississippi River and across Glaze Creek.

14. What are the gradients of the Mississippi River and of Glaze Creek?

15. Describe the nature of both of the streams mentioned on Exercise 14. Be complete.

16. What is Moredock Lake and how was it formed?

17. What are the curved or horseshoe-shaped features outlined by contour lines on the flat area east of the Mississippi River?

18. What is the origin of the small depressions that occur between the Meramec and Mississippi Rivers on the upland surface? This is an area of limestone bedrock. Why do some contain water whereas others do not? Where does the surface water drain in these areas?

19. Why does Fountain Creek flow some 10 mi south along the floodplain before entering the Mississippi River? What is this feature called?

20. Why don't the country roads follow the section lines as they do in many parts of northern Indiana?

21. What engineering problems might you expect in this area (building foundations, road construction, etc.)?

Lexington, Nebraska; 15-min quadrangle map

22. What is the scale, size, and contour interval of the quadrangle?

23. What is the regional stage of the area?

24. About how much rainfall is indicated in the area? Explain.

25. Describe the nature of the Platte River as observed on this map (including its gradient). What is the geomorphic name for this channel pattern?

26. Is the Platte River aggrading or degrading its bed? Explain.

27. Would you expect to find bedrock at a shallow depth beneath the town of Lexington? Explain.

28. Would water supply be a problem for Lexington? How would water be obtained?

29. What engineering problems would you expect along the Platte River?

Harrisburg, Pennsylvania; 15-min quadrangle map

This quadrangle is located in the Ridge and Valley or Folded Appalachian province. The structure is a folded sequence of sedimentary rocks; sandstones, shales, and limestones primarily.

30. What is the scale, size, and contour interval of the quadrangle?

31. What type of drainage pattern is present in the mountain regions? Why is it developed rather than some other type?

32. What is the landform (geomorphic) name for the type of pass where the Susquehanna River crosses the mountains? What would this be called if the river abandoned the channel?

33. Describe the configuration of Conodoquinet Creek.

34. Compare engineering problems in eastern and western Harrisburg.

Written Questions

35. If 1% infiltrates into the subsurface in an arid region, how much water per acre is supplied when the annual rainfall equals 10 inches? Calculate this in acre feet of water per acre.

36. Explain why a sudden warming trend that melts snow-cover is particularly prone to cause flooding with regard to soil infiltration.

37. Explain why urban runoff yields a stream pollution problem. Why is this an expensive problem to solve?

38. Give a low, moderate, and high value for the common parameters with regard to U.S. streams. Parameters should include velocity, gradient, and discharge.

39. The surface velocity of a large stream was found to be 2.6 ft/sec. Approximately what would be the average velocity? If the cross-sectional area of the stream was 450 ft^2 what value for the discharge is obtained? Give the answer in cubic feet per second and millions of gallons per day.

40. Distinguish between the three views of a stream. Name them and discuss each in turn.

41. Given the Manning equation, $v = \frac{1.5}{n} d^{2/3} s^{1/2}$, how much will the velocity increase in a stream if the depth increases from 5 ft to 12 ft, assuming the other variables do not change? How much is the erosive force increased when this occurs? Show calculations.

42. At zero flood stage, a certain river is 12 ft deep at its deepest point. If the stream rises 8 ft during a flood period, how much scour of the stream bottom is likely to occur? What would be the maximum water depth for this situation? Show your calculations.

43. What does the term *competency* mean? Why does competency decrease downstream although the average velocity increases slightly? Explain in terms of turbulent velocity and average velocity. Why is the decrease in competency important with regard to the supply of aggregates for construction?

44. What diameter particle is easiest for a stream to transport? Give the numerical value. Why is this particle easiest to move? Explain.

45. High sediment load in a stream coming from plowed fields or construction sites is considered a type of pollution. Why is this a reasonable conclusion? What effect is it likely to have on fish and other fauna in the stream? How can this be controlled for large construction sites? What farm practice is being implemented to reduce this effect?

46. What is meant by the term *base level*? Why was the ultimate base level lower in the geologic past than it is today? What is the evidence for this? Explain.

47. What is meant by the term *drainage pattern*? How does this differ from channel pattern? Explain. Which drainage pattern is indicative of folded sedimentary rocks? Which one of a resistant central peak?

48. Refer to Figure 11.13. Draw a cross section across the valley showing the various landforms in cross section view rather than in map view. Include some older backswamps deposits in the subsurface.

49. What does a braided channel pattern indicate? What climatic change could cause this to occur?

50. New Orleans, Louisiana, is located on the extensive Mississippi River delta. What engineering construction problems are associated with the city? Base your answer on knowledge gained about problems of wide alluvial floodplains.

51. Why would an area adjacent to a water gap be subject to flooding problems? Why would runoff be high in such areas?

52. Northern Missouri has been described as a maturely dissected till plain. What would be the regional stage for this area? Would you expect the secondary roads to run north-south, east-west as they do in northern Indiana and northern Illinois? If not, what orientation would they have?

53. Why is encroachment on a floodplain by buildings a particularly poor policy? If the structures are open to pass water to a height above the 100-year flood, would this solve the problem? Explain.

54. Is the construction of a golf course in a floodway a good use of the land? Explain.

Additional Readings

BLOOM, A.L., 1978, *Geomorphology, A Systematic Analysis of Late Cenozoic Landforms,* Prentice Hall, Englewood Cliffs, NJ.

LEOPOLD, L.B., WOLMAN, M.G., and MILLER, J.P., 1964, *Fluvial Processes in Geomorphology,* W.H. Freeman and Co., San Francisco.

MORISAWA, M., 1968, *Streams: Their Dynamics and Morphology,* McGraw-Hill, New York.

SCHUMM, S.A., 1979, *The Fluvial System,* John Wiley & Sons, New York.

12

The Work of Glaciers

GLACIAL ICE

Glaciers today cover about 10% of the land surface, some 6,875,000 mi^2 (17.9 × 10^6 km^2), most of which is in Antarctica and Greenland. They form in regions of the world where more snowfall occurs in winter than is melted away in summer. When the ice volume becomes sufficiently great, glaciers flow outward from the accumulation area under the influence of gravity.

A glacier is a mass of ice, formed by recrystallization of snow, which may flow across the land surface. Years of accumulation and burial are required to convert snow to glacial ice. Icebergs and pack ice do not qualify as glaciers because both float across the oceans, and pack ice is frozen seawater.

The specific gravity of newly fallen snow lies between 0.08 and 0.10, indicating a high degree of porosity. This provides for the estimate of 10 in. of snowfall yielding an equivalent of 1 in. of rainfall that is commonly used by meteorologists. After snow settles, its density increases significantly to a specific gravity between 0.2 and 0.3.

On further densification, aided by melting and refreezing plus the effects of sublimation, snow turns to firn, a granular material with a specific gravity greater than 0.4 but generally less than 0.8. Glacial ice is achieved when this value through densification increases to 0.83 but it may reach a value of 0.91 deep in the glacier. The natural density of 0.917 for ice frozen directly from water cannot be achieved for glacial ice. Air voids incorporated in the snow continue to persist to some degree in the glacial ice.

Glacial ice behaves like a metamorphic rock. In the metamorphic process the crystal size increases from 0.1 mm for snow to 4 mm or more for the ice. Air content is reduced and specific gravity increased.

The boundary between firn and glacial ice occurs at a considerable depth within a glacier. Ice temperature is a factor because colder ice retards the growth of crystals. For valley glaciers this depth may be about 100 ft (30 m) but in the Greenland glacier the boundary is closer to 1000 ft (300 m) deep.

TYPES OF GLACIERS

The three primary types of glaciers are 1) valley glaciers, 2) piedmont glaciers, and 3) continental glaciers. Piedmont glaciers occur when two or more valley glaciers coalesce on the plains below the alpine region where glaciers originate. Piedmont glaciers are not of major significance and a detailed evaluation is not warranted in this basic discussion on the subject.

Valley glaciers are rivers of ice that flow down existing stream valleys from mountainous regions above. They are also referred to as mountain or alpine glaciers because of their origin in the mountains. The term *valley glacier* seems more descriptive because the glacier occupies steep valleys, which extend throughout their length and determine their direction of flow.

Continental glaciers are great mound-like volumes of glacial ice that advance under their own weight from the accumulation area. They cover the entire landscape, not just valleys and adjacent plains as do other glaciers. They may exceed 10,000 ft (3000 m) thick in their central portions and are perhaps half that thick near the advancing ice front.

Modern glaciers move slowly, only a few centimeters to a meter (3 ft) per day. They move by internal flow within the ice, much like stream flow, with the central portion moving the fastest and the boundary zones at the edges moving the slowest. Much of the movement occurs within the individual grains along atomic slip planes, and growth in crystal size is a consequence. Crystals may increase from the 4-mm size of newly formed ice to several inches in diameter by the time they reach the snout or furthest extent of a valley glacier.

RESULTS OF GLACIATION

The effects of modern-day valley glaciers can be examined to learn the manner in which they erode, transport, and deposit materials. Many similarities exist between these glaciers and their continental counterparts and, through comparison, considerable detail can be learned about continental glaciers as well.

Valley Glaciers
Erosion and Transportation

Valley glaciers obtain debris through frost action of the adjacent rock and from landslides or avalanches onto the ice itself. They also pluck blocks of rock from the jointed bedrock surface at their base by scouring of the ice and along the sides of the valley.

At the head of a valley glacier a bowl-shaped feature or amphitheater known as a cirque (pronounced "sirk") is carved from the rock by plucking and frost action. A major crevasse develops where the ice and bedrock meet at the head of the glacier. This is termed a bergshrund and it is a consequence of the ice moving by gravity away from the headwall.

Meltwater streams flowing away from the front of the glacier carry finely ground rock particles called rock flour. This provides a grayish color to the water and demonstrates the grinding power of the ice. Glacial lakes, which contain rock flour in suspension, take on a turquoise color caused by the refraction of light.

A glacier drags materials of all sizes (ranging from boulders to clay) along its base. These materials abrade the bedrock surface, sometimes yielding a smooth polish on fine-grained massive rocks or forming scratch marks in other instances. These fine scratches or striations can develop both on the bedrock and on those rocks held by the glacier that provide the grinding action. After glacial melting these striated rocks provide evidence of glaciation. Continental glaciers may also abrade deep trough-like features on a bedrock surface, known as grooves. Both striations and grooves show the bearing of ice movement but they do not indicate the direction from which the ice came. Because only a line on the rock surface is provided, one cannot distinguish the direction of origin. So for an east-west striation, the ice could have come either from the east or from the west.

Valley glaciers sharpen or steepen the topography by erosive processes because they are concentrated in stream valleys and in snow accumulation areas. This is in contrast to the erosional effects of continental glaciers, which tend to smooth the topography as they override the entire land surface. The distinction between these effects is discussed further when continental glaciation is considered in a subsequent section. The common erosional features of valley glaciers include the following: cirque, arete, horn, col, U-shaped valley, rock basin, hanging valley, and fjord.

As previously mentioned, a cirque is the bowl-shaped depression at the head of the valley glacier where snow accumulates and forms into glacial ice. An arete is the saw-toothed or serrated ridge that forms the common boundary between the sides of two cirques. Where three cirques intersect, the steep central peak or spire that remains is called a horn, which is bounded by the side walls of these cirques. The Matterhorn of Switzerland (Figure 12.1) is the most famous example of this feature. A col occurs when two cirques on opposite sides of a ridge intersect to form a gap in the ridge (Figure 12.2).

Glaciated valleys have a characteristic U-shaped or trough-shaped cross profile, in contrast to the characteristically V-shaped narrow mountain valleys formed by stream action alone. When the ice moves down the preexisting stream valley it broadens the base of the valley by eroding the weathered and fractured rock and accumulations of soil, yielding a trough shape. The regular gradient of the stream valley is disrupted as well. The main glaciated valley, long after the ice has melted, typically shows ponded and swampy areas adjacent to the newly established stream channels.

Higher up in the mountains, above this lower portion of the main valley, rock basins and hanging valleys are observed. Glacial action selectively quarries

FIGURE **12.1** The Matterhorn, the Alps, Switzerland.

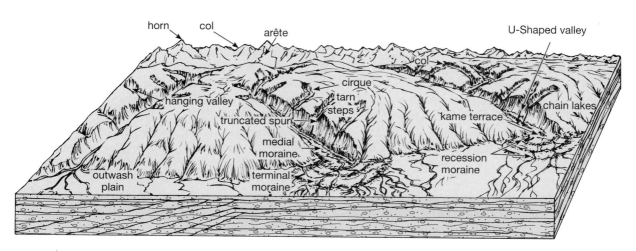

FIGURE **12.2** Features of valley glaciation.

fractured rock, yielding a series of depressions in the rock surface. These rock basins in the base of a cirque when filled with water are called tarns, whereas a series of lakes filling rock basins at a lower elevation are known as chain lakes. Hanging valleys are formed when tributary valleys are left high above the main valley that carried the glacier and underwent extensive erosion. Some hanging valleys are locations for waterfalls such as those observed in Yosemite National Park in the glaciated Sierra Nevadas. These features are illustrated in Figure 12.2.

Fjords occur in coastal areas where bedrock valleys were extensively glaciated and are now covered by the sea. The valleys were glaciated when sea level was considerably lower than today, but glaciers can erode solid bedrock even when it is below sea level. The fjords of Norway, Greenland, Labrador, Alaska, and New Zealand today are partially filled by the sea, yielding steep-sided, deep-water passages.

Depositional Features

A number of the landforms deposited by valley glaciers are similar to features formed during continental glaciation. These deposits will also be considered later in our discussion of continental glaciers. Features generally limited to the effects of valley glaciation and those that serve to introduce the subject are discussed in this section.

The debris transported by a glacier is eventually deposited after the ice has melted. Debris may be deposited directly by the glacier when released on melting or carried some distance away by the meltwater from the ice. The general term *drift* or *glacial drift* includes both modes of deposition. Unstratified drift occurs by deposition directly from the glacier; stratified drift is a consequence of meltwater deposition.

The unstratified drift, which is laid down directly by the glacier, consists of material known as till. It consists of a mixture of materials ranging in size from clay to large boulders greater than a ton in weight (>900-kg mass). Till is composed mostly of silt and clay with occasional pebbles, giving it an overall texture likened to raisin pudding. These deposits, known as clay tills, are prominent in the north central United States (including Ohio, Indiana, Illinois, Wisconsin, and Michigan) where continental glaciation prevails. In parts of New England and in the state of Washington there are extensive deposits consisting of large rock fragments and boulders, which are known as boulder tills.

Till is deposited by the receding glacier to yield landforms collectively known as moraines. Depending on their location and configuration, different types of moraines are distinguished. For valley glaciers, the sides of the glacier adjacent to the valley wall are locations for lateral moraines. Here material that falls from the valley walls accumulates, yielding a ridge along each side of the valley when the ice melts. When two glaciers join, adjacent lateral moraines unite to form a medial moraine near the center of the widened glacial flow. These are marked as dark zones in the active glaciers but they are usually reworked and disappear when the ice melts. Water running along the valley wall of a melting glacier may deposit coarse material to build up the lateral moraines. After melting, these stratified deposits stand above the valley floor and are called kame terraces. They may yield supplies of aggregate materials for construction. Glacial features are illustrated in Figure 12.2.

Common to deposition by both valley glaciers and continental glaciers are end moraines and ground moraines. An end moraine is a ridge of till that marks the place where the ice stagnated for some time while the advance equaled the retreat caused by melting. Debris accumulates in a high, arc-shaped landform with the concave portion pointing in the direction from which the ice advanced. The terminal moraine is that individual end moraine marking the greatest extent of ice advance. Other end moraines are recessional moraines, which develop when the ice recedes to another stagnation point as it melts headward. Much of the debris is laid down directly below the ice as melting occurs to provide a rolling terrain along the valley floor. The landform is called ground moraine and it may be hundreds of feet thick in deposits of continental glaciers. These various moraines are illustrated in Figure 12.3.

Outwash sand and gravel are deposited by meltwater flowing away from the face of a melting glacier. Streams of water loaded with material eroded from the till form braided channel patterns on the land surface beyond the glacier. As these streams lose their velocity, they drop the transported material, largest particles first, to form a wide apron of stratified material. This is called an outwash plain. If the deposit continues down a narrower valley beyond the

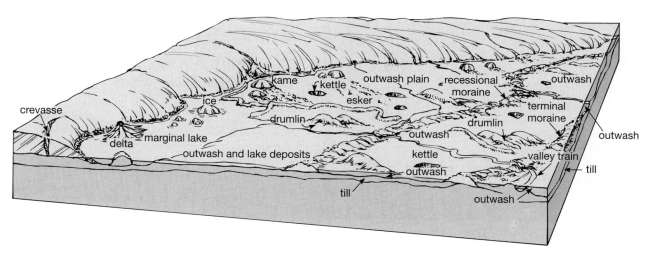

FIGURE **12.3**   Features of continental glaciation.

outwash plain, a valley train deposit is the consequence as more material is dropped by the flowing water. An outwash plain and a valley train are illustrated in Figure 12.3. A photograph of lateral and medial moraines forming on an active glacier is shown in Figure 12.4.

Continental Glaciers

Continental glaciers completely override the terrain and therefore tend to smooth the land surface rather than sharpen it in the fashion of valley glaciers. Both the erosional and depositional features of continental

FIGURE **12.4**   Lateral and medial moraines of a valley glacier, shown also is a lake occupying a crevasse in the ice.

glaciers provide a general smoothing effect because the deposits tend to fill in old stream valleys to the level of the surrounding hills. Some less common exceptions to this are the erosive effects and enlargement of stream valleys by the ice sheet as evidenced by the Great Lakes and by the Finger Lakes of New York State. These zones of less resistant rocks were selectively quarried by the glaciers to yield closed depressions now filled with water. The Great Lakes are discussed more fully in a subsequent section.

Continental glaciers may erode exposed bedrock hills to yield a unique knob-form known as a roche moutonnée. This is a sculptured hill that has a gently sloping face in the direction toward the advancing glacier and a steep face on the opposite slope. The ice rides along a bedding plane or other weakness zone and plucks the rock off a joint plane that is nearly perpendicular to the bedding. Consequently, the gentle slope marks the direction from which the ice advanced. A roche moutonnée is illustrated in Figure 12.5.

Continental glaciers yield depositional features somewhat similar to those of valley glaciers. Table 12.1 provides a comparison between the features of the two types of glaciation.

Erratics are boulders or cobbles of a certain rock type carried by glaciers a sufficient distance from their place of origin so that they are foreign to the type of bedrock where they are deposited. They may be found imbedded in till, lying directly on the bedrock surface, or located within outwash formed

FIGURE **12.5**   Roche moutonnée, or glacier-carved bedrock hill, showing direction of ice movement.

TABLE **12.1**   Comparison of Features for Valley and Continental Glaciation

Features	Valley	Continental
Till	Common	Common
Ground moraine	Common	Common
Terminal moraine	Common	Common
Recessional moraine	Common	Common
Lateral moraine	Common	Common
Medial moraine	Common, but quickly eroded	Absent
Drumlins	Rare or absent	Locally common
Stratified Drift	Common	Common
Outwash plain	Common	Common
Valley train	Common	Common
Kames	Common	Common
Kame terraces	Common	Hilly country only
Eskers	Rare	Common
Lake beds	Rare	Common
Erosional Aspects		
Striations, polish	Common	Common
Cirques	Common	Absent
Horns, aretes, cols	Common	Absent
U-shaped valleys, hanging valleys	Common	Rare
Fjords	Common	Absent
Roche moutonnée	Absent	Common
Loess	Rare	Common
Glacial erratics	Common	Common

from glacial meltwater. In the north-central United States, erratics typically consist of igneous or metamorphic rocks from the Canadian Shield. As such, they contrast greatly with the sedimentary bedrock in the area of deposition.

Kettles are pits or depressions in the drift. They can occur in outwash plains or in ground moraines as well as end moraines. Formed from the delayed melting of ice blocks in the drift, these depressions range from 30 ft to a mile in diameter (10 m to several kilometers) and may be from 3 to 100 ft deep (1 to 30 m). Outwash deposits intensely marked by kettles are called pitted outwash plains. Kettles may be filled with water to form the lakes so common in the northern United States and Canada or may become filled with organic deposits to yield the extensive peat and muck deposits. Kettles are shown in Figure 12.3.

Eskers are sinuous ridges of stratified sand and gravel formed by meltwater streams at the base of stagnant ice near the margin of the glacier. They are steep sided, from 10 to 50 ft (3 to 15 m) high, and may extend for 0.5 to several miles (1 to 10 km) in length. These ridges of sand and gravel usually occur as discontinuous features.

A common feature superimposed on ground moraines of continental origin is the small, conical-shaped hills known as kames. They consist of stratified drift formed in the crevasse openings of the melting ice. They may appear as clusters or in isolated mounds and contain a wide range of sand and gravel-

sized materials. By contrast, drumlins are cigar-shaped hills, typically composed of till, that were streamlined by the readvance of glacial ice over a rolling till plain. The blunt end indicates the direction from which the ice came and the narrow end shows the direction of ice movement. A drumlin is shown in Figure 12.6.

Kame terraces form when ice moves down a stream valley then melts back to yield stratified terraces against the valley wall. They occur in the lower reaches of valleys traversed by valley glaciers and in those valleys partially filled by continental glaciers.

Lake beds form as bottom sediments in lakes receiving siltation over a period of years. Because many marginal lakes form as a consequence of glacial melting, lake beds are a common variety of stratified drift deposits. Sandy beach ridges showing the location of a previous shoreline are associated with the lake beds. They are collectively referred to as lacustrine deposits.

Many lake beds consist of pairs of thin sedimentary layers, each doublet known as a varve. The two layers represent a single year's deposition with the darker, clayey layer providing the winter contribution. A varve is usually less than 1 in. (2.5 cm) thick but may reach several inches (more than 5 cm) in thickness under some circumstance.

Associated with many depositional features related to continental glaciation are silt deposits known as loess. These buff-colored materials in North America consist of small angular grains and are generally considered to be wind deposited and linked to glaciation. It is concluded that wind blowing typically from

west to east across newly formed, broad outwash areas picks up silt-sized particles only to deposit them some distance downwind. Shells from air-breathing snails, which abound in the loess, suggest that the deposits are not water deposited.

Five major areas of loess accumulation occur in the United States: lower Mississippi River, central interior lowlands, Great Plains, Snake River plain, and Columbia River (or Palouse Area). Refer to Figure 17.14, a map of the United States, showing the loess areas, for details.

Greater detail on deposits formed by continental glaciation is given in Table 12.2. Both the constituents of the deposits and the topographic detail are supplied in the table for purposes of comparison.

Another consequence of continental glaciation is the enlargement of stream valleys that lie immediately beyond the glacial margin. Water from the melting glacier erodes the stream channel, both widening and deepening it. When glacial melting ceases, the discharge dwindles and a small volume stream is found within a larger valley. Such a feature is called an underfit stream. A prime example of this is the Mill Creek Valley located in the center of Cincinnati, Ohio. The valley is quite wide and deep but the stream (Mill Creek) today is only of small size.

Two landform features of continental glaciation are shown in Figures 12.7 and 12.8. The first is an end moraine viewed from an area beyond the glacial ice advance. Figure 12.8 shows a till plain or a ground moraine typical of the midwestern United States.

GLACIAL HISTORY OF THE UNITED STATES

Glacial Stages

In North America four major advances of continental glaciers occurred during the Great Ice Age of the Pleistocene epoch. These four glacial stages are named after the midwestern state where deposits are extensive or were first studied. In chronological order starting with the oldest they are the Nebraskan, Kansan, Illinoian, and Wisconsin glacial stages. Each major glacial advance is estimated to have persisted about 100,000 years with interglacial episodes of warming lasting a duration of several hundred thousand years apiece. A schematic diagram showing ice advance versus time is presented in Figure 12.9.

FIGURE 12.6 Drumlin field showing direction of ice movement from left to right.

TABLE 12.2 Deposits Associated with Continental Glaciation

Type of Deposit	Lithology	Topographic Form	Special Features and Remarks
Till	Unsorted, nonstratified, heterogeneous. Contains angular rock fragments ranging from clay to boulders. Layer of till deposited by single advance of ice called "till sheet." Larger rock fragments may contain glacial striations or polish.	Ground moraine	Gently rolling; 5- to 50-ft relief. When flat, called till plain. When rolling, called "swell and swale." Drumlins, stream-lined hills, characterize some ground moraines. Formed beneath the ice.
		End moraine	Varies from extremely rugged to subdued forms. Commonly called "knob and kettle" topography. Usually occurs as a belt or range of hills marking the edge or margin of a former glacier lobe. Commonly given local geographic name.
Outwash	Stratified, sometimes cross-bedded, sand, gravel; less frequently silt. Wide range of rock types and mineral suites. Grains and pebbles more rounded than till cobbles due to water transport.	Outwash plain	Flat to irregular, depending on number of stagnant ice masses present when formed. Highly "pitted" outwash plains may appear as some end moraines.
		Valley train	Preglacial valley filled with extensive glacial outwash. May extend beyond glacial boundary.
		Esker	Subglacial stream deposit, winding ridges, sometimes compound, often discontinuous. Cross-bedded and slumped along sides where former ice walls collapsed.
		Kame	Irregular-shaped knoll believed to form in depressions in stagnant ice or near ice margins. Commonly associated with end moraines.
Lacustrine	Sands and gravel	Beach ridges	Gentle ridges of low relief that mark former shorelines of extinct lakes. May be accentuated by dune deposits. Can be traced for miles in many Great Lakes states.
	Silt and clay	Lake beds	Flat, featureless plains, covered with varying thicknesses of stratified silt and clay. Common in the Great Lakes region. Includes varves and varve-like deposits; fine sands also.
Eolian	Sandy	Sand dunes	Associated with beach sands, either ancient or modern; also found in areas of extensive outwash. Many Pleistocene dunes are now stabilized with vegetation.
	Silty	Loess	Nonstratified, buff-colored, calcareous, up to 90% silt fragments, believed by many to have been derived from valley train deposits associated with advancing or retreating ice front.

FIGURE **12.7** End moraine feature looking from a direction beyond the ice advance.

FIGURE **12.8** Till plain or ground moraine feature of the midwestern United States.

The distribution of glacial deposits according to geologic age is shown in Figure 12.10, in which the surface materials are indicated by different symbols. Distinctions are made between Kansan, Illinoian, and Wisconsin deposits. Isolated patches of glaciation (not shown) also occurred in the mountainous regions of the western United States where valley glaciers prevailed during the Wisconsin glacial stage. In Figure 12.10 an island of unglaciated terrain is shown in an area of southwestern Wisconsin at the Illinois, Iowa, and Minnesota boundaries, which is known as the "Driftless Area." This territory of resistant Cambrian and Ordovician sandstones marked the extent of ice

advance for several glacial stages and substages in the upper Midwest. Never completely encircled by ice at any time, the area experienced an ice advance from the north and west, which then retreated without covering the area. In a later stage glaciers came from the north and east also to fall short of covering the area before melting back. Consequently all the land surrounding the Driftless Area was glaciated at different times but leaving it as an untouched island in the glaciated terrain.

The general extent of continental glaciation in the United States can be observed by viewing Figure 12.10. In a general way, we can state that the boundary

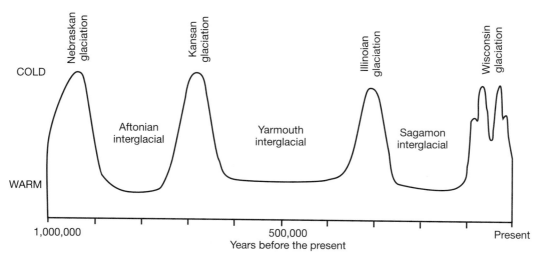

FIGURE **12.9** Pleistocene glacial advances of the United States.

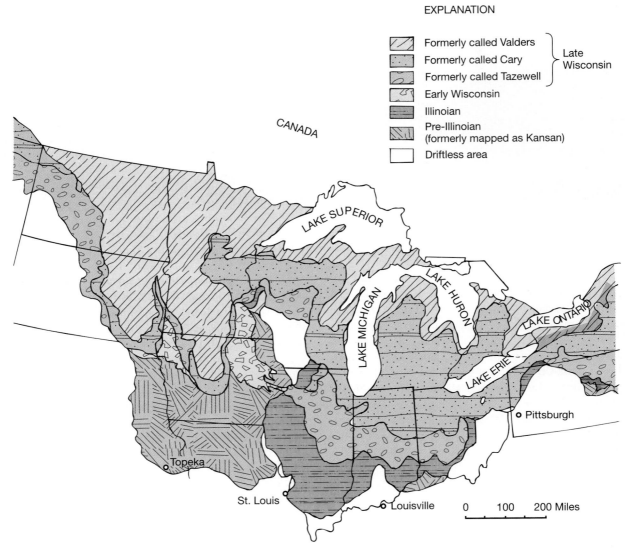

EXPLANATION

▨	Formerly called Valders
▨	Formerly called Cary
▨	Formerly called Tazewell
▨	Early Wisconsin
▨	Illinoian
▨	Pre-Illinoian (formerly mapped as Kansan)
☐	Driftless area

Late Wisconsin

CANADA

LAKE SUPERIOR

LAKE MICHIGAN

LAKE HURON

LAKE ONTARIO

LAKE ERIE

○ Pittsburgh

○ Topeka

St. Louis   ○ Louisville

0   100   200 Miles

FIGURE 12.10 Distribution of glacial deposits in north-central United States.

showing maximum extent of glacial advance extends from Long Island, New York, across northern Pennsylvania to Pittsburgh down along the Ohio River to its confluence with the Mississippi River, up the river to St. Louis and the confluence of the Missouri River, and then up the Missouri River into North Dakota and straight across the country to Seattle, Washington. There are exceptions to this general boundary, such as southeastern Ohio, south-central Indiana, and the southern tip of Illinois, which are not glaciated. However, this easily remembered boundary serves as an approximate indication for those sections of the United States that were overrun by continental glaciers during the Pleistocene epoch.

The elevation of the ocean's surface is intimately associated with the amount of glacial ice that covers the land surface. If all glaciers were suddenly melted today, the sea level is estimated to rise some 100 to 200 ft (30 to 60 m). Such a rise would obviously have a devastating effect on numerous coastal cities throughout the world. Indications are that glacial ice has been retreating in the last century with an attendant increase in sea level of several inches.

During the glacial epochs of the Pleistocene, water locked up in glacial ice was much more extensive than that amount involved with glaciers today. Therefore, during glaciation, sea level stood at a much lower elevation. In like manner, during the interglacial periods, sea level was significantly higher than it is today.

It is of interest to estimate how low sea level dropped during the maximum ice advance of Wisconsin glaciation. To calculate the volume of ice involved, both the area of ice coverage and average thickness of the ice are needed. The area of the Earth's surface covered by ice can be estimated with reasonable accuracy, but determining the average ice thickness is much more difficult. Thickness obviously decreased near the margins. Also the pressure of the ice load can be established for some locations. In all, an estimate of up to 330 ft (100 m) has been suggested for the drop in sea level that occurred during the maximum advance of the Wisconsin glacier.

History of the Great Lakes

Geological evidence indicates that the Great Lakes were formed in late Wisconsin time by the deepening of weak rock lowlands that in preglacial time contained streams draining eastward into the St. Lawrence River. The Great Lakes formed in the time period from about 14,000 to 2500 years b.p. (before the present) by glacial scour yielding ice-marginal water bodies when the ice front receded. Lake evo-

lution was influenced by five contributing factors: 1) oscillating ice fronts, 2) topographic irregularities uncovered by retreating ice, 3) variations in directions of retreat and advance of the ice, 4) lowering of lake outlets by erosion, and 5) differential uplift of the land adjacent to the ice following glacial melting.

There were no preglacial (or pre-Pleistocene) Great Lakes, but there would surely have been sequences of lake evolution during the final phases of the Illinoian, Kansan, and Nebraskan glacial stages. No evidence of a pre-Wisconsin lake system of the magnitude of the Great Lakes has been recognized to date.

The history of the Great Lakes has been determined primarily by tracing topographic and geologic features showing former shorelines and outlets, along with radiocarbon dating of specific deposits. During certain phases of glacial melting, the lake levels stood at considerably higher elevations than the present-day shorelines of the Great Lakes and at considerably lower levels. This yielded lacustrine deposits that extend well beyond the confines of the current lakes. These lacustrine plains along with the lake outlets that drained the high-level water bodies are shown in Figure 12.11. The area of marine subsidence along the St. Lawrence River is also depicted.

ENGINEERING CONSIDERATIONS OF GLACIAL MATERIALS

Glacial deposits are used both as construction materials and for foundation support of engineering structures. Because of the great variety of glacial materials, the

FIGURE 12.11 Lacustrine areas surrounding the Great Lakes.

contrast in value or usefulness in the first case and associated problems in the second is quite considerable.

Construction Materials

Sand and gravel are used for many construction purposes including aggregates for concrete, base course materials under roads, aggregates for bituminous pavements and overlays, cohesionless backfill material, filter blankets for drainage, and for slope protection from wave action and mass movement. Certain types of stratified drift are sources of sand and gravel supplies.

Outwash plains and valley trains yield good supplies of aggregate. These are typically associated with stream valleys and wide plains adjacent to drainage ways. The maximum size and distribution of sizes may vary considerably throughout the deposit depending on the velocity of the water carrying the material. Typically, an excess of sand and fine pebbles exists compared to the amount of coarse gravel material, which is needed for aggregates used in concrete and bituminous mixes. Deposits with less than a 20% to 25% gravel fraction are not economical to process because there is little market for the excess sand, which must be wasted.

Eskers, kames, and kame terraces are composed of stratified drift but the range of sizes is usually extreme. A range from large boulders to silt may occur with fine sand and silt sizes predominating. These landforms are also prone to extreme chemical weathering of the boulders and cobbles because these hills lie above the general terrain and are affected by the downward seepage of water through time. Bank run sand and gravel for general-purpose fills or base courses for secondary roads are easily obtained from these deposits. Consequently, such features often disappear from the landscape because of surface mining soon after urbanization approaches a region.

The economic value of gravel in a glaciated area is determined by its location relative to the point of use. If it must be transported a significant distance, the cost to provide the gravel increases significantly. Such material is said to have a high "place value" because location (or place) is so significant in determining its value. High-volume, low-unit-cost materials (such as gravel) are particularly subject to such considerations. In glaciated areas the effect of place value for gravel

is magnified owing to the common occurrence of gravel deposits in these areas. Because of this, the market for gravel must generally be located within about 15 mi (25 km) of the source if its price is to be competitive.

Ground moraines and drumlins are not good sources for sand and gravel. The unstratified drift consists primarily of silt and clay particles in much of the north-central United States. Kames may be associated with end moraines so that some sand and gravel deposits can be found within the end moraine complex. Careful study of the landforms is needed to ensure that coarse materials are contained within the feature of interest.

Loess Deposits

Loess, the wind-deposited material consisting primarily of silt, has special engineering characteristics that warrant consideration. It is yellow to buff in color with a uniform particle size in the vicinity of 0.01 to 0.005 mm in diameter. The material is without obvious stratification and consists mainly of fresh, angular grains of quartz and feldspar. When it is not saturated with water, loess has a strong tendency to split along vertical joints and to maintain vertical faces when excavated in that fashion.

Loess is deposited in the windward direction from major valley trains of glaciers, that is, glacial sluiceways. It is thickest adjacent to these features and thins out rapidly from the source. Loess blankets the entire landscape; hills, slopes, valleys and plains alike. As discussed previously its origin is linked to a retreating phase of continental glaciation.

Like all silty materials, loess dries quickly when exposed to the air. This leads to the problem that loess is very sensitive to changes in moisture during earth-moving operations when the material is being compacted in an earth fill. In the past, slopes were cut vertical in loess deposits because of its ability to maintain this configuration when dry. Deeper cuts related to interstate highway construction were found to be unstable in a vertical excavation and slumping was extensive. Ground-water seepage contributed greatly to this instability. In such situations, the loess must be cut at a much gentler slope (as low as 3 : 1, horizontal to vertical) and trees are established on the slopes to reduce seepage effects and to stabilize the slope.

Foundation Problems

Engineering properties of glacial deposits vary greatly with regard to the problems associated with foundation support for structures. With this in mind, engineering applications for the following categories are examined: 1) sandy outwash plains and valley trains, 2) ground moraine or till plains, 3) lacustrine plains, and 4) peat and muck deposits.

Outwash Plains

Sandy outwash plains and valley trains typically contain high ground-water tables and cohesionless materials, which will settle under static loads and as a result of vibrations. Problems include settlement in sand because of footing contact pressures and vibrations during pile driving. Dewatering of the sands to accomplish dry working conditions may also lead to settlements and loss of ground on adjacent properties. Layers of compressible silts and clays may occur within the outwash materials so that settlements due to consolidation must also be considered.

Ground Moraines or Till Plains

Ground moraine in many areas of the Midwest is extremely dense because of the great thickness of the overriding ice. Some layers are locally referred to as "hard pan" because of the difficulty experienced in drilling or in driving a split spoon sampler through the material. This soil sampler is used in the Standard Penetration Test (SPT) where the number of blows per foot to drive the sampler is recorded. This exploration technique is discussed in Chapter 19.

These tills provide quite suitable support for most building foundations. Glacial till also provides an excellent material for construction of small dams of uniform cross section. These impound water for farm ponds under the Soil Conservation Service (SCS) program. For larger dams glacial till may provide the core material for the dam, whereas the shell material may be somewhat coarser in size. Dams are discussed in Chapter 20.

Lacustrine Plains

A lacustrine plain is the location of stratified drift formed at the base of glacial lakes. These layers of silt and clay are highly compressible. They have low shear strengths and consolidate under imposed loads. The extent of lacustrine deposits associated with the Great Lakes is shown in Figure 12.11. Extensive areas occur in 1) Chicago and the southern end of Lake Michigan, 2) southeast of Lake Erie in the Toledo, Ohio, area extending along the lake to Cleveland, 3) southern end of Lake Huron in the Detroit area, and 4) Buffalo, New York, and eastward on the south side of Lake Ontario. These are all populous areas and the problems associated with the compressible clays have contributed greatly to the study and applications of geotechnical engineering.

Underground construction in compressible lake beds leads to other difficulties. The low shear strength of the clays causes loss of ground into tunnels during excavations by tunneling shields. The Chicago subway system built in the 1940s was constructed in the lake clays of the Chicago Loop area. The problem of weak shear strength and high compressibility had to be dealt with during this construction.

Peat and Muck Deposits

Peat and muck deposits are formed in glacial depressions that fill with plant remains as vegetation grows into shallow lakes or in swamps. Kettles provide these depressions and they can occur in ground moraines, end moraines, and outwash plains. Peat deposits may comprise an area of several square miles ($\sim$7 km^2) or can occur in small patches of less than an acre (0.004 km^2 or 0.4 hectares) in size.

The vertical extent of a peat deposit is generally 10 ft (3 m) thick or less but some large deposits may extend for depths greater than 25 ft (7.5 m). For construction purposes it is more expedient to avoid peat deposits by relocating buildings or transportation routes and power transmission lines than to deal with the problems they involve. Avoidance is not always possible, however, and for road construction, excavation of the material or bridging over it with a displacement fill may be the only feasible solution. Pile foundations through the muck and peat may be used to support structures that need to be located at the particular site. As more construction and housing developments occur in the United States fewer options of changing the alignment of roads or the location of proposed buildings will prevail. The solution of difficult problems for supporting buildings and other engineering structures on challenging geologic materials will be required of trained specialists with increasing frequency in the future.

EXERCISES ON THE WORK OF GLACIERS

Map Reading

Mt. Rainier National Park, Washington; 30-min quadrangle map

This map shows mountain or valley glaciers in existence on Mt. Rainier.

1. On which side of Mt. Rainier are the glaciers the largest where they extend to the lowest elevations? How do you account for this? (*Hint:* Consider weather patterns.)

2. Locate examples of the following features using names shown on the map:
   (a) tarn
   (b) medial moraine (glacier name)
   (c) end moraine (glacier name)
   (d) cirque
   (e) U-shaped valley
   (f) hanging valley

3. Describe the shape of Cathedral Rocks just southeast of the summit of Mt. Rainier. Account for the shape you have just described. What is this feature called?

4. What do the brown dots on Cowlitz glacier represent? Does this material contribute significantly to the transport of debris by a glacier? Explain.

5. Locate a possible source of glacial gravels for highway construction. What *type* of glacial deposit have you located?

Waterloo, Wisconsin; 15-min quadrangle map

6. One inch on the map equals how many miles in the field?

7. What is the regional stage of this quadrangle? During which glacial stage was the surface material formed?

8. What direction of ice movement can be inferred from the alignment of the drumlins of this quadrangle?

9. What is the average length, height, and width of drumlins in this area? Describe their overall shape. What type of material probably would be found in the drumlins?

10. What does this general area consist of: an outwash plain, till plain, terminal moraine, or what?

11. Describe the nature of the natural drainage in the area.

12. What engineering problems would be expected concerning the Chicago and Northwestern railroad in this area?

Otterbein, Indiana; 15-min quadrangle map

The northern part of this quadrangle is ground moraine west of Lafayette. Note the location of High Bridge and its topographic setting. Also note the Granville bridge setting. What are the bridges' respective distances from the Wabash River at West Lafayette?

13. How would you describe the topography of the northern part of this quadrangle? (Include its regional stage.)

14. How well drained is the northern part of this map? Has this area undergone much stream dissection to date?

15. What is the geologic age of the glacial drift in this area? Can you tell this by the amount of stream dissection? Explain.

16. Why is Little Pine Creek (near High Bridge) flowing in such a deep, wide valley in spite of the limited dissection of the uplands? What is such an oversized valley called?

17. Why does Lost Creek terminate before reaching the Wabash River? What kind of deposits would you expect to find along the river?

Flora, Illinois; 15-minute quadrangle map

18. What surficial features shown by the topography suggest an older age of glaciation than that shown on the Otterbein quadrangle?

19. From what glacial stage would you expect the drift in this area to have originated? Explain.

20. What is the regional stage of development for the quadrangle?

21. Account for the meandering of the Little Wabash River.

22. Are there any evidences of drumlins, eskers, and recessional moraines? What glacial landform is present in this area?

23. What do the little circles indicate? (*Hint:* See back of map.)

24. Suggest a source for sand and gravel in the area.

25. What engineering problems might be expected in this area? Explain your answer.

Old Speck Mountain, Maine; 15-min quadrangle map

26. What is the landform name (or physiographic province) for this general location of the United States? How does it differ from the Appalachian Mountains in Pennsylvania and Virginia?

27. What evidence of glaciation appears on the map? (*Hint:* Consider details of surface erosion to arrive at your answer.)

28. If the region has been glaciated, what explanation accounts for the smoothly rounded contours of the hills?

Kingston, Rhode Island; 7.5-min quadrangle map

This area is located in the Appalachian orogenic belt. Structurally it consists of intrusive igneous rocks overlain in many places by glacial drift.

29. What evidence of glaciation appears in the highland area in the northern portion of the quadrangle?

30. Name the type of landform for the irregular belt of topography just south of Worden's Pond.

31. How is Worden's Pond related to this irregular landform?

Written Questions

32. A cubic foot of "frozen water" was obtained from an Antarctic glacier. The sample weighed 30.1 lb and had a crystal size of 3.1 mm. What is its specific gravity and what name best applies to the sample? Explain. What would be its approximate air content?

33. Why are the terms *valley glacier* and *mountain glacier* used interchangeably for one of the major types of glaciers? Where in the United States can the effects of valley glaciers be observed?

34. If sea level was 100 m (328 ft) lower during the maximum glacial advance, what volume of water (in mi^3) from the oceans does this represent? What volume of ice does this represent (assume a SpG of ice = 0.90)? If the glacial ice averaged 2100 m (6890 ft) thick on the Earth, how much area would this volume of ice cover?

35. Approximately how long did the Yarmouth interglacial period last? What would be the effect of this extensive time interval on the glacial deposits of Kansan age? Explain.

36. What is the significance of the Wisconsin Driftless Area? Why was the area not overridden by glaciers? About how large an area is involved?

37. Was the Pittsburgh, Pennsylvania, area glaciated? Was Cincinnati, Ohio? Was Rapid City, South Dakota? What is the approximate percentage of Indiana that is unglaciated?

38. In what areas located around the Great Lakes in the United States do lacustrine deposits prevail? What major cities are found in these areas?

39. What glacial deposits are best exploited for sand and gravel to be used as construction materials? Are these stratified or unstratified deposits? Why?

40. What is meant by the phrase *place value* with regard to geologic deposits (see Construction Materials)? If transportation costs for gravel are $0.25 for the first ton-mile and $0.05 per ton-mile thereafter, how much does it cost to transport gravel 15 miles? What percentage increase in cost does this represent for gravel that sells at $6.00 per ton at the processing plant?

41. What types of concrete deterioration can occur when glacial gravels are used as coarse aggregates? Refer to Chapter 6, Engineering Properties of Rocks, for details.

42. What are the five major areas of loess accumulation in the United States? In which of these is Vicksburg, Mississippi, located? Walla Walla, Washington? East St. Louis, Illinois? (See also Figure 17.14).

43. In the United States, why would deep loess cuts on the northside of an interstate highway be more stable than cuts on the southside of the highway? Consider climatic factors in your answer.

Additional Readings

Flint, R.F., 1971, *Glacial and Quaternary Geology*, John Wiley & Sons, New York.

Paterson, W.J.B., 1969, *The Physics of Glaciers*, Pergamon Press, London.

Sharp, R.P., 1960, *Glaciers*, University of Oregon Press, Eugene, Oregon.

13

Physiographic Provinces and Engineering Considerations

In applied geology and engineering construction a significant advantage accrues from the consideration of geologic features on a region-by-region basis. Construction problems and geologic concerns in one area of the United States will differ from those of another area. One major reason for these differences is the contrast in geologic setting and geographic details that occur in one place versus another. Many examples can be cited where outside construction contractors, coming into an unfamiliar area, overlook a local geologic condition and bid a major job too low. After winning the contract, they subsequently must

fight to minimize their losses or attempt to break even on the project.

Physiography or landform features can be used as a basis to isolate specific construction problems or concerns for an individual region. The geologic formations that make up an area have a direct effect on the construction problems. These formations are a consequence of geologic history, from their origin, through time and the processes that have acted on them.

In this chapter regional physiography is examined first followed by the details of geologic history that produced these landscapes. Next, the problems that are common to certain physiographic regions are considered. Finally, a review of the geologic and physiographic characteristics of two sample locations is accomplished.

PHYSIOGRAPHIC PROVINCES

Continental areas such as the United States have been divided into regions, each having a particular type of landscape. These subdivisions of the continental landmass are called physiographic provinces.

In the United States, the physiographic provinces are large, ranging from one to several states in size. Within a province, the landscape is generally similar because it has developed under similar climatic conditions on rocks of the same general type and struc-

ture, depicting a common geologic history. That is to say, the landscape is a consequence of 1) the structure of the rock units present, 2) the geologic processes under which the rocks were weathered and eroded, and 3) the stage or degree to which the geologic processes have acted to wear away those geologic materials. These factors are discussed further in a subsequent section.

The physiographic provinces of the United States are shown on the Lobeck physiographic Diagram as Figure 13.1. Mountainous provinces in general are underlain by orogenic belt structures, that is, by various combinations of folded and faulted sedimentary rocks; intrusive igneous bodies; and regional metamorphic rock masses—all of which are associated with mountain building processes. The lowland and plateau provinces are underlain by stable region

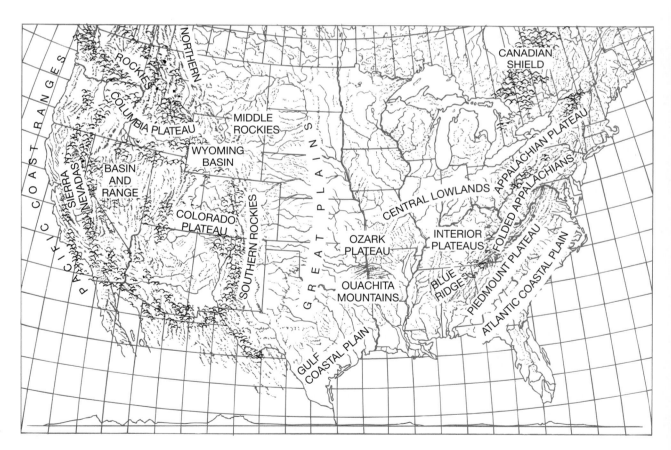

FIGURE 13.1 Physiographic provinces of the United States. (Reprinted from Columbia University, New York. Copyright 1932 by A.K. Lobeck.)

structures. These consist of horizontal to gently dipping sedimentary rocks or broadly domed sedimentary features that extend over areas several states in size. Thick layers of lava flows and volcanic cones form a final group of landscape features. These typically formed at a later time on top of preexisting orogenic belt (mountain forming) structures. Most Precambrian orogenic belts have been generally reduced to an erosion surface of low relief.

A brief discussion of the major physiographic provinces of the United States is presented in Table 13.1. These descriptions are, by nature of their brevity and by the large extent of the various provinces, greatly simplified. In reality, each province yields a chapter of discussion in books that specialize in this subject. The descriptions do provide a general indication of the features that prevail in an individual province.

TABLE 13.1 Descriptions of the Physiographic Provinces of the United States

The Canadian Shield Province

This is the foundation on which the North American continent was built. It has undergone many periods of uplift and erosion. Most of the units are hard metamorphic and igneous rocks of Precambrian age, highly resistant to erosion. Surface relief on the Canadian Shield has been greatly reduced by erosion since Precambrian time, from a mountainous terrain to one of gently rolling hills and valleys.

Two parts of the province are located in the United States, the Superior Upland of Northern Minnesota and Wisconsin, and the Adirondacks of New York. Both are composed of rocks with extremely complex structures, have been exposed to erosion for nearly two billion years, and have undergone, most recently, extensive glaciation. The soil cover is thin because of glacial erosion and numerous lakes occur in the massive bedrock areas.

Gulf and Atlantic Coastal Plain Province

This province is composed entirely of undisturbed sedimentary strata consisting mostly of loosely consolidated sands and gravels. Much more area than is visible today was exposed during Pleistocene glaciation when sea level stood considerably lower. The plain is relatively flat, sloping gently to the ocean, with some low, hilly sections.

Ridge and Valley Province

The folded Appalachian belt consists of upturned, strongly folded Paleozoic sedimentary rocks extending from the New England Province to Alabama. It is centrally located between the Appalachian Plateau to the west and the Piedmont Plateau and Blue Ridge to the east. The structure is not nearly as complex as the belt to the east, nor are the mountains as high. The belt consists of parallel, elongated, steep-sided ridges, with valleys between them. Drainage is almost entirely confined to these valleys. The province contains major coal deposits.

Appalachian Plateau Province

This is an area of very gently folded rocks that has been uplifted like a table and then dissected by streams. The northern half was glaciated and contains many lakes and other glacial features. Numerous coal deposits are found here.

Blue Ridge Province

This is a massive ridge complex that extends some 500 mi (800 km) from southern Pennsylvania to northern Georgia. It consists of metamorphosed sedimentary rocks of the Lower Cambrian and Upper Precambrian ages plus Precambrian gneisses and granites. The Great Smoky Mountains occur in the southeastern portion of this province.

(Continued)

TABLE 13.1 Descriptions of the Physiographic Provinces of the United States—Continued

Piedmont Plateau Province

This province extends from southern New York state to Alabama and lies as a low plateau above the Coastal Plain Province to the east but at a much lower elevation than the Blue Ridge Province to the west. It consists of Precambrian metamorphics and igneous intrusives; hence the name "Crystalline Appalachians." Deep residual soils known as saprolites occur in the province as well as occasional resistant bedrock domes such as Stone Mountain, Georgia.

New England Province

This is the northern extension of the easternmost mountain belt, that is, of the Piedmont, Blue Ridge, and Ridge and Valley Provinces. The rocks are very complex having been subjected to folding, faulting, and intrusions by igneous rocks, with most episodes in early Paleozoic time. Extensive areas of gneiss, schist, slate, marble, and quartzite indicate the widespread metamorphism. The entire region has been recently glaciated, leaving a landscape dotted with thousands of lakes and numerous streams.

Central Lowlands Province

This extensive province is underlain by limestones, sandstones, and shales of Paleozoic age. It is gently warped into domes and basins. The northern half was recently glaciated and contains many lakes and glacial features. The province embraces almost the entire Mississippi Valley and the Great Lakes area. It is one of the greatest food-producing areas of the world.

Interior Low Plateaus Province

This is a transition area between the Appalachian Plateau and the Central Lowland with the Coastal Plain to the southwest. It consists of gently dipping sedimentary rocks ranging in age from Ordovician to Cretaceous. Extensive erosion has formed a series of rock escarpments in the province.

Ozark and Ouachita Provinces

These are sometimes collectively referred to as the "interior highlands." They are remnants of an extensive mountain range that can be correlated with the Appalachian system. The Ozark Plateau consists of Paleozoic sedimentary rocks, predominantly cherty limestones, whereas the Ouachitas are strongly folded and faulted rocks with sandstones and shales predominating.

Great Plains Province

The U.S. portion of this province extends from the Rio Grande River on the south to the Canadian border on the north (it actually continues into northern Canada) with the Central Lowlands to the east and the Rocky Mountains to the west. The rocks are mainly Mesozoic or Cenozoic in age and at the surface consist commonly of poorly lithified rock units formed on the weathering of the Rocky Mountains, or of glacial deposits. Beds are mainly horizontal. Only the northern section has undergone continental glaciation.

Southern Rocky Mountains Province

This province consists of a north-south trending mountain system extending from southern New Mexico to southern Wyoming. It is comprised of a series of anticlinal ranges with cores of Precambrian igneous or metamorphic rocks, and it yields the highest peaks in the Rockies.

(Continued)

TABLE 13.1   Descriptions of the Physiographic Provinces of the United States—Continued

Wyoming Basin Province

An extension of the Great Plains, the Wyoming Basin lies between the Southern and Middle Rocky Mountains. Tertiary age formations occur at the surface, attesting to its relationship to the Great Plains.

Middle Rocky Mountain Province

Separated from the Southern Rockies by the Wyoming Basin, this province consists of two distinct mountain groups. The eastern group includes the Bighorn, Wind River, Bear Tooth, and Owl Creek mountains, which are similar to the Southern Rockies with their cores of Precambrian igneous and metamorphic rocks. The western group includes the Gros Ventre, Hogback, Wyoming, Snake River, Salt, and Teton Ranges, which were formed mainly by thrust faulting. The origin of these ranges is more closely related to that of the Basin and Range Province than to the Southern Rockies. Jackson Hole and Yellowstone Park are located in the western group of the Middle Rockies Province.

Northern Rocky Mountain Province

The Northern Rockies are located in the northern two-thirds of Idaho and the western part of Montana. They are composed of a major "granite batholith," the Idaho Batholith, plus prominent folded mountains in western Montana and other prominent structures elsewhere.

Columbia Plateau Province

This plateau, located adjacent to the Columbia River in Idaho, Oregon, and Washington, consists of an accumulation of thousands of feet of lava flows. The northern portion was glaciated and several lakes were formed by this process and by lava flows transecting the rivers. It is an arid region and not heavily populated.

Colorado Plateau Province

Located in the southwestern U.S., this plateau system formed on gently dipping sedimentary rocks; it includes the Grand Canyon. It is an area where the slow uplift has been matched by the downcutting of the rivers to form deep canyons. In the southern section, extensive lava flows and episodes of volcanism, some as recent as 900 years ago, have occurred. The sharp topographic expression is due largely to the arid climate of the province.

Basin and Range Province

The geologic history of this region is complex, but block faulting forming large basins and ranges has superimposed a certain similarity to the province. This has yielded deep, closed basins, surrounded by relatively flat-topped ridges. The area is one of the most arid in the world and poorly populated.

Sierra-Cascade Province

This province consists of the Sierra Nevada, which is a prominent block-faulted mountain system located along the California–Nevada border and the Cascade Range, a series of volcanic peaks extending from northern California through Oregon and Washington. Included in the Cascades are such prominent peaks as Mt. Shasta, Mt. Rainier, Mt. Lassen, and Mt. St. Helens. Mount Whitney, the highest point in the 48 contiguous states of the United States, is located in the Sierra Nevada.

(Continued)

TABLE 13.1   Descriptions of the Physiographic Provinces of the United States—Continued

Pacific Coast Ranges Province

These are the westernmost mountains on the continental United States, extending along the coastal area. They are also the most recently formed mountain belt and many of the mountains consist of partially unconsolidated materials. Los Angeles and San Francisco are located in this area. To the north (in Washington), the mountains form offshore islands. In the south, rainfall is sporadic, yielding a semiarid condition. Because of the young nature of these mountains, there are many very active faults in the area.

Pacific Troughs Province

This belt of deep valleys divides the Sierra-Cascades from the Pacific Coast Range. The troughs are formed by large down-faulted blocks. They have been accumulating sediments from the mountains on both sides and some are very fertile. The San Joaquin Valley lies south of San Francisco with the Sacramento Valley to the north. In Oregon, the Willamette Valley forms the trough and in Washington it is the Puget Sound. In the north the troughs are water filled and form the island passage up the coast to Alaska.

As previously stated, there is a concern for the nature of problems that could develop in a specific geographic area. The concept of physiographic provinces can be applied relative to highways, dams, bridges, and building foundations. Table 13.2 presents a list of possible construction problems or considerations that could pertain to different parts of the United States.

FACTORS AFFECTING PHYSIOGRAPHY

The appearance of the land surface, that is, the landscape, is determined, as previously stated, by three considerations: 1) geologic structure, 2) the dominant geologic process that has operated in the region, and 3) the stage or degree to which the geologic process has been able to carve the land surface.

Structure

Structure involves the type of rock present, and its attitude or position in space, along with the extent or lack of faulting and folding of the rock mass. For example the horizontal sedimentary rocks of the Central Lowlands yield a different landscape than do the massive igneous intrusives that form the Sierra Nevada. Of the three factors just mentioned, structure is the dominant feature in determining the overall appearance of the landscape.

Process

The geologic processes of major significance relative to the appearance of the land surface are glaciation, weathering and erosion in humid regions, weathering and erosion in arid regions, subsurface solution in limestone terrains, and development of coastal features along the oceans and the Great Lakes. Although less significant than structure in determining the overall appearance of the landscape, process is none-theless important. As an example the New England Province is equivalent structurally to the combined area of the Ridge and Valley plus the Blue Ridge. Extensive continental glaciation in the New England area, as contrasted to no glaciation in the two other provinces, provides a major difference in appearance and therefore the separate designations.

Stage

Stage can be thought of as the extent to which the process has run its full course. Youthful topography in a humid region shows only the initial effects of stream erosion, leaving extensive interstream divides, which are flat and untouched by channel development. A contrast can be seen between the appearance of the till plains in central Indiana and the same type of feature in northern Missouri. The till plain of Wisconsin age in Indiana (only 14,000± years old) is in early youth with only a small network of streams developed. In Missouri the topography is mature because the Kansan age till has been undergoing dissection for

TABLE 13.2   Engineering and Geological Factors of Concern for Various Regions of the United States

1. Tunnels required	18. Earthquakes
2. Frost action in soils	19. Fires
3. Water supply	20. Icing of roads
4. Permafrost	21. Water pollution
5. Aggregate shortage	22. High ground-water table
6. Landslides and slope stability	23. Storms near seacoast
7. Location–bridges, foundations, etc.	24. Beach erosion
8. Excavation, rock cut	25. Soil erosion
9. Sinkholes	26. Reservoir leakage
10. Irregular bedrock surface	27. Saltwater intrusion (Coastal)
11. Sanitation	28. Expansive soils
12. Embankment construction	29. Compressible soils
13. Water for compaction	30. Poor bearing capacity
14. Flash floods	31. Reactivity of concrete aggregates
15. Snow removal, snow avalanches	32. Presence of faults
16. Accessibility during exploration	33. Subsidence
17. Blowing sand	

Note: This list is intended to be a general yet somewhat complete list of highway and construction problems due to natural conditions that can occur in various parts of the United States. There is no preferred order of presentation.

nearly a half-million years. The rolling terrain of Missouri looks obviously different from the Indiana region.

The concept of stage was discussed in a previous chapter on running water and stream development (Chapter 11). Figure 11.17 shows the effect of different stages of development on the appearance of the landscape. The existence of flat interstream divides is a criterion for youthful topography, with mature topography occurring when these flat areas are destroyed.

SPECIFICS ON GEOLOGIC STRUCTURE

An outline of the major geologic structures of continental regions is shown in Table 13.3. These are the primary geologic structures that strongly influence the appearance of the landscape. There are three subdivisions of these geologic structures: 1) undisturbed, 2) disturbed, and 3) volcanoes and lava flows. The undisturbed structures underlie most of the interior areas of the United States, extending from the Appalachian Plateau on the east to the Great

Plains on the west. This is the stable interior of essentially horizontal sedimentary rocks, which is a common feature for all the continents of the world. The rocks have not been deformed by mountain building stresses.

Disturbed structures include folded and faulted sedimentary rocks along with igneous intrusives and complex metamorphic configurations. These are the areas of active mountainous regions and of such locations now worn down through the great expanse of geologic time. The Appalachian Mountain region, the Canadian Shield, the Rocky Mountain system, the Pacific Coast Ranges, and the Basin and Range Province, which lies between these last two continental subdivisions, make up the vast area of disturbed structures in the United States. Details on the different varieties of structure included in this large group of landforms are discussed later in this chapter.

Volcanoes and lava flows comprise the third major subdivision of disturbed structures. These are extensive areas where volcanic mountain chains dominate the terrain or where great expanses of flood basalts cover large areas of the landscape. The extensive areas in the world where flood basalts occur were

TABLE 13.3   Geologic Structure of the Continents

The main types of structures found on the continents can be divided into three groups. The structures in each group owe their general similarities to a similar geologic history.

A. Undisturbed Structures (Stable Region)

These structures are developed on layers of sedimentary rocks, mostly shale, sandstone, and limestone or on thick glacial deposits. These layers are horizontal, gently dipping, or very broadly warped into domes or basins, and they form a relatively thin veneer over the Precambrian basement complex, ranging in thickness from a few hundred to a few thousand feet. The broad expression of a stable region structure is that of a plateau if the relief is moderately high; or that of a plain if the relief is low.

1. Horizontal or very gently dipping sedimentary layers including coastal plains
2. Broadly domed sedimentary layers.

B. Disturbed Structures (Orogenic Belt)

These structures are formed on sedimentary, metamorphic, and igneous rocks. If layered, these

rocks are seen to be folded and faulted from their original positions, and they form a thick, though long and narrow, section above the basement complex. Post-Cambrian orogenic belts are typically expressed as long narrow mountain chains, having extensive amounts of granitic rock exposed in their central part. Also within the belt, and usually bordering the granitic rocks, are areas of belted metamorphic rocks. Most ancient (Precambrian) orogenic belts have generally been reduced by erosion to a surface of low relief.

3. Folded or tilted sedimentary layers (striped, canoe-shaped, and zigzag patterns)
4. Major faults (may be expressed as an abrupt change in topography along a more or less straight line)
5. Homogeneous crystalline rocks: granite, gneiss, some schists
6. Complexly related igneous, sedimentary, and metamorphic rocks.

C. Volcanoes and Lava Flows

7. Volcanic cones
8. Layered flows of lava.

---

discussed in the chapter on igneous rocks (Chapter 3). In the United States the province comprising extensive lava flows is the Columbia Plateau Province, located in the northwest portion of the country. The Cascade system, located in central Washington and Oregon and extending some distance into northern-California, is the prominent volcanic mountain system in the United States. Mount St. Helens, which violently erupted in 1980, is located within the Cascades in southern Washington.

Block Diagrams
The geologic structures described in Table 13.3 are illustrated in Figure 13.2 in block diagram form. The eight diagrams in Figure 13.2 are keyed to the specific subdivisions shown in Table 13.3. The two upper drawings refer to A1 structures, coastal plain

features for the upper left diagram and horizontal strata for the upper left. The A2 structure of broadly domed sedimentary layers is shown in the drawing directly below.

The disturbed structures of the B subdivision are illustrated in four diagrams below the A structures in Figure 13.2. B3 shows the folded sedimentary rocks common in the Ridge and Valley Province of the eastern United States. B4 illustrates major faulting and B5 shows the rounded topography of homogeneous crystalline igneous and massive metamorphic rocks. B6 is a diagram depicting complex combinations of igneous, metamorphic, and folded sedimentary rocks including faulting.

Terrain illustrating volcanic cones is presented in diagram C7. This is typical of the Cascades region. Layered volcanic rocks typical of flood basalts formed

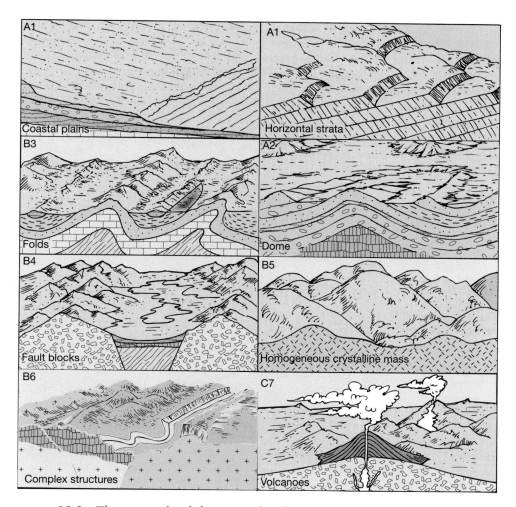

FIGURE **13.2** The geographical diagrams of geologic structures related to landscape.

by fissure eruptions are not included in Figure 13.2. These have a generally similar appearance to the dissected plateau of horizontal sedimentary rocks shown in A1, upper right drawing.

The various features depicted in Figure 13.2 can be observed on the topographic maps of selected areas of the United States. By observing the configuration of the contour lines, the relief, drainage pattern, gradients, and other factors, details can be ascertained about the geologic structure that prevails in the area. This information is used in turn to predict the construction problems, concerns, and geologic constraints that should be considered. The interpretation of topographic maps to determine the underlying geologic structure is required in the map exercises presented at the end of this chapter.

DETAILS ON PROCESSES AND STAGES

Additional details of geologic processes are considered in other chapters of this text; running water, glaciation, coastal processes, ground water, and wind. They are not considered in a major way, however, in the exercises for this chapter. Differences in weathering effects of humid versus arid conditions can be observed for the eastern versus the western areas of the United States, respectively. These differences are discussed in detail in Chapter 8 on Rock Weathering and Soils.

The regional stage for an area is discerned by comparing the appearance of the topographic map to Figure 11.17, which depicts the different types of

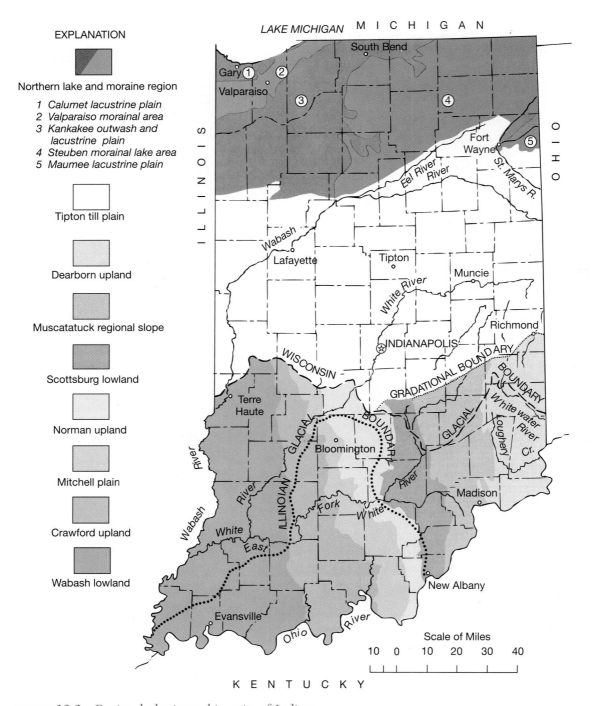

EXPLANATION

Northern lake and moraine region

1 *Calumet lacustrine plain*
2 *Valparaiso morainal area*
3 *Kankakee outwash and lacustrine plain*
4 *Steuben morainal lake area*
5 *Maumee lacustrine plain*

Tipton till plain

Dearborn upland

Muscatatuck regional slope

Scottsburg lowland

Norman upland

Mitchell plain

Crawford upland

Wabash lowland

FIGURE 13.3   Regional physiographic units of Indiana.

stage development. Youth, late youth, early maturity, and late maturity are possible designations. If an area did not begin as essentially a flat sloping plane onwhich stream drainage systematically developed, then the comparison is not applicable.

SUBDIVIDING PHYSIOGRAPHIC PROVINCES

Sections

All of the physiographic provinces described in Table 13.1 can be subdivided into smaller portions based on physiography. These smaller units, typically called physiographic sections, have more features in common and a less diverse range of landscape. In the case of provinces consisting of mountainous terrain, the sections may be individual mountain ranges and intermountain basins. This is the case for the Middle Rockies. Another example, the Ozark Province, consists of individual plateau areas and higher mountain terrains, which have distinct structural differences. On a site-specific basis, analysis of the individual sections is a must, preceded by a study of the regional conditions, which is accomplished by a review of the one or more physiographic provinces.

Divisions on a State Basis

Another means for subdividing physiographic provinces or large parts of them is accomplished by studying the landscape detail on a state-by-state basis. This proves useful because state geological surveys approach the studies of surface geology for the state in this fashion and the citizens of that state can best appreciate these subdivisions. Many states can be studied using this approach. Subdivisions of a state physiographic map into these areas based on a very similar landscape appearance are referred to as regional physiographic units.

Indiana

As an example, the state of Indiana with its individual units is shown as Figure 13.3. The extent of Wisconsin glaciation and Illinoian glaciation is also provided on the map. The landscape of those units with a north-south elongation, located in the central to southern portion of the state, is influenced by the bedrock characteristics in these areas. For northern Indiana the landforms are dictated primarily by the effects of glacial deposition.

Table 13.4 provides the description for the physiographic units of Indiana. They are keyed to the map of the state presented in Figure 13.3. A generalized geologic map of Indiana is supplied as Figure 13.4.

Oklahoma

Oklahoma has been selected as the other example to illustrate how physiographic divisions can be related to highway construction problems. Figure 13.5 shows the physiographic units of the state and Table 13.5 provides a description of these physiographic subdivisions. Figure 13.6 is a generalized geologic map of Oklahoma.

COMMON HIGHWAY CONSTRUCTION PROBLEMS IN OKLAHOMA The following highway construction problems have been described by Hayes (1971). Listed by frequency of occurrence rather than by severity of hazard or by economic cost, they are 1) seepage, 2) landslides and slumps (slope stability), 3) rippability, 4) expansive soils, 5) rock excavation, 6) construction materials location, 7) erosion-sedimentation, 8) sinkholes and cave-ins, 9) corrosion of metals underground, and 10) land use.

Some of these problems are more prevalent and severe in certain parts of the state. For instance, landslides are almost always restricted to the eastern third of the state because of the higher annual rainfall (45+ in. or 112+ cm), steep slopes, and thick sequences of shale and/or presence of colluvium.

Seven of the nine physiographic units described in Table 13.5 are included in this comparison of highway construction problems to the different landscape regions. This is summarized in Table 13.6.

Additional details follow concerning highway construction problems in Oklahoma:

- Difficult to locate riprap or coarse aggregate for paving in High Plains area.
- Severe sinkholes occur in gypsum areas. Gypsum beds thicker than 10 ft (3 m) cannot be ripped.
- Seepage problems occur where the granular terraces overlie less permeable shale.
- Salt springs typically are corrosive to culverts.
- Seepage is a frequent problem on east-west roads that must cross east-facing escarpments.
- Some clay shales are expansive.
- Nonrippable limestones are common; seepage in these areas is also common; underground mine areas yield cave-in problems.

TABLE 13.4   Description of Regional Physiographic Units of Indiana

1. *Calumet Lacustrine Plain*
   Lake sediments of glacial Lake Chicago. Series of beach ridges and sand dunes from former and present shorelines.
2. *Valparaiso Morainal Area*
   Wide terminal moraine of a substage of Wisconsin glaciation. It is the prominent relief feature in the area.
3. *Kankakee Outwash and Lacustrine Plain*
   Sandy glacial lake deposits developed on outwash from the Valparaiso moraine. Many low scattered sand dunes on the flat lake plain.
4. *Steuben Morainal Lake Area*
   Composed of recessional moraines from the ice lobe that entered the state from the northeast (Saginaw lobe).
5. *Maumee Lacustrine Plain*
   Lake sediments, an extension into Indiana from an extensive lake plain in Ohio.
6. *Tipton Till Plain*
   Flat Wisconsin age till sheet underlain by locally rugged bedrock topography giving rise to a great range in thickness of glacial deposits.
7. *Wabash Lowland*
   Area of moderate to low relief developed on Pennsylvanian age shales and sandstones.
8. *Crawford Upland*
   Most rugged topography in Southern Indiana, results from erosion of alternating massive sandstones, shales, and limestones of Late Mississippian age.
9. *Mitchell Plain*
   Area of moderate to low relief developed by solution of Middle Mississippian limestones; many caves.
10. *Norman Upland*
    Gently westward-sloping surface on resistant sandstones and siltstones of Early Mississippian age. The Knobstone escarpment along its eastern margin rises 400 to 600 ft above the adjoining Scottsburg lowland.
11. *Scottsburg Lowland*
    Narrow lowland area developed on Devonian and Early Mississippian age shales.
12. *Muscatatuck Regional Slope*
    A westward-sloping surface held up in the east by resistant cherty Silurian limestones along the border of the Dearborn upland.
13. *Dearborn Upland*
    Flat upland surface with deeply entrenched valleys developed on Upper Ordovician limestones.

- Acid seepage water is corrosive to culverts.
- Expansive soils are common in Gulf Coastal Plain.
- Landslide problems can be related to certain rock formations, members, and beds in the state. Johns-Valley Formation in southeast Oklahoma (Ouachita Mountains) almost always yields landslides unless a special design is used.

The two states used as examples in this application of physiography to engineering construction represent an area where glacial deposits and horizontal rocks prevail (Indiana) and another where a variety of sedimentary structures and igneous rocks are found (Oklahoma). These provide the needed detail to illustrate how this relationship between physiography and engineering construction is shown. Similar engineering applications can be made to physiographic divisions of the other states.

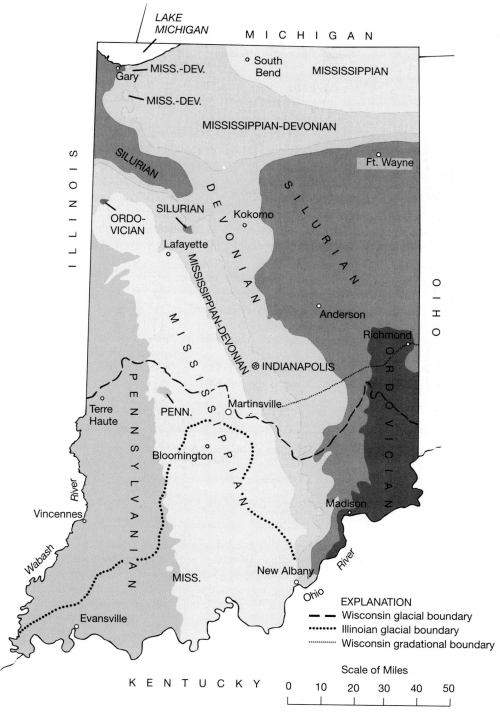

FIGURE **13.4** Generalized geologic map of Indiana. (Compiled by J.B. Patton, January 1, 1952, Indiana Department of Conservation, Geological Survey.)

TABLE 13.5 Descriptions of Regional Physiographic Units of Oklahoma

High Plains

Composed principally of nearly flat-lying weakly to unconsolidated calcareous sandstones and conglomerates with some soft, chalky, sandy limestones present locally. Tertiary in age, these plains were formed essentially by Rocky Mountain outwash. Steep slopes adjacent only to large streams.

Prairie Plains Homocline

Permian and Pennsylvanian shales and sandstones that dip westward 10 to 50 ft/mi (2 to 9 m/km). Shales are red to the west (Permian) and gray to the east (Pennsylvanian). Region consists mostly of gently rolling hills and broad, flat plains. Resistant beds of cemented sandstone or of limestone are more common in the Pennsylvanian strata. Hence, relief is greater in the east where long cuestas of east-facing escarpments overlook broad shale plains.

Ozark Uplift

A deeply dissected plateau formed on gently dipping (westward) cherty limestones of Mississippian age. Caves, solution cavities, and other karst features are more prevalent than in other sections of Oklahoma.

Andarko-Hollis-Marietta Basins

Flat-lying red Permian shales and sandstones. Few rock units are hard and the topography is mostly gently rolling hills and broad flat plains. Locally 100-ft- (30-m)-thick beds of gypsum occur along with thin (5-ft or 1.5-m) dolomite beds. Resistant gypsum layers in Andarko Basin and western Hollis Basin locally cap high escarpment. Dips are mostly gentle (usually less than 15 ft/mi or 3 m/km), although a narrow area 5 mi (8 km) at the southern edge of Andarko Basin has a dip of 250 ft/mi (50 m/km). In some areas thick beds of soft sandstone are deeply dissected into steep-walled canyons. Extensive granular terrace deposits occur adjacent to major streams.

Arkoma Basin

Pennsylvanian strata consisting of gray shales and hard, resistant sandstones. Strata are gently folded into a series of many local anticlines and synclines. Dips typically less than 15° and many faults are evident locally. Resistant sandstones yield broad hills and mountains rising 300 to 2000 feet above wide, hilly plains and valleys.

Ardmore Basin

Lowland of folded Mississippian and Pennsylvanian shales and sandstones lying between the Arbuckle Mountains and Gulf Coastal Plain. General structure is synclinal, but a large number of anticlines are also present. Dips are steep particularly along margins of the basin, with angles of 45 to 90° being common. Dark gray shales are from 100 to several thousand feet thick (30 to more than 600 m) and separated by resistant sandstones, limestones, and conglomerates only 10 to 15 ft (3 to 5 m) thick. These latter units form conspicuous subparallel ridges rising up to 100 ft (30 m) above adjacent lowlands.

Ouachita Mountains

Along with the Wichitas and Arbuckles they comprise Oklahoma's three mountain regions. Located in the southeast the Ouachitas form an arcuate fold belt consisting mostly of Mississippian and Early Pennsylvanian sandstones and shales locally about 30,000 ft (9100 m) thick deposited in an elongated trough. During the Pennsylvanian age these strata were folded into broad anticlines and synclines and thrust northward along a series of major thrust faults. Resistant, steeply dipping sandstones form long sinuous ridges and hogbacks towering 1000 to 1500 ft (300 to 450 m) above adjacent shale valleys.

TABLE 13.5 Descriptions of Regional Physiographic Units of Oklahoma—Continued

Arbuckle Mountains

Located in south-central Oklahoma this is a small area of low to moderate hills containing 15,000 ft (4500 m) of folded and faulted sedimentary rocks ranging in age from Cambrian to Pennsylvanian. About 80% of this thickness is comprised of limestones and dolomites with the remainder being shale and sandstone. Rocks were folded and thrust faulted during several mountain building episodes of the Pennsylvanian. Sedimentary rock cover has been eroded to expose Precambrian granites in a 150 mi^2 (380 km^2) area in its southeastern portion, which is the largest exposure of Precambrian rocks in the state.

Wichita Mountains

Located in southwest Oklahoma, granite, rhyolite, and gabbro are the dominant rocks, being Middle or Early Cambrian in age. These igneous rocks are flanked by scattered outcrops of Cambrian and Ordovician limestones and dolomites much like the folded rocks of the Arbuckle Mountains. Thrust faulting during several episodes in the Pennsylvanian uplifted the igneous rocks and erosion removed the sedimentary cover. Igneous rocks now form mountains 500 to 1000 ft (150 to 300 m) high rising above a surrounding plain of Permian red beds.

Gulf Coastal Plain

Located in southeast Oklahoma, loose sands, gravels, limestones, and clays, Cretaceous in age, dip gently southward toward the Gulf of Mexico. Sediments, only slightly dissected by streams, form gently rolling hills and plains.

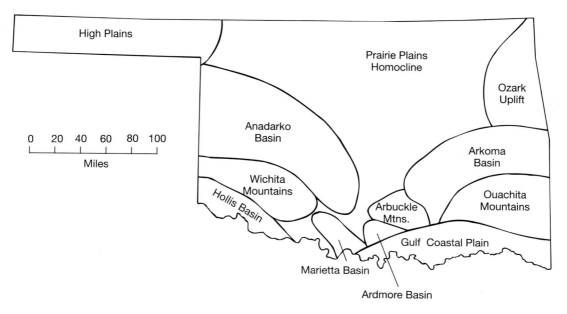

FIGURE 13.5 Physiographic divisions of Oklahoma.

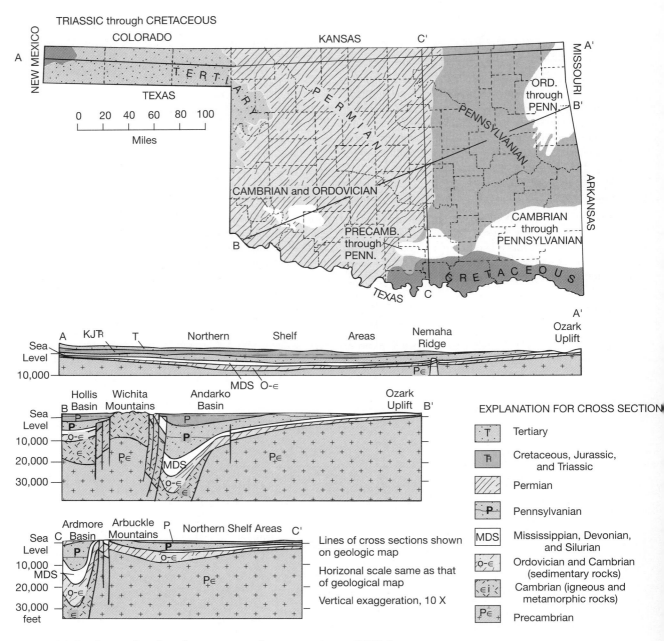

FIGURE 13.6   Generalized geologic map and cross sections of Oklahoma.

TABLE 13.6 Highway Construction Problems Related to Oklahoma Physiographic Units

	1	2	3	4	5	6	7
Seepage	SN,G	SN,G	M,G	S,G	S,G	S,G	M,G
Slumps	SN,G	SN,G	M,L	M,L	S,L	S,L	M,G
Rippability	SN,G	M,L	M,G	S,G	S,G	S,G	M,G
Expansive soils	SN,L	SN,G	M,G	SN,G		M,G	S,G
Rock excavation	SN,G	M,L	M,G	S,G	M,G	S,G	SN,G
Materials location	S,G	M,G	SN,G	M,G		M,G	SN,G
Erosion	M,G	M,G	M,G	SN,G		M,G	M,L
Sinkholes	SN,L	S,L	SN,G	S,L	M,L	SN,G	SN,G
Corrosion	SN,G	SN,G	SN,G	S,G	M,G		SN,G
Land use	SN,G	SN,G	M,L	SN,G	SN,G	SN,G	SN,G

1. High Plains
2. Andarko-Hollis-Marietta Basins
3. Prairie Plains Homocline
4. Ozark Uplift
5. Arkoma Basin
6. Ouachita Mtns.
7. Gulf Coastal Plain

KEY: S—severe; M—moderate; SN—slight to none; G—general, L—local.

EXERCISES ON PHYSIOGRAPHIC PROVINCES AND ENGINEERING CONSIDERATIONS

1. *Exercise on Physiographic Provinces Maps:* Refer to the physiographic diagram of Figure 13.1 to complete this exercise. Draw the boundary lines around the various physiographic provinces. It is suggested that this be done in pencil first, checked carefully, and then outlined using red pencil or red pen.

   The provinces are listed for your convenience.

   Southern Rockies
   Northern Rockies
   Middle Rockies
   Basin and Range
   Sierra-Cascades
   Pacific Coast Range (indicate also Great Valley of California and the trough areas in Oregon and Washington)
   Coastal Plain, Gulf and Atlantic
   Canadian Shield

   Piedmont Plateau
   Blue Ridge Province
   Ridge and Valley Province (Folded Appalachians)
   Central Lowlands
   Interior Low Plateaus
   Ozark and Ouachita Province
   Great Plains
   Columbia Plateau
   Colorado Plateau
   New England Province
   Appalachian Plateau
   Wyoming Basin

2. After reviewing the brief description of the physiographic provinces (Table 13.1), indicate to the left of each province in the list provided in Exercise 1 whether the province is basically a stable region structure (SRS), orogenic belt structure (OBS), a volcanic cone (VC), or a volcanic lava flow structure (VLS).

Where does the Canadian Shield fit into this classification since it is not a mountainous region? Why is the Appalachian Plateau considered to consist of geographic "mountains" but not mountains in a structural geology sense?

3. Refer to the list of engineering problems for various regions of the United States (shown in Table 13.2). Which of these problems would be associated with the various physiographic provinces listed in Exercise 1? To the left of the list in Exercise 1, indicate by number which problems would be associated with which province. In addition, list the problems (again by number) that would be associated with 1) mountainous regions, 2) wide alluvial flood plains, 3) arid portions of southwestern United States, and 4) limestone regions in Kentucky and southern Indiana.

Map Reading

The following topographic maps (Exercises 4 through 33) relate to the physiographic provinces of the United States.

Kassler, Colorado; 7.5-min quadrangle

4. To what physiographic province does the eastern part belong? The western part? To which block diagram in Figure 13.2 does each refer?

5. What is the regional stage of each part?

6. Give the maximum relief of the eastern and of the western parts (use mountain to valley).

7. In what direction do the hogbacks dip? Explain the difference between a hogback and a questa.

8. What evidence indicates that the basement complex rocks underlying the western portion are homogeneous crystalline rocks (rather than being folded sedimentaries)?

Allens Creek, Indiana; 7.5-min quadrangle

This quadrangle is underlain by sedimentary rocks (shale, limestone, and sandstone) that dip gently westward and form part of the western flank of a very broad gentle arch that has its center near the Indiana–Ohio line.

9. To what physiographic province does this area belong?

10. Allens Creek is located a few miles north of Bloomington; in which Indiana physiographic section is it contained?

11. What is the approximate maximum relief?

12. What evidence on the map suggests that the beds are nearly horizontal? Is there any evidence of sinkhole formation? How does your answer relate to the nature of the rocks present?

13. What evidence of recent changes in the course of Salt Creek may be seen on the map? Where is Salt Creek currently widening its floodplain?

14. What depth to bedrock is suggested in this area? What problem in constructing sewer lines does this indicate?

Orbisonia, Pennsylvania; 15-min quadrangle

15. In what physiographic province is this quadrangle located? To which block diagram in Figure 13.2 does this refer?

16. What evidence on the map indicates that the rock layers are folded rather than being only tilted?

17. Note the two water gaps and the trellis drainage. These are indirect evidence of folded rocks. How are water gaps formed?

18. What rock type is likely to make up the mountain tops; what kind makes up the valleys?

Mt. Whitney, California; 15-min quadrangle

The scarp forming the western edge of Owens Valley marks the eastern front of the Sierra Nevada, which is formed of a great block of the Earth's crust upfaulted and evidently tilted toward the Pacific Ocean. The Sierra Nevada is formed mostly of granite; the Inyo Mountains of granite and disturbed sedimentary rocks. Movement along the fault is essentially vertical and did not take place all at one time, but rather in small increments of a few feet or tens of feet over a long period of geologic time.

19. In what physiographic province is this quadrangle located? Does this appear similar to diagram B4 in Figure 13.2? Explain.

20. What evidence of faulting appears along the front of the Alabama Hills?

21. If the front of the Sierras is a fault scarp, what is the total vertical movement along the fault, as measured between the east front of Alabama Hills and the top of Mt. Whitney?

22. What evidence indicates that Owens Valley cannot be a stream-cut valley such as is the case for the Grand Canyon?

23. What engineering problems would be expected to occur in this area?

Ewing, Kentucky–West Virginia; 7.5-min quadrangle

The northern part of this quadrangle is underlain by nearly horizontal sedimentary beds and the southern part by beds that are folded.

24. To what physiographic province does the northern part belong? The southern part?

25. Why does the rock layer forming Cumberland Mountain yield a ridge, whereas the adjacent layers form valleys? What rock types are involved? What is the strike of these beds?

26. Name one stream on the map that has a trellis drainage pattern.

27. What is the regional stage of the northern part? Why is that area regarded as a plateau?

28. What engineering problems would you suspect in this area?

Thousand Springs, Idaho; 7.5-min quadrangle

This area is underlain by layered flows of basalt that have a total thickness of several thousand feet.

29. To what physiographic province does this quadrangle belong? Refer to Table 13.3. What subdivision of this table relates to this area?

30. What is the regional stage in the northeast part of the map?

31. What evidence of layered bedrock structure appears on the map?

32. What properties of basalt would account for the fact that there is an aquifer from which many large springs emerge?

33. What engineering problems would you envision in this area? Consider this also relative to the major dams built on the Columbia River.

Exercises 34 through 45 pertain to the physiographic regional units of Indiana.

34. In what U.S. physiographic province is Indiana located?

35. Examine the geologic map of Indiana (Figure 13.4). Describe in a general way the age and configuration of the bedrock in the state.

36. Omitting glacial deposits, to what part of the state would you go to find the oldest rocks? The youngest rocks?

37. Compare the physiographic map of Indiana (Figure 13.3) to Figure 12.10 showing the extent of glaciation in north-central United States. To what glacial stage do the units 1 through 6 listed in Table 13.4 belong?

38. In which Indiana physiographic province would you expect to find caves? Explain why.

39. Describe the possible ground-water problems in the following provinces:

Kankakee Outwash and Lacustrine Plain
Tipton Till Plain
Crawford Upland
Scottsburg Lowland

40. What kind of foundation material (specifically the bedrock or glacial material type) would be found in the following cities?

Gary　　　　　Fort Wayne
Lafayette　　　Bloomington
Indianapolis　 Evansville

41. List the provinces in which swamps would be common and why.

42. What would you expect the most common bedrock type in the state to be? What bedrock type is most commonly used as a crushed stone aggregate for Portland cement concrete and for bituminous mixes in Indiana? Why would it not be basalt?

43. What highway engineering problems might be expected in the area of Wisconsin glaciation in Indiana?

44. What highway engineering problems would be expected in the unglaciated portion of Indiana that would not occur elsewhere in the state?

45. A number of years ago a Midwest site for a linear accelerator for the U.S. Nuclear Regulatory Commission had been proposed for the Indianapolis area—the Eagle Creek site. (The accelerator was subsequently located near Batavia, Illinois.) From your knowledge of Indiana geology, what problems would you expect in the foundation for such a structure in the Indianapolis (north side) location?

Exercises 46 through 49 pertain to the physiographic units of Oklahoma.

46. A list follows of the 10 physiographic units of Oklahoma. After reading Table 13.5 and reviewing Figure 13.2 and Table 13.3 indicate whether each of the 10 units belongs to a stable region structure (SRS) or an orogenic belt structure (OBS).

High Plains
Prairie Plains Homocline
Ozark Uplift
Andarko-Hollis-Marietta Basins
Arkoma Basin
Ardmore Basin
Ouachita Mountains
Arbuckle Mountains
Wichita Mountains
Gulf Coastal Plain

47. Only seven physiographic sections are considered in Table 13.6 but 10 are described in Table 13.5. Which three were omitted? Why were they probably omitted? (*Hint:* See Figure 13.5 and note the size and location of the omitted units.) Which of the highway construction problems given in Table 13.6 is likely to be associated with these other three physiographic sections? Explain why.

48. Examine the geologic map of Oklahoma on Figure 13.6. Where are Mesozoic age rocks found? With which physiographic division(s) are these associated? (*Hint:* See Figure 13.5.) Are there any Cenozoic age rocks present? Explain.

49. What are the three mountain regions in Oklahoma? Which of the three seems to be the most rugged or highest? Explain your answer.

Additional Readings

FENNEMAN, N.M., 1931, *Physiography of Western United States,* McGraw-Hill, New York.

FENNEMAN, N.M., 1938, *Physiography of Eastern United States,* McGraw-Hill, New York.

HAYES, C.J., 1971, "Engineering Classification of Highway-Geology Problems in Oklahoma," in *Proc. 22nd Highway Geology Symposium,* Norman, OK.

HUNT, C.B., 1967, *Physiography of the United States,* W.H. Freeman and Co., San Francisco.

JOHNSON, K.S., and MANKIN, C.J., 1971, "Geology of Oklahoma—A Summary," *Proc. 22nd Highway Geology Symposium,* Norman, OK.

LOBECK, A.K., 1950, *Physiographic Diagram of North America* (with text). The Geographic Press, Columbia University Press, New York.

THORNBURY, W.D., 1965, *Regional Geomorphology of the United States,* John Wiley & Sons, New York.

WOODS, K.B., and LOVELL, C.W., 1960, "Distribution of Soils in North America," in *Highway Engineering Handbook,* K.B. Woods, Ed., McGraw-Hill, New York.

14

Landslides, Subsidence, and Slope Stability

CHAPTER OUTLINE

Mass Movement Classifications
Types of Slope Failure
Plastic Outflow
Underground Opening Collapse
Falls
Topples
Movement by Slip
Failure by Flow
Slope Failure by Lateral Spread
Complex Slope Failures
Causes of Slope Failure
Prevention and Correction of Slides

Mass movement or mass wasting is the bulk transfer of soil or rock debris downslope under the direct influence of gravity. This is also known as gradation by gravity. Landslides, subsidence, and slope stability are phenomena dependent primarily on gravity.

Water may or may not be present in appreciable amounts for the various types of mass movement since water occurs as an accessory constituent, not as the primary agent of transport. Stream flow, by contrast, requires abundant water relative to the volume of transported solids to accomplish movement.

Gradation by running water and by gravity play the major role in reducing the land surface in both arid and humid climates. Other agents of gradation are glacial ice, wind, and wave action (on lakeshores and seacoasts). These typically play a lesser role than water and gravity when viewed as geologic agents working throughout geologic time.

Typically, running water and mass wasting work hand in hand to erode the land surface. Running water establishes the channels and continues to downcut the terrain with time. Mass wasting accepts the local base level set by channel erosion and transports the bulk of soil and rock toward this lower elevation under gravity's influence. An analysis of the Grand Canyon, Arizona illustrates this twofold erosion. The Colorado River has cut a narrow gorge during the last million years or so and mass wasting has moved the remaining large volume of material to the base of the channel for transport away by the Colorado River.

MASS MOVEMENT CLASSIFICATIONS

Classifications of mass movement provide an overview of the total, extreme variation in slope movement and make it possible to compare and relate the different types of failure. Because of the diverse nature of the failure parameters (that is, of material type, rate of movement, and geomorphic description), a classification is needed to associate the various features of slope failure and appreciate their differences.

Sharpe's classification of 1938 provides a valuable guide to the overall nature of mass movement and also gives a useful historical review of slope analysis. It is presented as Table 14.1. Starting from the left, the classification is subdivided according to the manner of movement, yielding two categories, the first involving downward and outward displacement, which forms a new surface or free side along the failure boundary. The second does not develop this new free surface but consists mostly of downward, vertical movement of the land surface. Known as subsidence, it is a common occurrence in human-related mass movement. Specifics about these failures are presented later in this chapter.

The first category as seen in Table 14.1 is subdivided on the basis of the failure mechanism involved. Two such mechanisms are recognized, flow and slip. Flow involves failure of the mass along many internal planes of slippage, much in the fashion of flow in a

viscous liquid such as molasses or hot taffy. There is no single plane of failure but countless internal ones, along which small amounts of slippage have occurred.

Slip involves the movement of a solid mass downslope along a single failure plane or a well-defined failure zone. In this way, the center of gravity of the mass is translated downslope under gravity forces. This type of failure lends itself to analysis using the limit equilibrium approach, which states that at failure the driving forces on the failure mass are equal to the resisting forces along the surface of failure. Prior to failure, the resisting forces must be greater than the driving forces. As shown in Table 14.1, the types of mass movement that fail by slip are slump, debris slide, debris fall, rock slide, and rock fall.

The rate of movement for both flow and slip is used to further subdivide mass movement. Creep failures of soil and rock occur slowly, whereas avalanches are rapid failures. Slip failures can also be considered according to failure rates as shown in the classification.

The materials carried by mass movement provide the final basis for subdividing Sharpe's classification. The range of constituent materials includes chiefly ice, earth or rock plus ice, earth or rock, earth or rock plus water, and chiefly water. This last category is stream flow and does not qualify as a part of mass wasting *per se*. Its inclusion in the chart is meant to illustrate the transitional relationship from mass movement to stream flow, which occurs as water becomes the predominant constituent. The specific mass-movement varieties included in this classification are discussed in a later section of the chapter.

The Transportation Research Board (TRB) Classification of Slope Movements provides another example of a useful overview of slope failure. This entails the 1978 revision of the Highway Research Board Classification of Landslides presented in HRB Special Report #29, which has been used by geologists and engineers alike since its publication in 1958. A somewhat simplified version of the TRB classification is presented in Table 14.2. In the TRB classification, six major types of movement are considered: I) falls, II) topples, III) slides, IV) lateral spreads, V) flows, and VI) complex. In comparison with the earlier classification of Sharpe, two types of movement, topples and lateral spreads, have been added in the TRB classification. Topples refer to blocks of

TABLE 14.1   Sharpe's Classification of Mass Movement

Movement			Material				
Direction of Movement	Type	Rate	Chiefly Ice	Earth or Rock Plus Ice	Earth or Rock	Earth or Rock Plus Water	Water
With free side (movement downward and outward)	FLOW	Imperceptible (usually)	GLACIAL TRANSPORT	Rock glacier solifluction	Rock creep Talus creep Soil creep	Solifluction Earth flow Mud flow	Stream flow
		Perceptible, slow, rapid					
	SLIP	Perceptible, very slow to rapid, rapid		Debris avalanche	Slump Debris slide Debris fall Rock slide Rock fall	Debris avalanche	
No free side (vertical movement primarily)	SUBSIDENCE						

bedrock or massive soil that topple over because the block becomes top heavy through the reduction in block size owing to the development of closely spaced fractures. This feature is discussed fully in the subsequent section on specific slope failures. Lateral spreads include liquefaction of soils during earthquakes when such soils lose their shear strength because of the buildup of excess pore pressure. This too is discussed fully in the next section. Falls are included under slip failure in Sharpe's classification but in the TRB counterpart, they are considered separately. Subsidence is not included in the TRB classification. Examples of types of mass movement are shown in Figure 14.1.

The TRB classification also considers two material types: bedrock and engineering soils. In the second category, two subgroups are 1) debris, that is, a coarse collection of broken pieces of rock plus soil, and 2)

finer grained soil materials consisting of various combinations of sand, silt, and clay.

TYPES OF SLOPE FAILURE

The various types of slope failure or mass movement varieties are discussed and compared in a detailed manner in the following section. To initiate this review of slope failures, several key definitions must be considered. These are mass movement, slope movement, slope stability, landslide, and subsidence.

Mass movement and slope movement are similar terms but mass movement carries a geologic emphasis implying that slope failure is part of the natural geologic process of gradation, which continues to occur. Slope movement and slope stability usually involve a shorter term consideration, with the primary concern being to prevent or minimize the effects.

TABLE 14.2 Types of Slope Movements (after Transportation Research Board Classification, 1978)

Type of Movement	Type of Material (before movement)		
	Bedrock	Predominantly Coarse (Engineering Soils)	Predominantly Fine (Engineering Soils)
I Falls	Rock falls	Debris fall	Earth fall
II Topples	Rock topple	Debris topple	Earth topple
III Slides			
A. Rotational	Rock slump	Debris slump	Earth slump
B. Translational	Rock block slide	Debris slide (very slow to rapid)	Earth block slide
IV Lateral spreads	Rock block spread		Earth lateral spread (very rapid)
V Flows	Rock creep	Debris flow	Wet sand or silt flow (rapid to very rapid)
	Valley bulge	Debris avalanche	Rapid earthflow (very rapid)
		Block stream	Mudflow
		Solifluction lobes	Dry sandflow (rapid to very rapid)
		Creep mantle	Dry loess flow
			(Due to earthquakes, extremely rapid)
VI Complex	Rock fall–debris flow or rock fall avalanche (extremely rapid)	Cambering spreading plus bending overflow	Slump–earthflow
	Slump and topple		Mudslide (sliding mudflow)
	Rock slide–rock fall		

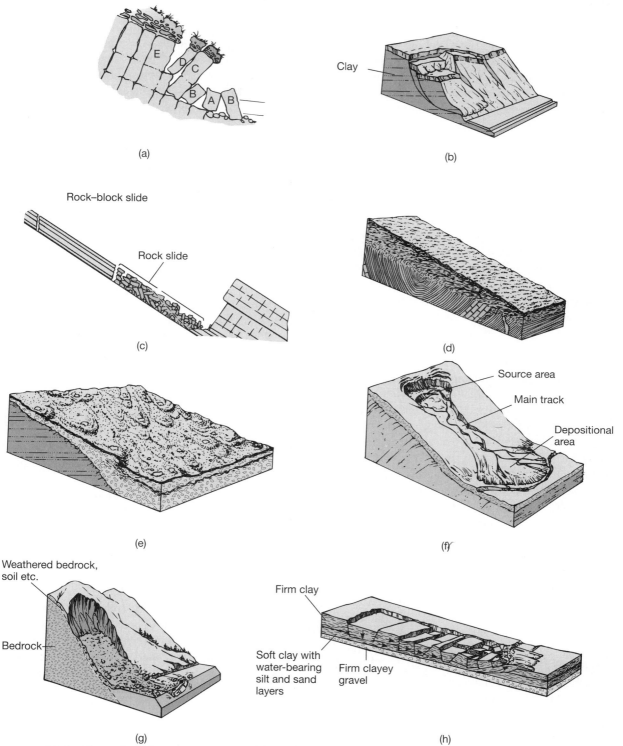

FIGURE 14.1 Examples of mass movement. (a) Rock topple, (b) earth slump, (c) rock-block slide, (d) soil creep, (e) solifluction, (f) earthflow, (g) debris avalanche, and (h) earth lateral spread. (After Transportation Research Board, 1978).

Slope movement studies involve both materials and the mechanisms of failure, with the knowledge that a better understanding of these factors should reduce the effects of future slope problems. Slope stability is obviously concerned with prevention of slope failure or of extensive movement. It also involves the short-term concept of slope control, rather than a long-term geologic consideration, and therefore falls within the scope of engineering geology.

Commonly, an accurate stability evaluation of a previously intact slope can best be made shortly after it has failed. This is accomplished by back-calculating the stability following such failure, assuming the original slope conditions can be determined. This is possible because prior to failure, the slope obviously had a factor of safety greater than 1, which was reduced to unity at the time of failure. How much greater was the factor of safety prior to failure is the question. Conditions at failure are determined first and then, based on other considerations prior to failure, the initial factor of safety is calculated. Before failure, the available resisting and driving forces can be estimated but it is difficult to determine just how these forces were mobilized at the onset of failure.

The term *landslide* is commonly used as a collective term to include most if not all slope failures. By simple inspection, however, the word *slide* forms part of the term, which suggests failure by slide or slip. Consequently, the term *landslip* can be used in a strict application of slip failure, but landslip excludes failure by flow, and possibly by fall, toppling, and lateral spreading. It also does not include subsidence, the vertical downward movement of the land surface. Therefore, to provide an overall term, *landslide* has come to encompass all of the slope failures included in Sharpe's Classification of Mass Movement plus those in the TRB classification with only the omission of subsidence. This collective nature of the term is illustrated by the title of the TRB publication that contains the classification presented in Table 14.2. "Landslides, Analysis and Control," TRB Publication, Special Report 176, pertains to slip, flow, toppling, and lateral spreads.

Subsidence is the lowering of the land surface by a vertical, downward movement or downwarping. Typically a subsidence hollow develops, that is, a dish-shaped feature with the greatest displacement in the center. Subdivisions of subsidence according to the mechanisms of failure are 1) compaction, 2) consoli-

dation, 3) plastic outflow of weak layers, and 4) collapse of underground openings. Subdivision based on the location where the subsidence originates yields two varieties: shallow subsidence and deep subsidence. These types of subsidence are discussed in the following paragraphs.

Compaction and Consolidation

With regard to engineering construction, the downward movement of the ground surface is termed *settlement*. Geologically speaking, this downward movement is called subsidence. Within the soil mass, settlement or subsidence is typically of two types: compaction and consolidation.

Compaction,[1] in both a construction sense (geotechnical engineering) and an engineering geology sense, is the reduction in volume of a soil mass under load caused by reorientation and realignment of soil particles to form a denser packing, but without drainage of water from the soil. Such reorientation can be caused by applying compressive loads (rollers used on earth-moving projects), vibration (vibrocompactors, earthquakes, pile-driving operations), or the downward movement of water through a dry, loose soil.

Consolidation,[2] in the geotechnical sense and engineering geology sense, is the volume reduction under load that occurs as water flows from the sample, and is related to the reduction in excess pore water pressure. Therefore, saturated, compressible clays consolidate under load whereas all types of soils, cohesive and noncohesive alike, can be compacted by moving the grains closer together. Geotechnical engineering details of consolidation and compaction are discussed in Chapter 7.

One feature of compaction that is of considerable interest is the collapse that occurs when water flows through loosely packed, dry soils. It is sometimes called hydrocompaction. The motion of the water provides sufficient energy to move the soil particles to a tighter packing arrangement. This occurs to some soils when water is first impounded in a reservoir or when water leaks from conduits or canals into the

---

[1] Compaction in a strictly geologic sense is the volume reduction and loss of water that occurs when sediments (soil) becomes more rock-like under geostatic loading. It includes the engineering aspects of both compaction and consolidation.

[2] Consolidation in the geologic sense is the process by which sediments become rock-like or become lithified. It may include consolidation and compaction in the engineering sense plus, in some cases, cementation.

loosely packed material below. An example of interest is related to the construction of the California Water Project, which supplies water to southern California from the northern part of the state. Mudflow deposits in the San Joaquin Valley were particularly susceptible to such settlement. Water leaking through cracks in the concrete canal lining caused large-scale settlement below the canal and the subsequent collapse of the lining. Prewetting of susceptible soils prior to construction is a viable means for preventing the subsequent failure of the canal.

Consolidation occurs when heavy loads are placed on compressible soils such as plastic clays and organic materials. The foundation loads from footings, piles, and mats can initiate such settlement by inducing consolidation. These loads increase the pore pressure or neutral stress on the soil, causing water to flow from it. Hence, as the effective stress increases, the soil reduces in volume. Details regarding effective stress are provided in Chapter 7.

Another cause of consolidation is the removal of the subsurface fluids (water or oil) usually because of human activities. Lowering the water table reduces the buoyancy force contributed by the water and in effect increases the downward pressure on the soil at the former level of the water table. This initiates consolidation and settlement. Since it takes time for water to flow through a fine-grained soil, consolidation is time dependent. This is in contrast to compaction, which occurs much more quickly in response to the applied load.

Consolidation commonly occurs in oil or ground-water supply areas. Interbedded sands and clays provide the features needed for the consolidation process. Drainage of the sand layers leads to greater loads on the clays, which then consolidate. This problem has occurred in Mexico City where ground-water removal from the volcanic ash beds and clays has been extensive. It also occurred in Long Beach, California, where great volumes of oil and brackish water were removed, and in the Houston, Texas, area where ground-water removal from the coastal plain sediments has caused extensive consolidation. In some Houston suburbs surface areas are now below the high-tide elevation and are subject to ocean encroachment at high tide.

A bit of reflection on this subject will bring the concept of shallow and deep subsidence into perspective. Compaction due to water movement, vibration, or applied load typically occurs near the land surface because the soil is loosest there and surface loads have the greatest effect. Therefore, shallow subsidence is typically related to compaction. Consolidation may occur hundreds or thousands of feet below the land surface if the fluids are removed from these depths. Settlement is transferred upward from a considerable depth, therefore, it is termed deep subsidence. Most ground-water pumping for irrigation causes deep subsidence, whereas shallow subsidence is provided by initial wetting near the surface.

Subsidence above underground openings is particularly destructive to surface structures because of the differential settlement it invokes, which occurs because the center settles much more than the edges of the trough or dish. In addition, it provides tensile stresses near the edge of the trough, which can be more destructive to buildings than compressive stresses. Soil tunneling, which yields a subsidence trough at the surface, can be extremely destructive to buildings, particularly relatively stiff buildings such as some old masonry, wall-bearing structures. This is why the loss of ground into a tunnel must be minimized in areas of urban development. Loose, cohesionless soils or weak, cohesive ones are particularly prone to loss of ground during tunneling when a tunnel shield is used. Compressed air, dewatering, and possibly grouting ahead of the advancing face may be needed to strengthen the soil to minimize soil movement into the tunnel.

PLASTIC OUTFLOW

Plastic outflow occurs when an organic or silty layer near the surface is subjected to loading by a heavy structure and the compressible layer exits to the surface yielding subsidence under the loaded area. Before final site selection is made, this problem can be alleviated by exploring the subsurface to determine the presence of such layers close to the surface.

UNDERGROUND OPENING COLLAPSE

The collapse of underground openings can involve natural openings underground or those due to human activities. Natural openings subject to collapse include caverns in soluble rock such as limestone, dolomite, gypsum, and rocksalt, plus lava tubes and

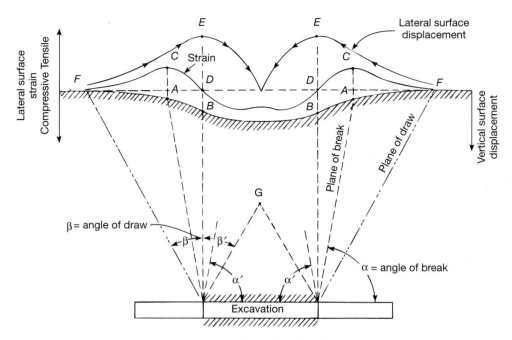

FIGURE 14.2 Idealized representation of trough subsidence. (From Rellensmann, O., 1957, *Rock Mechanics in Regard to Static Loading Caused by Mining Excavation,* Second Symposium on Rock Mechanics, Quarterly, Colorado School of Mines, 52.)

tunnels in extrusive igneous rocks. When the roof of the opening collapses, a stable arch may form if the cavern width is sufficiently small or the rock above the cavern is massive, containing only widely spaced joints. Under these conditions, the void will not migrate upward but instead forms a cathedral-shape extending upward to a point above the opening. Stable openings as large as 100 ft wide have been found in large limestone caves with a stable roof of massive rock. The value for the inward sloping angle is a function of the rock and its joint spacing.

When the roof opening is wide compared to the strength of the rock mass, roof failure will occur as blocks of material fall into the opening. The side boundaries of these blocks tend to slope outward from the vertical so that a wedge of rock moves downward from above. The angle measured from the horizontal is known as the angle of break. The relationship between the angle of break and the subsidence trough at the surface is shown in Figure 14.2.

The angle of break, α, can be estimated for soils and rocks using ϕ, the angle of internal friction, and a relationship based on the Mohr circle representation for the state of stress on the plane of failure. This yields $\alpha = \phi/2 + 45°$, which is derived in Chapter 7. A listing of ϕ values, calculated α values, and observed α values for various soil and rock types is presented in Table 14.3.

In Figure 14.2 the plane of draw and angle of draw, β, are also shown. As depicted in this figure, β is measured from the vertical whereas α is measured from the horizontal. The angle β is related to the boundary on the surface of the subsidence hollow and, typically as a matter of convenience, this boundary is taken at a point where only 5% of the maximum vertical displacement occurs. A similar designation must be made or the bounds of the trough will become a function of surveying accuracy rather than actual subsidence. The plane of draw always extends further away from the underground opening than does the plane of break [that is to say, $\beta > (90 - \alpha)$].

TABLE 14.3   Measured Angles of Internal Friction, Calculated Angles of Break, and Observed Angles of Break for Various Soils and Rock

Material	Measured Angles of Internal Friction ϕ (deg)	Calculated Angle of Break α (deg)	Observed Angle of Break α (deg)
Clay	15–20	52.5–55	
Plastic strata,			60–80
Clays and shales			60
Sand	35–45	62.5–67.5	
Unconsolidated strata			40–60
Sand			40
Moderate shale	37	63.5	
Hard shale	45	67.5	
Sandstone	50–70	70–80	
Coal (average)	45	67.5	
Rocky strata			80–90
Hard sandstone			85
Limestone			

Note in Figure 14.2 that point B lies vertically above the boundary of the opening. Point B is the inflection point on the subsidence curve, the point where the shape changes from concave up (in the center of the trough) to concave down (on the two limbs of the subsidence hollow). When the subsidence trough shape is compared to a normal probability curve (the Gaussian curve used in statistics), the inflection point occurs about 1/2.5 of the distance from the center of the trough to the edge, again the edge being at the 5% vertical displacement point. If the half width is designated w, point B occurs at $w/2.5$.

The plane of break on Figure 14.2 intersects the ground surface at A and the strain curve at C. It can be seen that C is the point of maximum strain and, therefore, the likely place for the break to occur. In the drawing, the plane of break occurs about one-third of the distance between D and the edge of the trough or AB $\cong$ BF $\times$ 1/3. Recall that Figure 14.2 is an idealized representation of the subsidence relationship so values for angles of draw, obtained from angles of break using the geometry of this drawing will be, by their very nature, only approximate.

Values for the angle of draw are measured from actual cases of subsidence, both for tunnels driven through soil and for underground mining in rock. Values of β as low as 9° and as high as 58° have been determined for different soils and geologic conditions. For massive rocks, β values are in the range of 11 to 20°. Table 14.4 provides a list of estimated angles of draw based on field observations.

Referring again to Figure 14.2, point G occurs at the intersection of lines inclined inward from the vertical at an angle of β or β', the angle of draw. As the excavation is widened equally on both sides (outward along the darkened bed), point G moves up vertically toward the surface. As long as G remains below the ground surface, the subsidence is termed

TABLE 14.4   Estimated Angles of Draw for Various Soils and Rock

Material	Angle of Draw, β (deg)
Rock, hard clay, sands above the ground-water table	11–26
Stiff to soft clays	26–50
Sands below the ground-water table	<50

subcritical, with the critical point occurring when G reaches the surface. This marks the situation when maximum subsidence occurs at the center of the trough. As further widening ensues, G rises above the surface and the trough will widen but no longer deepen. Movement will cease in the center of the trough. When G lies above the surface, the subsidence is termed *supercritical*. Total vertical movement in the center of the trough will be from 70% to 90% of the height of the mine, that is, of the thickness of the extracted zone.

The cathedral effect that forms above some small cavities is related to the location of point G and to the strength of the rock mass. If G lies too far above the excavation, progressive subsidence will occur and eventually reach the surface. Below that distance, a stable arch can develop and no further subsidence ensues. The width of the opening and the angle of draw influence the location of G. This holds true for underground mines as well as caverns. For coal mines, the critical width may be 10 to 15 ft, whereas for massive carbonate rocks (limestones, dolomites) this width may be 40 ft or greater.

Human-made openings underground are the result of mining activities and of engineering construction, such as tunnels and underground power plants. Mining activities include extraction of coal, limestone, gypsum, salt, or metallic ores such as iron, copper, lead, zinc, gold, silver, and molybdenum. Mining methods include 1) room and pillar, 2) stoping, 3) block caving, and 4) longwall methods. In block caving and longwall methods, most or all support is eventually removed so that subsidence is imminent. In the stoping and the room and pillar methods, sufficient support is commonly left behind to prevent roof failure and general subsidence. However, in room and pillar mines when a higher extraction ratio is desired and subsidence is acceptable, the pillars are removed (robbed) during a retreating operation from the far reaches of the mine section working back toward the permanent haulage ways. Surface subsidence is almost immediate when the support pillars are removed.

Solution mining of salt (NaCl or halite) leads to complicated subsidence problems. In this mining technique, two boreholes into the salt body are connected by hydrofracturing, that is, by fracturing between the holes using water pressures that exceed the overburden stresses. Hot water is then circulated down one of the holes and brine is taken out of the other. On the surface, this water is allowed to evaporate to yield the salt for further processing. Cavities thus formed are commonly asymmetrical in shape, because the flow paths take on preferred orientations and remove salt along these directions. When a sufficiently large cavity is formed, subsidence occurs in the rock layers above the salt and is propagated to the surface. Areas near Detroit, Michigan; Windsor, Ontario; Rochester, New York; and central Kansas have major subsidence problems related to such salt extraction.

A final point for consideration of subsidence has to do with the development of sinkholes. In this context, the term *sinkhole* refers to the conical-shaped surface depressions that occur in carbonate and other soluble rock terrains (usually limestone, dolomite, and marble, sometimes gypsum and salt). The depression marks a point above a vertical joint, enlarged by solutioning, which leads to the underground drainage system. Soil and weathered rock collapse into the enlarged joints to form the sinkhole.

Sinkholes are part of the natural, geologic, erosional process. In a natural setting it takes a considerable span of time for downward migrating waters moving through narrow openings to dissolve away the rock volume needed to develop sinkholes. Terrains exposed during the Pleistocene age (1 to 2 million years) had sufficient time to develop sinkholes. However, this natural development is not expected to occur in a human lifetime or during the life of an engineering structure. Therefore, when sinkholes suddenly occurred at the rate of many per year in areas of Alabama and Florida, human involvement in the situation was logically suspected.

One reason for rapid sinkhole development is the lowering of the ground-water table. This commonly occurs as a consequence of human activities such as removal for water supply, dewatering for construction, and construction of permanent road cuts or surface mines—all of which bring down the ground-water table. Figure 14.3 illustrates one of the possible situations. In some areas deep weathering of previously exposed limestone terrain is covered with sand or clay deposits from a later stage of development. Such deposits serve as a cap for the vertical openings below. When the water table is lowered, the buoyancy force is lost, yielding an increase in effective stress. This stress acts downward on the clay and may force

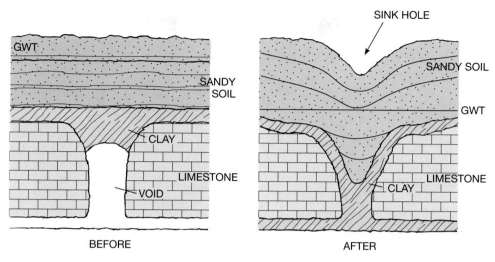

FIGURE 14.3   Possible mechanism for rapid sinkhole formation, human-induced failure.

it to fail (punch) into the opening below. When this occurs, a sinkhole forms quite suddenly on the surface. An example of rapid sinkhole failure was the formation of "December Giant," which occurred overnight (December 1972 in Alabama). The sinkhole measures 400 × 300 ft across and 150 ft deep (125 × 94 × 47 m). Fortunately, it occurred in an undeveloped, wooded area adjacent to an expanding residential tract. The ground-water table had been dropping in this general region for several years. A photograph of this large sinkhole is shown in Figure 14.4.

A question for consideration is could this sinkhole have occurred suddenly because of purely natural (nonhuman-related) causes? Obviously the ground-water table could lower for strictly geologic reasons alone. The answer involves the rate of change, not its magnitude. For the ground-water table to lower such a significant amount to trigger this failure, many years of reduced rainfall would be required. A long-term climatic change would be needed to cause the failure. But for the widespread development of new sinkholes in a short period of time in areas occupied by homes or other human activities, the problem is placed squarely on human involvement. Lowering of the water table because of increased pumping is the most logical answer.

This is not to say that humans should stay away from all sensitive areas. Instead, cognizance of the problems is required and the development of proper

solutions to prevent such catastrophic failures from occurring. If the water table level is maintained by balancing withdrawal and recharge, the problems will be alleviated. Other situations require different solutions but humankind must engineer these solutions to fit the potential problem.

FALLS

In this category of failure, the mass of soil or rock travels most of the distance through the air. It involves free fall, which is sometimes followed by bouncing and rolling of fragments. Bedrock containing steeply dipping weakness planes that dip into an opening or valley are particularly prone to this problem. Weathered bedrock zones may be most susceptible to rock falls as weathering opens joints and bedding planes. Debris and earth masses on steep slopes may also fail as a unit by falling into the opening. They would be less likely to bounce and roll after dislodgment than would the intact blocks of rock.

TOPPLES

Topples occur mostly on rock blocks and are a result of an overturning moment about a pivot point below the center of gravity of the block. The opening of joint planes perpendicular to the bedding or to another weakness plane that dips into an opening is

FIGURE **14.4**   Collapse of large sinkhole in Alabama. (United States Geological
Survey).

primarily responsible for toppling in bedrock. It is a
function of the height and width of the block as
shown in Figure 14.5 and is also illustrated in Figure
14.1(a).

With increased weathering, the joint planes become more closely spaced until the block is unstable
and toppling occurs. Loose blocks and weathered and
jointed blocks that would likely yield a toppling
failure within the proposed lifetime of a structure
should be removed from slopes during construction.

Application of the concept shown in Figure 14.5
provides information for such judgments about removal.

An example problem is provided next.

EXAMPLE PROBLEM _____

Bedding planes in a dense limestone dip at 20° into
a road cut. The open bedding planes are spaced
6.5 ft apart. What joint spacing perpendicular to
bedding will cause toppling to occur?

ANSWER

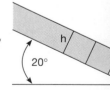

$$\frac{b}{h} = \tan\theta; \quad b = 6.5 \tan 20°$$
$$= 2.4\,ft$$

If the joints are spaced at closer than 2.4 ft the
blocks will topple into the opening.

MOVEMENT BY SLIP

In slopes that fail by slip (or sliding), movement
occurs as a displacement along one distinct surface of
failure or within a relatively narrow zone of failure.

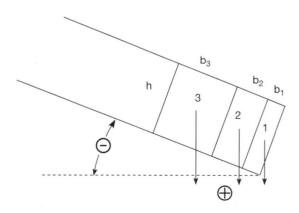

FIGURE **14.5**   Toppling of rock blocks.

The rock or soil unit fails as a mass by moving down the failure surface. This configuration is well suited to analysis by equating the driving forces to the resisting forces on the block at the time of failure. Known as the limit equilibrium approach, this method is discussed and examples are supplied in this section.

The failure surface ideally is considered to be of two general shapes, circular or planar. The circular failure plane is commonly considered to occur in essentially homogeneous materials such as soil and weak rock; the slope failure itself is called a slump. The TRB classification (Table 14.2) indicates that slumps occur in earth, debris, and weak rock materials alike.

Slump Failures

Slumps are extremely common slope failures in humid areas. They occur on natural slopes, cut slopes, and embankments. A buildup of ground water in the slope (pore pressure development) is commonly involved with such failures so that the prime time for slumps to occur is shortly after the spring thaw and attendant rainfall. The various parts of a slump failure are illustrated and defined in Figure 14.6.

Analysis of slump failures is accomplished by use of the Swedish circle method of analysis or modifications of that procedure. This method involves a limiting equilibrium method equating resisting forces to the driving forces that tend toward failure. The parameters of the soil (or weak rock) that resist movement are cohesion c and the coefficient of internal friction, or when related to a friction angle, $\tan\phi$. Values for c and ϕ are measured in the field, determined in the laboratory, or simply estimated, and these values are used to determine the resisting forces. The specifics regarding slump failure analysis, though outside the scope of this text, are considered in soil mechanics or geotechnical engineering textbooks. The contributing factors for the resisting and overturning forces are weight of the soil, weight of any surcharge, internal seepage forces, and pore water effects, plus the c and ϕ values. The problem may be complicated by the fact that not all of the available c and ϕ strength are mobilized at failure.

Typically, slope stabilities are determined for cut slopes and embankments by finding the critical failure circle through the slope cross section by plotting the factors of safety for each circular failure and locating the minimum value. This can now be accomplished by computer programming techniques.

Three general locations for slump failure surfaces are known to occur. Illustrated in Figure 14.7, they are slope failure, toe failure, and base failure. Slope failure is a common slump feature that is limited to the near-surface materials. These failures may involve mostly corrective maintenance for highways and pits. Toe failures are more deeply seated and hence involve greater volumes of material. They sometimes develop when the slope is extended by additional excavation. Base failures involve extremely large volumes of material and because of the considerable depth of the failure plane, it may not be intercepted by borings during exploration prior to construction or during analysis after slope failure. Typically, it involves a weak layer at depth along which failure is concentrated to yield a flat-bottomed shape to the failure surface. Methods to prevent or remedy slope failures are presented at the end of this chapter.

Translational Failures and Rock-Block Slides

Planar failure surfaces yield movement of a translational nature by sliding down the inclined plane into an opening or valley. This failure is commonly associated with movement along weakness planes in a bedrock mass; in addition, debris slide and earth-block glide are also identified as other possible failures in the TRB classification (Table 14.2). For the debris slide, the surface of failure is commonly along the debris-bedrock interface where ground water would accumulate because of the contrasting permeability of the bedrock (much less) and the debris above. Earth-block glides occur along a planar feature that has low shear strength compared to that of the material above it. The failure surface connecting this planar feature with the surface is typically very steep, yielding a block shape for the total failure surface. In this respect it is not a great deal different from a deep-seated slump feature, which also has a flat-bottomed shape for its failure surface.

Rock slides (and/or rock-block slides) are of particular interest within the overall category of failure by slip. Many of the large-scale and destructive slope failure events of the past have involved rock-block slides. Several of the spectacular slope failures are discussed at the end of this chapter but as an intro-

MAIN SCARP—A steep surface on the undisturbed ground around the periphery of the slide, caused by the movement of slide material away from undisturbed ground. The projection of the scarp surface under the displaced material becomes the surface of rupture.

MINOR SCARP—A steep surface on the displaced material produced by differential movements within the sliding mass.

HEAD—The upper parts of the slide material along the contact between the displaced material and the main scarp.

TOP—The highest point of contact between the displaced material and the main scarp.

TOE OF SURFACE OF RUPTURE—The intersection (sometimes buried) between the lower part of the surface of rupture and the original ground surface.

TOE—The margin of displaced material most distant from the main scarp.

TIP—The point on the toe most distant from the top of the slide.

FOOT—That portion of the displaced material that lies downslope from the toe of the surface of rupture.

MAIN BODY—That part of the displaced material that overlies the surface of rupture between the main scarp and toe of the surface of rupture.

FLANK—The side of the landslide.

CROWN—The material that is still in place, practically undisplaced and adjacent to the highest parts of the main scarp.

ORIGINAL GROUND SURFACE—The slope that existed before the movement of interest occurred. If this is the surface of an older landslide, that fact should be stated.

LEFT AND RIGHT—Compass directions are preferable in describing a slide, but if left and right are used they refer to the slide as viewed from the crown.

SURFACE OF SEPARATION—The surface separating displaced material from stable material, but not known to be a surface of failure.

DISPLACED MATERIAL—The material that has been moved from its original position on the slope. It may be in a deformed or undeformed state.

ZONE OF DEPLETION—The area within which the displaced material lies below the original ground surface.

ZONE OF ACCUMULATION—The area within which the displaced material lies above the original ground surface.

VC—Vertical component of slump.

HC—Horizontal component of slump.

L—Length of displaced zone along the slope.

LC—Length of slump along the slope.

D—Depth to rupture surface.

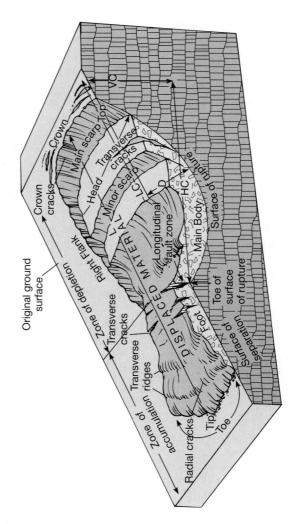

FIGURE 14.6 Nomenclature of the parts of a slump.

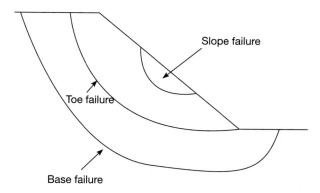

FIGURE 14.7 Locations for slump failure surfaces in soil.

ductory statement it is significant that three major slides, the Gros Ventre Wyoming failure (1925), the Frank, Alberta, slide (1903), and the Vaiont, Italy, slide (1963) all involved major movements by rock-block sliding.

Rock-block sliding is illustrated in Figure 14.8, which also includes details on toppling. Assuming a dry slope, if the slope angle θ is less than the angle of sliding resistance φ, then the block will not slide down the inclined plane. This would be the case for zones A and C in Figure 14.8. The φ angles are discussed later in this section.

In zones B and D, θ is greater than φ and sliding will occur. The φ angle for many competent rocks is about 30 to 35°. For weak rocks such as shale, however, the φ value may be only 15° or lower.

Topples are also considered in Figure 14.8. Recalling from Figure 14.5, if $b/h <$ tanθ the blocks will topple forward as depicted in zone D for Figure 14.8. Topples will occur in both zones C and D where the curved boundary is for $b/h =$ tanθ and φ is assumed equal to 35°.

Summarizing the situation depicted in Figure 14.8, zone A is stable both for toppling and sliding. Zone B shows sliding but no toppling, and zone C has toppling but no sliding. For zone D, both sliding will

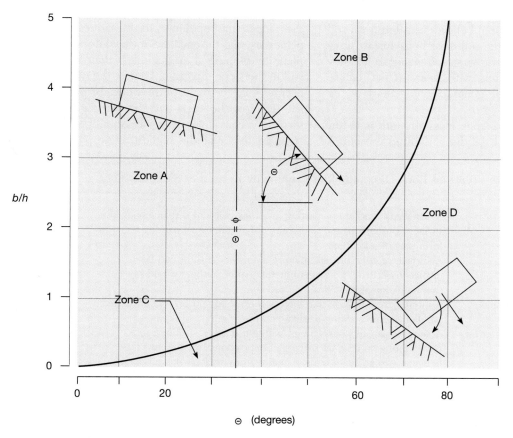

FIGURE 14.8 Stability fields for sliding and toppling.

occur because $\theta > \phi$ as well as toppling because $b/h < \tan\theta$.

Weakness Planes in Rock

The stability of rock masses is not dependent on the strength of the intact rock, but on the nature of the weakness planes contained within the mass. This is in contrast to most masses of soil in which the mass strength is determined by the basic material strength itself. Consequently, laboratory tests on small soil samples, if they approximate field conditions, can be used to indicate field behavior. For a rock mass, however, material strength is much greater than that of soils and the brittle nature of the rock induces fractures; it is these weakness planes that determine the rock mass strength. Relative to the slope stability of a rock mass, shear strength along the fractures or discontinuities determines the resistance to movement.

If the compressive strength of a material is greater than about 100 psi (700 kN/m^2 = 700 kPa), the material behaves as a rock, that is, its mass strength is a function of the weakness planes present. Below 100 psi (700 kPa) the material behaves as a soil, with soil parameters depicting the soil mass behavior. Relative to slope failure geometry, when soil materials fail they typically develop slumps with near-circular failure surfaces because of the general homogeneity of the material.

As previously stated, rock mass strength is determined by the nature of the discontinuities present. A discontinuity is defined as a structural weakness plane or surface along which movement can take place. Table 14.5 contains a list of common rock discontinuities.

Significant Features of Discontinuities

The nature of the discontinuities in the rock mass, that is, their physical properties and characteristics, is of major importance to rock slope stability. The significant features are 1) geometry, 2) continuity, 3) spacing, 4) surface irregularities, 5) physical properties of the adjacent rock, 6) infilling material, and 7) ground water.

GEOMETRY   The geometry of a discontinuity involves its orientation in space plus its position relative to potential slope failures. The steepness of discontinuities and how they intercept the slope in question are critical factors. The strike and dip of the weakness planes and of the slope faces are used to relate these features. Blocks can slide more easily on steeper weakness planes, but these planes must dip toward the slope (the opening) to be critical. Weakness planes are said to "daylight" if they intersect the slope plane by dipping at an angle less than the slope but still in the slope direction. Hence, the worst situation occurs when the strike of the weakness plane parallels the strike of the slope and the two dip in the same direction. That is why roadways should be oriented at right angles or at the highest angle possible to the

TABLE 14.5   Types of Rock Discontinuities Commonly Related to Slope Stability

Joint	A surface of fracture or parting in a rock, without displacement; the surface is usually planar and commonly occurs with parallel joints to form part of a joint set.
Fault	A surface or zone of rock fracture along which there has been displacement from a few centimeters (some authors would say 1 m) to a few kilometers in scale.
Shear zone	A tabular zone of rock that has been crushed and brecciated by many parallel fractures owing to shear strain.
Bedding surface	A surface, usually conspicuous, within a mass of stratified rock, representing an original surface of deposition. It is the interface between two adjacent beds of sedimentary rock. If the surface is more or less regular or nearly planar, it is called a bedding plane.
Foliation	A general term for the planar arrangement of textural or structural features in metamorphic rocks, such as rock cleavage in a slate, schistosity in a schist, and gneissic structure in a gneiss.

strike of steeply dipping beds in order to minimize slope failure effects.

Resistance against sliding along a discontinuity is determined by the steepness of the plane of weakness as well as the resistive forces of cohesion and coefficient of sliding friction along that plane. These items are discussed here because they relate directly to the basic understanding of rock-block slides.

Failure of rock blocks by sliding is commonly analyzed by use of the limit equilibrium relationship. This states that at the moment of failure the forces resisting movement are equal to or slightly less than the forces tending to move the block (known as the driving forces). Therefore, at any time prior to failure, the summation of the resisting forces is greater than that of the driving forces, $(\Sigma F_{resisting} > \Sigma F_{driving})$.

The resisting forces are typically evaluated for rock-block movement using the Mohr-Coulomb relationship shown in Figure 14.9. In the Mohr-Coulomb theory, the shear resisting stress along a plane is equal to cohesion c along that plane plus a friction stress σ_n tanϕ. This stress is a function of the normal stress across the plane, σ_n, and the coefficient of sliding friction of the block on the substrata, that is, f = tanϕ.

If the weight force W is considered now, it is observed that the normal force F_n is a function of the slope of the inclined plane, that is, F_n = W cosθ. Combining these details, the resisting force is $\tau \cdot A$ = CA + W cosθ tanϕ.

For most discontinuities the cohesive force is relatively small, from less than 100 up to 800 psf ($<$ 4.9 to 38.9 kPa). In many situations, the cohesion contribution is neglected in the analysis. This conservative approach provides an extra margin of safety against sliding for most calculations.

The ϕ value for rock sliding on rock is primarily a function of the mineralogy of the constituent materials. This information is presented in Table 14.6.

Referring again to Figure 14.9 it can be observed that, by applying similar reasoning as that used for the resisting forces, the driving force F_d, due to the weight of the block, equals W sinθ. Hence, we now have F_r = CA + W cosθ tanϕ and F_d = W sinθ. Using the commonly applied means for determining the factor of safety (FS) by dividing the resisting force by the driving force, we have

$$FS = \frac{\text{Resisting force}}{\text{Driving force}} = \frac{CA + W \cos\theta \tan\phi}{W \sin\theta}$$

If CA is considered to be zero, as is assumed in certain cases, and canceling W above and below,

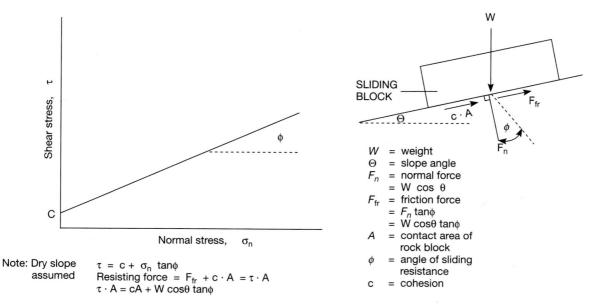

Note: Dry slope assumed

τ = c + σ_n tanϕ

Resisting force = F_{fr} + c $\cdot$ A = $\tau \cdot$ A

$\tau \cdot$ A = cA + W cosθ tanϕ

W = weight
Θ = slope angle
F_n = normal force
= W cos θ
F_{fr} = friction force
= F_n tanϕ
= W cosθ tanϕ
A = contact area of rock block
ϕ = angle of sliding resistance
c = cohesion

FIGURE 14.9 Mohr-Coulomb plot for rock-block sliding resistance.

TABLE 14.6 Angles of Sliding Resistance According to Rock Type, Smooth Contacts

Rock Name or Varieties	Approximate ϕ Value
Quartzite and quartz-rich rocks	30°
Schists and related rocks	20–25°
Carbonates	30–35°
Clays and clay shales	8–20°

$$FS = \frac{\cos\theta \, \tan\phi}{\sin\theta} = \frac{\tan\phi}{\tan\theta}$$

A final consideration can be made by allowing the ϕ angle to equal the slope angle θ; then

$$FS = \frac{\tan\theta}{\tan\phi} = 1$$

From this it follows that if $\phi > \theta$ and neglecting any contribution from the cohesion (also assuming dry slopes or zero pore pressure) then the FS will be greater than 1, yielding a stable condition. This is the basis for the sliding stability considerations given in Figure 14.8.

CONTINUITY Returning to the significant features of bedding planes, continuity or the extent of the weakness planes is of considerable importance. Major discontinuities such as faults extend for considerable distances but smaller features (for example, joints) may extend for much shorter distances. Some joint sets will extend through a number of beds or even a formation, whereas other sets will be terminated after several bed thicknesses or even a single bed. If failure is to occur in a rock mass that contains discontinuities that terminate within the mass, some intact rock must fail in order to propagate the fracture. This involves considerably larger forces for failure to occur than for simple sliding along through-going discontinuities. The shear strength of intact rock will be many times that of the fractured rock. It is sometimes possible to estimate the length of intact rock breakage needed based on the existing fracture discontinuities in a rock mass. This information can be used as a further refinement of rock slope stability. Typically, only 1 or

2% of intact rock along a boundary can nearly double the factor of safety against sliding.

SPACING The spacing of discontinuities contributes to rock mass strength. Joints develop in parallel sets or combinations of sets with a more or less regular spacing between the joint planes. In addition, bedding planes have a regular spacing, which should be included as part of the descriptive information provided in an engineering geology field report. Spacing ranges from thinly bedded, a few centimeters apart, to widely spaced or massive, which would have a separation of a meter or so. The spacing of discontinuities is directly related to the amount of intact rock that must break before through-going fractures can develop. Closely spaced discontinuities, though each individually may not fully traverse the rock mass, lead to a weaker situation than to widely spaced discontinuities, which also presumably do not fully cut the mass. Simply stated, closely spaced discontinuities indicate a weak rock mass with a more continuous fracture surface.

SURFACE IRREGULARITIES Surface irregularities along a discontinuity in rock can provide additional resistance not available to a smooth fracture surface. Two mechanisms allow these irregularities to contribute to increased strength: 1) overriding of the irregularities or 2) fracture through them. Figure 14.10 illustrates these two possibilities.

The overriding of irregularities in effect increases the angle of sliding resistance ϕ by an amount equal to the irregularity angle i. This yields the equation $\tau = \sigma_n \tan(\phi + i)$. Here the small contribution of cohesion along the fracture surface has been neglected or assumed equal to zero. Figure 14.10 illustrates both the relationships for overriding of irregularities and shearing through them based on a Mohr-Coulomb plot. The equation for shearing through rock in Figure 14.11 is $\tau = c_r + \sigma_n \tan\phi_r$. C_r is the cohesive strength of intact rock and $\tan\phi_r$ is the value for internal friction for intact rock (approximately 50° for massive rocks). The value of C_r is equivalent to the S_o values obtained from triaxial tests for rock, listed in Table 5.1. These values range from 1000 to 10,000 psi (7000 to 70,000 kPa) and are sufficiently large that C_r completely dominates the equation except for depths greater than several thousand feet (hundreds of meters) where the confining stress (σ_n) becomes

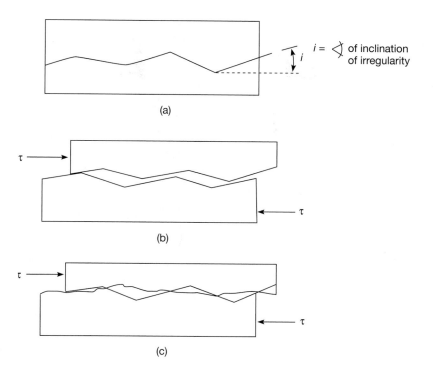

FIGURE 14.10   Cross-section view of surface irregularities related to movement along a discontinuity. (a) Prior to movement, (b) movement by riding over irregularities, and (c) movement by shearing through irregularities.

more significant. The dashed lines on the diagram show the locations where that relationship would not control slope failure as the other failure mode occurs at a lower stress level and thus prevails.

The lowest plot, $\tau = \sigma_n \tan\phi_{jr}$ is for continued movement along a surface of prior displacement. The value of ϕ_{jr} is somewhat less than the ϕ_j value that controlled when slope movement first occurred. For a massive limestone a typical value for ϕ_j would be 35°, but the ϕ_{jr} value, after block displacement, would be about 25°.

From the preceding discussion, we can conclude that surface irregularities increase the shearing resistance along failure surfaces. These features must be noted in the field to allow for recognition of the additional contribution of resisting force. As a matter of experience it is considered that the effective angle for these irregularities, angle i, will range from 10 to 15° in actual practice, although field and laboratory measurements may suggest higher values.

EXAMPLE PROBLEM _____

A series of quartzite beds dips into an open pit mine. How steep can the beds dip if they have an irregularity value of $i = 12°$, assuming a dry slope, no cohesion, and a required factor of safety of 1.2?

ANSWER

$$FS = \frac{CA + W\cos\theta\tan(\phi + i)}{W\sin\theta} \quad \text{dry slope}$$

If $C = 0$:

$$FS = \frac{\tan(\phi_j + i)}{\tan\theta}$$

The value of ϕ_j for quartzite is estimated at 30° (Table 14.6):

$$\tan\theta = \frac{\tan(\phi_j + i)}{FS}$$

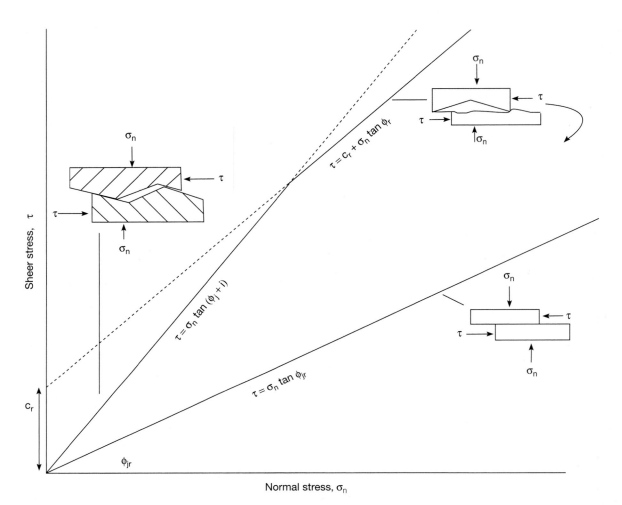

FIGURE 14.11   Failures on discontinuities in rock.

$$= \frac{\tan(30 + 12)}{1.2}$$

$$= 0.7503$$

$$\theta = 36.9°$$

CHARACTER OF ADJACENT ROCK   The next feature of discontinuities for consideration with regard to the resistance of rock-block sliding relates to the physical properties of the adjacent rock. Different rock types and their alteration products develop different shearing resistances. This is suggested by the range in ϕ values shown in Table 14.6 and for the cohesion values given on the previous page. For irregular

contacts between rock units of different strength, failure through the irregularities, if it were to occur, should involve more of the weaker rock than of the stronger. Consequently, the value of C_r to include in the equation represents that of the weaker rock.

INFILLING MATERIALS   The infilling materials of a discontinuity are assumed here to include all soil-like material that occurs between the walls of the fracture. This includes gouge developed previously during movement between the walls of the discontinuity, as well as any sediment that has accumulated in the opening. The resistance to shearing along the discontinuity is related to the properties and thickness of the infilling material plus the shear strength of the rock walls.

Four possible situations exist with regard to the infilled discontinuity:

1. Failure surface is contained entirely within the infilling material and shearing resistance is dependent on the properties of the infilling material alone.
2. Failure surface passes partially through both the infilling and the intact wall rock. Shearing resistance is complex and both materials will contribute to strength.
3. Infilling is present but very thin so that the properties of infilling material will only modify the shearing resistance developed between the rock walls of the discontinuity.
4. Infilling is absent and the shearing resistance is dependent only on properties of the wall rock.

In practice, it is extremely difficult to determine the nature and extent of infilling material for a large rock mass volume even if surface exposures are good. One concern is that infilling material may be washed from rock cores or ground up during the drilling operation. Therefore, it is impossible to obtain truly quantitative information for the total rock mass so that values may have to be extrapolated over considerable distances. Data on infilling material may be used simply to adjust slightly those calculations made assuming the existence of smooth, clean discontinuities.

Another factor concerned with infilling of discontinuities, not mentioned previously, may greatly increase rock slope strength. Secondary mineralization, in the form of quartz, calcite, and possibly iron oxide, can greatly increase the rock mass strength. In some cases the joint planes become as strong or stronger than the intact rock. This is another reason why careful examination of rock cores and rock exposures is necessary for engineering applications.

GROUNDWATER The final consideration concerning important features of discontinuities and rock slope stability is a major one, ground water. The considerations related to ground water are explored first and then a quantitative evaluation is presented.

Ground water contributes to the instability of rock slopes in the following ways:

1. A reduction in shear strength occurs because of fluid pressure (pore water pressure) acting along potential failure surfaces. Along discontinuities, water causes an uplift pressure perpendicular to the weakness planes and in effect reduces the downward normal stress. When the discontinuities are numerous with diverse orientations, the neutral stress (pore water pressure) increases yielding hydrostatic pressure ($h\gamma_w$), like that occurring in soils. When discontinuities have only a few preferred orientations and the spacing between them is large, abnormal distributions of water pressure (usually higher than hydrostatic) can result.
2. The presence of water increases the bulk unit weight of the rock mass (W) and thus increases the driving force ($W \sin\theta$). The term W is present in the resisting force also, but the reduction in normal stress caused by pore water pressure more than offsets the increase in W caused by the addition of that same water.
3. Ground water freezing in discontinuities can force open these planes, which increases water access to the slope and causes downslope displacement because of the volume increase when water turns into ice. Another feature sometimes overlooked is that the freezing of water along the downslope surface can block drainage to the surface from the rock mass, thus increasing the water pressure within the slope itself.
4. Erosion of infilling material can occur to open up discontinuities as a result of high-velocity groundwater flow. Removal of these materials may reduce the shearing resistance along the discontinuity. Increases in pressure caused by impounding reservoirs or raising the ground-water level by other means may lead to this condition.

The major concerns relative to ground-water effects can be best considered by returning to the basic equation shown in Figure 14.12. Resisting force = $CA + F_n \tan\phi$ along with the relationship for the driving force of Driving force = $W \sin\theta$. Figure 14.12 illustrates the problems that develop.

For purposes of an example, assume $L = 10$ ft, $h = 4$ ft.

$$\gamma_{rock} = 170 \text{ lb/ft}^3, \quad c = 200 \text{ lb/ft}^2, \quad \phi = 35° \quad \text{and} \quad \theta = 20°$$

$$\gamma_w = 62.4 \text{ lb/ft}^3$$

Assume first that the slope is dry. The U and V in Figure 14.12 would equal zero in this situation.

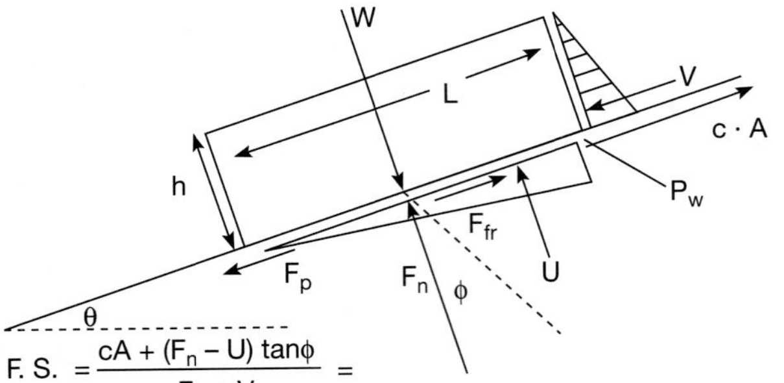

W = weight
θ = slope
φ = angle of internal resistance
F_n = normal force = W cosθ
F_p = component force parallel to plane = W sinθ
A = contact area
c = cohesion
L,h = length and height of block
U = uplift force due to pore water pressure
V = downslope force due to pore water pressure
γ_w = unit weight of water

$$\text{F. S.} = \frac{cA + (F_n - U)\tan\phi}{F_p + V} =$$

$$= \frac{cA + (W\cos\theta - U)\tan\phi}{W\sin\theta + V}$$

Calculating this, $P_w = h\cos\theta\,\gamma_w$

$$U = P_w \times L \times \frac{1}{2} = \frac{hL\gamma_w\cos\theta}{2}$$

$$V = P_w \times h \times \frac{1}{2} = h\gamma_w\cos\theta \cdot h \times \frac{1}{2} = \frac{h^2\gamma_w\cos\theta}{2}$$

Assume unit thickness (equal to 1) into the sheet of paper.

Volume of block = h × L (1) = hL
Weight of block = vol. · γ_{rock} = hL γ_{rock}
Contact area = A = L(1) = L

$$FS = \frac{cL + \left[\left(hL\gamma_{rock}\cos\theta \frac{-hL\gamma_w\cos\theta}{2}\right)\right]\tan\phi}{hL\gamma_{rock}\sin\theta + \frac{h^2\gamma_w\cos\theta}{2}}$$

FIGURE **14.12** Rock-slope stability, pore pressures included.

$$FS = \frac{CA + W\cos\theta\tan\phi}{W\sin\theta}$$

$$= \frac{200(10)(1) + (170)(10)(4)(1)\cos 20°\tan 35°}{170(10)(4)(1)\sin 20°}$$

$$= \frac{2000 + 4474.26}{2325.7} = 2.78$$

Next, the pore pressure situation is considered.

The pressure in the water at the bottom of the fracture is P_w. It is equal to the vertical height of the water column, $h\cos\theta$, multiplied by the unit weight of water γ_w, which equals 62.4 lb/ft^3. Therefore, $P_w = h\cos\theta\gamma_w = 10\cos 20°(62.4) = 586.4$ lb/ft^2. The U and V are triangular pressure diagrams, which equal hP_w × 1/2 and LP_w × 1/2, respectively. This yields the equation:

$$FS = \frac{200(10) + (4 \times 10 \times 170)\cos 20° \frac{-4(10 \times 62.4)\cos 20°}{2}\tan 35°}{4(10 \times 170)\sin 20° + \frac{(4^2 \times 62.4)\cos 20°}{2}}$$

$$= 2.02$$

The factor of safety is greater than 1 but substantially less than it was for the same slope when dry, FS = 2.78. The effect of the pore pressure buildup in a slope is quite significant and therefore the slope should be drained or one must consider the water pressure effect in the slope design.

Figure 14.13 shows a final detail for ground-water and rock slope stability. In this diagram, downslope drainage has been blocked by ice on the slope face. The factor of safety for this condition is now reduced to 0.86. Therefore, from the dry case to the one with the crack full of water under free drainage, the FS was reduced from 2.78 to 2.02. It is further reduced when drainage is blocked, to a value of 0.86, yielding a failure condition. Although $\phi >> \theta$ (35° versus 20°), failure will occur under these high, pore water pressure conditions.

Landslides that fail by slip represent a major portion of the most significant types of mass movement. Several photos of rock slides and slumps are shown in Figure 14.14.

FAILURE BY FLOW

Slope failure by the mechanism of flow occurs by internal slippage within the slope mass along many planes of failure similar to downslope movement of a viscous liquid. Rather than a single plane of failure or a definite failure zone, which is characteristic of slip movement, flow failure consists of internal slippage of the failure mass at many locations at different rates, somewhat analogous in that regard to the flow of water.

The materials affected by flow failure include a wide variety of soils and bedrock plus ice. Various types of mass movement by flow are designated in Sharpe's classification (Table 14.1) and in the TRB classification (Table 14.2), and some are depicted in Figure 14.1. The terms included in the Sharpe classifications are rock glacier, solifluction, debris avalanche, rock creep, talus creep, soil creep, earthflow, and mudflow. For TRB, the terms are rock creep and valley bulge, debris flow, debris avalanche, block stream, solifluction lobes, creep mantle, wet sand or silt flow, rapid earthflow, dry sand flow, dry loess flow, and mudflow. These terms are discussed briefly in the following paragraphs.

Rock Glacier

A rock glacier is a mixture of ice and rock that flows from a steep mountainous area onto an adjacent valley or plain along a definite channel. Such channels can be mapped by visual observation or from aerial photographs. Rock glaciers can be destructive features and must be avoided as building sites for homes, industrial structures, or public buildings.

(a)

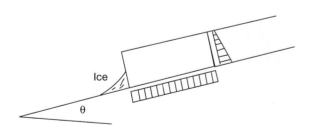

$$FS = \frac{CL + (hL\gamma_{rock}\ \cos\theta - hL\gamma_w\ \cos\theta)\ \tan\phi}{hL\gamma_{rock}\ \sin\theta\ +\ h^2\ \dfrac{(\gamma_w\ \cos\theta)}{2}}$$

FIGURE **14.13** Rock-slope stability, pore pressures with blocked drainage.

FIGURE **14.14** (a) Rock slides along bedding planes.

Figure 14.14 (b) Rock slides forming landslide dam, Thistle, Utah. (c) Slump in soil and rock with damage to residence.

Solifluction

Solifluction involves soil flow down gentle slopes, mostly in arctic and subarctic areas, that occurs because of surface melting of the frozen soil. Water is unable to penetrate the frozen soil below and the mixture of water, soil, and sometimes ice will flow downslope at angles of as low as 2 to 3°. This is the process by which the flat tundra areas are developed. It is also thought that till plains owe much of their expression of flat terrain to the effects of solifluction in the years following glacial retreat. The TRB classification uses the term *solifluction lobes,* which relates to the lobate features that depict an area actively undergoing solifluction.

Debris Avalanche and Avalanche

A debris avalanche is an extremely rapid failure of broken rock and soil that is sometimes accompanied by ice. It may be relatively dry but in some instances contains enough water to saturate the debris. In general, they are long and narrow with characteristic V-shaped scars tapering uphill to the head. Commonly they occupy a preexisting stream channel and may advance well beyond the foot of the slope in an extensive run-out zone, an area with a much lower gradient, that extends for hundreds or thousands of feet beyond the foot of the slope, over which the debris expends its final energy.

The term *avalanche,* if unmodified, refers strictly to slope movements of snow or ice. These failures are also quite destructive to property and their locations can be mapped based on broken and uprooted areas along narrow paths on forested hillsides. These areas stand out particularly well during the summer season. Runout zones from avalanches can also be depicted and obviously building permits should not be granted for areas in avalanche runs and runout zones.

Debris Flow

The TRB classification also includes debris flow, which is similar to debris avalanche but the flow is classified as a rapid failure rather than a very rapid one. Debris flows are generally restricted to channels and they commonly result from unusually heavy rainfall or from rapid thawing of snow. These features, in a manner similar to debris avalanches, have a considerable runout zone beyond the base of the slope, a fact that must be made known to planners, architects, and contractors who may attempt to initiate construction in such destruction-prone areas.

In the TRB classification, a distinction is also made between debris flows and mudflows. Debris flows involve material containing a high percentage of coarse fragments, whereas a mudflow consists of soil of which 50% or more is sand, silt, and clay. The mudflow also must be wet enough to flow rapidly.

Mudflow

Mudflows are the rapid failures that are common in arid and semiarid regions and are confined to a definite channel. Soil and vegetative debris will accumulate at the heads of dry channels until an infrequent rainfall event triggers the high runoff so typical of these dry, poorly vegetated areas, and the material becomes saturated. With increased moisture comes decreased shear strength, and the wet mass flows rapidly down the channel obliterating any trees or structures in its path. Active mudflow zones can be recognized during field studies or from aerial photographs, and old mudflow deposits can be identified by their diagnostic landforms. Active mudflow areas should be properly zoned to prevent construction of housing and industrial buildings, but golf courses or parks without permanent structures may be quite suitable for these locations.

A particularly serious set of conditions periodically gives rise to damaging mudflows in California. During a prolonged drought, brush fires burn off the vegetative cover. Then, if a period of intense rainfall occurs, runoff is particularly extreme because of the denuded nature of the landscape, and it triggers the onset of numerous mudflows. This and the lack of restrictive building codes in the past allowed for widespread destruction of homes by mudflows. Proper zoning of mudflow areas and a program to remove excess debris from the heads of mudflow-prone channels can greatly reduce the potential financial loss associated with these slope failures. A schematic of a mudflow is shown in Figure 14.15.

Rock Creep

Rock creep is the movement of bedrock downslope by the slow, flow-failure mechanism of creep. It occurs in bedrock that dips downslope due to the slow slippage along bedding planes and in rock units that dip into the slope by creep along joint planes that parallel the slope. An extreme case of rock creep into an existing valley can lead to a valley bulge (also known as cambering) or the inward movement of the valley sidewalls because of slow, flow failure. This feature, recognized by the TRB classification, has occurred in England where massive dolomite overlies marine shales, and rock creep of the dolomite units into the valley has been documented.

Recently, a flow phenomenon in massive bedrock that is manifested by the folding, bending, or bulging of rock has been recognized during field studies. These features are found in many areas of high relief throughout the world. The term *sackung* is used for

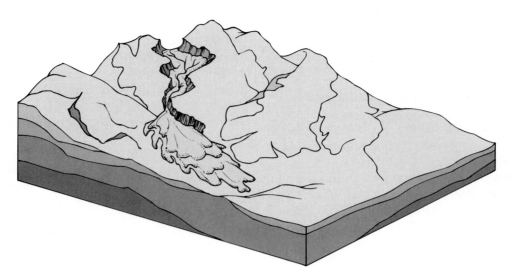

FIGURE **14.15**   Diagram of a mudflow.

the ridge top depression that develops because of large-scale creep toward the valley walls.

Talus Creep and Soil Creep

Talus creep and soil creep refer to the slow movement downslope of either talus (accumulations of broken rock along the base of mountain slopes) or of topsoil and subsoil. The TRB classification combines the two categories of creep referring to them collectively as mantle creep (the mantle is the loose, weathered material lying above bedrock). Soil creep is evidenced by tilted fence posts and other shallow-buried markers that are moved progressively downslope. This creep can lead to surveying problems and property disputes if the fence line also serves as the property line.

Earthflows

Earthflows are common failures in cohesive soils within humid areas. They involve slope failures that are not limited to a definite channel and the soil flows in a relatively rapid fashion (although usually not rapid enough to be perceptible to the eye). This action yields a spoon-shaped failure hollow, not too dissimilar from a slump feature. Typically, the soil is moist but not saturated with water. The evidence, or lack thereof, of internal shear planes (flow) within the failed mass determines whether an earthflow (flow failure) or a slump (slip failure) has occurred.

Block Streams

Narrow channels of rock debris on steep slopes that move extremely slow and are typically fed by talus accumulations at the head are known as block streams. Rain wash removes the fines from the surface so the resulting appearance is more coarse than the overall nature of deposit.

Wet Sand Flow and Wet Silt Flow

Several types of failure listed in the TRB classification have not yet been discussed. These are wet sand or silt flow, dry sand flow, and dry loess flow. Wet sand or silt flow is a particular problem in dunes along the shores of large lakes or the ocean. Rapid failures can occur during changes from partial saturation to full saturation. Partially saturated sand has an apparent cohesion because of capillary forces caused by numerous water-air interfaces in the pores. With saturation, not only is this apparent cohesion lost, but pore water pressure further reduces the effective stress and rapid slope failure can occur.

Dry Sand Flow

Dry sand flows are common along shores or embankments underlain by dry granular material. They occur as both channelized flows and as sheet-like features.

Dry Loess Flow

Dry loess flow failures are induced by earthquakes on slopes that are generally quite steep because of the characteristic of loess to stabilize on a near-vertical slope. Earthquake shaking provides the vibration needed to mobilize the grains of silt in the porous loess. Apparently, the internal structure is destroyed by the shaking and the loess becomes a fluid suspension of silt in air, which flows down into the valleys, filling them and overwhelming any villages or other human-made structures in its path.

Loess flows mobilized by earthquake shocks are associated with two major instances of extreme loss of life in this century. In Kansu Province, China, following the 1920 earthquake, about 100,000 lives were lost when loess flowed into the valleys, burying villages. On July 10, 1949, in Tadzhikistan, south-central Asia, earthquake-triggered loess flows buried or destroyed 33 villages as the flow covered the valleys to depths of several tens of meters for a distance of many kilometers.

SLOPE FAILURE BY LATERAL SPREAD

Lateral spreads are the next category of slope failures in the TRB classification. As previously stated, this failure mechanism was not considered in Sharpe's Classification of Mass Movement of 1938 or in the Highway Research Board landslide classification in 1958. It has only been recognized as significant since the Anchorage, Alaska, earthquake of 1964 where the mechanism of lateral spread was the dominant mode of failure.

The most significant detail of lateral spreading involves fracturing and extension of intact material, either bedrock or soil, because of liquefaction[3] or plastic flow of material at a shallow depth below the surface. The coherent upper units may subside, slide, rotate, disintegrate, or simply liquefy and flow downhill. The total process may involve rotation, translation, and flow and therefore lateral spreading failures could well be included under the complex category of the TRB classification. They are such a distinctive group, however, and because they are related to specific geologic conditions, a special designation for them is warranted.

[3] Liquefaction is the rapid loss of shear strength that occurs in saturated fine sands and silts as a result of earthquake shocks because of a sudden increase in pore water pressure and the attendant decrease in effective stress.

Laterally spreading slope failures form in fine-grained earth materials on shallow slopes, particularly in sensitive silts and clays that lose most or all of their shear strength when disturbed or remolded. Failure is commonly progressive; it starts locally and spreads from that point into the slope. The initial failure may be a slump along a stream bank or along a shore line with the failure progressing further and further into the bank. The primary surface movement is translation and not rotation. Movement commonly begins suddenly, without warning, and it proceeds with rapid to very rapid velocity from the initial point of failure.

If the underlying mobile zone is thick, blocks at the head of the failure typically drop downward as grabens, which may or may not involve a backward rotation. Upward and outward extrusion and flow may occur at the toe of the failure. Lateral spreads seem to include a gradational series of landslides in surficial material, ranging from block slides at one extreme when the mobile zone is very thin, to earthflows or completely liquefied mudflows at the other extreme in which the zone comprises the entire mass.

Most spreading failures in the western United States involve less than total liquefaction and are mobilized by seismic shock. This holds true for the failures during the San Fernando Valley earthquake of 1969 and in the failure of the Bootlegger Cove clay below the Turnagain Heights residential district of Anchorage, Alaska, in the major earthquake of 1964. It also occurred in some areas during the San Francisco earthquake of 1906 where spreading failures caused direct damage to structures and also severed water supply lines to reduce fire-fighting capabilities. In the Loma Prieta earthquake of October 17, 1989, lateral spreading in the Marina District of San Francisco caused major damage to the existing structures. Lateral spreads were negligible during the Los Angeles earthquake of January 17, 1994, because of the dry nature of the valley-fill soils in the city of Northridge.

For lateral spreading failures in Pleistocene age glacial and marine sediments, several common characteristics of the failed material have been recognized: Movement commonly occurs without apparent reason, failure is generally sudden, even gentle slopes are unstable, dominant movement is by translation, the materials are sensitive, and pore water pressure is involved in the stability.

COMPLEX SLOPE FAILURES

These failures typically include a combination of two types of the failures previously discussed that act together or in sequence to cause slope movement. The specific combinations included in the TRB classification are:

1. Rock fall–debris flow (or rock fall avalanche)
2. Slump and topple
3. Rock slide–rock fall
4. Cambering spreading plus bending overflow
5. Slump–earthflow
6. Mudslide (or sliding mudflow).

Rock fall–debris flows or, similarly, rock slide–debris flows are most common in rugged mountainous regions. The Elm, Switzerland, failure of the 1800s, which took 115 lives, started with a small rock slide adjacent to a quarry on the mountainside but quickly became a flow of high velocity. About 13 million cubic yards (10 million cubic meters) of rock descended an average of 1540 ft (470 m) vertically in an elapsed time of 55 sec. The much publicized Frank, Alberta, slide of 1903 began as a rock slide and turned into an avalanche, which killed about 70 people in the outlying portions of the town. Yet these failures were minor by comparison to the rock fall–avalanche that followed the May 31, 1970, earthquake in Peru, which buried the city of Yungay and part of Ranrahirca, causing a loss of life greater than 18,000. The movement started at an elevation of 21,100 ft (6400 m), coming to rest 12,000 ft (3660 m) below, and involved 65 to 130 million cubic yards (50 to 100 million cubic meters) of rock, ice, snow, and soil that traveled 9 miles (14.5 km) from its source to Yungay at a velocity between 175 to 210 mph (280 to 335 km/hr).

The combined flows of extreme magnitude described seem to be lubricated at least in part by a cushion of air trapped beneath the debris. The enormous size of these flows is required to develop this means of transfer, which fortunately does not occur in the commonplace smaller slope failures. The large prehistoric Blackhawk landslide in the San Bernadino Mountains, Southern California, which measures 2 miles (3.2 km) across is also thought to have slid on a layer of compressed air.

The specific combinations of slump and topple, rock slide–rock fall, and slump–earthflow are fairly straightforward relationships between the two failure modes involved. Cambering and valley bulging were described previously under the mechanism of flow failure.

CAUSES OF SLOPE FAILURE

Slope failures or landslides occur as short-lived phenomena that have a profound effect on the land surface and commonly on human activities. The stress applied to a slope must exceed a certain critical level before the equilibrium conditions, which may have endured for many years, are upset and failure proceeds. The causes of major landslides can be considered under three primary groups: 1) excessive precipitation, 2) earthquakes, and 3) human activities.

Excessive Precipitation

The existing moisture conditions in a slope mass determine the extent of rainfall necessary to trigger a landslide. If the slope contains significant moisture from prior rainfall, the contribution of water from a new storm can be relatively small and still induce slope failure. Other factors being equal, the intensity and duration of the storm are the significant factors that contribute to instability of the slope.

Addition of ground water to the slope increases the total downward-acting force on potential failure surfaces, and reduces the frictional resistance along such surfaces by increasing the pore water pressure (with a resulting decrease in effective stress). The geologic conditions of the slope mass have a great effect on the means and degree in which water enters the slope and, in turn, drains from it. A high permeability of soils or bedrock provides easy ingress into the slope except under frozen conditions. Low permeability or a frozen surface prevents the exit of water at the base of the slope, yielding an increase in pore water pressure and greater instability.

Exceptionally heavy rains have caused massive sliding to occur in various parts of the world. In 1966–67 very heavy rainstorms in Brazil initiated massive landslides and flooding, which killed more than 2700 people. The landslides included a slump–earthflow, debris slides and avalanches, debris flows and mudflows, and rocks and rock slides.

Japan is subject to numerous landslides related to high-intensity storms because of the high relief of the islands and the high annual rainfall. Kobe was hit in 1938 by rainstorms that produced rock–mudflows killing 461 people and destroying 100,000 houses. In 1945 the Makurazaki typhoon hit Kure and produced mudflows causing 1154 deaths. The Kenogarva typhoon of 1958 produced the heaviest rainfall ever recorded for Tokyo: 15.4 in., or 392.5 mm, in 24 hr. This killed 61 people and caused more than 1000 landslides.

In June 1966 Hong Kong received a series of rainstorms that amounted to nearly 15 times the normal intensity. Near mid-month, 15.7 in. (400 mm) of rain occurred in 24 hr, which was added to the previous 12.3 in. (314 mm) from earlier that month. This was the trigger for wholesale landsliding, the most disastrous and spectacular being large-scale, deep-seated slumps. Also included were debris avalanches, boulder falls, and rock slides. Bedrock in Hong Kong is predominantly volcanic and intrusive rocks on which a deep residual soil has formed by weathering.

The United States has also suffered extensive landslides associated with high-intensity storms. In August 1969 Hurricane Camille produced an 8-hr deluge of 30 in. (710 mm) in central Virginia. In Nelson County alone, property damage was more than $116 million and 150 people were killed by the combination of debris avalanches and flooding. The debris avalanches followed preexisting drainages on hillsides steeper than 35°; produced head scars at the steepest part of the hill; were prevalent on north-, northeast-, and east-facing slopes; caused rapid surges of water and sediments into stream channels with devastating effects; and typically left scars from 200 to 800 ft (60 to 240 m) in length, 25 to 75 ft (7.5 to 23 m) wide, and 1 to 3 ft (0.3 to 0.9 m) deep.

In California many landslides have been linked to the effects of rainfall. The distribution, amount, and pattern of rainfall strongly influence slope failure. Heavy rainfall that occurs when the slopes are still wet from previous storms causes the largest number of landslides. On this basis, the pattern of rainfall during the rainy season is more important than is the total amount received.

Earthquakes

History suggests that the most catastrophic landslides that occur are related to earthquakes. Earthquake shocks have profound effects on loose and sensitive soils and even on bedrock to the extent that major terrain changes are associated with the world's most severe earthquakes. The Lisbon earthquake of 1755 produced widespread effects throughout Portugal, causing many failures along sea cliffs and in the valley walls of rivers. The massive Indian earthquake of 1897 profoundly changed and scarred valleys for distances of 20 mi. The shock was felt as a strong motion in an area of 160,000 mi^2 and was noticeable in a region more than 10 times as great.

In discussions concerning major earthquakes and related landslides in the United States, the specific earthquakes commonly considered are 1) New Madrid, Missouri, 1811; 2) San Francisco, 1906; 3) Madison Canyon, Montana, 1959; 4) Anchorage, Alaska, 1964; and 5) San Fernando Valley, California, 1971.

The New Madrid, Missouri, earthquake caused widespread landslides in the loess river bluffs as far north as the Ohio River. On the east side of the Mississippi River a continuous strip of landslides prevailed for 35 mi (56 km). Large depressions, 100 ft in diameter, 100 ft deep (30 m) were still visible as late as 1869.

The San Francisco earthquake, which occurred at 5:12 A.M. on April 18, 1906, was felt over a land area of 175,000 mi^2 (453,000 km^2). The landslides that occurred were divided into four categories based on field studies conducted soon after the earthquake. The categories were earth avalanches, earthflows, earth slumps, and earth lurches (presumably liquefaction or lateral spreading failures). Most earth avalanches occurred along the sea cliffs of the coastal areas. The largest single landslide scar measured about 1 mi (1.6 km) wide and 0.5 mi (0.9 km) long. One of the largest earth slumps had a width of 1500 ft (450 m), extending 400 ft (120 m) downslope with a 50-ft- (15-m)-high scarp. The largest earthflow was 2700 ft (810 m) long, 100 ft (30 m) wide, and 3 ft (1 m) deep involving about 90,000 yd^3 (67,000 m^3) of material.

In the Madison Canyon of Montana, an earthquake triggered a rock slide–rock avalanche that killed 28 campers in August 1959. The rocks were dry because there had been no rain for six weeks prior to failure, but the earthquake shocks were sufficient to weaken the rock mass. Paleozoic dolomites, dipping 40° into the valley, overlying a Precambrian basement complex of gneiss and schist, provided the structure that led to instability. In addition to the main slope failure, numerous other landslides including rock falls, earthflows, slumps, debris slides, and debris avalanches were initiated by the earthquake. Many involved weathered bedrock atop the steep cliffs. Several were debris slides into nearby Hegben Lake, which may have been helped in part by water in the slopes around the lake. Others involved massive slumps of colluvium.

The Alaska earthquake of March 27, 1964, caused significant damage over a 50,000-mi^2 (129,000-km^2) area, killing 114 people. It had a Richter magnitude of 8.5 and the main portion of the earthquake lasted up to 7 mins. Aftershocks followed the main earthquake for 69 days with some as high as 6.7 on the Richter scale. The greatest damage of property and loss of life occurred in Anchorage where 14% of the city (700 acres involving 750 homes) was destroyed by lateral spreading failures. The failures occurred in the Bootlegger Cove clay, a glacial estuary-marine deposit that underlies much of the Anchorage area.

Another type of failure induced by the Alaska earthquake was the rock avalanche known as the Sherman landslide. Rock blocks broke loose along bedding planes that dipped 40° into the valley. The sandstone-argillite sequence was strongly jointed, which also lead to instability.

The Alaska earthquake also initiated extensive subaqueous slides in the adjacent harbor areas. In the seaport of Seward, the waterfront was altered when deltaic sediments slid oceanward, which increased the depth of the pier area from an original depth of 20 to 30 ft (6 to 9 m) to that of 130 to 180 ft (39 to 54 m). Subaqueous slides in the Passage Canal near Whittier, Alaska, created 104-ft (31-m) waves. One 52-ft (15.6-m) wave struck Whittier destroying the harbor, pier, docks, associated installations, and nearby homes.

The San Fernando Valley earthquake of 1971, with its Richter magnitude of 6.5, produced more than 1000 landslides in an area of 100 mi^2 (260 km^2) in the mountainous region nearby. Landslides included rock falls, soil falls, debris slides, debris avalanches, and slumps. Landslides also occurred in human-made structures such as earth embankments for dams and roadway embankments.

Two other severe earthquakes that induced major slope failures were already discussed in the classification of landslides section of this chapter. The 1920 failure in Kansi Province, China, that killed from 100,000 to 200,000 people owing to the failure of loess slopes into the stream valley marks the greatest loss of life involving any single slope failure event. An area of 30,000 mi^2 (76,000 km^2) was devastated.

The other staggering catastrophe was associated with the Peru earthquake of May 1970. The earthquake (7.7 Richter magnitude) initiated a rock fall–debris flow that took more than 18,000 lives when it destroyed the city of Yungay and parts of Ranrahirca.

Human Activities

An increase in the number of slope failures has occurred during the 20th century because of human activities. The problem is two-fold in nature: 1) the construction of engineering structures composed of earth materials to create embankments for roads, bridge abutments, dams, railroads and tunnel grades which may encroach upon the limits of the equilibrium conditions for slope stability. This also includes the stacking of refuse material following mining or other excavations. Also the impoundment of water behind some of these structures has changed the ground water regime of adjacent areas. 2) The alteration of natural slopes by excavation and urbanization. This includes the steepening of slopes, removal of the slope base, changing surface drainage conditions, removal of vegetation and disruption of the subsurface drainage characteristics. Regarding property damage, in some cases humans have not altered the natural setting, but simply encroached on an active area of erosion and must pay the consequences for such folly. This would include construction in snow avalanche and debris avalanche-debris flow areas as well as on active mudflows. Human error may be only that of choosing the wrong area while not inducing any natural changes by such action. In the longer term, this too may not hold true as the human presence and simple activities continue to alter the natural environment to some extent with roads, water wells, buildings, sewage disposal and cultivation.

This is not to say that humans must not encroach on the natural setting. Preservation of nature in an untouched fashion is not the answer; instead the current definition of conservation serves best. It involves management of the land for the greatest benefit to humankind.

The real challenge to the engineering geologist and construction engineer is to "Design with Nature," taking advantage of the natural setting and minimizing the impact of human activities on it. Humans have the capability of building structures which advanced-society demands and yet induce few effects to the landscape, particularly regarding slope stability.

Countless specific examples of slope stability failures related to human activity could be presented here to demonstrate various concerns. These would include failures in dam embankments, in reservoir areas, or natural slopes; embankments or cut slopes due to mining or oil production; and for embankment and cut slopes for highways. Only a few specific cases are presented to illustrate the overall diversity of these slope failures and to provide discussion on some of the most quoted, human related landslides.

Examples

Gros Ventre Slide, Wyoming

This is an example of the effects of natural slope failures on human activities through secondary involvement. A rock slide involving 50 million cubic yards (38 million cubic meters) occurred on June 23, 1925. The slide debris moved 1.5 miles (2.5 km) down the dip-slope of Sheep Mountain in northwestern Wyoming, across the Gros Ventre River and 305 feet (107 m) up the opposite slope. The mass, mostly Tensleep Sandstone, slid along the 18 to 21° bedding probably at the top of a saturated clay shale unit. The rock slide formed a dam 225 to 250 ft (69 to 76 m) high, which in turn created a lake nearly 5 mi (8 km) long. Two years later, May 1927, the lake overtopped the dam, causing serious flooding downstream.

Frank Slide, Alberta

For many years it was thought that coal mining played an important role in the initiation of this slide but more recent mapping and calculations suggest that other factors such as structural control were more dominant. On April 29, 1903, about 40 million cubic yards (30 million cubic meters) of rock broke loose from Turtle Mountain in southern Alberta, rushing down the mountain side to kill 70 people and destroy much of the outlying portion of the town of Frank.

The Frank slide was considered to be a classic example of a joint-controlled rock slide with the mass moving along joint planes parallel to the slope and perpendicular to the bedding, which dipped steeply into the mountain. Heavy spring rains and settlement due to mining of a coal seam at the base of the mountain, adjacent to the town, were thought to have triggered the slide.

Recent mapping in the area has shown that the movement was primarily along bedding planes. The mountain had an anticlinal feature prior to failure with the axial plane running through the mountain toward the town. Fractures parallel to the axial plane could have provided ingress for the heavy spring rains, which in turn caused a buildup of excess pore pressure in the slope and the subsequent sliding along the bedding planes.

Vaiont Reservoir Slide, Italy

This example illustrates the extreme need to prevent large-volume landslides from occurring along the rim of a reservoir. On October 9, 1963, the most disastrous landslide in European history occurred, the Vaiont Reservoir slide in northeastern Italy. A volume of rock of about 325 million yards^3 (250 million meters^3) suddenly slid into the reservoir, displacing most of the water and sending a wave 800 ft (260 m) up the opposite slope and at least 300 ft (100 m) over the top of the dam into the river valley below. This wall of water destroyed five villages and killed between 2000 and 3000 people. A remarkable fact is that the thin arch dam, the world's largest at the time of construction, sustained little structural damage. The dam remains intact today, sitting astride the valley at the base of the empty reservoir as a monument to this catastrophic event.

The area was known to be landslide-prone well before the dam was constructed, and landslides were experienced in the reservoir some years before the major failure. The slopes were well instrumented and it was thought by the designers that the reservoir could be lowered if slope movement became excessive and that slope failures of the size anticipated could be accommodated without undue risk. In 1960 a large slide mass (900,000 yd^3 or 700,000 m^3) slid into the reservoir a short distance upstream from the dam but this was only about 3% of the volume of the massive slide that ultimately occurred in 1963. It would appear in retrospect that the major error committed was the underestimation of the likely volume of a single slide mass. The failure occurred after a period of sustained, heavy rainfall, so extreme weather conditions should have been considered along with the design details.

An adverse set of structural features led to the slide. The Vaiont gorge forms the axis of a syncline with the bedding planes dipping inward from both sides of the valley. The sliding surface was largely along bedding surfaces that also contained well-defined joints parallel to the beds.

Alberfan, Wales

In 1966 a pile of coal mine spoil from a coal processing plant failed by moving downslope to kill 144 people, mostly children in the town below. The mine spoil had become saturated at depth previous to failure and the slope angle was approximately 30°. Intense rainfall added the needed driving force to cause a flow failure within the material.

Buffalo Creek Dam, Saunders, West Virginia

A fairly recent major catastrophe in the United States involving slope failure was the 1972 failure of the Buffalo Creek Dam. Heavy rains induced failure of three coal-refuse impounding structures. The major debris flow that resulted, consisting of water, coal

TABLE 14.7   Landslide Disasters

	Date	Number Killed	Remarks
Brenno Valley, Switzerland	1512	600	Rockslide dammed valley; dam broke after 2 years causing destruction
Tour d'Ai, Switzerland	1584	300	Landslide devastated village of Yvorne in Rhone Valley
Mount Conto, Switzerland	1618	2430	Rockslide
Goldau, Switzerland	1806	457	Landslide destroyed village
Mt. Ida, Troy, New York	1843	15	Sediment slump and flow
Elm, Switzerland	1881	115	Rock avalanche also demolished 83 houses
Trondheim, Norway	1893	111	Liquefaction flow in marine clays
Frank, Alberta, Canada	1903	70	Rock avalanche destroyed most of town
Kansu Province, China	1920	100,000–200,000	Earthquake caused loess flows
Nordjord, Norway	1936	73	Rockfall created 74-m wave
Kobe, Japan	1938	461	Rocky mudflows
Kure, Japan	1945	1154	Rocky mudflows
Yokahama, Japan	1958	61	Rocky mudflows
Madison, Montana	1959	28	Rock avalanche buried campers
Vaiont, Italy	1963	2000	Rockslide into reservoir created wave that flooded below dam
Anchorage, Alaska	1964	114	Combined toll from landslides and earthquake
Aberfan, Wales	1966	144	Human activity, mining spoil hill; landslide buried mostly children
Brazil	1966–67	2700	Combined toll from landslides and floods
Nelson County, Virginia	1969	150	Combined total from debris avalanches and flood
Huascaran area, Peru	1970	21,000	Combined rock avalanche and debris flow buried two cities
St. Jean Vianney, Canada	1971	31	Slab flows buried people and houses

TABLE 14.8   General Considerations for Preventing or Correcting Slope Failures

1. Weight at the head of a slope is a driving force. Never load the head unnecessarily. Relieve the head by removing material. Tie the surface to stable material deeper within the slope if needed.
2. Weight at the toe is a resisting force. Never excavate material at the toe without being sure that the slope will be stable afterward. Load the toe with berms or support the toe with piles, retaining walls, cribbing, etc., if more support is required.
3. Keep surface water off the slope. Infiltration may occur or erosion and gullying may develop depending on the nature of the slope. Water should be channeled and drained away from the slope.
4. Provide internal drainage for the slope when needed. This reduces the pore water pressure, increasing the frictional component of shear strength. It also increases cohesion of plastic materials by lowering the water content. Horizontal drains perpendicular to the slope direction may be required.
5. Reduce the slope angle as needed. This will decrease the driving force but greatly increases excavation costs. The slope should be graded to obtain a stable surface profile.

wastes, and sludge, traveled 15 mi (24 km) down-stream, killing 125 people and leaving 4000 home-less.

The major landslide disasters of the world since 1500 are listed in Table 14.7. These include a number of the examples discussed in the text.

PREVENTION AND CORRECTION OF SLIDES

The prevention of landslides or the correction of the slope after failure involves only two basic concepts: increase the resisting force or reduce the driving force or accomplish both, to the extent that the slope becomes stable. The specifics of how this is accomplished lie in the realm of slope design procedures and are beyond the scope of this text. The general principles can be outlined in a descriptive way, however, as shown in Table 14.8.

Additional information is supplied in Table 14.9 and Figure 14.16 on slope design procedures. Table 14.9 pertains to slopes in soil, and Figure 14.16 pertains to slopes in rock.

TABLE 14.9 Summary of Slope Design Procedures for Soils

Category	Procedure	Best Application	Limitation	Remarks
Avoid problem	Relocate highway	As an alternative anywhere	Has none if studied during planning phase; has large cost if location is selected and design is complete; also has large cost if reconstruction is required	Detailed studies of proposed relocation should ensure improved conditions
	Completely or partially remove unstable materials	Where small volumes of excavation are involved and where poor soils are encountered at shallow depths	May be costly to control excavation; may not be best alternative for large slides; may not be feasible because of right-of-way requirements	Analytical studies must be performed; depth of excavation must be sufficient to ensure firm support
	Bridge	At sidehill locations with shallow-depth soil movements	May be costly and not provide adequate support capacity for lateral thrust	Analysis must be performed for anticipated loadings as well as structural capability to restrain landslide mass
Reduce driving forces	Change line or grade	During preliminary design phase of project	Will affect sections of roadway adjacent to slide area	
	Drain surface	In any design scheme; must also be part of any remedial design	Will only correct surface infiltration or seepage due to surface infiltration	Slope vegetation should be considered in all cases
	Drain subsurface	On any slope where lowering of ground-water table will affect or aid slope stability	Cannot be used effectively when sliding mass is impervious	Stability analysis should include consideration of seepage forces
	Reduce weight	At any existing or potential slide	Requires lightweight materials that are costly and may be unavailable; may have excavation waste that creates problems; requires consideration of availability of right-of-way	Stability analysis must be performed to ensure proper use and placement area of lightweight materials

314

			Remarks
Increase resisting forces	Drain subsurface	At any slide where water table is above shear plane	Requires experienced personnel to install and ensure effective operation	
	Use buttress and counterweight fills	At an existing slide, in combination with other methods	May not be effective on deep-seated slides; must be founded on a firm base	
	Install piles	To prevent movement or strain before excavation	Will not stand large strains; must penetrate well below sliding surface	Stability analysis is required to determine soil-pile force system for safe design
	Install anchors	Where rights-of-way adjacent to highway are limited	Involves depth control based on ability of foundation soils to resist shear forces from anchor tension	Study must be made of *in situ* soil shear strength; economics of method is function of anchor depth and frequency
	Treat chemically	Where sliding surface is well defined and soil reacts positively to treatment	May be reversible action; has not had long-term effectiveness evaluated	Laboratory study of soil-chemical treatment must precede field installation
	Use electro-osmosis	To relieve excess pore pressures at desirable construction rate	Requires constant direct current power supply and maintenance	
	Treat thermally	To reduce sensitivity of clay soils to action of water	Requires expensive and carefully designed system to dry out subsoils artificially	Methods are experimental and costly

SOURCE: Gedney, D.S., and Weber, W.G., Jr. 1978, Design and construction of soil slopes. In R.L. Schuster and R.J. Krizek (Eds.), *Landslides, Analysis and Control*, Special Report 176, Transportation Research Board, Washington, DC.

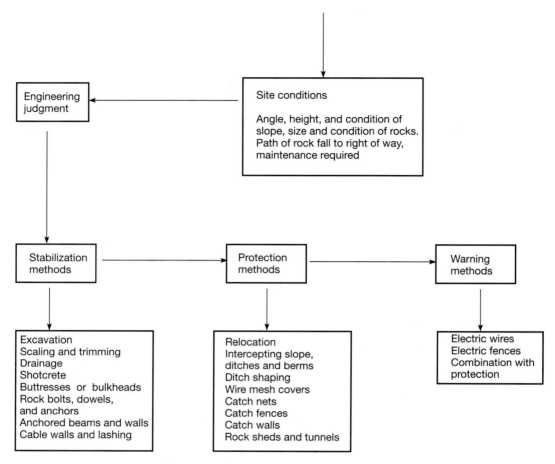

FIGURE **14.16** Order in which slope treatment methods should be considered for design and construction of rock slopes. (From Piteau, D.R., and Peckover, F.L., 1978, Engineering of Rock Slopes, In R.L. Schuster and R.J. Krizek (Eds.), *Landslides, Analysis and Control*, Special Report 176, Transportation Research Board, Washington, DC.)

EXERCISES ON LANDSLIDES, SUBSIDENCE, AND SLOPE STABILITY

1. Discuss the various ways in which Sharpe's Classification of Mass Movement differs from the TRB Classification of Slope Movement. Refer to Tables 14.1 and 14.2 plus the appropriate text material for details.

2. Explain how an earthflow would be distinguished from a slump in the field. What distinction is made between mudflow and debris flow in the TRB classification?

3. Discuss the difference between the terms *compaction* and *consolidation* as used by civil engineers and by traditional geologists (not including engineering geologists). What geologic materials would be particularly subject to consolidation (in the engineering sense)?

4. What factors lead to shallow subsidence? To deep subsidence? How can the two types be alleviated or greatly reduced?

5. The load on a rock pillar for a room and pillar mine is defined by the following equation:

$$\text{Stress on pillar} = \frac{h\gamma_{\text{rock}}}{1 - \text{Fraction of material removed}}$$

where h is the height of overburden and γ_{rock} is the unit weight of the rock. If the pillar strength is determined by its compressive strength, the factor of safety is

$$FS = \frac{\text{Compressive strength}}{\text{Pillar stress}}$$

Show this in terms of the equation given for pillar stress. If a factor of safety of 3 is required and the pillars have a compressive strength of 5000 psi (for good coal), develop in graph form the relationship between depth and allowable extraction ratio (or fraction of material removed). Show the equation for depth versus extraction ratio and plot it on the drawing.

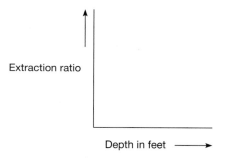

6. How is sinkhole terrain (karst topography) recognized in the field or on a topographic map? Why do some sinkholes contain water whereas others are free draining? What problems might develop if sinkholes are simply filled with earth and buildings placed on the leveled land surfaces? How might the fill be properly constructed to prevent such problems?

7. Stability against toppling of rock blocks is maintained when $b/h > \tan\theta$ (see Figure 14.5). For a bed dipping at 30° how closely spaced can those joints perpendicular to bedding be and still maintain stability if the beds are 1 ft apart?

8. The term *landslide* by common usage typically includes slope failures by fall or topple, flow and lateral spreading, and slide or slip. How did this usage develop? Why is it important to distinguish between flow failures and slip failures?

9. Why is it important that the potential failure surface of a slump be intercepted during subsurface exploration? Describe two geological settings in which a deep-seated slump could occur. A labeled cross section should prove helpful in this regard.

10. Why is the continuity of fractures in a rock mass related to the spacing of these weakness planes relative to slope stability? How are surface irregularities of bedding planes related to slope stability? Explain.

11. Refer to Figure 14.11 in the text. Note the point where the upper solid line changes slope.
    (a) Calculate the σ_n value for this point given the following information for a rock mass of quartzite: $C_r = 6000$ psi (42,000 kPa), $\phi_r = 40°$, $\phi_j = 32°$, and $i = 15°$.
    (b) Having calculated σ_n, assuming this to be a vertical stress at depth in a rock mass, calculate the depth required to obtain this σ_n value given a unit weight or unit density for quartzite of 170 lb/ft³ (2720 kg/m³).

12. A block of limestone, generally similar to that illustrated in Figure 14.12, is resting on an inclined, smooth bedding plane. The block measures 8 ft long by 9 ft high by 10 ft thick (into the slope). The mass is dry. The unit weight of the rock is 160 lb/ft³ (= 2.56 g/cm³). Cohesion equals 300 lb/ft² (14.7 kPa) and $\phi = 32°$.
    (a) Calculate the factor of safety if the slope angle $\theta = 35°$.
    (b) More difficult question requiring iterative calculations: What is the maximum slope angle that the block can have to yield a factor of safety against sliding of 1.2?

13. Assume conditions in Exercise 12(a) except that the fracture behind the block is completely filled with water and free drainage occurs at the downslope end of the block (see Figure 14.12).
    (a) Calculate the factor of safety.
    (b) Assume next that the fracture is completely filled and drainage is blocked at the downslope end (see Figure 14.13). Calculate the factor of safety.

14. If those slopes that fail by flow (earthflows and lateral spreads in particular) cannot be analyzed by the limit equilibrium method, how is stability for these slopes determined?

15. How is a mudflow recognized in the field and on aerial photographs? How can an ancient mudflow be told from an active one? How is shallow subsidence related to mudflows? How can the damage caused by mudflows to residential areas be minimized?

16. What is meant by the term *liquefaction*? Why would this be of concern for earth fill dams constructed over alluvium in earthquake-prone areas? Explain.

17. If a major earthquake occurred in Evansville, Indiana, what areas in the state would be most prone to landslide failures? See Indiana map, Figure 13.3, for reference purposes. Explain. (Hint: Alluvial areas are most prone to earthquake effects.)

18. Examine Figure 14.16 on slope treatment for rock slopes. What is the purpose of berms (or benches) in a cut slope? Consider construction and maintenance primarily in your analysis.

19. What is the purpose for drilling horizontal drains back into a rock slope? Does this seem to be a difficult operation to accomplish? Explain.

20. In contour strip mining today, mining companies are required to put all stripped rock and soil back on the mining bench after coal has been removed. What slope stability problems might develop under such circumstances? If a failure on such a slope occurs, what methods could be used to explore the failed area so that a back calculation of stability could be made? What problems do you anticipate in the exploration?

21. Regarding construction of interstate highways in mountainous areas (such as I-70 west of Denver), sometimes it is more economical to make cuts and fill embankments that might fail locally rather than to prevent all such failures from occurring by more costly construction procedures. What are the economical considerations involved in this decision. Other than economics, what other factors should be considered?

22. In a room and pillar limestone mine, the pillar dimensions are 20 ft by 20 ft and the distance between the pillars is 30 ft, offset in the manner shown in the following drawing. The pillars are numbered for reference purposes.
    (a) What is the extraction ratio for this mine? (That is, the volume removed, divided by the volume of the mine for the level in question.)
    (b) If the limestone weighs 160 lb/ft^3, its compressive strength is 15,000 psi, and the mine depth is 500 ft, what is the factor of safety against crushing for the pillars? When pillars 8 and 9 are completely removed, a subsidence trough develops above the

opening, with its long dimension located between pillars 7 and 10 and its shorter dimension between 3 and 15.
    (c) Give a representative angle of break for this limestone. Give a representative angle of draw for it. Make a cross section from pillars 3 to 15 extending to the ground surface showing these reference angles.
    (d) Would the subsidence be subcritical, critical, or supercritical? Explain what this means and show calculations.
    (e) How wide would the subsidence hollow be if it reached the surface?
    (f) What is the factor of safety now against crushing for pillars 2, 3, 4, 7, 10, 14, 15, and 16? As an aid for this problem draw in the new areas of influence for the remaining pillars on the mine plan given. Are additional pillar failures likely to occur? Show calculations.

Additional Readings

COATES, D.L., Ed., 1977, *Landslides, Reviews in Engineering Geology*, Vol. III, Geological Society of America, Boulder, CO.

CORDING, E.J., Ed., 1972, *Stability of Rock Slopes, Proc. Thirteenth Symposium on Rock Mechanics*, American Society of Civil Engineers, New York.

ECKEL, E.B., Ed., 1958, *Landslides and Engineering Practice*, Highway Research Board, Special Report 29.

HOEK, E., and BRAY, J.W., 1977, *Rock Slope Engineering*, 2nd ed., Institute of Mining and Metallurgy, London.

RELLENSMANN, O., 1957, *Rock Mechanics in Regard to Static Loading Caused by Mining Excavation*. Second Symposium on Rock Mechanics, Quarterly, Colorado School of Mines, 52.

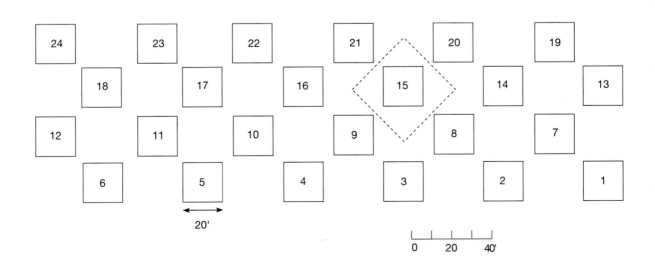

Schuster, R.L., and Krizek, R.J., Eds., 1978, *Landslides, Analysis and Control,* Transportation Research Board, Special Report 176.

Sharpe, C.F.S., 1938, *Landslides and Related Phenomena: A Study of Mass Movements of Soil and Rock,* Columbia University Press, New York.

Zaruba, Q., and Menel, V., 1969, *Landslides and Their Control,* Elsevier, New York.

Ground-Water Geology

G round water is a subject of major significance in the field of engineering geology because it is inherently valuable as a primary source of water for human use, and its presence has a pronounced effect on engineering construction and engineering works, including slope stability, surface and subsurface excavation, and foundation support.

The great quantity of water needed for human use is obtained in only two basic ways: from surface supplies (streams, lakes, reservoirs) and from ground water. More than 100 billion gallons of water per day are used in the United States at the current time and of this total about 20% is taken from ground-water supplies. Each year more than a half million ground-water wells are drilled in the United States alone. The major categories for ground-water use are rural, both domestic and farms; public water supply; industry; and irrigation.

Ground water provides a major storage reservoir for freshwater. Only 5% of the total freshwater supplies are contained in the Earth's atmosphere or on its surface. The remaining 95% occurs below the surface where it remains as a largely untapped reservoir.

The use of ground water rather than surface water for water supply purposes results in a number of advantages:

1. It is commonly free of pathogenic organisms and hence requires no purification for domestic or industrial use.
2. The temperature is nearly constant, which is important for heat exchange purposes.
3. Color and turbidity effects are usually minimal.
4. Chemical composition for a single source is essentially constant.
5. Ground-water reservoirs are generally larger than those for surface water and therefore are not affected by droughts of short duration.
6. Biological and radiological contamination of ground water is more difficult and less likely.
7. Because ground water has accumulated in the Earth during long years of recharge, it is sometimes found in areas that do not have significant surface supplies today.

The disadvantages for ground-water development, which apply to only some locations, may negate its use.

1. It may not be available in sufficient quantities to match anticipated needs, because some soils and rocks have too low a permeability to transmit much water to a well.
2. In most cases ground water has more dissolved solids present (for example, hardness) than does surface water in the same region.
3. The cost of developing wells, particularly in humid areas, may likely be greater than that of impounding reservoirs on small streams for surface supplies.
4. Ground-water contamination by industrial chemicals has become prevalent in some urban areas.

ENGINEERING CONSIDERATIONS

Ground water is related to many facets of civil engineering and construction. As previously mentioned, water supply is a major area of interest. Many of the smaller cities and towns in both the eastern and western United States rely on ground water as their only water source. Water treatment of ground water typically involves only chlorination to kill bacteria.

In addition to water supply and sanitation, ground-water considerations also involve land drainage, irrigation, seepage related to dams and levees, control of water during soil and rock excavation, foundation support, slope stability, and settlement. The need to determine ground-water conditions for a site is one of the major reasons for performing a subsurface investigation. Numerous complications can occur during construction projects when ground-water levels are misinterpreted or when that phase of the investigation is overlooked. Dewatering systems or other methods of ground-water control must be designed to provide dry conditions for construction. Settlement in the form of subsidence caused by the withdrawal of water from the subsurface must be anticipated or provisions made to minimize subsidence by maintaining the same water level by recharging the aquifer. Slope failures related to pore water pressure are a particularly prevalent problem. This subject, in addition to subsidence, is discussed in Chapter 14 on landslides.

ORIGIN OF SUBSURFACE WATER

Most of the water in the subsurface has been supplied via the atmosphere by way of the hydrologic cycle. This diagram appears as Figure 11.1 in the chapter on running water. Some water in volcanic areas may be juvenile water, that is, new water supplied from igneous fluids. The hydrologic cycle demonstrates the way in which water circulates from the oceans through the atmosphere and back to the sea by a number of different paths, either over the land surface or underground. These various paths may range from short to long, both in terms of time and distance traveled.

Most moisture in the world is evaporated directly from the oceans, an amount estimated at 80,000 mi^3

(328,000 km³) per year. Another 15,000 mi³ (61,400 km³) of water per year are evaporated from the land surface, that is, from lakes, streams, and the soil, and as a result of transpiration by plants.

Total precipitation from the atmosphere will equal the amount of evaporation supplied to it, so approximately 95,000 mi³ (389,000 km³) of water each year fall on the Earth. The continental land masses receive about 24,000 mi³ (98,300 km³) of freshwater annually. In turn, a portion of this, 15,000 mi³ (61,400 km³) of water per year, is returned once more to the atmosphere with the remainder shared by runoff and infiltration. In that way the cycle continues from year to year.

Connate Water

Therefore, most ground water is meteoric in origin. Some ground water in sedimentary formations, however, is a carryover from water present during deposition of the formation. For marine sediments, which are the most common, connate water originates as seawater. If the fluids are not flushed out by freshwater after the beds are uplifted, the water remains as a salt-rich, pore fluid, unsuitable for most uses as a water supply. In the case of oil production, considerable quantities of brine water must be disposed of on the Earth's surface in oil fields because brine water accompanies the crude oil pumped from the ground. This can be a serious environmental problem.

By definition, an aquifer consists of a formation or of strata from which ground water can be obtained for beneficial use. Formations containing salty, connate water are not considered to be aquifers because the water is not suitable for human consumption. Such formations may be potential sites for waste disposal of objectionable fluids by injection wells if other conditions of containment are met. This subject is discussed in more detail later in the chapter.

Magmatic Water

Magmatic water is another source for subsurface water because water is one of the most abundant fluids absorbed by magma moving upward toward the Earth's surface. When extrusive rocks flow onto the surface a great release of steam and other gases into the atmosphere occurs. Consequently, in volcanic areas and areas of high heat flow it is likely that some of the subsurface water has been released by the magma; this is properly termed *magmatic* water.

Juvenile Water

A point sometimes raised by geologists is whether such magmatic water is new (juvenile) water. This water could be an entirely new addition to the Earth's supply, implying that such water has not gone through the hydrologic cycle before. It has been determined in fact that most magmatic water is acquired from the rocks through which the magma intrudes so that the water is actually meteoric or connate and is not new *per se*. Many areas of hot springs and geysers such as Yellowstone Park are known to spray forth meteoric water or "recycled water" rather than juvenile water.

Water in the Oceans

Another point commonly considered in this context is the origin of water in the oceans. The oceans were formed early in the Earth's history, apparently from clouds of water vapor given off by the cooling Earth. There is no evidence to suggest that the oceans have been slowly growing in volume since the Precambrian despite the fact that volcanic eruptions have occurred continuously since that time. This suggests that very little volume of juvenile water has been contributed by volcanic eruptions since the early history of the Earth; instead, magmatic water by and large marks the return of deep-seated water to the hydrologic cycle.

SUBSURFACE DISTRIBUTION OF WATER

Not until about 1700 was it first demonstrated in a convincing way that ground water and surface water supplies were provided as a direct consequence of precipitation. Prior to this, many mistaken ideas prevailed, for example, that the sea moved underground, lost its salt by distillation or some unexplained process, and reappeared as freshwater in springs, rivers, and lakes. Even after the general public became aware of the direct relationship between precipitation and steam flow, the interdependence between ground water and surface water remained obscure. This lack of knowledge was reflected in laws concerning water supply wherein ground water and surface water were considered to be independent entities. Some of this confusion still prevails in our water law today. The legal details of ground water are considered in a later section of this chapter.

Pierre Perrault (1608–1680) demonstrated in France during the latter part of his lifetime that rainfall in a drainage basin (watershed) provided the corresponding runoff for that area. He showed that the Seine River system above Arnay-le-Duc had a runoff volume equal to only one-sixth of the precipitation it received during his three-year period of study. This showed that rainfall could easily account for stream flow, water used by plants, and water that infiltrates the ground.

Other contemporary scientists of Perrault were able to verify his findings; one was a fellow Frenchman named Mariotte (1620–1684). Mariotte suggested that if one-third of the precipitation evaporated and one-third remained in the Earth there would still be enough water left to sustain the flow of rivers. Only an extension of this statement was needed to reach the conclusion that ground water could feed streams to sustain their flow in dry weather and that streams could in turn feed the ground-water regime under other circumstances. Despite this situation, the general public did not become cognizant of the interrelationship of surface water and ground water for more than 200 years.

Subsurface Distribution

Understanding the occurrence and nature of subsurface water requires a knowledge of its vertical distribution in the Earth's crust. The outer portion of the Earth, which is somewhat porous to a fairly great depth, is known as the zone of interstitial water and may extend as deep as 6 mi (9.5 km)—at least in sedimentary rocks. In this zone the pores are either partially or completely filled with water. In its deepest portion pores are isolated from one another so that no water movement occurs. Below this thick interstitial portion the zone of combined water occurs, a zone where water is chemically attached in the form of hydrated minerals or those with hydroxl or OH groups. All fractures and open pores have been closed by the confining pressure from the overlying rock.

The subdivision of subsurface water into various zones is shown in Figure 15.1. There are no sharp boundaries between the subdivisions; instead, the changes are more gradual. Soil water, for example, is distinguished from the water directly below it in the intermediate vadose zone by its greater fluctuations in quantity in response to evaporation and transpiration.

In a forest environment, trees may have roots that penetrate to depths of 30 ft (9 m) or more. Also, moisture levels retained in the upper few inches near the ground surface are affected by the temperature, pressure, and movement of the air.

Vadose Zone

The vadose zone, sometimes referred to as the zone of aeration, is divided into three portions: 1) soil moisture, 2) intermediate vadose, and 3) the capillary fringe. In this third portion, capillary-rise prevails yielding saturation at its greatest depths. Here the term *aeration* (pores containing some air) does not apply, so that *zone of aeration* is not an appropriate term for the total zone. Vadose or suspended water provides a more appropriate term for this zone.

The intermediate vadose zone commonly separates the soil water from the saturated zone below. This intermediate portion may be 1000 ft (300 m) thick or greater in arid regions, or absent in some humid environments. Water is moving downward under the influence of gravity in the zone and has been referred to as *gravitational water*. Such terminology is not recommended because almost all water movement is controlled or strongly influenced by gravity, not merely the vadose water.

The capillary fringe, which occurs at the base of the vadose zone, is a band of soil directly above the water table where capillary size openings lift water upward because of surface tension. The boundary between the intermediate vadose and the capillary zones is abrupt in coarse-grained, cohesionless soils (sands and gravels) but is quite gradational in silts and clays. The upper surface of the capillary fringe is irregular and it changes depending on the amount of recharge. Contained in its upper part are pockets of air that slow the movement of water, but in the lower part of the fringe, saturation is just as complete as it is below the water table. This movement of water is similar to that below the water table. The capillary fringe in silt and clay materials is sometimes as great as 8 ft (2.4 m) thick, whereas in coarse sand or gravel it may extend upward only a fraction of an inch (about 1 cm) above the water table.

Water Table

The phreatic zone is separated from the capillary fringe by the water table or phreatic surface. Theoretically, this is the surface depicted by the elevation

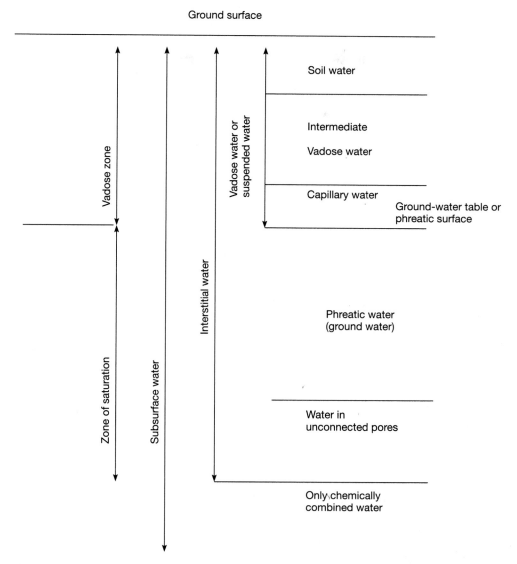

FIGURE **15.1**   Subdivisions of the Earth's crust based on the nature of subsurface water.

of water in wells that penetrate a short distance into the saturated zone. If ground-water flow is essentially horizontal, then such water levels in wells will correspond quite closely to the water table. The shape of the water table is controlled partly by the topography of the land. It tends to follow, in a general way, the outline of the land surface so the shape of the water table is commonly considered to be a subdued replica of the Earth's surface.

Common definitions for the water table indicate that it is the upper level of the zone of saturation or it is the surface of separation between the zone of saturation and the capillary fringe. A more precise definition states that the water table is the surface in an unconfined aquifer along which the hydrostatic pressure (pressure in the water or pore pressure) is equal to the atmospheric pressure (or zero gauge pressure). Figure 15.2 illustrates this relationship. In

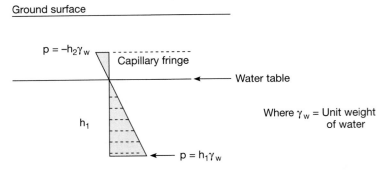

FIGURE 15.2 Hydrostatic pressure relative to the water table.

the capillary fringe the water is in tension rather than in compression. (This condition is known as capillary tension.)

Figure 15.2 also indicates the positive pore water pressure, which increases with depth below the water table, and the negative pore water pressure, which increases upward into the capillary fringe.

Zone of Phreatic Water

In specific geological terms, water below the water table is designated as ground water and the zone below the water table is called the zone of saturation. Both terms lead to confusion because in nonscientific terms ground water suggests any water below the ground surface and the zone of saturation should include all saturated materials. For this reason, subsurface water is a better general term for all water below the land surface. In addition, the lower part of the capillary fringe is saturated and water there travels at much the same velocity as water just below the water table. Therefore, a substitute for the term *zone of saturation* is needed, with a more acceptable one being *zone of phreatic water*. This is defined as water that will flow freely into a well, thereby indicating the level of the water table. Water in the capillary fringe will not drain freely into a well and will not contribute to the static water level. A pumping well with its intake below the water table, however, will pull water from the capillary fringe as it draws down the water table.

CONFINED AND UNCONFINED WATER

Water in direct vertical contact with the atmosphere by way of interconnecting pores in a permeable material is called unconfined water. The water table marks the phreatic surface or top of a saturated zone in an unconfined aquifer. There is zero fluid pressure in the water at the water table, as shown in Figure 15.2.

In many areas the first unconfined water zone encountered in the subsurface lies above the general or regional zone of phreatic water and thus indicates a more or less isolated body of water. Its position is dictated by permeability differences related mostly to the stratigraphy of the deposit, yielding a permeable zone lying above a small-sized impermeable one. The upper surface of this saturated zone is called a perched water table (Figure 15.3).

Confined water is separated from the atmosphere by an impermeable layer, which allows for a buildup of pressure in excess of atmospheric at the top of the saturated zone. Such water when intercepted by a well will rise to a level higher than that at which it was intercepted, that is, higher than the top of the saturated aquifer. This is referred to as an artesian system and it also is shown in Figure 15.3.

The surface to which the artesian wells will rise is called the piezometric surface, and the rise in feet above the point of interception (top of the aquifer) is the artesian head. If this head is sufficiently great that the water flows out onto the Earth's surface this yields what is known as a flowing, artesian well. Apparently the term *artesian well* in the early days of study referred specifically to a flowing artesian well, but in modern-day usage *artesian* is synonymous with confined water, and flowing artesian wells are those special ones in which the water reaches the surface.

The conversion of water pressure in the confined aquifer to feet of head in the well or standpipe is related to Bernoulli's equation, which states:

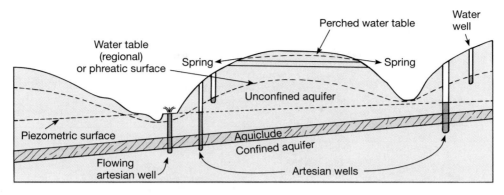

FIGURE 15.3 Pressure associated with subsurface water.

Total pressure = Pressure head + Velocity head + Elevation head

or

$$H_t = \frac{P}{\gamma} + \frac{V^2}{2g} + H$$

where

H_t = total pressure in feet
P = water pressure
γ = unit weight of water
V = velocity of water
g = gravity
H = elevation head.

Because the velocity of subsurface water is very low unless the water moves through open conduits in caverns or lava tubes, the velocity head is assumed equal to zero. When the confined aquifer is intercepted, the pressure head P/γ is converted to elevation head and the water rises in the well. For example, if a confined aquifer has a pressure head of 20 psi and it is intercepted by a well,

$$\frac{20(144)\ \text{lb/ft}^2}{62.4\ \text{lb/ft}^3} = 46.2\ \text{feet of rise will occur}$$

above the aquifer boundary. This yields an artesian well. If the Earth's surface is reached within the 46.2 ft, a flowing artesian well results.

The material overlying an aquifer may be semipermeable, allowing some water movement upward under high pressures. Under these conditions high pressures are not sustained and water will rarely rise more than 10 ft or so above the aquifer. Most artesian water in recent alluvial deposits would be semiconfined.

A number of other geologic conditions can lead to confined aquifers: alternating beds of dipping sandstones and shales; fault zones in crystalline igneous rocks; folded beds with alternating layers, recharged along joint fractures; horizontal sedimentary rocks involving unconformable relationships; glacial deposits consisting of till and outwash; and stabilized sand dunes displaying impervious deposits in the interdune area. These geologic conditions are illustrated in Figure 15.4.

POROSITY AND RELATED PROPERTIES

Porosity is a measure of the total void space in a rock and it equals the void volume divided by the total volume. The porosity of a soil or rock depends on 1) the shape and arrangement of the particles, 2) gradation or range of grain size, 3) the degree of compaction and cementation, and 4) the portion, if any, of soluble rocks removed by solution. A list of porosities for common soils and rocks is presented in Table 15.1. These are given as ranges of values and include most geologic situations.

The terms related to porosity, such as degree of saturation, void ratio, and water content, can best be defined by means of a comparison through the use of a solid–water–air diagram (Figure 15.5). This information is presented in Chapter 7, Elements of Soil Mechanics, but is repeated in part in this chapter

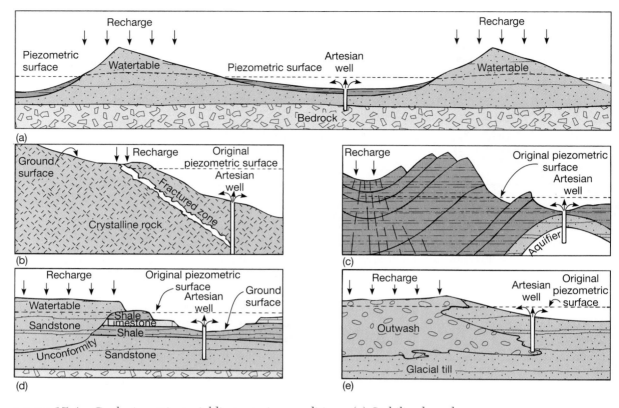

FIGURE 15.4   Geologic settings yielding artesian conditions. (a) Stabilized sand
dunes, (b) crystalline rock, (c) complexly folded and fractured sedimentary rocks,
(d) horizontal sedimentary rocks, and (e) glacial deposits.

because of its significance to the subject of ground-water geology.

A soil or rock can be considered in terms of the three components or phases that comprise it: solid, water, and air. If each phase were isolated, the three would yield the diagram shown in Figure 15.5. On the mass side, M_s and M_w designate the mass of the solid and of the water, respectively. Because M_a, the mass of the air, is so small compared to the mass of water and solid, M_a is considered to be zero. Therefore, the total mass, $M_t = M_w + M_s$.

On the volume side, V_v, the volume of the voids, equals the volume of the air plus the volume of the water. The total volume V_t equals $V_s + V_v$ or $V_s + V_a + V_w$ as indicated.

Using the volumes and masses as designated in Figure 15.5, the following relationships are developed; weight instead of mass is used in the British engineering system.

Porosity, $n = \dfrac{V_v}{V_t} \times 100$ on the % basis

void ratio $e = V_v/V_s$, as a fraction

w = water content or moisture content,

$= M_w/M_s \times 100$, as a percent;

ρ = mass density = M_t/V_t;

ρ_d = dry density = M_s/V_t

Using weights instead of mass

Unit Weight, $\gamma = \dfrac{W_t}{V_t}$

Dry Unit Weight, $\gamma_d = \dfrac{W_s}{V_t}$

$W = W_w/W_s \times 100$

TABLE **15.1** Porosities of Soil and Rock

Material	Maximum Porosity (%)
Soil, loam	Up to 60
Chalk	Up to 50
Sand and gravel	25 to 30
Sandstone	10 to 25
Limestone	5 to 30
Chalky limestone	30
Oolitic limestone	10 to 20
Compact limestone	5
Dolomite	5 to 30
Fossiliferous, reef dolomite	30
Crystalline, porous	10
Interlocking mosaic	5
Shale	2 to 25
Siliceous shale	2
Marble	5
Slate	4
Granite	1.5
Other dense rocks	0.5

$$S = \text{degree of saturation} = V_w/V_v \times 100,$$

as a percent.

This analysis can be extended by applying the relationship $M_s = V_s G_s \rho_w$ where G_s is the specific gravity of the solid and ρ_w is the density of water, $1 \text{ gm/cm}^3 = 1000 \text{ kg/m}^3$ and $W_j = V_s G_s \gamma_w$. In like manner, $M_w = V_w \rho_w$ and $W_w = V_w \gamma_w$. $\gamma_w = 62.4$ lb/ft^3.

These various relationships can be illustrated by way of an example problem using the British engineering units.

EXAMPLE PROBLEM

A soil sample in the field had a volume of 0.02 ft^3 and weighed 2.45 lb under natural moisture conditions. After oven drying, the sample weighed 2.05 lb. The specific gravity of the solids was found to be 2.67; calculate the answers to all the relationships listed.

ANSWER:

To begin, draw the solid–water–air diagram as shown in Figure 15.6.

$$V_s = \frac{W_s}{G_S \gamma_w} = \frac{2.05}{2.67\,(62.4)} = 0.0123\,ft^3$$

$$V_w = \frac{W_w}{\gamma_w} = \frac{0.40}{62.4} = 0.0064\,ft^3$$

$$n = \frac{0.02 - 0.0123}{0.02} \times 100 = 38.5\%$$

$$V_a = 0.02 - 0.0123 - 0.0064 = 0.0013$$

$$e = \frac{0.02 - 0.0123}{0.0123} = 0.626$$

$$w = \frac{0.40}{2.05}\,100 = 19.5\%$$

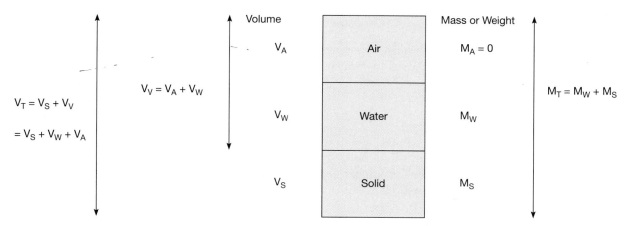

FIGURE **15.5** Solid–water–air diagram.

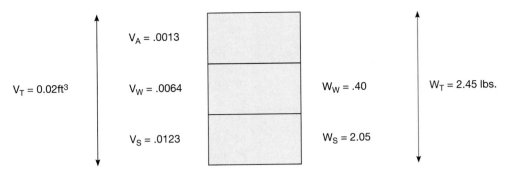

FIGURE **15.6**   Calculations using solid–water–air diagram.

$$\gamma = \frac{2.45}{0.02} = 122.5 \text{ lb/ft}^3$$

$$S = \frac{0.0013}{0.0064} \times 100 = 20.3\%$$

$$\gamma_d = \frac{2.05}{0.02} = 102.5 \text{ lb/ft}^3$$

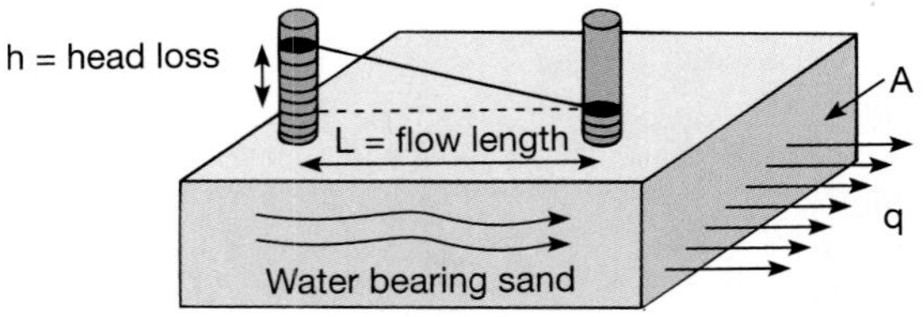

FIGURE **15.7**   Head loss and hydraulic gradient.

PERMEABILITY

The rate at which water is able to flow through soil or rock is of major significance in the fields of construction, civil engineering, and engineering geology. This flow rate is a major factor in situations such as the following: 1) the rate and volume of water supplied by pumping wells, 2) the rate of leakage through or beneath dams, 3) the ease with which soils can be dewatered, and 4) the rate of consolidation of a saturated soil under load.

Darcy's Law

The governing equation for flow through a porous medium has come to be known as Darcy's law, which is given by:

$$q = kiA$$

where
  q = rate of flow through the cross section
  k = coefficient of permeability
  i = hydraulic gradient, a dimensionless number equal to the loss of head over a given flow distance = h/L
  A = cross-sectional area of the soil through which the water flows.

This is illustrated by Figure 15.7. By substitution $v = q/A$, the equation becomes

$$v = ki$$

where v and k have the same units of L/t (or length/time) typically in centimeters per second for engineering geology and geotechnical engineering.

In a general way, at least some of the voids in a particulate system are interconnected and continuous. This holds true for most soils and as a consequence the smaller the soil particles, the smaller the water-conducting passageways. Permeability is the capacity of a material to transmit a fluid, which in the case of geotechnical engineering is usually water. The coefficient of permeability k, therefore, decreases with decreasing particle size for soils. Table 15.2 gives the effective size of D_{10}[1] of the soil. Hazen's approximation has been used to estimate the coefficient of permeability when the effective size is known. It states that $k = 100(D_{10})^2$, where D_{10} is given in centimeters and k is in cm/sec. This approximation works reasonably

---

[1] D_{10} is the size designation on a gradation curve such that 10% of the material is finer than that size.

TABLE **15.2** Soil Permeabilities and Effective Size

Material	k (cm/sec)	Effective Size, D_{10} (mm)
Uniform coarse sand	0.4	0.6
Uniform medium sand	0.1	0.3
Clean, well-graded sand and gravel	0.01	0.1
Uniform, fine sand	4×10^{-3}	0.06
Well-graded, silty sand and gravel	4×10^{-4}	0.02
Silty sand	10^{-4}	0.01
Uniform silt	5×10^{-5}	0.006
Sandy clay	5×10^{-6}	0.002
Silty clay	10^{-6}	0.0015
Clay (30% to 50% clay size)	10^{-7}	0.0008
Colloidal clay (minus 2 μm ≥ 50%)	10^{-9}	$40 \overset{\circ}{\text{A}} = 4 \times 10^{-6}$

well for the sandy gravel to sand range but is less applicable for finer grained soils.

Several assumptions are involved for Darcy's law with regard to ground-water flow: 1) the flow is laminar, 2) the temperature is constant and in the range of 60°F (15.5°C), and 3) steady-state flow under saturated conditions prevails. Laminar flow conditions are violated only in clean, coarse gravels where the actual permeability would be somewhat less than the *k* value would suggest because of the turbulent flow conditions.

Measures of Permeability

A different unit of permeability has been used in the past in the field of hydrogeology in the United States. Rather than *k* in centimeters per second, the unit in gallons per square foot per day (the Meinzer unit) was used previously. At first, the unit of gal/ft²/day may not seem compatible with cm/sec but more careful inspection will show that both have units of *L/t* or length divided by time. In hydrogeology the term *hydraulic conductivity* rather than *coefficient of permeability* is commonly used, which tends to underscore the fact that permeability relative to water is involved. The relationship between Meinzer and cm/sec units is shown for different soils and rocks in Table 15.3.

In more recent years, since the emphasis on metric units developed in scientific literature in the United States, meters per day have been employed for hydraulic conductivity in hydrogeology studies. The conversion is

$$1 \text{ cm/sec} = 864 \text{ m/day} = 21.2 \times 10^3 \text{ Meinzers}$$

The third major field involved with permeabilities of soils and rocks (in addition to geotechnical engineering and hydrogeology) is petroleum engineering. In this specialty the flow of crude oil through porous media under reasonably high pressures (high gradients) is involved. The viscosity of crude oil is not only different from that of water, it also varies from one crude to another and temperatures are typically elevated well above 60°F (15.5°C). For this reason, the direct application of hydraulic conductivity to the flow of crude oil is not appropriate. Therefore, another unit, the darcy, is used instead.

For the reasons indicated, in petroleum engineering intrinsic permeability is used rather than hydraulic conductivity. It is a measure of the medium (rock or soil) alone and not related to a specific fluid such as water or oil. A darcy is defined as the flow of 1 cm³ of fluid having a viscosity of one centipose in one second under a pressure drop of one atmosphere over a length of one cm through a porous medium of 1 cm² in cross-sectional area. Substitution into this equation and simplifying common terms shows that a darcy has the units of length squared or L^2. Specifically,

$$1 \text{ darcy} = 0.987 \times 10^{-8} \text{ cm}^2 = 1.062 \times 10^{-11} \text{ ft}^2$$

Because the density and viscosity of water are functions of temperature, one can convert from darcies to Meinzers, to centimeters per second, or to meters per day. Water at 60°F (15.5°C) is typically the conversion

TABLE 15.3   Comparable Values of Effective and Intrinsic Permeabilities for Common Rocks and Soils

Material	k cm/sec	k(gal/ft^2/day or Meinzers)	K (darcies)
1. Ranges of Values Gravel	$1 - 10^2$	$10^4 - 10^6$	$10^3 - 10^5$
Clean sands (good aquifers)	$10^{-3} - 1$	$10 - 10^4$	$1 - 10^3$
Clayey sands, fine sands (poor aquifers)	$10^{-6} - 10^{-3}$	$10^{-2} - 10$	$10^{-3} - 1$
2. Specific Values Argillaceous limestone, 2% porosity	8.6×10^{-8}	1.8×10^{-3}	10^{-4}
Limestone, 16% porosity	1.2×10^{-4}	2.5	1.4×10^{-1}
Sandstone, silty, 12% porosity	2.23×10^{-6}	4.74×10^{-2}	2.6×10^{-3}
Sandstone, coarse, 12% porosity	9.4×10^{-4}	19.9	1.1
Sandstone, 29% porosity	2.1×10^{-3}	43.6	2.4
Very fine sand, very well sorted	8.4×10^{-3}	1.8×10^2	9.9
Medium sand, very well sorted	2.23×10^{-1}	4.7×10^3	2.6×10^2
Coarse sand, very well sorted	3.69×10^1	7.83×10^5	4.3×10^4
Montmorillonite clay	$\simeq 10^{-8}$	$\simeq 10^{-4}$	$\simeq 10^{-5}$
Kaolinite clay	$\simeq 10^{-6}$	$\simeq 1.0^{-2}$	$\simeq 10^{-3}$

3. Equivalencies

1 darcy = 18.2 Meinzer units for water @ 60°F, or 8.58×10^{-4} cm/sec for water @ 60°F; 1 Meinzer = 0.134 ft/day = 4.72×10^{-5} cm/sec = 5.49×10^{-2} darcies, water @ 60°F; 1 cm/sec = 1.165×10^3 darcies for water @ 60°F = 21.2×10^3 Meinzers; 10^{-6} cm/sec = 1.165 millidarcies for water @ 60°F; 1 cm/sec = 1.03 10^6 ft/yr = 864 m/day; 1 millidarcy = 0.001 darcy = 0.858×10^{-6} cm/sec for water @ 60°F.

point used. Table 15.3 provides equivalent values of permeability for common soils and rocks.

SPRINGS

Water above the water table moves downward in permeable material toward the saturated zone. If its downward movement is restricted by layers of low permeability, water will flow down gradient along the interface. Eventually this contact between the permeable layer above and the impermeable layer below will intersect a valley wall or other vertical exposure, and water will exit from the slope. If there is enough volume to show noticeable water movement, such supplies are called springs. Less discharge than this yields surface seepage or, simply, a seep.

The geologic conditions yielding permeable layers that overlie impermeable ones are quite common so that specific, different types are numerous. Sequences of 1) sandstone above shale, 2) sand above silt or clay, and 3) broken (talus) or weathered rock lying above massive bedrock are examples of stratigraphic sequences leading to spring development.

Concentrations of fractures in fault zones can provide the permeable portion, with the surrounding rock being its impermeable counterpart.

Spring water has the reputation, which is perpetuated by various advertisements, of being pure or therapeutic or of having some other attribute not available to surface water or well water supplies. Facts do not substantiate those claims, but spring water typically does have more dissolved solids and consequently a stronger taste. It also can be easily polluted. Pollution occurs because springs flow near the Earth's surface and the flow is in open conduits in the ground without the benefit of the filtering characteristics of flow through soils.

Springs can be classified according to origin or according to size, with size based on the amount of discharge. First-magnitude springs, those with the greatest flow, discharge more than 100 ft^3/sec (2.83 m^3/sec).

Second-magnitude springs have a discharge ranging from 10 to 100 ft^3/sec (0.28 to 2.83 m^3/sec). These designations extend down to eight-magnitude springs at less than 1 pint/min (7.9 ml/sec) (see Table 15.4).

TABLE 15.4 Meinzer's Classification of Springs According to Discharge

Magnitude	British Units	Metric Units
1st	>100 ft³/sec	>2.83 m³/sec
2nd	10 to 100	0.283 to 2.83
3rd	1 to 10	28.3 to 283 l/sec
4th	100 gal/min to 1 ft³/sec	6.31 to 28.3
5th	10 to 100	0.631 to 6.31
6th	1 to 10	63.1 to 631 ml/sec
7th	1 pint/min to 1 gal/min	7.9 to 63.1
8th	<1 pint/min	<7.9 ml/sec

Big Springs, Missouri, located in the Missouri Ozarks about 100 mi southwest of St. Louis, is a first-magnitude spring with an average discharge of 380 ft³/sec (250 mgd or 10.8 m³/sec). A first-magnitude spring is illustrated in Figure 15.8.

At best, only a few hundred first-magnitude springs exist throughout the world. This fact is a

FIGURE 15.8 Big Springs, located in the Missouri Ozarks 100 miles south of St. Louis, has a discharge of about 175,000 gallons per minute. (Pierre Studio, Sullivan, MO.)

result of the rare combination needed to produce such high-volume flows: large amounts of infiltration, a large drainage area, and favorable geologic conditions to localize the discharge. Almost all first-magnitude springs are located in lava, limestone, boulder, or gravel aquifers with flow from cavern-like openings. The common aquifers of sandstone, conglomerate, and sand generally lack the extremely high permeability and hence the enormous discharge to exhibit first- or even second-magnitude springs.

Small springs may occur in all types of soil and rock. Loess, dolomite, graywacke, gypsum, and serpentine may contain springs of seventh or eighth order. Shales may contain the smallest of springs, issuing from joints or small layers of more silty or sandy rock.

GROUND-WATER MOVEMENT

Ground water (or the water table) can either provide water to surface streams or derive water from them. A gaining stream is one that receives water from the water table; such streams tend to be permanent, that is, they flow year round. Typically, they occupy well-entrenched valleys located in well-developed floodplains. A losing stream is one that supplies water to the water table; quite commonly they dry up when the rainy season is past. They are usually small, fairly steep tributaries to the main stream. Older terminology, no longer preferred by most geologists or geological organizations, includes the terms *effluent stream* for a gaining stream and *influent stream* for a losing one. Older ground-water literature makes extensive use of these terms.

The term *aquifer* was introduced previously to describe a unit that can yield significant quantities of usable water. A rock that neither transmits nor stores water is called an *aquifuge* and one that stores water but does not transmit significant quantities of it is called an *aquiclude*. The term *aquitard* is sometimes used to mean a rock unit that transmits enough water to have a regional effect on flow but not enough to supply a well.

The level of the water table is of special interest to the engineering geologist because it shows the elevation of saturation of soil or rock. When the elevation of the land surface is known, the depth to the water table is easily determined. Maps depicting the water table are used to estimate the direction of flow. Movement occurs at right angles to the water table contours in the downslope direction.

Figure 15.9 is a water table map showing both surface contours and water table contours. Observe that the stream flows southwestward with a tributary stream entering from the north. The water table intercepts the main stream, showing it as a gaining stream, whereas the tributary stream loses water to the water table. The pond also intercepts the water table in part and obtains some water from it. The pumping well, by contrast, has depressed the water table surface (the cone of depression is outlined by the water table contours).

The piezometric surface of a confined aquifer can also be depicted on a topographic map. The contours of the piezometric surface show the levels to which the water would rise, not the depths at which it would be intercepted. Piezometric surfaces exist for each confined aquifer, and sometimes misleading data are obtained when two aquifers are intercepted by the same well, which experiences pressure effects from both. Well construction needs to take into account the contribution of multiple layers of artesian head in the

subsurface. Maps of the piezometric surface can be used to estimate where pump intakes should be placed in an artesian well and how much lift will be required to bring the water to the surface.

The velocity of ground-water movement is determined based on Darcy's law using the relationship $v = ki$, where v is the average velocity through the cross section and k and i are the coefficient of permeability and hydraulic gradient, respectively. The velocity of flow along the path of seepage, V_s, is appropriately called the seepage velocity, where $V_s = V/n$ with n equal to the porosity represented as a fraction. When the seepage velocity is known, the distance or time of travel through the subsurface can be determined. Quantities of flow can also be calculated. This is best illustrated by an example problem.

EXAMPLE PROBLEM

Two ponds of water are 1000 ft apart with the water surface in one pond being 150 ft higher than in the other. Both ponds have a square shape, measuring 200 ft on a side. A 2-ft-thick sandy layer connects the ponds and the sand has the permeability $k = 5 \times 10^{-2}$ cm/sec with a porosity of

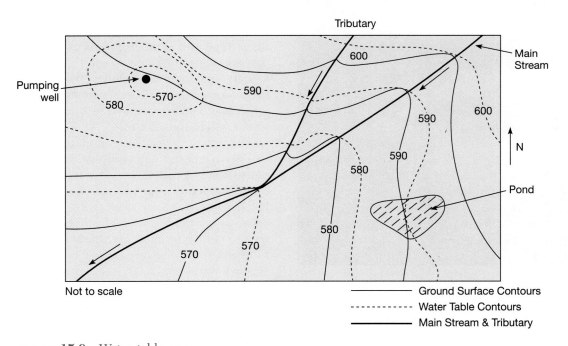

FIGURE **15.9** Water table map.

25%. How long does it take for water to travel from the upper pond to the lower one? How much water seeps from the upper pond into the lower one on a yearly basis?

ANSWER

$$v_{avg} = ki = 5 \times 10^{-2} \text{ cm/sec} \times i = 5 \times 10^{-2} \text{ cm/sec}$$

$$\times \frac{150}{1000} = 7.5 \times 10^{-3} \text{ cm/sec}$$

Knowing 0.134 ft/day = 4.72×10^{-5} cm/sec (conversion equivalent), then

$$v_{avg} = \frac{7.5 \times 10^{-3}}{4.72 \times 10^{-5}} (0.134 \text{ ft/day}) = 21.29 \text{ ft/day}$$

$$v_s = \frac{v_{avg}}{n} = \frac{21.29}{0.25} = 85.17 \text{ ft/day}$$

$$t = \text{Distance}/v_s = 1000 \text{ ft}/85.17 \text{ ft/day}$$

$$= 11.74 \text{ days}$$

$$Q = kiAt, ki = 21.29 \text{ ft/day},$$

$$A = 200 \text{ ft} \times 2 \text{ ft} = 400 \text{ ft}^2,$$

$$t = 365 \text{ days}$$

$$Q = 21.29 (400) (365) = 3,108,340 \text{ ft}^3 \text{ of water}$$

$$= 3,108,340 \times 7.48 \text{ gal/ft}^3 = 23.2 \times 10^6 \text{ gal/yr}$$

PRODUCTION OF GROUND WATER

Two criteria must be met to obtain a usable water supply from the subsurface: 1) The rock or soil must be saturated and 2) it must have sufficient permeability to deliver water to the well. If both conditions are not met simultaneously, a dry well will result. A well in dry, coarse gravel or one in saturated clay will produce no flow of water.

In most discussions, two types of aquifers are considered, those in soil materials and those in bedrock. Soil or unconsolidated materials (geological definition) offer sites for wells that will have the greatest rate of production. These are considered first.

A related question is how much water production is needed for a ground-water source to be called an aquifer. The lower limit of production is about 1 gal/

min to be termed an aquifer but in areas of very low production, water-bearing zones producing 100^+ gal/day (0.07 gal/min) have been developed to provide a domestic water supply. Typically, a minimum of 50 to 100 gal/person/day is required for domestic water use.

The distinction between an aquifer and a water-bearing zone has recently taken on added significance with regard to ground-water contamination. For aquifers it is typically required that they be cleaned up to background ground-water parameters; the requirement is less critical for water-bearing zones that cannot support well production.

Production from Unconsolidated Materials

Soils that produce meaningful amounts of ground water for supply purposes must be coarse grained. Gravel, sandy gravel, gravelly sand, and coarse sand are the usual producers. Fine sand may be too fine in size to be effectively screened out during pumping, which can lead to problems with the pump and the well because of the sand removal required. In any case, the search for ground water in unconsolidated deposits is really a quest for sand and gravel. An outline of the geomorphic (landform) features that consist of coarse, cohesionless material is presented in Table 15.5.

Landforms Composed of Sand and Gravel

Stream-related features provide the major supply of ground water. These include items 1, 2, and 3 in Table 15.5 and generally 4 and 5 as well. Outwash plains and valley train deposits occur in conjunction with stream deposits; they are formed during meltwater flow from the glacial ice. So despite their glacier-related origin, outwash plains and valley trains are associated with major-stream development, at least for many of the major deposits related to continental glaciation.

Outwash plains of a smaller areal extent may form during the local melting episodes of a continental glacier. Such plains may exist at the current surface of the Earth in glaciated terrain or they may have been buried by readvancing and subsequent retreating ice to provide a sand and gravel layer at depth within a deposit of till. Small valley trains are formed under similar circumstances and today can be found as channels of permeable material within the mass of till.

TABLE 15.5 Ground-Water Aquifers,
Unconsolidated Materials by Origin

Stream-Related Features
    1. Floodplains
    2. River terraces
    3. Alluvial fans
Glacial (Glacial-Fluvial) Features
    Most reliable
    4. Outwash plains
    5. Valley train deposits
    6. Kame terraces
    Less reliable
    7. Eskers and kames
    8. Selected portions of end moraines
    9. Beach ridges
Coastal Features
    10. Coastal plains
    11. Sea terraces

Alluvial fans are discussed in Chapter 11. They form along an intermittent stream course where a sudden reduction in gradient causes the deposition of coarse-grained sediments to occur. Alluvial fans most commonly occur at the base of mountains or of sufficiently high hills. They will be covered by successive deposition during the retreat of the mountain front due to continued erosion. Water supplies are obtained from wells drilled in the basins associated with retreating mountain fronts.

Kame terraces form during glaciation of steep valleys by alpine glaciers or by continental glaciers that partially fill such valleys. The permeable layers are formed as terraces on opposite sides of the valley by meltwater flowing at the edge of the glacier along the valley wall.

Eskers and kames provide isolated deposits of sand and gravel and local supplies of water. Perched-water aquifers may be the consequence, in which case the associated reservoir of stored water is severely limited in volume. Such water supplies run short during dry summer periods and periods of drought.

End moraines are dotted with glacial deposits of sorted, coarse-grained material. These features are usually easily identified in the field or from aerial photographs. They too tend to provide limited supplies of water because of their isolated nature.

Beach ridges occur in major lacustrine deposits where they mark the shoreline of the retreating lake water. They are linear features that stand above the terrain and typically contain medium to fine sand. Some can supply limited amounts of water.

Coastal plains commonly form areas of wide extent, which mark the flat ocean bottom after retreat or off-lap of the ocean has occurred. They are prevalent along the Atlantic Ocean and Gulf of Mexico in the United States where uplift of the landsurface during the late Cenozoic age exposed large land areas. This landform slopes gently toward the sea, and major supplies of water are contained within its sandy layers. Excessive drawdown from overpumping and drainage is a serious problem and is considered in a later section of this chapter under the discussion on water supply problems.

Sea terraces are found overlooking the ocean where fairly recent uplift has elevated the beach area some distance above sea level. Such terraces are prevalent along the California coast where active uplift of the Pacific Coast Range continues today. Freshwater is found at depth but typically at an elevation only a few feet above sea level.

Production from Bedrock

Ground water can also be obtained from bedrock but typically in lesser quantities than from unconsolidated materials. This is because of the lower permeability of bedrock (see Table 15.3). The permeability of bedrock can be primary, meaning that it is a consequence of the characteristics of the rock dating back to its formation, or it can be secondary, that is, due to secondary permeability induced by an occurrence subsequent to formation. Table 15.6 provides information on the permeability relationships of bedrock.

Primary Permeability

The best bedrock aquifers are those with primary permeability. Sandstones, particularly the poorly cemented, friable ones, provide the most water in the sedimentary rock areas east of the Rocky Mountains. Reef limestones and dolomites are also good aquifers and occur frequently in the central United States. Sandstones and carbonate rocks also comprise the oil producers in the oil fields of the world.

For basaltic and andesitic rocks, the flow tops of the lavas provide the major supplies of ground water.

TABLE 15.6 Bedrock Material and Conditions Yielding Adequate Water Supplies

Primary Permeability
  Sandstone
  Reef limestones and reef dolomites
  Flow tops of lavas
Secondary Permeability
  Jointing in massive rocks; includes limestone, dolomite, sandstone, intrusive igneous rocks, extrusive lava-type rocks, gneiss, marble, quartzite, massive slates and schist, plus other similarly massive units
  Fault zones and deeply weathered zones
  Caverns and solution channels in limestone, dolomite, and gypsum

There are only a few areas of the world where these rocks predominate, one being the Columbia Plateau in the northwestern United States. In this area, which is semiarid in nature, the withdrawal of water for domestic supplies and irrigation far exceeds the recharge of water into the aquifer. To this extent the removal of water is much like mining a valuable mineral from the subsurface.

Secondary Permeability

Secondary permeability tends to be sporadic in nature and therefore more difficult to locate in the subsurface and less reliable as a water supply. Low volumes of flow are also associated with such wells. Typical supplies are only a few gallons per minute or as low as a few hundred gallons per day. Intersecting fractures at depth typically supply the low-volume yields so that the capacity of the reservoir is severely limited. Extended dry periods will also cause such wells to run dry or to be seriously curtailed. Pollution of these wells is a major concern because the few localized joints and fractures tend to be the collection points for septic tank effluent and infiltration of runoff from animal waste areas, as well as the location of ground-water recharge.

WATER WELL TERMINOLOGY

The permeability of a saturated zone is not the only determinant for the quantity of ground-water supply. A small amount of reflection on the subject will suggest that a 50-ft-thick layer of saturated coarse sand should be able to deliver to a well more water than a similar layer only 10 feet thick.

The coefficient of transmissibility T, or just simply transmissibility, is a measure of this ability to deliver water to a well. It is the rate at which water will flow through a vertical strip of the aquifer, 1 ft wide extending through the full saturated thickness under a hydraulic gradient of 1. Written in equation form:

$$T = kHi = kH$$

where i, the hydraulic gradient, equals 1 and k is given in gal/ft²/day (Meinzers) or ft/day and H in feet. (Also k in meters per day and H in meters.)

The transmissibility value for an aquifer will indicate its water supply capabilities. Values of T range from 1000 to 1,000,000 gpd/ft (gallons per day per foot), or 135 to 135,000 ft²/day, or 40.8 to 40,800 m²/day. Transmissibility of less than 1,000 gpd/ft, (<135 ft²/day or 40.8 m²/day) is good for domestic supplies only, whereas transmissibility of 10,000 gpd/ft or more, (>1350 ft²/day or >408 m²/day) is adequate for industrial, municipal, or irrigation purposes. This can be best illustrated with an example problem.

EXAMPLE PROBLEM

A sandy unconfined aquifer is 40 ft thick with the water table located 8 ft below the ground surface. The coefficient of permeability is $k = 4.3 \times 10^{-2}$ cm/sec. What is the transmissibility of the aquifer and is this a good candidate for an industrial water supply?

ANSWER

$T = kiH,$ with $i = 1$

$k = 4.3 \times 10^{-2}$ cm/sec $= 4.3 \times 10^{-4}$ m/sec

$\quad = 3.71$ m/day $= 1.41 \times 10^{-4}$ ft/sec

$\quad = 12.18$ ft/day

$\quad = 91.1$ gal/ft²/day

$H =$ Saturated thickness $= 40 - 8$

$\quad = 32$ feet $= 9.76$ m

$T = 91.1 \ (32 \ \text{ft}) = 2915.2$ gal/day/ft

$\quad = 12.18$ ft/day $(32) = 389.8$ ft²/day

$\quad = 3.71$ m/day $(9.76 \ \text{m}) = 36.2$ m²/day

T lies between 1000 and 10,000 gal/ft/day.

Therefore, the acquifer is good for a domestic water supply, but marginal for an industrial water supply.

When pumping starts in a well, the water level nearby in the aquifer is lowered. The amount the water surface drops is called *drawdown* and the greatest amount will occur immediately near the well. Drawdown will be less at greater distances from the well until, at some distance, a point occurs where the water level is essentially unchanged. The slope of the unwatered zone increases toward the well and as the water flows inward from all directions, a cone of depression forms around the pumping well (Figure 15.10).

As pumping continues, the cone of depression continues to move outward from the well. At some point in time the cone of depression stops expanding, although the pumping rate holds constant. When this occurs, the condition of equilibrium has been reached and no further drawdown develops as pumping continues. In some wells, equilibrium develops within a

few hours after pumping is initiated. In others, however, it does not occur even after an extended period of pumping, continuing for a year or more. The distance from the well to the outer circle of zero drawdown is known as the radius of influence. In a water table aquifer the radius of influence is related to the transmissibility. Assuming equal pumping rates, a well with very high transmissibility (100,000 gpd/ft) will have a small drawdown and a small radius of influence compared to a well with moderate transmissibility (10,000 gpd/ft), which will show a considerably greater drawdown and also a much greater radius of influence. Hence, a highly permeable material has a steeper drawdown curve and smaller drawdown than does a lower permeability material. In addition, wells in artesian aquifers have a much greater radius of influence (in the range of 5000 ft or about 1500 m) than do water table aquifers (400 ft or 120 m extent more common).

Another useful term regarding ground-water supply is the *specific yield*. This is the quantity of water that a unit volume of the aquifer will give up when drained by gravity. It is expressed as a portion of the porosity. The remainder of the water, that portion not

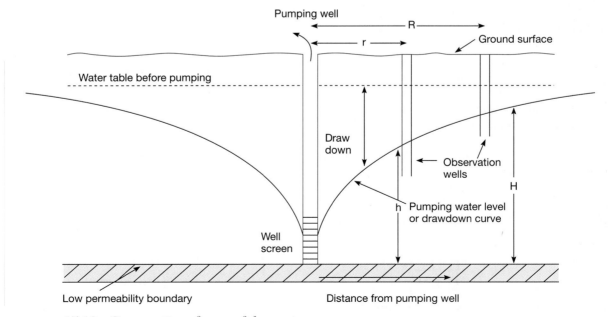

FIGURE 15.10 Cross section of cone of depression.

given up under gravity drainage, is called the *specific retention*. The specific yield plus the specific retention is equal to the porosity. For example, if 0.15 ft^3 of water is drained from 1 ft^3 of saturated coarse sand, the specific yield of the sand is 0.15 or 15%. If the porosity of the sand is 35%, its specific retention is 0.20 or 20%.

The coefficient of storage S of an aquifer is the volume of water released from storage (or taken into storage) per unit of surface area of the aquifer per unit change in head. This is simply the volume of water removed per unit volume of material unwatered during pumping. In water table aquifers S is the same as specific yield of material in regard to pumping. In artesian aquifers, S is somewhat more complicated. It is the result of two elastic effects, compression of the aquifer and expansion of the contained water when the pressure (or head) is reduced during pumping. As with porosity and specific yield, S is a dimensionless term. For water table aquifers S ranges from 0.01 to 0.35 and for artesian aquifers the range is from 1×10^{-5} to 1×10^{-3}. The coefficient of storage is used along with the transmissibility to determine the permeabilities of aquifers that have not reached steady-state conditions. This aspect of ground-water supply is beyond the scope of this text. Pumping wells under steady-state conditions are discussed in the next section.

MEASUREMENTS OF PERMEABILITY

The permeability of soils can be measured or, perhaps more appropriately, "estimated" by means of several different testing procedures. One such estimate, mentioned previously, is Hazen's approximation, which is used for clean sands and gravels. Hazen's approximation is $k = 100D_{10}{}^2$, with k in centimeters per second and D_{10} in millimeters. Recall that D_{10} is the effective grain size and represents the diameter at which 10% by weight of the grain size distribution is smaller than the diameter. Hazen's approximation is an empirical relationship because the units on opposite sides of the equation are not the same.

Some test procedures are conducted on laboratory samples, whereas other tests are performed in the field. Those performed in the field consist of two types: 1) the field pumping test with monitoring of

the drawdown in observation wells nearby and 2) in-hole pump tests of several different varieties but without observation wells. These are considered later in our discussion.

Table 15.7 contains information on the various types of soil permeability tests. Permeabilities are shown in terms of centimeters per second presented as a log scale. Five specific tests are included in the table and can be grouped collectively as field tests on *in situ* material or laboratory tests on prepared samples. Under the laboratory category the following tests are included: constant-head permeameter, falling-head permeameter, grain size distribution (Hazen's approximation), and consolidation.

Field Pumping Test

The field pumping test with monitoring by observation wells will be considered first. This test measures preferentially the horizontal permeability in a deposit rather than the vertical. This is desirable in most cases because seepage under dams and levees is horizontal and flow to pumping wells is also in that direction.

Two conditions should be met to apply the equation for the field pumping test that we next derived. First, steady-state conditions should be established. In an attempt to accomplish this, the well is pumped for 24 hr or longer at a constant rate prior to conducting the test. Steady-state conditions are determined by measuring the drawdown values in the monitoring wells at specific time intervals, to determine when water levels stabilize. Second, the aquifer should be fully penetrated, that is, the well should extend to the bottom of the aquifer with the well screen placed in the lower one-half to one-third of the aquifer. This ensures that the flow is radial to the well, which is assumed in the analysis with which we derive the equation.

Figure 15.10 shows a field pumping test for an unconfined or water table aquifer. Two observation wells are needed in addition to the pumping well. The drawdown is measured at these observation wells. In practice, the distance that these wells are located from the pumping well is important because in permeable aquifers the drawdown may be so small that it cannot be detected as close as 100 ft away. In such cases observation wells at 10- and 25-ft distances should prove satisfactory.

TABLE **15.7** Permeability and Drainage Characteristics of Soils

Coefficient of Permeability k (cm/sec, log scale)

10^2	10^1	1.0	10^{-1}	10^{-2}	10^{-3}	10^{-4}	10^{-5}	10^{-6}	10^{-7}	10^{-8}	10^{-9}

Drainage

Good — from 10^2 to about 10^{-3}

Poor — from about 10^{-3} to about 10^{-6}

Practically Impervious — from about 10^{-6} to 10^{-9}

Soil types

Clean gravel (high k range)

Clean sands; clean sand and gravel mixtures

Very fine sands; organic and inorganic silts; mixtures of sand, silt, and clay; glacial till; stratified clay deposits; etc.

"Impervious" soils, e.g., homogeneous clays below zone of weathering

"Impervious" soils modified by effects of vegetation and weathering

Direct determination of k

Direct testing of soil in its original position—pumping tests; reliable if properly conducted; considerable experience required

Constant-head permeameter; little experience required

Falling-head permeameter; reliable; little experience required

Falling-head permeameter; unreliable; much experience required

Falling-head permeameter; fairly reliable; considerable experience necessary

Indirect determination of k

Computation from grain-size distribution. Applicable only to clean cohesionless sands and gravels

Computation based on results of consolidation tests; reliable; considerable experience required

SOURCE: After Terzaghi and Peck, 1948.

340

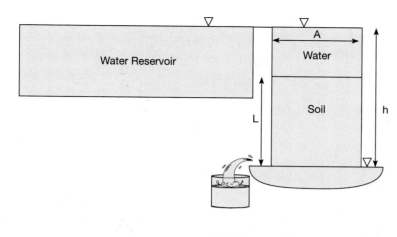

Q = total quantity of flow

A = cross sectional area of conduit

t = elapsed time

k = coefficient of permeability

i = hydraulic gradient

$$Q = kiAt \qquad \text{Darcy's law}$$

$$i = h/L$$

$$Q = \frac{khAt}{L}$$

$$k = \frac{QL}{hAt} \qquad \text{k measured from } 10^2 \text{ to } 10^{-3} \text{ cm/sec}$$

FIGURE **15.11**   Constand-head permeameter test.

The equation for the water table aquifer, with full penetration of the section is derived next where

$$q = kiA \text{ (Darcy's law)}$$
$$i = dy/dx$$
$$A = 2\pi xy$$
$$q = 2\pi k\, xy\, dy/dx.$$

Separating variables,

$$\frac{dx}{x} = \frac{2\pi k}{q} y\, dy$$

Integrating between the limits of two points on the drawdown curve (or at two observation well locations) as shown in Figure 15.10:

$$\int_r^R \frac{dx}{x} = \frac{2\pi k}{q} \int_h^H y\, dy$$

$$\ln \frac{R}{r} = \frac{\pi k}{q}(H^2 - h^2)$$

where ln is the natural logarithim

or

$$\log \frac{R}{r} = \frac{\pi k}{2.3q}(H^2 - h^2)$$

where log is log $_{10}$ or the common log

$$k = \frac{2.3q\,\log R/r}{\pi\,(H^2 - h^2)}$$

or for Meinzer units

$$P = \frac{1055q\,\log R/r}{(H^2 - h^2)}$$

where

k = ft/sec with all units consistent, that is, q in ft^3/sec, R, r, H, and h in feet

P = permeability in gpd/ft^2 with q in gpm

r = distance to nearest well (ft)

R = distance to farthest well (ft)

H = saturated thickness, farthest well (ft)

h = saturated thickness, nearest well (ft)

In a similar analysis for a confined or an artesian aquifer the equation is

$$k = \frac{2.3\,q\,\log R/r}{2\pi\,m(H - h)}$$

$$= \frac{0.366q\,\log R/r}{m(H - h)} \qquad \text{with all units consistent}$$

$$P = \frac{528q\,\log R/r}{m(H - h)}$$

except H and h = heights for the piezometric surface above the base of the aquifer during pumping, m = aquifer thickness and the same units for P in gpd/ft² as in the water table case.

Also note for the artesian case that $T = km$ or transmissibility = permeability multiplied by aquifer thickness. Therefore,

$$T = km = \frac{0.366q \ logR/r}{(H - h)} \quad \text{with all units consistent.}$$

It is possible to calculate the coefficient of permeability in a field pumping test even though steady-state conditions (or equilibrium conditions) have not been reached. This is accomplished through the application of the Theis equation developed in 1935. This concept is beyond the scope of this text but can be found in reference books on ground-water geology and ground-water seepage. A nonequilibrium well shows a change in drawdown with time rather than reaching a stable steady-state condition.

EXAMPLE PROBLEM: FIELD PUMPING TEST ___

A sandy, unconfined aquifer is 50 ft thick with hard, glaical till below it. The depth to the ground water table is 3 ft. Under steady-state conditions, the well discharge was 480 gal/min; and drawdown in a monitoring well 50 ft away from the pumping well was 1.5 ft and at a monitoring well 100 ft away was 0.2 ft. The well fully penetrated the aquifer thickness. Calculate the coefficient of permeability and the transmissibility of the aquifer.

ANSWER

$$k = \frac{2.3q \ logR/r}{\pi \ (H^2 - h^2)}$$

The saturated thickness before pumping is $50 - 3 = 47$ ft. At $r = 50$ ft the drawdown is 1.5 and, therefore, $H = 47 - 1.5 = 45.5$ ft. At $R = 100$ ft, the drawdown is 0.2 ft and, therefore, $h = 47 - 0.2 = 46.8$ ft.

$q = 480$ gal/min $= 480/7.48 \times 1/60 = 1.069$ ft³/sec

$$k = \frac{2.3(1.069) \ log\frac{100}{50}}{\pi (46.8^2 - 45.5^2)} = \frac{0.7401}{376.9} = 0.00196 \text{ ft/sec}$$

$$= 0.05984 \text{ cm/sec} = 5.98 \times 10^{-2} \text{ cm/sec}$$

$T = kH_{sat}$, where $H_{sat} = 47$ ft

$\quad = 1.96 \times 10^{-3}$ ft/sec (47 ft)

$\quad = 9.21 \times 10^{-2}$ ft²/sec

$\quad = 7960$ ft²/day

This transmissivity is adequate for an industrial supply (>1350 ft²/day).

Constant-Head and Falling-Head Permeameter Tests

The constant-head permeameter test and the falling-head permeameter test are laboratory methods based on Darcy's law as well. Figure 15.11 illustrates the constant-head permeameter device. The falling-head permeameter test is illustrated in Figure 15.12. This laboratory procedure is performed on finer grained soils than those tested by the constant-head procedure.

In the falling-head permeameter test, the flow volume is not measured directly so that very low flows through low-permeability soils can be accommodated. The head is measured in the standpipe before and after the test, so with the elapsed time of flow plus the dimensions of the soil sample known, the permeability can be determined.

Consolidation Test

Permeability is a factor in the consolidation test because settlement of the soil is related to the drainage of the sample under load. The permeability is calculated from the coefficient of consolidation C_c, which is obtained during consolidation testing. For further details on that subject the reader is referred to textbooks on geotechnical engineering. The consolidation testing procedure is discussed in this text in Chapter 7, Elements of Soil Mechanics.

Open Hole Permeability Tests

Open hole permeability tests, ones in which only a single well is monitored, can be performed in soils and weakly cemented or porous rocks with reasonably accurate results. These are small-scale, *in situ* tests as compared to the large-scale tests involving a pumping well and adjacent monitoring wells as discussed earlier.

Three possible tests may be used: 1) the open auger hole method for shallow depths, 2) the open tube method, and 3) the piezometer method. The auger method is performed in a shallow uncased hole

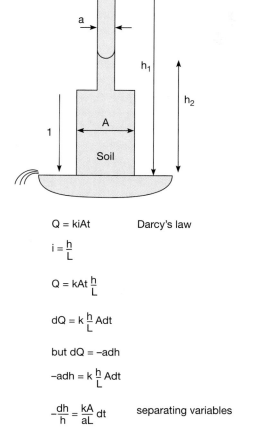

$$Q = kiAt \qquad \text{Darcy's law}$$

$$i = \frac{h}{L}$$

$$Q = kAt\frac{h}{L}$$

$$dQ = k\frac{h}{L}Adt$$

$$\text{but } dQ = -adh$$

$$-adh = k\frac{h}{L}Adt$$

$$-\frac{dh}{h} = \frac{kA}{aL}dt \qquad \text{separating variables}$$

$$-\int_{h_1}^{h_2}\frac{dh}{h} = \frac{kA}{aL}\int_0^t dt$$

$$-\ln\frac{h_2}{h_1} = \frac{kA}{aL}t$$

$$k = \frac{aL}{At}\ln\frac{h_1}{h_2}$$

$$= \frac{2.3\,aL}{At}\log\frac{h_1}{h_2}$$

FIGURE 15.12 Falling-head permeameter test.

below the water table. The open tube and piezometer methods are accomplished in cased holes, which are open at the bottom. The piezometer method has a sand-filled zone around the piezometer itself. The three methods are illustrated in Figure 15.13. As

shown, the auger hole method and piezometer method measure primarily horizontal flow, whereas the open tube method measures vertical or effective flow.

For the three test methods, depending on soil permeability, a constant-, falling-, or rising-head can be accommodated in the well. The appropriate equations, derived for these similar laboratory tests, can be applied except that the geometry of the "sample" differs, that is, it has a different shape factor. The dimensional details of the three open hole tests are shown in Figure 15.14.

Referring to the open auger hole test, water is pumped from a shallow open hole, which lowers the surface to a depth h below the water table, and a rate of inflow q equivalent to the rate of change in h is measured immediately, that is, without waiting for steady-state conditions to develop. In a similar manner as in the laboratory falling-head test

$$k = \frac{\pi a}{(A/a)h}\frac{dh}{dt}$$

where a is the radius of the well, h is the drawdown, and A is the shape factor, a function of the borehole and the boundaries of the flow medium.

Integrating,

$$k = \frac{2.3\pi a}{(A/a)t}\log H/h$$

Values for the shape factor are:

horizontal flow and $D \simeq 0$,

$$A/a = 0.75\left(\frac{H}{a} + 10\right)\left(2 - \frac{h}{H}\right)$$

and for $D = \infty$, allowing for considerable vertical flow,

$$\frac{A}{a} = 0.68\left(\frac{H}{a} + 20\right)\left(2 - \frac{h}{H}\right)$$

The second relationship holds for intermediate values of D to as low as $D/H=1$ and for high H/a values. This technique can best be illustrated by an example problem.

EXAMPLE PROBLEM

In an open hole, 4 in. in diameter, the water level dropped 10 ft in 1 min of pumping (Figure 15.15). Determine the k value for the material.

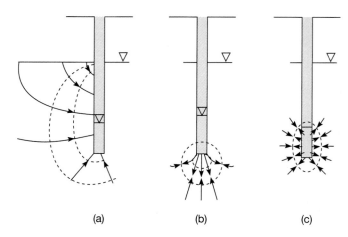

(a)   (b)   (c)

FIGURE **15.13** Flow patterns for permeability tests, open
   bore holes. (a) Open auger hole, (b) open tube, and
   (c) piezometer.

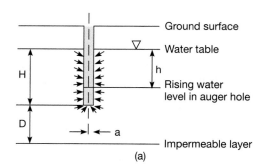

FIGURE **15.14** Dimension details, open hole
   permeability tests. (a) Auger hole test and
   (b) open tube or piezometer permeability test.

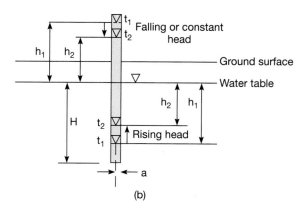

ANSWER

$$\frac{D}{H} = \frac{15}{15} = 1$$

Use $A/a = 0.68 \, (H/a + 20) \, (2 - h/H)$

$$A/a = 0.68 \left(\frac{15}{0.1667} + 20\right)\left(2 - \frac{10}{15}\right) = 99.7$$

$$k = \frac{2.3\pi \, (1/6)}{99.7 \times 1 \text{ min}} \log \frac{15}{10}$$

$$= 0.0021 \text{ ft/min}$$

$$= 3 \times 10^{-5} \text{ ft/sec}$$

$$= 1.06 \times 10^{-3} \text{ cm/sec}$$

For the open tube and piezometer methods the
shape factors have also been developed. They are
particularly complicated for the piezometer method
and are not included here. The reader is referred to a
paper by Hvorslev (1949) for these details.

An open tube (also a piezometer) can either be
tested under constant-head or changing-head conditions during pumping. For the constant-head situation the water surface is raised to a level h above the
water table and held there during pumping. The
equation for the constant head case is

$$k = q/Ah_1$$

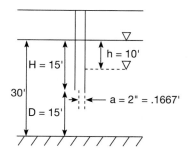

FIGURE 15.15 Open auger in-hole pump test.

where q is the pumping rate, h_1 is the height above the water table, and A is the shape factor.

For the changing water table case

$$k = \frac{2.3\pi a}{A} \frac{\log(h_1/h_2)}{t_2 - t_1}$$

where a is the diameter of the well, h_1 is the initial distance between the water table and the level measured at time t_1, and h_2 is the final distance between the water table and the level measured at time t_2. The shape factor for the open tube case is shown in Table 15.8. This also can best be represented by solved problems. Given the situations shown in Figures 15.16 and 15.17, the solutions to these problems are provided here. Based on Figure 15.16,

$$q = 0.5 \text{ ft}^3/\text{min}$$
$$2a = 4''; a = \text{radius of well} = 2''$$
$$= 0.1667 \text{ ft}$$

$$k_e = \frac{q}{5.5ah_1}$$

$$= \frac{0.5}{5.5 \, (0.1667) \, (50)} = 0.0109 \text{ ft/min}$$

$$= 3.32 \times 10^{-1} \text{ cm/sec}$$

Also, based on Figure 15.17,

$2a = 4''$, $a = $ radius of well $= 2'' = 0.1667$ ft. For one minute of pumping, $h_1 = 50$ ft and $h_2 = 40$ ft

$$k_e = \frac{2.3\pi a}{4(t_2 - t_1)} \log \frac{h_1}{h_2}$$

$$= \frac{2.3\pi \, (0.1667)}{4 \, (1)} \log \frac{50}{40}$$

$$= 0.02918 \text{ ft/min}$$

$$= 8.89 \times 10^{-1} \text{ cm/sec}$$

Field Testing of Rock Permeability

As a final category of permeability measurements, *in situ* permeability of rock will be considered. A common in-hole method used in bedrock borings is the packer permeability test. In this test, water is injected into a zone of rock through a perforated tube located between two inflated packers. Also, for testing the bottom section of the hole, a single packer can be used. An advantage of this test is that a selected portion of rock can be checked for permeability.

Packer permeability tests are quite commonly expressed in lugeon units. One lugeon is equal to a flow

TABLE 15.8 Shape Factors for Open Tube Method of Permeability Determinations

Drawdown Condition	Open Tube Situation	
	Penetrates Full Aquifer	Within Infinite Aquifer
Constant head	$A = 4a$	$A = 5.5a$
Variable head	$A = 4a$	$A = 5.5a$
Equations Developed		
Constant head	$k_e = \dfrac{q}{4ah_1}$	$k_e = \dfrac{q}{5.5ah_1}$
Variable head	$k_e = \dfrac{2.3\pi a}{4(t_2 - t_1)} \log \dfrac{h_1}{h_2}$	$k_e = \dfrac{2.3\pi a}{5.5(t_2 - t_1)} \log \dfrac{h_1}{h_2}$

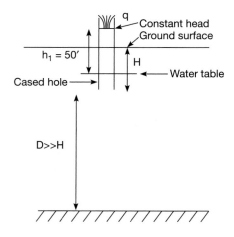

FIGURE **15.16**   Open tube, constant head, D = ∞.

of one liter per meter per minute at a pressure gradient of 10 kilograms per square centimeter per meter of hole. It is equivalent to a permeability coefficient of 10^{-7} m/sec or 10^{-5} cm/sec. Packer tests can be applied both above and below the water table, but greater success is obtained below the water table. The test length should be at least five times the hole diameter.

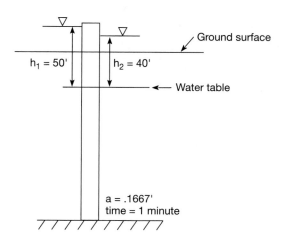

FIGURE **15.17**   Open tube, constant head, D = O.

Rock Fissure Permeability

In massive rock the permeability of fissures is many times greater than that of the solid rock between them. In tunnel and dam design it is desirable to determine the actual spacing of fissures and their thickness (width of opening). This relates to rock mass strength, permeability, and ease of grouting the fractures. Spacing of fissures can usually be determined from site investigation, including field mapping and a subsurface drilling program.

Permeability is directly related to the cube of the fissure width and directly related to the frequency of fractures. Therefore, fracture frequency has much less effect on permeability than does fissure width and, in the same way, fracture frequency has a relatively small affect on fissure width for a given permeability. Consequently, a reasonably accurate fissure frequency survey, coupled with good rock mass permeability measurements can predict fairly accurately the fracture width of the rock mass. The equation for this relationship is:

$$d^3 = \frac{12\eta k_r}{N\rho_w}$$

where d = fissure width (mm) and η = coefficient of viscosity of water = 1 centipose = 0.01 dyne/sec/cm², k_r = radial permeability (m/sec), N = frequency of fissures in number/m, and ρ_w = density of water.

$$d^3 \cong 1520 \frac{k_r}{N}$$

$$d \cong 11.5 \sqrt[3]{\frac{k_r}{N}}$$

EXAMPLE PROBLEM _____

The fracture spacing in a massive rock unit was found to be 1.58 feet (0.481 m) and the rock mass permeability is 5×10^{-8} m/sec. Calculate the fissure width for this rock mass.

ANSWER

$$d \cong 11.5 \sqrt[3]{\frac{k_r}{N}} \cong 11.5 \sqrt[3]{5 \times 10^{-8} \times \frac{0.481}{1}}$$

$$\cong 0.072 \text{ mm}$$

LEGAL DETAILS OF GROUND-WATER OWNERSHIP

Ground-water ownership in the United States can best be understood by considering it in relation to the legal details of surface water. Legal rights to surface water are usually obtained either by land ownership or by appropriation of specific water resources that are located close to or a short distance from the place of use.

Ground-Water Rights Based on Land Ownership

Originally, by virtue of land ownership, an owner had a raparian right to unlimited use of water from a stream or lake that bordered the owner's land. Known as the English rule, it developed as common law in England when people knew little or nothing about ground-water movement. When applied to ground water, this rule leads to complications because the law is based on the Doctrine of Capture, that is, anyone who owns the overlying land also owns the water beneath it. In regard to ground water, the law seems flawed because it does not recognize that adjacent owners (and sometimes distant ones) take water from the same aquifer and thereby it encourages individual use rather than some plan of sharing. It also encourages increased usage by virtue of individual, unlimited use rather than any form of conservation.

Because of malicious abuse of the English rule a modification known as the American rule developed in many states with raparian rights. This rule adds the qualification that the ground water must not be removed for a malicious or wasteful reason but that some positive purpose must exist for pumping water from the subsurface. This reduces water use somewhat but not necessarily a great amount. Dewatering for construction or for surface and subsurface mining is an acceptable purpose. Raparian rights for surface water and ground-water ownership prevail in much of the eastern half of the United States.

Ground-water law develops in response to the needs and the demands of citizens in individual states of the United States. For example, in Indiana, a problem developed during the 1980s in the northwestern portion of the state where ground water is obtained from an artesian bedrock aquifer. This is in the area of the Kankakee Outwash and Lacustrine Plain (see Chapter 13).

Known as the Fair Oaks Farms case, it involved a high rate of pumping of wells for irrigation purposes, which drastically lowered the piezometric surface on the adjacent farmer's land. A drop of more than 50 ft was noted close to Fair Oaks Farms and the cone of pressure relief with greater than 1 ft of drawdown extended for a radius of more than 10 miles. Consequently, the pumping affected an area of about one county in the state of Indiana.

Neighbors wells went dry if the drawdown extended below the depth where their pump intake was placed. This problem was not easily solved because in-hole pumps were now needed because the water had to be lifted more than 25 ft (the piezometric surface prior to the pumping was only 10 ft or so below the ground surface). A suction pump can only lift water 1 atm, or 32 ft, which reduces to about 25 ft based on pump efficiency. Some older, small-diameter wells would not accept in-hole pumps, which necessitated the drilling of a new, larger diameter well and purchase of the new pump. This was an expensive endeavor that the adjacent landowners felt was unjustified. They appealed to the Indiana legislature and also initiated litigation against Fair Oaks Farms.

The result of this controversy was that Fair Oaks Farms settled out of court for damages to the landowners, and the Indiana legislative changed the state law. They replaced the American rule, which had allowed Fair Oaks Farms to obtain as much water as they desired as long as it was not wasted. Now all large-capacity wells (greater than 250 gal/min) must be registered with the Indiana Department of Natural Resources and if the piezometer surface is drawn down more than 25 ft in an area, pumpage is reduced by all parties to allow the levels to be maintained.

Ground-Water Law Based on Appropriative Rights

In regions of extreme water scarcity, the doctrine of appropriative rights or the "Doctrine of Prior Appropriation" has always prevailed. In the United States, this is the region generally to the west of the 100th meridian. Simply worded, the doctrine states that the first person or group of people to appropriate the water gets first rights to the use of water. That is, the order of priority is based on the order in which the people have established the right to the supply. Reductions start with the last allocation and work back from there. The appropriation can be for water on the owner's land or it can be for water located hundreds of miles away from where it will be used. In the western states where appropriative rights prevail, surface water and ground-water sources are well integrated under the law to protect the prior appropriations.

Legal Complications

Complications do develop in the application of ground-water laws. It is difficult to apply laws developed for surface water to the ground-water regime. For example, it is extremely difficult to define the boundaries of ground-water supplies, or "ground-water streams" as one would define surface drainage basins, to yield distinct regions to which the doctrine of raparian rights can be strictly applied.

Another problem is related to the large number of people affected by ground-water use in a region. An aquifer may extend beneath the property of hundreds to many thousands of individual owners and include parts of several states. Water removed from one part of the aquifer must eventually affect the water level and possibly the pressure in other parts of the aquifer. An equitable division of water rights under such an involved situation is virtually impossible.

SAFE YIELD   The total amount of water to be removed from the subsurface under the doctrine of prior appropriation is another difficulty. The annual safe yield, which is determined by the amount of ground-water recharge, is the maximum water withdrawal that can be sustained for extended periods of time without an ultimate shortage. Pumpage greater than this amount is simply the mining of water. In California the safe annual yield is used to determine the amount of water to be appropriated. In other Western states this is not the case and with ever-increasing population in the southwestern Sun Belt this problem of safe yield must soon be faced.

INDUSTRY AND MUNICIPAL USE   In many states with raparian rights for water supply, individuals have legal rights not extended to industry or municipalities seeking ground-water supplies. As previously stated, individuals can withdraw as much ground water as they wish as long as it is removed for a positive purpose (American rule). A municipal water supply, however, does not have that same right without certain conditions. For example, a water supply company or utility could not purchase a small lot in a housing area or subdivision and compete with surrounding landowners for subsurface water by pumping without limit. The landowner accepts certain implied risks when buying property with regard to ground-water supply, but the possibility that a major water user such as a water works or industrial plant locating nearby is beyond the normal level of expected risk. Therefore, the industrial plant or municipal water works would have to buy a sufficiently large piece of property so that the water supply of their neighbors would not be altered by ground-water pumping.

Another consideration would come into play regarding the use of subsurface water by the municipal water company. Implied in the common law relationship is that the water would be used on the property from which it is obtained. If used off site, as in the water works example, the common law rule would no longer apply. The rights of the water works in this situation would not be covered by the rule.

SALTWATER INTRUSION   Along the eastern seaboard saltwater intrusion from the ocean into freshwater aquifers is a major concern. The technical details of this subject are discussed in a later section. In some mining operations, typically those for potash and phosphates, the area adjacent to the ocean is dewatered to allow for a dry, surface-mining procedure. Such dewatering accelerates saltwater encroachment. It was determined, however, that this dewatering does not violate the American rule because the water is removed for a productive reason, more economical mining of a necessary raw material. A law limiting pumping or permanent lowering of the water table in regard to saltwater intrusion had to be enacted to control this problem. The apparent technical solution is to return the water to the subsurface in an area located between the operating mine and the ocean, but to continue dewatering operations at the mine.

APPROPRIATIVE RIGHTS   In the western United States, it may develop that an owner with second appropriative rights can cause a lowering of the water table that affects the owner with first appropriative rights. The net effect is that the first owner must deepen his or her well, which costs money. The second person should compensate the first for this expense but not for all of it because part of the depletion is due to the first person's pumping as well as that of the second. This percentage of compensation would be determined legally or by mutual agreement.

In many states that operate under raparian rights, special legislation is required to move water from one drainage basin to another. This applies to ground water and surface water and involves major consequences even for states that apparently have more than adequate water supplies, those in the eastern half of the United States. For the western states where appropriative rights prevail and water supply is acute, even greater consequences are involved. Typically, federal and state legislation is involved in such projects as the Arkansas–Frying Pan project, which brings westslope water from Colorado east of the Rockies into the Arkansas River.

The present-day arrangement is to vest the power for ground-water regulation in governmental agencies at either the district, state, regional, or federal level. In many cases the individual states have appropriate agencies, which have the authority to act in this capacity. Other states have local agencies that set limits of ground-water production and levy taxes on owners of large wells to pay for the cost of recharging ground-water aquifers.

WATER WITCHING

Water witching, divining, or dowsing is the nonscientific means by which certain people attempt to find water in the subsurface by use of a forked stick, coat hanger, or metal rod. It is believed that some un-

known force attracts the stick or other object and causes it to point downward where water supposedly occurs in the subsurface. Laypersons in the United States apparently do not reject this method of water exploration because they are so poorly informed about the details of water-bearing zones in the subsurface.

Lack of knowledge regarding ground water is revealed by considering two questions on the subject: 1) Do people generally know about how ground water is stored in the Earth and how it moves through it? The answer is no. 2) Do people generally know about water witching or dowsing and how one goes about it? The answer is yes.

Some proposed explanations by the proponents of water witching attempt to suggest why it should work. Two possible explanations are 1) extrasensory perception (ESP) and 2) electromagnetic force. Little is known about ESP at the present time but regardless of how ESP works, if in fact it does, it does not help explain the process of dowsing. A scientific explanation is needed if we are to give any credibility to the subject.

With regard to electromagnetic effects, the claim is made that flowing water sets up an electromagnetic force, which attracts the stick or metal rod. This has no basis in fact. Water does not flow through the ground at anything approaching a moderate velocity nor does it occur in veins or narrow conduits in the ground. The assumptions of laminar flow for Darcy's law are met by ground-water movement, and low velocities are involved, well below 10^{-2} cm/sec or considerably less than 1 ft/hr. A beginning course in geology will convince the serious student that saturated, highly permeable earth materials are found in many geologic environments which do not occur as veins or tubes of porous material that could host high-velocity water movement. Just as the layperson must trust that a surgeon or other professional understands his or her specialty better than does an untrained, though enthusiastic individual, the same holds true that ground-water geologists are well versed in the details of ground-water flow.

GROUND-WATER POLLUTION

Ground-water pollution can occur in many different ways and has an equal number of causes. Contamination by heavy metal cations; organic chemicals in the form of pesticides, herbicides, and solvents; and biological constituents can occur. If the dissolved solids or suspended solids become too great, thereby lowering water quality to an unacceptable level, then water pollution has also occurred. The problems to be considered here are the following sources of pollution: 1) septic tanks, 2) sanitary landfills, 3) saltwater encroachment, and 4) miscellaneous causes.

Septic Tank Fields

A septic tank system is a simple sewage treatment method used by individual households or a small group of homes or apartments. Sewage is held in the septic tank under anaerobic conditions for a period of time after which the processed liquid is transferred to the ground through an underground tile system. Figure 15.18 provides the details of a septic system.

Septic tank systems must have free drainage into the subsurface (and not onto the Earth's surface) if they are to operate on a long-term basis without causing pollution problems. Two features of the soil can prevent a septic tank from operating properly: 1) too low a permeability at the contact area of the septic tank field to allow drainage into the soil and 2)

a perched water table, which prevents gravity flow of water from the tiles in the septic field into the subsurface.

Soil areas can be rated relative to septic tank use based on two different approaches: 1) soil percolation and 2) details from agricultural soils maps, which indicate relative permeability and depth to the highest seasonal water table. Percolation tests and the design of septic tank fields are examined first. This is sometimes referred to by contractors and other construction people as the "perc" test.

Percolation Tests

To perform the percolation test the following method is used. Four-inch-diameter auger holes are drilled to the appropriate depth at the site of the proposed septic tank field. Six or more holes are required, with an average value obtained from them. Two inches of coarse sand or fine gravel is placed on the bottom of the hole. The hole is filled to at least 12 in. above the gravel. Water is left in the hole overnight. Before the test begins the next day water is added to a level 6 in. above the bottom and the water drop measured at 30-min intervals for 4 hr, refilling to the 6 in. level

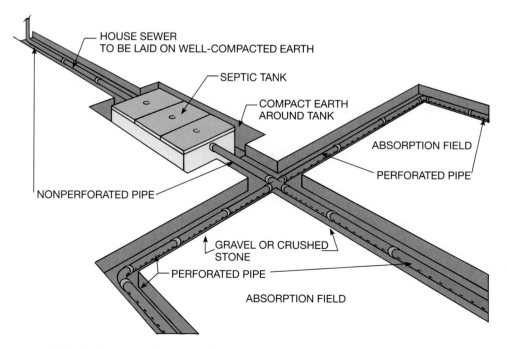

FIGURE 15.18 Septic tank sewage disposal system.

above the bottom as needed. The drop that occurs in the final 30-min period is used to determine the percolation rate. In sandy or other soils in which the water drains away in less than 30 min, time intervals between readings should be 10 min with the test run for 1 hr. The drop occurring in the last 10 min is used to calculate the percolation rate in this situation. The final reading is used to calculate the percolation rate in terms of the number of minutes required for the water surface to fall 1 in.

The size of the septic leach field is determined on the basis of the necessary area for the base of the septic tank trench or area base for a seepage pit. The trench area is related in a seemingly strange manner to the number of bedrooms in the planned house and the percolation rate of the soil. Table 15.9, for example, indicates that a percolation rate of 10 min/in. requires 165 ft^2 per bedroom as the area for the bottom of the septic tile field trench. If a three-bedroom home is involved, 3 × 165 = 495 ft^2 of trench bottom is required. If the trench is to be dug 2 ft wide, then 495/2 = 248 feet of trench length will be needed.

Table 15.10 indicates the restrictions for distances between the septic tank field and other portions of

TABLE 15.9 Percolation Rate Versus Trench Area and Number of Bedrooms in House

Time Required for Water to Fall 1 Inch (hr)	Required Absorption Rate (ft²/bedroom, standard trench and seepage pits)
1 or less	70
2	85
3	100
4	115
5	125
10	165
15	190
30	250 (unsuitable for seepage pits if <30)
45	300
60	330 (unsuitable for leaching systems if <60)

TABLE 15.10 General Restrictions—Distance Between Septic Tank Field and Property Aspects

Property Aspect	Minimum Distance to Closest Part of Septic Tank System (ft)
Water well	100
Stream, lake, or other water course	50
Dwelling or property line	10

the property. Table 15.11 deals with the final task to be considered: selecting the size of the septic tank.

A discussion about why the relationship between *number of bedrooms* versus the size of septic tank field and versus the size of the septic tank is important is in order. Obviously, the more sewage, the larger the septic tank should be and, in turn, the larger the absorption field. The size of the house or the number of occupants is related to this, but so is the number of appliances such as dishwashers and garbage disposals, and this is related to the general affluence or standard of living of the residents. One measure of affluence is the number of bedrooms, because houses with more bedrooms are generally more expensive and contain more water-using appliances. Therefore, the number of bedrooms is a reasonable measure of house size and affluence and is more appropriate than simply using the square footage of the house. Also this information is usually easily accessible.

TABLE 15.11 Capacity of Septic Tanks Versus Size of House

Number of Bedrooms	Recommended Minimum Tank Capacity (gal)
2 or less	750
3	900
4	1000
Each additional bedroom, add 250.	

Contrary to popular belief, septic tanks do not accomplish a high degree of bacteria removal. Although some treatment occurs, not all infectious material is removed. Also the effluent released from the septic system may be more objectional (and more odiferous) than was the sewage entering it. This does not detract from the primary, valuable service performed by the septic tank, which is to condition the sewage so that less clogging of the disposal field will occur. Further treatment of the effluent including removal of pathogens is accomplished by percolation through the soil. The disease-producing bacteria die in time (one to three months) as a result of the unfavorable conditions prevailing in the soil. An example problem on septic tank absorption field design is in order.

EXAMPLE PROBLEM

The average of six auger holes indicated that 10 min were required for the water level to fall 1 in. in the hole. The house will have four bedrooms. What size septic tank absorption field is required?

ANSWER

A reading of 10 min/in. of drop requires 165 ft^2 per bedroom for the standard trench. For a four-bedroom house, 4(165) = 660 ft^2 is required. Using a trench 2 ft wide, the trench should be 660/2 = 330 ft long. Three "fingers" 110 ft long each, would be workable. See the sketch in Figure 15.19.

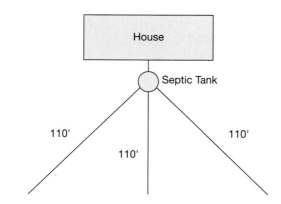

FIGURE 15.19 Plan view of septic tank absorption field construction.

EXAMPLE PROBLEM

A related problem concerning soil retention time is considered next. Assuming that k for the soil is 1×10^{-3} cm/sec and the gradient of the water table is 1%, how long does it take for water to travel 50 and 100 ft through this soil?

ANSWER

$$v = ki = 1 \times 10^{-3} \text{ cm/sec} \times 0.01$$

$$= 1 \times 10^{-5} \text{ cm/sec}$$

$$v_s = \frac{v}{n_e}$$

where n_e is the effective porosity for the soil. For a silty material, assume $n_e = 15\%$. Therefore,

$$v_s = \frac{1 \times 10^{-5} \text{ cm/sec}}{0.15} = 6.66 \times 10^{-5} \text{ cm/sec}$$

$$= 2.187 \times 10^{-6} \text{ ft/sec}$$

$$d = v_s t \quad t = \frac{d}{v_s} = \frac{50 \text{ ft}}{2.187 \times 10^{-6} \text{ ft/sec}} =$$

$$22.88 \times 10^6 \text{ sec} = 0.725 \text{ yr}$$

For 100 ft, $t = 1.45$ years.

Agricultural Soils Maps

Agricultural soils maps are typically large-scale maps (1:4800), which depict the different soil series as delineated by a soil scientist based on field study including interpretation of aerial photographs. A soil series consists of a location name and a texture name (for example, Russel silt loam), and is subdivided according to the slope of the land surface. The soil textural terms used in the agricultural soils classification (based on USDA designations) are much like engineering soils classification in which the terms sand, silt, clay, and loam are commonly used. (A loam is a combination of nearly equal amounts of silt and sand.)

Agricultural soils maps are prepared on a county-by-county basis. They are based on the upper 3 or 4 ft of the Earth's surface. Each soil horizon or zone is described on the basis of color, texture, consistency, depth, porosity, stoniness, moisture aspects, and perhaps other features. The top soil (zone A), subsoil (zone B) and parent material (zone C) are the commonly occurring soil zones present. Soils are subdivided into slope subclasses with the following typical ranges for each:

0% to 2%, level to nearly level;
2% to 6%, gently sloping (undulating);
6% to 12%, sloping (rolling);
12% to 20%, strongly sloping (hilly);
20% to 35%, steep;
35+%, very steep.

Modern agricultural soils maps (those produced in the last 15 years) also contain information relating engineering soils data and engineering uses to the basic agricultural soils data and description. The subclass (for example, Russel silt loam, 2% to 6% slope) is given a rating relative to engineering use. Such uses may include building foundations, ponds, homesites, roadways, septic tank fields, cemeteries, basements, sanitary landfills, and excavations. The various subclasses are rated good, moderate, and severe relative to these various proposed uses. Relative to the subject under discussion here, septic tank absorption fields, the factors of slope, permeability, and depth to seasonal perched water table are the important determinants for the rating.

Subclasses with too steep a slope (76% usually), too low a permeability, or too high a seasonal perched water table (12 in. or less from surface during wet season) are rated severe regarding septic tank fields. When these categories are less extreme, a moderate rating is received, and a gently sloping, permeable soil (sands and silts) with no perched water table yields a rating of good.

In some counties of the Midwest (good agricultural producing areas primarily) construction permits for houses with septic tanks are granted on the basis of the soils map or soil subclass rather than on the basis of percolation tests. There is cause to question percolation test results in some cases, because the tests may be run in the dry season when the perched water table is absent and high permeability readings will prevail because the soil is cracked from drying out. Percolation tests should illustrate the average or perhaps the worst condition rather than the best. Building permits may be denied on the basis of the agricultural soils class alone in some areas.

LIMITATION OF RATING PROCEDURE   Rejection of sites for septic tank use focuses attention on a critical concern about the procedure for rating soils relative to specific engineering use. This rating practice causes certain areas to be bypassed as housing sites when the real need is not to prevent construction in marginal areas but instead to ensure that proper construction procedures are used as needed. With proper design and construction, severe sites can be made acceptable. This may require drainage of perched areas by field tile, the addition of compacted soil at the site to bring the surface above the perched water, or dual septic tank fields, with alternate use allowing a time for each to recover. The proposed solution is to require that construction in severe areas be approved by a competent professional engineering geologist, civil engineer, or agronomist who has established expertise in this specialty. Simple exclusion is not the answer; proper design and construction provide a better approach.

Sanitary Landfills

A sanitary landfill is a method for disposing of refuse or solid waste on the land in a manner that protects the public health and environment. This is accomplished by placing the waste in layers or cells, densifying it through compaction, and covering with compacted soil at the end of each working day. This prevents spontaneous combustion of the material. Bales of compacted trash can also be placed in layers in the fill and covered with soil each day.

The two primary ways to develop landfills are 1) the area method and 2) the trench method. In the area method the refuse is placed on level ground, or the land surface is excavated to a designated depth and refuse placed above the soil base. Cover material is placed over the trash, generally over the face of an on-moving pile, after each working day. Low-permeability material is required below the base of the landfill and also for the cover material. In the trench method, a vertical cut 15 to 20 ft deep is made and the refuse is placed back into the trench using the soil excavated from the advancing trench as cover material for the landfill. The area method and trench method are illustrated in Figure 15.20.

The refuse in the landfill undergoes some aerobic decomposition prior to the time it is covered with compacted soil at the close of the work day. After

(a)

(b)

FIGURE 15.20 Methods for constructing sanitary landfills. (a) The trench method and (b) the area method.

burial, anaerobic decomposition takes over. Meteoric water, either vadose water or ground water containing CO_2, moves through the fill, dissolving soluble constituents as it proceeds. Weak acids thus formed increase the solvent power of the water and more leaching occurs. This liquid is referred to as leachate. The principal gases that form are carbon dioxide

(CO_2) and methane (CH_4), but landfill gas also contains water vapor. Production of gas is at a maximum early in the decomposition process but the intensity decreases markedly with age.

Common cations and anions in addition to some heavy metal cations are commonly incorporated in the leachate. Ca^{++}, Mg^{++}, Na^+, and K^+ are common cations along with the anions of Cl^-, SO_4^{-2}, and NO_3^-. Cadmium, lead, zinc, chromium, mercury, and other heavy cations also may be carried in solution by the leachate. Organics including human-made chlorinated hydrocarbons are also present. These toxic constituents pose a potential ground-water pollution problem of great significance to the environment. Health hazards include the concern for aquatic life and for human life. A number of organics are thought to be carcinogenic (cancer causing) in quite low concentrations: in the parts per billion range in water and parts per million range in soil.

Contaminant transport in ground water is complicated by the nature in which constituents behave in a water environment. For example, some organic substances such as gasoline and fuel oil float on the water surface, whereas other constituents, heavier than water, will sink. A third option is for the constituents to dissolve in the water and travel along with it.

Fine-grained soils such as clay and silty clay provide a more desirable containment for the leachate than do coarser grained soils. The advantage is twofold. Finer soils have a lower permeability and will pass the leachate through less readily. Also, the clay minerals present in finer grained soils are able to exchange ions with the leachate removing the heavy toxic cations and replacing them with Na^+, Ca^{++}, or Mg^{++}.

The criteria for containing the leachate within the landfill site involve the nature of the surrounding material (type of soil or bedrock), the thickness of the soil, the nature of the bedrock (type and structural defects), rainfall, infiltration, and the details of site topography. These considerations are summarized in Table 15.12.

Miscellaneous Ground-Water Pollution

There are numerous other means by which ground-water pollution can occur. Industrial wastes can cause ground-water pollution if the wastes are not properly contained. In some cases toxic heavy metal cations can be transmitted to the ground water. High concentrations of cadmium, chromium, lead, zinc, and other cations have been supplied to the ground water from corroded containers of industrial wastes.

A concern about chloride contamination gained attention a few years ago. Large supplies of $CaCl_2$ are stored in the summer and fall for use during the winter to deice snow-covered, icy roads. Typically some of the stockpile remains year round unless supplies are depleted during a severe winter. In the past, such piles were left outdoors uncovered. Rainfall onto the piles dissolved the salt and runoff carried heavy concentrations of Ca^{++} and Cl^- into surface streams and into the ground-water regime. Today, permanent, tall, pointed structures are built to house the salt stockpiles. These structures prevent rainwater accumulation and the subsequent salt-charged runoff.

Ground-water pollution caused by open fissures and caves in limestone and dolomite, and to a lesser extent in gypsum and rocksalt, provides nearly an endless list of examples. The high permeability of fractures and caves is the basis for the problem, which, rather than yielding water velocities of a few hundred feet per year, can be hundreds of feet per minute. No exchange with the soil or rock mass occurs during such rapid transport so there is virtually no reduction in toxicity, bacteria count, or suspended solids during the process.

Countless cases of ground-water pollution by livestock in limestone terrain have been recorded and pollution of distant wells from industrial waste through cavernous limestones is far too commonplace. The flow of septic tank effluent to an adjacent well can be short circuited by open fissures, yielding pathogenic bacteria to the water supply. Special care to keep pollution away from open fissures must be practiced in areas with karst topography.

Joint-controlled ground-water flow in massive rocks of all kinds must be carefully scrutinized. In massive quartzites, granites, gneisses, schists, and similar rocks, preferred joint directions may focus much of the flow through a major bedrock region. Joint intersections and shear zones may carry much of the water flow for a bedrock mass measuring many tens of feet in dimension. Water that supplies domestic wells along with water from septic systems may commonly be directed through this dominant flow system. If the fissures are fairly open and the travel time is short, pollution problems can develop in such terrains.

As a case in point, the city of Ottawa, Ontario, experienced problems some years ago with pollution of ground water from an aquifer in jointed quartzite, which is encountered at some depth below the city.

TABLE 15.12   Geologic Considerations for Proper Siting of Landfills, Midwestern U.S. Application

Type, Nature, and Stratigraphy of Soils
  Glacial till comprises most of ground moraine, which is mostly clayey silt and silty clay with occasional sand lenses. Place monitoring wells in water-bearing sand lenses.
  End moraine features; contain glacial till plus thicker sand lenses in more contorted shapes than ground moraine. Need to determine geologic detail before placing monitoring wells in these sands.
  Outwash sands; challenging areas to locate landfills. May have layers of low-permeability materials which subdivide the cross section into separate water bearing zones. Need to determine the subsurface details before placing monitoring wells.
  River floodplains; similar to outwash sands.
  Residual soils; weathered material above bedrock. Determine the thickness and nature of this material.
Thickness of Unconsolidated Material
  Trench method of landfill; excavate material before placing trash. Must have adequate thickness of soil after excavation, 20 ft or more needed between base of trench and an aquifer.
  Concept; want to prevent movement of leachate through the water-bearing zone.
  Monitoring wells need to be placed in the appropriate water-bearing zones.
Type and Nature of Bedrock Below Unconsolidated Material
  Sandstone; permeable
  Limestone, particularly weathered surface of bedrock; permeable
  Shale; nonpermeable
  Other areas outside of the Midwest may consist of igneous or metamorphic rocks, which have different permeabilities.
  Concept; may need monitoring well in bedrock aquifer if overburden soil thickness is not very great or if movement through bedrock aquifer a concern.
Ground-Water Supplies in Vicinity
  Protection of existing water supply is crucial. Where do neighbors obtain their water, what depth, what geologic material, what artesian pressures? Geologic detail is extremely important. There may be no connection between water-bearing zones at landfill site and for adjacent areas. Need to study landforms and subsurface information. Pressure relationships are important. Water flows from high pressure to low pressure through permeable materials. Must study local ground-water supplies before placing monitoring wells.
Topography of Site Including Flood Potential
  Upland area away from tributary streams minimize the effects of runoff from landfills. This relates to regional water flow as well. Consider during monitoring program.
  Flood potential; generally need to stay away from 100-yr flood elevation. Site can be protected by levee built to this elevation.
Ground-Water Table Location and Water-Bearing Zones at Site
  Water table in permeable material; easy to observe during drilling program.
  Water table in low permeability soil; more difficult to determine. Check water levels in borings without sand layers after extended periods of time.
  Need to locate monitoring wells in water-bearing zones that can be affected by leachate migration.
  Need to determine geologic details before placing monitoring wells. Perform subsurface investigation first, using split spoon sampling and Shelby tube sampling. Develop geologic cross sections for site. Return to site and place monitoring wells after evaluating data.
  Perform geologic study first, then place monitoring wells based on geologic findings.

The problem was traced to the farm belt a number of miles away where the jointed quartzite crops out. Runoff from animal feed lots was getting direct access to the open quartzite joints and little or no improvement in water quality had occurred during the miles of travel underground.

Returning to the discussion on landfills, there are two different procedures by which designs have attempted to control the leachate effects generated by a landfill: 1) attenuation and 2) containment.

Attenuation landfills were constructed in the past with the understanding that clayey soils would attenuate or remove the constituents in the leachate and ground water by exchanging ions or filtering out dissolved and suspended constituents. Clayey soils have a capacity to remove exchangeable cations. A rule of thumb has developed that a clay layer can neutralize about 2 to 2.5 times its thickness of landfill debris. For example, 100 ft of trash should have a clay base of from 40 to 50 ft thick to attenuate the cations and neutralize the leachate constituents.

Recent concerns have occurred regarding this procedure because a number of constituents rely on dilution to decrease their concentration, which is a risky approach to the problem. Nature cannot dissipate a continuous influx of chemical constituents through dilution. In addition, dilution is not an acceptable procedure according to the United States Environmental Protection Agency for ground-water remediation. Leachate constituents that rely on dilution to decrease concentration are Cl^-, F^-, SO_4^{-2}, and NO_3^-, which are not removed from ground water moving through soil.

Containment is the second way to control landfill leachate. In this procedure, the leachate is prevented from moving away into the environment because the leachate is collected for processing much like sanitary wastewater. Liners of either compacted clay or synthetic geomembranes, such as PVC (polyvinyl chloride) or HDPE (high-density polyethylene), are used to prevent leachate migration and collection systems below the waste are used to drain off the leachate.

Containment procedures have been required for hazardous waste landfills since the early 1980s. This technology of containment liners is now being applied to conventional waste landfill construction. Double liners with detection systems located between them reduce the liklihood of leachate migration from the site.

Another desirable procedure is to minimize the amount of leachate that is generated. Infiltration of precipitation into the landfill generates leachate. To minimize this production, landfill caps are made more impermeable and sandy or other drainage layers are placed above the trash to intercept any infiltration. Caps are constructed of compacted clay layers or of geomembranes much like the landfill liner. The top of

the completed landfill is sloped steeply enough to encourage runoff but not so steep as to cause increased erosion. Slopes of 3% to 6% on the top of landfills and 20% to 33.3% on the sides are generally considered. As an illustration a liner base and cover configuration for a hazardous waste landfill are shown as Figure 15.21.

Saltwater Encroachment

Along coastal areas freshwater can be obtained from the subsurface because its lower density allows it to float above saltwater. This balance can be upset by overpumping of the ground water. When that occurs saltwater encroaches or moves into the freshwater aquifer.

Saltwater encroachment has become a major problem for the populated areas along the eastern seaboard of the United States. Some of the areas are heavily populated year-round, such as Long Island, New York, and others are popular tourist areas, which experience large increases in population during the tourist season. Atlantic City, New Jersey, is an example of a coastal area that typically experiences heavy increases in population during the summer months and had to allow for such needs in their water supplies. With the advent of casino gambling there, this increase in tourism may be sustained over much of the year, putting a further strain on their water supply. Some coastal areas of California and Florida have also undergone problems with saltwater encroachment.

Recharge of an aquifer after encroachment of saltwater is not easy to accomplish. In the lowering of the water table, some permeability is lost because of compaction and consolidation so that greater pressures are required to put the freshwater back into the aquifer. This collapsing of the aquifer prevents adequate recovery of the freshwater aquifer even when high pump pressures are used.

The effect of ground water withdrawal can best be illustrated for an island observed in cross section in Figure 15.22. Since pressure under the ocean must equal the pressure under the island,

$$d(1.03) = H(1.00)$$

$$\text{but } H(1.00) = (h+d)(1.00)$$

$$d(1.03) = (h+d)(1.00)$$

$$d = 33h$$

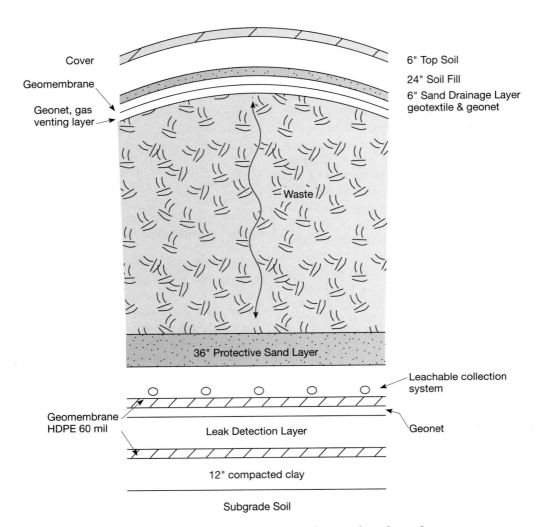

FIGURE 15.21   Cross section showing liner base and cover for a hazardous waste landfill.

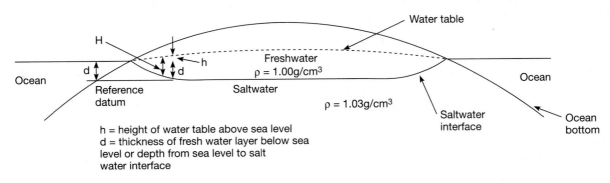

h = height of water table above sea level
d = thickness of fresh water layer below sea level or depth from sea level to salt water interface

FIGURE 15.22   Island with saltwater intrusion.

Hence if h is decreased by 1 ft, that is, if the water table is lowered by 1 ft, d is decreased by 33 ft or, indeed, the interface moves up 33 ft as a consequence of the 1-ft drop in the water table.

This illustrates why it is so critical that permanent lowering of the water table not be allowed in coastal areas either by overuse or simply to dewater the sediments to allow mining. As previously stated, this can be prevented in mining areas if the water is recharged into the ground between the mining site and the seacoast, thus replenishing the saturated zone in that location.

FORMATION OF CAVES IN CARBONATE ROCKS AND ENGINEERING SIGNIFICANCE

The origin of caves in limestone and dolomite is a subject of much discussion in the field of geology. The primary elements of discussion are the ways in which solutioning occurs, how structural features of the rock affect dissolution, and where in the ground-water regime solutioning develops.

Karst topography is common in a limestone or dolomite region. Sinkholes, rounded surface depressions, are prominent. Figure 15.23 is a photograph of a karst area with trees growing out of the sinkholes. Figure 15.24 is a topographic map of karst terrain; sinkholes appear as closed circles on the map.

Calcite ($CaCO_3$), the primary constituent in limestone, and to a lesser extent dolomite [$CaMg(CO_3)_2$], the primary constituent in dolomite rock (or dolostone), are soluble in dilute carbonic acid, which occurs in subsurface water. This is shown in the following equation:

$$H_2O + CO_2 \rightarrow H^+ + HCO_3^-$$

$$(HCO_3^- \text{ is carbonic acid})$$

$$CaCO_3 + 2H^+ \rightarrow Ca^{++} + H_2O + CO_2 \uparrow$$

The CO_2 gas yields the sudden effervescence of bubbles that occurs when HCl is dropped on a piece of limestone. Dolomite "fizzes" much less vigorously and it may require that the rock be powdered before any action is observed. A limestone cave is pictured in Figure 15.25. Stalagmites extend from the ground up and stalactites extend from the ceiling down.

FIGURE **15.23**   Photograph of a karst area on limestone bedrock.

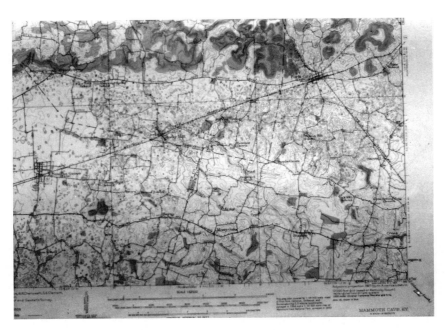

FIGURE **15.24**   Topographic map of Karst terrain.

Water movement in massive rocks such as carbonates is controlled by the discontinuities, the joints, and the bedding planes. Near the surface the joints are open and water moves freely into and along them. Some dissolution undoubtedly occurs. Between the joints, high pinnacles of unweathered rock remain, in contrast to the vertical or nearly vertical joints where the carbonate rock has dissolved. The insoluble residues of chert, iron oxide, and clay minerals remain. This yields an extremely uneven bedrock surface in limestone terrain with pinnacles and pits located in adjacent positions. Excavation and foundation support problems can result from this uneven bedrock surface.

The dissolution occurs over long durations of time by human standards, typically hundreds of thousand to millions of years. This is a rather short time geologically, but it does illustrate that short-term solutioning of carbonate rocks is seldom an engineering problem. The problem occurs instead where the solution channels already exist.

Dissolution of carbonate rocks is controlled mostly by the jointing. The two features of joints involved in this regard are 1) the attitude or orientation of the joints and 2) the continuity or extent to which the joints are through-going in the formation.

Typically, there are two prominent vertical or near-vertical joint sets in gently dipping sedimentary rocks. These are the strike joints, those running parallel to the strike of the rocks, and the dip joints, those parallel to the direction of dip. The two sets are therefore mutually perpendicular or nearly so. Water will move along these joints, intersecting a prominent bedding plane and then continuing along this nearly horizontal direction, at right angles to the joint. Hence, vertical and nearly horizontal openings are formed by enlargement along the rock discontinuity. Figure 15.26 shows the joint sets in limestone that give rise to solutioning and cave formation.

Not all joints run entirely through a geologic formation. Many terminate at prominent bedding planes. This can be observed both in outcrops and in underground openings in rock. Truncated joints not only provide increased rock mass strength but also disrupt the primary conduits for water movement through the mass.

One set of joints is typically more extensive than the other, either the strike joints or the dip joints. The strike joints, for example, may be more continuous, actually penetrating the entire formation. If this is the case, water movement will be concentrated along this vertical joint direction.

FIGURE **15.25** Limestone cave showing cave deposits or drip stone. Stalactites from the ceiling and stalagmites from the floor can join to form a column as shown near the center. Missouri Cave connected to Onandoga Cave. (Pierre Studio, Sullivan, MO.)

Dissolution occurs as the water moves through the joint. Also, the purity of the carbonate comes into play after an open joint and conduit system develops. For argillaceous (clayey) limestones and dolomites, less carbonate must be dissolved per unit volume of rock than for purer carbonates, and the clay can be carried away by the moving water. Therefore, a combination of carbonate purity, jointing pattern,

availability of water, and topography determines the extent of solution development.

The location within the ground-water regime where the major dissolution occurs is a primary consideration. Various hypotheses have been proposed, suggesting that a major cave system is formed 1) above the water table, 2) at the water table, or 3) below the water table.

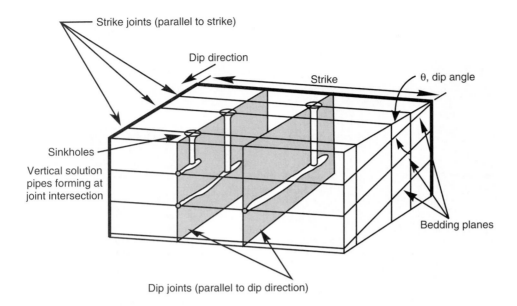

FIGURE **15.26** Diagram of strike joints and dip joints in limestone that provide pathways for solution channels.

The argument that dissolution and erosional effects to the limestone (or dolomite) would be greater for water flowing under pressure suggests an origin for the major cave system either right at or somewhat below the water table. Above the water table, in the vadose zone, water moves downward under the direct influence of gravity, but no fluid pressure (or positive pore pressure) develops. The dissolution and erosional capabilities would seemingly be less because of this situation.

Some distance below the water table the volume of flow would be less than that at and immediately below this surface. The flow and pressure relationships are shown in Figure 15.27. Depicted are flow lines (solid, with arrows) and equipotential lines (dashed and numbered). Between each pair of flow lines an equal volume of water flows. This shows that the upper part of the phreatic zone (immediately below the water table) carries the major portion of flow, and a progressively greater cross section is needed for the same flow as the depth increases.

For the reasons just discussed and based on field evidence, J Harlan Bretz (1953) proposed the following sequence for cave system formation.

1. Cave system formed at water table when the major stream system was moderately well entrenched and located adjacent to the limestone in the uplands.
2. Deeper entrenchment of the stream lowered the water table and drained the cave system.
3. Caves filled with red mud carried by downward movement of vadose water from the residual soil above.
4. Mud excavated from caves by water from surface flows when erosion works its way down toward cave level.
5. Dripstone, stalactites, stalagmites, etc., formed by precipitation of calcium carbonate in an air-filled cave.

Perhaps the only controversial sequence of this outline is the filling and excavating of the red mud. It does not seem likely that all cave systems would necessarily go through that sequence. The other details of the formational process seem quite reasonable.

Thornbury (1954) suggested that some cave systems in Indiana and Kentucky were likely formed in part above the water table. He cites evidence relating sinking streams and subsurface passageways to support this conclusion.

Based on these discussions it is generally accepted that most cave development occurs at or relatively close to the water table when the major stream

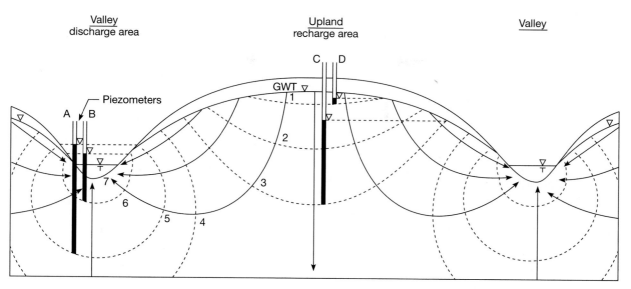

FIGURE 15.27   Simplified regional flow system in uniformly permeable material. (Adapted from Hubbert, 1940.)

entrenchment nearby is situated at the proper level to establish caves in the limestone unit in question. The base level elevation for the stream must persist for a sufficient time to develop the cave system at a specific elevation or narrow range of elevations. This requires a special geologic setting.

Carbonate units not associated with entrenched streams have little likelihood of developing caves of this type. Consequently, a regional study is necessary to determine the likelihood of cave development. Aided by available data on dissolution details of the strata from published reports, maps, and well records a determination can be made regarding the likelihood of cave development in a region containing carbonate rocks. This information is generally well established for areas in which formations are cavernous.

The information gathering process is crucial in the study of large dam sites in sedimentary rock areas. The extent of solution cavities has major significance for dam foundations and water impoundment, and much information must be obtained well before a decision is made to proceed with subsurface investigation.

The question of cave development at depth, that is, below the bottom of the entrenched stream, is sometimes raised. Referring again to Figure 15.27, we can see that at least one flow line in the drawing extends through an area well below the stream. In a uniformly permeable system, most of the water would flow above this level, but in areas of nonuniform materials water flow may be concentrated in lower zones of higher permeability. Therefore, it is concluded that caves can form in some cases not only well below the water table, but well below the base of the entrenching stream that dictates the water table level. This would, by nature of the list of conditions required for it to occur, seem to be an exceptional case rather than a common one.

EXERCISES ON GROUND-WATER GEOLOGY

1. Information supplied in this chapter indicates that the continental landmass of the Earth receives 24,100 mi^3 of precipitation per year. What average value, in inches per year, for rainfall on the continental landmasses is suggested by this? Recall that the Earth is essentially a sphere 8000 mi in diameter, and the continents constitute about 21% of the surface area.

2. The following information pertains to the subsurface at a specific location:

- The soil is 30 ft thick and it is a silt.
- A thick sandstone formation 1000 ft thick lies below the soil.
- The depth to the water table is 18 ft.
- The silt has a capillary fringe 5 ft thick.

Based on this information, show the subdivisions of this cross section relative to the subdivisions of subsurface water. Draw an appropriate hydrostatic pressure diagram for the cross section in keeping with Figure 15.2.

3. (a) If an artesian aquifer is intercepted by a well drilled 120 ft deep and the water rises in a standpipe to a height of 10 ft above the ground surface, how many feet of artesian pressure exist for this well?
   (b) What type of well is it?
   (c) What was the pressure in psf and psi for the water in the aquifer before entering the well?
   (d) If the piezometric pressure in the aquifer is reduced 15 ft for every mile of horizontal distance, what is the hydraulic gradient for the aquifer?
   (e) If the hydraulic conductivity of the sandstone is 15.3 ft/day, what is the average velocity of flow through the sandstone?
   (f) Assuming a porosity of 18% for the sandstone, what is the seepage velocity for that unit?
   (g) How long does it take for water to travel a mile in the sandstone aquifer assuming the effects of pumping are minimal?

4. A farmhouse is located above a perched aquifer that has a volume of 4000 yd^3. The sandy aquifer has a porosity of 28% and a specific yield of 12%.
   (a) What is the specific retention of the sandy aquifer?
   (b) What is the maximum number of gallons of water that can be obtained from the 4000-yd^3 aquifer if no recharge occurs because of a drought and all the aquifer is saturated at the start?
   (c) If the household uses 600 gal of water per day (four people at 150 gal/person/day) how many days can the family go before all available water is used?

5. An oil-bearing sand is known to have an intrinsic permeability of 120 millidarcies. What is the equivalent hydraulic conductivity for this sand? What would be a typical description for such a sand based on this hydraulic conductivity value?

6. Big Springs in the Ozarks of Missouri has a discharge of about 175,000 gal/ min.
   (a) How many ft^3/sec does this equal? How many MGD? (*Hint:* 1 cfs = 0.646 MGD.)
   (b) What magnitude spring would this qualify for in Meinzer's classification? (Table 15.4)

(c) The Missouri Ozarks consists of sedimentary rocks that are Lower Paleozoic in age. In what specific rock type is the spring most likely to occur?

7. What are the five most productive landforms relative to ground-water supply? What makes them so productive? What problems occur in wells in fine sand and how can they be minimized?

8. Why are shales typically very poor water-producing formations? Consider both concepts of primary permeability and secondary permeability in your answer.

9. A field pumping test is to be performed in an unconfined aquifer of gravely sand. This aquifer is 30 ft deep with a hard clay zone below it. The water table is found at a depth of 4 ft.
   (a) How deep should the pumping well be drilled and where should the well screen be placed?
   (b) How long should the well be pumped in an attempt to reach steady-state conditions?
   (c) What is meant by *steady-state conditions* in this regard?
   (d) Calculate k for the following test situations. For the pumping test assume that steady-state conditions exist and that the pumping well and observation wells are properly completed.

   $$q = 250 \text{ gal/min}$$
   $$r_1 = 25 \text{ ft where drawdown} = 5 \text{ ft}$$
   $$r_2 = 100 \text{ ft where drawdown} = 1.5 \text{ ft}$$

10. A sandy gravel with 50 ft of saturated thickness has a permeability of 5×10^{-1} cm/sec.
    (a) Calculate the transmissibility of the aquifer.
    (b) Rate this transmissibility relative to the needs for domestic or industrial use.

11. An artesian aquifer 100 ft thick was evaluated using a field pumping test. The artesian head for the aquifer prior to initiating the test was 120 ft. At a distance of 1000 ft, the drawdown was 40 ft, and at a distance of 5000 ft it was 20 ft. Find the permeability of the aquifer assuming that steady-state conditions prevail. The pumping well was sustained at a rate of 500 gal/min during the test. What is the permeability of the aquifer in ft/sec, cm/sec, Meinzers, ft/day, and darcies?

12. A constant-head permeameter test was run on a sand sample that had a diameter of 3 in. and a length of 4 in. The head loss through the sample was 12 in. The quantity of water flow in 1 min was 300 cm^3. Calculate the permeability in cm/sec, ft/sec, and m/sec.

13. A falling-head permeameter test was run on a clayey silt sample. The sample had a diameter of 2 in. and a

length of 4 in. The diameter of the standpipe is 0.25 in. At the beginning of the test, the head was 100 cm and at the end 30 hr later it was 54 cm. Find the permeability in cm/sec.

14. Use the open auger hole procedure to determine a numerical value of permeability for a percolation test. Given that for the 4-in-diameter hole it took 5 min for the water surface to drop 1 in., what is the permeability of the soil in cm/sec? If it took 30 min to fall 1 in., what would be the permeability?

15. What is the significant difference between the American rule and the English rule as applied to raparian rights? Why are individuals protected by excessive withdrawal of water from the subsurface by industry or by a municipal water company but not by a typical neighbor's use of water? Explain.

16. A cottage is to be constructed on a lakefront lot and a well and septic tank system are needed. The cottage is to be built on a half-acre lot, having rectangular dimensions and 100 ft of lake frontage. The lot slopes gently toward the lake. The dimensions of the cottage will be 50 ft by 50 ft and it will have the equivalent of three bedrooms relative to use of water and production of sewage. Results of six percolation tests indicated that it took 13 min for the water level to drop 1 in. in the test hole. Based on this information, determine the size of the septic tank needed and lay out the location of the house, septic tank absorption field, and well. It is desirable to place the cottage as close to the lake as is possible. Show calculations and then the layout of the septic tank, field, etc., on a plan map of the lot. State assumptions and reasons for various decisions made. Assume the trench in the percolation field is 2 ft wide.

17. When the gases formed in a landfill are collected, water vapor plus what two gases are the principal ones obtained? How must this gaseous mixture be processed to get a highly combustible product for an energy source?

18. Why are those landforms that are the best for obtaining ground-water supplies also likely to be the worst as a location for the placement of a sanitary landfill? What types of unconsolidated materials and types of bedrock are most favorable as disposal sites for refuse?

19. What are the two advantages of a clayey soil in retarding leachate movement as compared to a sandy soil? Explain in detail.

20. What is meant by the term *saltwater encroachment* and where in the United States is it particularly a problem?

21. A small island on the seacoast has an area of 9.5 mi². Because of extensive development of vacation homes,

infiltration, which was 15% for the island in its natural state, was reduced to 8% after construction. The annual rainfall for the island averages 30 in.
(a) What is the maximum annual safe withdrawal of water for the island in acre-ft, ft³, and gallons if no depletion of the ground water is to occur?
(b) How many tourists and vacation people can this support assuming the water consumption is 150 gal/person/day? Calculate this in tourist-weeks and in tourist-years.
(c) How many tourists per square mile and per acre does this maximum number of tourists involve?
(d) If the water table on the island is lowered 1 in. by overpumping, how much water does this represent? What percent increase over the maximum allowable "safe" number of tourist-years does this represent?
(e) If the water table is lowered 1 in., how much higher is the freshwater-saltwater interface raised?

22. Why is movement of water in open joints in a rock a much greater threat of pollution than is water movement through a soil?

23. If dissolution of limestone and dolomite is a slow, geologic process, why are humans concerned about placement of dams and surface reservoirs in such terrains? How are potential problems in these terrains determined?

24. Cave formation is most prevalent near the water table but some caves are developed at much lower elevations. Why is this important relative to leakage from reservoirs? How can the presence of such deep-seated caves be indicated?

Additional Readings

ATTEWELL, P.B., and FARMER, I.W., 1976, *Principles of Engineering Geology*, Halsted Press, John Wiley & Sons, New York.

BRETZ, J HARLAN, 1953, "Genetic Relations of Cover to Peneplains and Big Springs in the Ozarks," *American Journal of Science*, Vol. 251, pp. 1–24.

DAVIS, S.N., and DEWIEST, R.J.M., 1966, *Hydrogeology*, John Wiley & Sons, New York.

DRISCOLL, F.G., Ed., 1986, *Ground Water and Wells*, 2nd edition. Johnson Well Screen Co., St. Paul, MN.

HUBBERT, M.K., 1940, "Theory of Ground-Water Motion," *Journal of Geology*, Vol. 48, pp. 785–944.

HVORSLEV, M.J., 1949, *Subsurface Exploration and Sampling of Soils for Civil Engineering Purposes*, U.S. Army Engineer Waterways Experiment Station, Vicksburg, MS.

TERZAGHI, K., and PECK, R.B., 1948, *Soil Mechanics in Engineering Practice*, John Wiley & Sons, New York.

THORNBURY, W.D., 1954, *Principles of Geomorphology*, John Wiley & Sons, New York.

16

Coastal Processes

CHAPTER OUTLINE

Wave Motion
Tides
Effects of Severe Storms
Erosion in Coastal Areas
Estuaries
Classification of Coastlines
Shoreline Protection Structures
Sensitivity of Shoreline Features to Human Activities

The zone along the coast of oceans or large lakes such as the Great Lakes is a site of active geomorphic processes. Wave action and the relative level of the water surface provide the erosive force that modifies and rebuilds coastal landforms. Waves are the undulations on the water surface, usually set in motion by the wind. Other causes are noted when tsunamis are discussed later in the chapter. The size of waves depends on wind speed, wind direction, and the length of open water.

Frictional drag of moving air currents (the wind) on the water generates waves. Only the water surface and its near subsurface are disturbed by the wind. The area over which wind acts to generate and promote waves is known as the *fetch*. This area is extremely large for the oceans and the Great Lakes and is much of the reason why storm activity can be so devastating to adjacent coastal areas. For smaller lakes and reservoirs the fetch is smaller; indeed, engineers designing earth dams take care to minimize the fetch where wind builds waves that will ultimately strike the dam. Slope protection of the upstream face of a dam can be reduced accordingly when the fetch for that shoreline is minimized.

WAVE MOTION

The motion of waves in the ocean is best described by the terms used to describe wave behavior. This terminology is illustrated in Figure 16.1. Wavelength L is

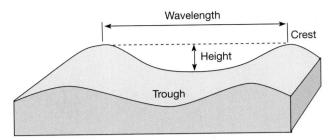

FIGURE 16.1   Terminology associated with waveforms.

the horizontal distance between successive repeat points on the wave trace, that is, between adjacent crests or adjacent troughs. Wave height *H*, or amplitude, is the vertical distance between the crest and the trough. The period *T* is the elapsed time between successive crests or troughs. Two additional terms associated with wave motion are *frequency* and *velocity*. Frequency *N* is the number of wavelets or cycles that pass in a unit of time and velocity *v* is the rate of movement of the waveform. Frequency is the reciprocal of the period (or *N = 1/T*) so it can be observed that *V = LN = L/T.*

The two primary types of ocean waves are oscillatory waves, which occur in the open ocean away from the coasts, and translational waves, which develop a short distance offshore.

Oscillatory Waves

For oscillatory waves, only the waveform advances, not the individual water molecules themselves. This is similar to stalks of grain bending in the wind as a breeze ripples across a field. Just as the stalk sways back and forth yielding a circular motion for a point on the stalk, a circular motion in the water is also generated, these circles decreasing in size with depth, as depicted in Figure 16.2. At the surface, the circular orbit has a diameter equal to the wave height with particles completing the circle in time *T.* Beneath the surface, the motion decreases rapidly (exponentially), and at a depth of about one-half the wave length (*L/2*) it is less than 5% of the velocity at the surface. This level at which negligible motion occurs is known as the *wave base.*

The velocity of the waveform represented previously in the equation *v = L/T* is not the same velocity as that of the water particles. Instead, it is the rate of propagation for the surface wave itself. Water par-

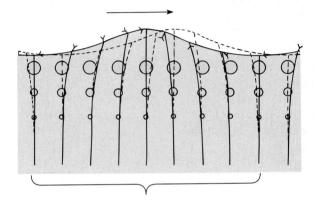

FIGURE 16.2   Movement of water molecules at depth as the wave travels along the surface.

ticles moving along the circular path travel at a considerably lower velocity.

Attenuation with Depth

Examples of the attenuating effect with depth have been documented. For a wave with a wavelength of 330 ft (100 m) and a height of 16 ft (5 m), traveling at 28 mph (45 km/hr), the orbital velocity of the surface particles is about 4.4 mph (7.0 km/hr). At a depth of 65 ft (20 m) the orbital motion is only about 1.2 mph (2 km/hr), whereas at 330 ft (100 m), movement is negligible. By contrast, for a wave with a wavelength equaling 1250 ft (380 m) and a height of 33 ft (10 m), traveling at 56 mph (90 km/hr), the surface particles have an orbital velocity of 4.6 mph (7.4 km/hr), but at a depth of 330 ft (100 m) the orbital motion is still nearly 1 mph (1.6 km/hr).

Wind Velocity Versus Wave Velocity

For oscillatory waves generated by the wind, periods from 10 to 20 sec have been observed. Wind supplies the energy to waves in two ways: 1) by pushing against the wave crests and 2) by friction of the air against the water surface. Wind blowing at 31 mph (50 km/hr) will produce a wave 22 ft high (6.7 m) with a 250-ft wavelength (76 m) traveling at 25 mph (40 km/hr). By contrast, wind at 69 mph (110 km/hr) provides a wave 48 ft high (14.5 m), with a wavelength of 1230 ft (376 m) traveling at 55 mph (88 km/hr). The energy of the resultant waves increases very rapidly with increasing velocity of the generating wind. It appears to vary as the fourth or fifth power of the wind velocity.

Waves from storm winds are transferred into waves called *swells* by interference of these waveforms to yield larger waves. These typically have periods in the range of 8 to 12 sec. Sea captains have estimated wave heights at more than 80 ft (24 m) in hurricanes and sailors aboard U.S. naval ships have sighted waves with maximum heights of 115 ft or so (35 m) during these storms.

Tsunami

Waves with greater periods are generated not by the wind but by other phenomena. Short-period waves have lower wave heights and less wave energy. Tsunami or seismic sea waves develop into long-period waves of from 800 to 3000 sec. Tsunami (from the Japanese, "tsu" for harbor, and "nami" for wave) are generated by fault displacement in the ocean basins some distance offshore. Sometimes referred to as "tidal waves," which is a misleading and improper term, tsunami typically develop wave heights up to 50 ft (15 m) or more when approaching the shore. Because the terms tidal wave and tides are too similar and are likely to lead to confusion, tidal wave should not be used. Tides as contrasted to tsunami are waves generated by the gravitational attraction of the Moon, and to a lesser extent the Sun, on the Earth's surface. A period of from 8000 to 9000 seconds is associated with the tides.

Tsunami have long wavelengths of from 35 to 125 mi (55 to 200 km) and travel across the ocean at speeds between 310 to 560 mph (500 to 900 km/hr). These waves differ from wind-generated waves because their energy is transferred to the water from the seafloor so that the entire water column is involved in the wave motion. In the open ocean the wave height of a tsunami is only 1 to 2 ft (20 to 60 cm) so that ships at sea may not perceive their passing. However, when the tsunami approaches the shore, significant changes occur. Energy distributed in a deep column of water now becomes concentrated in an increasingly shorter one, resulting in a rapid increase in wave height. This is why the waves increase to 50 ft (15 m) in height and in rare situations to more than 100 ft (30 m). Such waves can cause enormous damage to low-lying coastal areas.

Nearly all tsunami originate in the Pacific Ocean within active earthquake zones. Total evacuation of low-lying areas in the path of the tsunami is the only effective defense. Elaborate warning systems for the coastal areas of the Pacific Ocean have been devised to alert people in time to move to high ground.

Waves in Shallow Water

In shallow water oscillatory waves change their mode to become translational waves. The waves begin to drag or "feel the bottom" when the water depth decreases to about one-half of the wavelength and the wave base is contacted. When this occurs, the wave height increases, the wavelength grows smaller, and the velocity of the waveform decreases. Progressively the wave front becomes steeper until the wave finally breaks. After the wave breaks, a translational wave takes over and water expends its energy running ashore and up the beach as wave runup or swash. This process of wave movement in shallow water is shown in Figure 16.3.

An empirical relationship between wavelength, water depth, and wave height is also shown in Figure 16.3. The wave reaches its peak at about $d = 2H$ where d and H are water depth and wave height, respectively. This occurs at about a water depth of $d = L/2$ with L being the wavelength. The wave breaks at about $d = 1.3H$ and either a predominantly plunging wave or a spilling wave develops depending on the configuration of the ocean bottom. A plunging wave is formed on steep, smooth bottoms where the crest is propelled ahead of the wave. A spilling wave occurs on smooth, flat bottoms with the crest spilling down in front of the wave. Most breakers actually fall somewhere within these two extreme cases.

The force of waves is an important agent for coastal erosion and sediment transport. Its magnitude is primarily a function of wave height and frequency, with winter storms typically yielding the greatest force. Such forces against a vertical wall have been measured for the New England area. The study showed that hydrostatic pressure against a vertical wall occurs as a consequence of an upward surge of

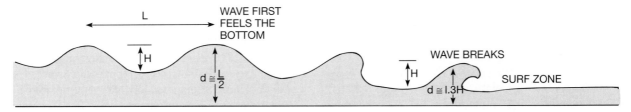

FIGURE 16.3   Waves breaking in shallow water.

water. A smaller, reflected wave moves away from the vertical wall. In this situation the water is sufficiently deep in front of the wall that no breakers develop prior to striking it, which of course would expend some energy. Values for this dynamic impact indicate a maximum destructive force of about 6000 lb/ft² (287 kPa) with an average of about 2000 lb/ft² (96 kPa) in the winter and 600 lb/ft² (29 kPa) in the summer.

Wave Paths and Erosive Forces
Wave Refraction
When waves approach the shore, they are *refracted,* or bent, in a natural effort to strike perpendicular to the shoreline. In this regard, waves can be depicted either by their crest line or by their ray path, the latter indicating the direction of propagation. Hence

the crest line tends to parallel the coast as refraction occurs. This is shown in Figure 16.4.

Wave refraction concentrates energy on the headlands and directs it away from the bay areas between them. As the wave moves toward the shore, the portion offshore from the headlands encounters or "feels" the bottom first and is slowed. Meanwhile, the rest of the wave continues forward without velocity reduction to be refracted later only when it drags the bottom. This causes progressive bending of waves toward the highlands, which finally receive the concentrated force of the breakers. As a result, the erosive force in the bays is minimized and sediment is commonly deposited there to form beaches. In this way, coastlines are straightened as the promontories are reduced by erosion and deposition builds up in

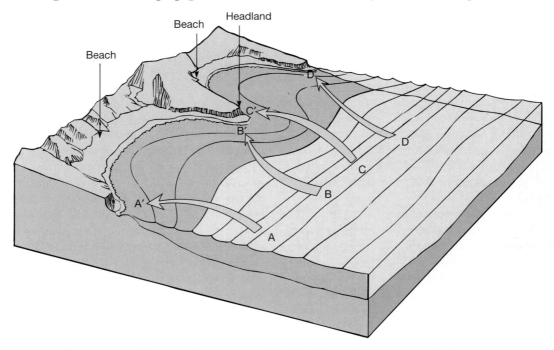

FIGURE 16.4   Wave refraction along an irregular shoreline, showing raypaths (A, B, C, D) and crestlines.

the bays. An aerial photograph showing wave refraction in a coastal area is shown in Figure 16.5.

Longshore Drift

Although wave direction is changed by refraction as the waves drag the bottom, as mentioned, they typically strike the shoreline at an angle to the perpendicular. This provides a component of movement parallel to or in the "downcoast" direction. Longshore drift (or littoral drift) of sediment is generated as waves strike the shore obliquely. Details of this process are shown in Figure 16.4.

As the wave strikes the shore at an angle, water and sediment are propelled obliquely up the beach. When the wave is spent, water and sediment are returned down the beach, perpendicular to the shore. Successive waves move the sediment obliquely up the beach and the backwash returns it again. In a series of

FIGURE **16.5** Aerial photograph of wave refraction along a coastal sand deposition shows the longshore drift move toward the observer; a jetty accumulates the sand.

steps the net transport is parallel to the beach in the component direction. This is known as beach drift, and the current that occurs in the breaker zone is the longshore current or littoral current. These two zones of movement provide transport parallel to the coast, the first along the beach as described and the other in the surf and breaker zone where material is transported by suspension and saltation. The combined action of these two comprises the longshore drift.

A great volume of sediment can be transported by the longshore drift, and for this reason beaches have been described as rivers of sand. As long as a continual supply of sediment occurs from the upcoast direction, the beach may show little change within a season. Increased or decreased sediment supply, typically brought about by human activities, may greatly disrupt this balance. More details on this subject are provided later in the chapter.

From one season to another or because of differences in direction between prevailing winds and storms, waves may approach the shore at different angles. This will change the intensity of the longshore drift or even reverse its direction. Longshore currents build their strength with increased distance down the shoreline. Ultimately they return seaward through the breakers in a narrow zone known as a rip current, which can be extremely dangerous to swimmers.

In a general sense, the longshore drift for coastal areas of the United States moves southward. This holds true for the Atlantic and Gulf Coasts generally, as well as for the Pacific Coast and for Lake Michigan. This is a consequence of the prevailing winds and storms, which come from a northwest or northeast direction. Locally, the direction of longshore drift may differ from this because of shoreline configuration, tributary streams, offshore islands, or other complications.

For Lake Michigan, the longshore current moves southward on both the western side of the lake (the Wisconsin–Illinois shorelines) and the eastern side (the Michigan shoreline). Indiana Dunes National Lakeshore is located at the southern end of Lake Michigan where the migrating sand has accumulated in large dunes along the shoreline (Figure 16.6).

Wave energy affects shoreline erosion along with the previously mentioned features of wave direction. High-energy waves that are very steep (that is, they have a high height-to-length ratio) cause beach erosion, whereas less steep, smaller waves provide beach

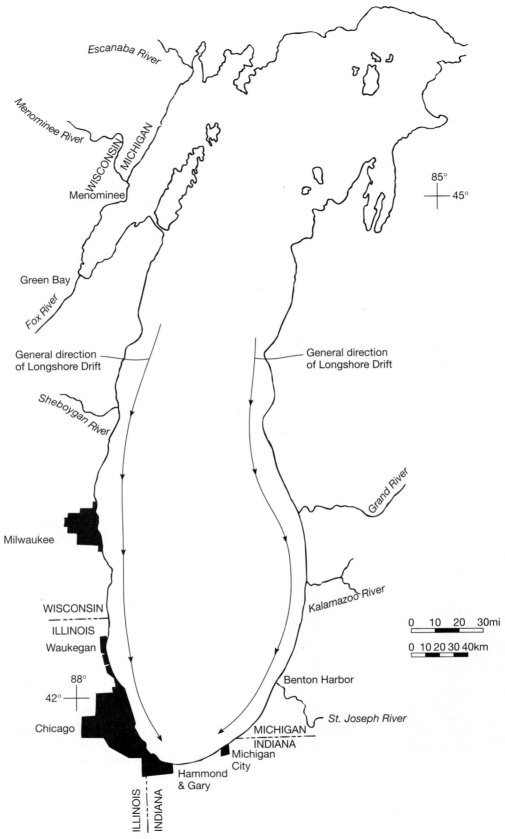

FIGURE 16.6 Map of Lake Michigan showing directions of longshore drift.

replenishment. Intense seasonal storms have resulted in the concept of a winter beach, where the beach is eroded, narrow, and deficient in sand, as opposed to a summer beach with a wide, sandy expanse. These extreme conditions do not occur along all beaches because of the different intensities of winter storms from one coastal area to another.

TIDES

Location and Frequency of Tides

The tides are daily fluctuations in the level of the sea caused by the attraction of the Moon and, to a lesser extent, the Sun on Earth's surface. Both the side of Earth facing the Moon and that directly opposite it will experience a high tide. As the Moon revolves around Earth, these bulges in the water surface follow along. At the intermediate locations, 90° away or one-fourth of the Earth's circumference, the coastal areas experience low tide. Because of this relationship, in a period of one day typically two high tides and two low tides are experienced. More accurately, the double cycle repeats itself once every 24 hr and 50 min not every 24 hours, as the Moon advances eastward in its orbit. The bulging of the tides relative to the position of the Moon is depicted in Figure 16.7.

Causes of Tidal Fluctuations

The bulge of water on the opposite side of Earth from the Moon can be explained in two ways. First, because the center of Earth is closer to the Moon, it is pulled more strongly toward the Moon than is the water on the opposite side of Earth. Hence, Earth in effect is pulled away from its water body on the side opposite the Moon. The other explanation is that Earth's centrifugal force, caused by its rotation, exerts the greatest influence on the opposite side of Earth as it pulls against the Moon's gravitational force at that point.

The Moon exerts more influence on the tidal fluctuations than does the Sun despite the Sun's much greater mass and greater attractive force. This is true because Earth's diameter of 8000 mi (12,800 km) is more significant when compared to the distance between Earth and the Moon—200,000 mi (320,000 km)—than when compared to the Sun's distance of 93×10^6 mi (149×10^6 km). The net result is that the Sun's effect on the tides is just less than half (0.46) that of the Moon.

Spring Tides and Neap Tides

Twice each lunar month, at new moon and full moon, the Sun and the Moon lie in a straight line and their influence on Earth is additive. These extremes are called spring tides. In like fashion, twice each lunar month, at first and third quarters, the Sun and the Moon are in opposite positions and the tides are their smallest. These are known as the neap tides.

Details on Tidal Extremes

The tides are not as simple a phenomenon as the previous discussion would suggest because 1) many parts of the world have only one tide per lunar day, 2) in other areas the high tide lags many hours behind the time when the Moon passes overhead, and 3) in still other locations the two daily tides are of greatly different heights.

In reality, several components comprise the tidal effect, and their relative contribution is a function of Earth latitude. The semidiurnal tide effect mentioned previously has its maximum contribution at the equator and a minimal one at the poles. Another diurnal effect has its maximum contribution at 45° latitude and none at the equator.

Theoretically the maximum fluctuation of the tides should only be about 20 in. (50 cm), but values of more than 33 ft (10 m) are experienced in some areas of the world. Tides are affected by the geometry of the coastline, the location of the landmasses, the variable water depth of the oceans, and the Coriolis force, which deflects tidal currents. This force, a

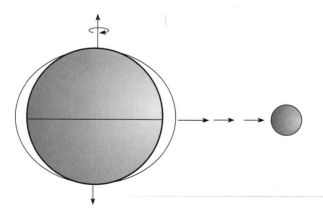

FIGURE 16.7 Earth and Moon system showing the bulges in the oceans toward and away from the Moon.

result of Earth's rotation, is an apparent force experienced by moving objects, which trends to the right in the Northern hemisphere and to the left in the Southern. Because of these complications, tide charts for the world are prepared based on empirical relationships involving past observations of tidal effects.

Locations with Maximum Tidal Effects

Mouths of estuaries and straights between bodies of water whose maximum water levels are not in phase with the local high tide, experience strong tidal currents. In the English Channel between England and France and into the North Sea, this effect is quite pronounced and tends to increase the extremes of high and low tide. In another extreme example, water flowing under the Golden Gate Bridge in San Francisco has a daily flow of 2.3×10^6 ft³/sec (6.5×10^4 m³/sec) or 3.5 times the discharge of the Mississippi River.

In some rivers the tide enters as a rushing wave of water called a bore. The bore of the Hangchow River in China can be as great as 16 ft (3 m) with speeds up the river at 16 mph (25 km/hr). Smaller tidal bores occur in Nova Scotia and New Brunswick at the head of the Bay of Fundy.

As previously mentioned, the configuration of the coastline has a pronounced effect on the tides. Irregular coastlines with long estuaries extending inland provide the greatest tides. As the tide advances, the wall of water funnels into the estuary and, due to the decreasing cross section, the water height must increase to accommodate a constant water volume. In long estuaries the tidal rise can be multiplied more than tenfold or even twentyfold by this aspect. Table 16.1 provides a list of the tidal differentials for a few selected areas of the world.

The Bay of Fundy, between New Brunswick and Nova Scotia, has the greatest tide range on Earth, 54 ft (16 m) as a spring tide. It is situated at the head of an estuary 62 mi (100 km) long, and one consequence is a reversal in the flow direction of tributary streams between high and low tide. Tidal currents traveling 9 mph (15 km/hr) during the rise and fall of the tides have scoured basins 150 ft (45 m) deep in the bay.

In La Ronce, France, on the Brittany Coast an estuary 62 mi (100 km) long has been developed for tidal power generation. At this location the maximum tide range is 45 ft (13.7 m). Currents reach 8 mph

TABLE 16.1 Maximum Tide Range for Selected Locations of the World

Location	Range (ft)	Range (m)
Bay of Fundy, Nova Scotia	54	16
La Ronce, France	45	14
London	20	6
Boston	9	2.7
New York City	6	2
Key West, FL, and Galveston, TX	2	0.6
Cook Inlet, Anchorage, Alaska	35	11
Honolulu, Hawaii	1.5	0.5
Seattle, Washington	16	5

(13 km/hr) during rise and fall of the tide. A concrete dam 1200 ft (366 m) long, 125 ft (38 m) wide at the base, and 85 ft (26 m) high above the foundation level has been constructed. It consists of 24 concrete bays. Each bay contains a 10,000-kw reversible pump turbine, the combination of these delivering some 624 million kilowatt hours per year at the present time. With the incoming tide the water is run through the turbines into the head of the bay to generate electricity. When the water level on the ocean side is only a few feet above that of the bay, generation is curtailed waiting for the outgoing tide. Water is also pumped from the ocean side to the bay side as the tide retreats, thus building a greater differential head. When the tide has sufficiently ebbed, the turbines are reversed and water flowing from the bay side to the ocean side again generates electricity. Obviously, a considerable time interval occurs within each cycle of the tides in which no power can be generated. The tidal power station is tied into a large, power-generating network so that other generating plants can supply the load when La Ronce is not producing.

An interesting difference in tidal fluctuations occurs for the Panama Canal. On the Atlantic (Caribbean) side the tides are less than 1 ft (0.3 m), whereas for the Pacific side they are 13 ft (4 m).

EFFECTS OF SEVERE STORMS

Two distinct types of processes act to modify the coastline. The first provides a gradual change and acts in the slow, continuous fashion of erosion and deposition. The tides and longshore currents provide this regular, continuous action. The second type causes rapid changes and acts in an irregular, intermittent fashion. These are the catastrophic storm events that yield flooding and severe erosion of coastal areas when they occur. Severe storms, hurricanes, and typhoons involve very high wave energies, which develop as a consequence of low-pressure cells in the atmosphere. They yield four related effects: 1) storm surge, 2) high wave energy, 3) extreme rainfall, and 4) high winds.

Storm surge is manifested by a rise in elevation of the water surface because of the low-pressure conditions. This surge commonly affects the shoreline long before the direct storm activity hits. As the storm approaches the wind on shore, increases in velocity and the wave heights increase as well. As the storm moves inland, the previously flooded coastal area is subjected to further flooding from the heavy rainfall. Erosion of the beach and dunes can be excessive during such a storm. In some instances the storm waves will breach the foredune and rapidly flood the inland area.

EROSION IN COASTAL AREAS

Coast is a general term that designates the broad area adjacent to the sea. It typically consists of a combination of features, which may include steep cliffs, low-lying beaches, bays, tidal flats, or marshes. The topography that develops is a function of the uplift or subsidence of the coastline relative to sea level; the geologic materials involved; the processes of erosion and deposition by the sea; and elapsed time since relative movement between land and sea has occurred.

Factors in Wave Erosion

Waves erode the coasts in two primary ways: 1) by impact and hydraulic pressure and 2) by abrasion—the grinding action of sand, gravel, and cobbles on the cliffs or across the foreshore. The massive impact of water against a sea cliff can exert pressures up to 3000 psf (144 kPa) during major storms with pressures of only one-quarter that on lesser storms. Water is also driven into every opening and fracture in the rock, compressing the air against the solid boundary. This acts as a wedge to widen fractures and to loosen blocks of rock.

Landform Features in Massive Rock

Features that form by this wave action include sea cliffs, wave-built terraces, and wave-cut benches, which are illustrated in Figure 16.8. In coastal areas composed of steeply sloping resistant rocks, the wave action cuts a horizontal notch into the rock at sea level. By subsequent erosion this may yield sea caves, sea arches, or sea stacks. Sea arches occur when notches encroaching from two sides of a rocky promontory connect to form a passageway through the rock. When the arch eventually collapses, an isolated pinnacle remains, known as a sea stack. In these same areas, water spouts and potholes in the rocky cliffs commonly provide pleasant tourist attractions.

Undercutting of the cliff also causes overhanging rocks to collapse, with the fallen debris soon carried away by wave action. With a new cliff face exposed the process is repeated and by this progressive action the sea cliff retreats. As a consequence, a wave-cut bench is produced, typically with the upper portion visible at low tide. Sediment transported by longshore currents can accumulate in deeper waters to form a wave-built terrace. These features are illustrated in Figure 16.9.

Equilibrium Conditions

As the bench is widened, the waves break farther and farther from the shoreline. Their energy is dissipated by friction as they travel across the shallow water on the beach. This greatly reduces the wave action on the cliffs, and beaches can develop. The cliff face is now subject only to weathering and mass wasting. In effect, an equilibrium condition has developed involving the length of wave-cut bench, the sea cliff, and the wave action of the sea for the established elevation difference between the landmass and sea level. If this relationship should change because of tectonic uplift of the land or a worldwide alteration in sea level, a renewed erosional sequence would be set in motion.

Impact on Human Activities

The extreme effects of coastal erosion are of particular concern to human activity because of the desire to prevent landform alteration. Because property and human activity are affected by landform change, humans strive diligently to minimize those alterations that natural processes bring about. The profound effect of wave action on shore protection structures is

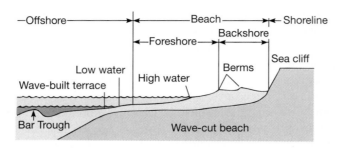

FIGURE 16.8   Cross section of features along a shoreline.

well documented. At Wick, in northern Scotland, storm waves have eroded away masses of concrete up to 2500 tons (about 2300 metric tons) from breakwaters built for shore protection. At Ymuiden, Holland, a 7-ton (6.3 metric ton) block in the breakwater was moved nearly a meter by the sea. The Chalk Cliffs of Dover are so rapidly undercut by the sea that occa-

sionally large landslides occur. A major landslide in 1810 caused earthquake vibrations felt in Dover, which is located several miles away.

The erosion rate of poorly consolidated materials by wave action of the oceans can be extreme, with an average retreat rate of 5 to 6.5 ft (1.5 to 2 m) per year a possible occurrence. Along parts of England's coast-

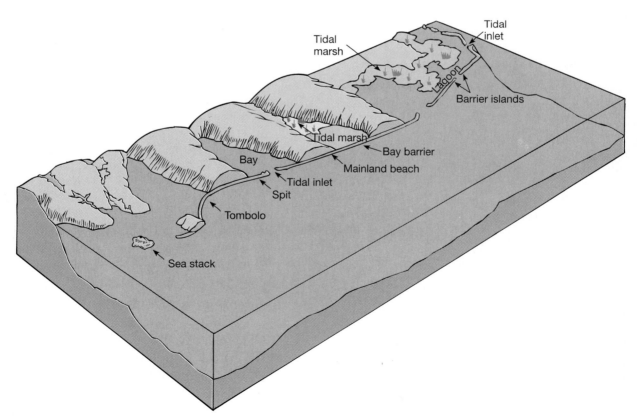

FIGURE 16.9   Depositional features along a coastline.

line, the shore has retreated more than 3 mi (5 km) since Roman times so that a number of coastal villages and landmarks have disappeared. The effect of erosion in Lake Michigan during the mid 1970s was also profound. In the suburban areas north of Chicago several meters of erosion occurred within a 2-year period. Many homes, beach cottages, and recreational areas were severely damaged during this period when the lake level was higher than average.

Deposition in Coastal Areas

Longshore currents and waves transport sediment along the coast until it is deposited in areas of low energy. These landforms include beaches, spits, tombolos, bars, and barrier islands. Small beaches may develop during the active erosional sequence in the bays between prominent rocky headlands, but the other depositional landforms are more common on smooth or gently curved coastlines. These coastlines occur when shoreline equilibrium has been established in the manner previously discussed.

Beaches

Beaches are the gently sloping portion of the shore consisting of unconsolidated sediment. Typically sand will dominate but some beaches are composed of cobbles and boulders, yet others are silt and clay. In general, beaches have a similar morphology, as illustrated in Figure 16.8. The characteristics of a beach— its slope, composition, and size—depend on wave energy and the supply of sediments. Beaches composed of fine-grained sediments are typically flatter than those made up of coarse sand and gravel.

Rivers that carry sediments from the landmass are the primary source of material for beach formation. This sediment is transported by the longshore current down the coast. If the source of sediment is curtailed by damming the river or by shore protection devices, this transport of sediment is interrupted and serious changes may occur downcoast. This problem is addressed in a later section.

The beach itself consists of two portions, the foreshore or the strand, which lies between low tide and high tide, and the back beach formed by two berms (see Figure 16.8). Both the lower and upper berms are the consequence of storm activity and mark the boundary of the beach. Landward are found cliffs, dunes, or limiting features that denote the extent of the erosive process of the sea.

Other Coastal Landforms

Sea terraces are found in regions with emergent coastlines, that is, those areas where the coast has recently moved upward relative to the sea. This marks the location of the beach and wave-cut bench, which were formed when that portion of the landmass stood at sea level. When such features are observed and the differential movement between the ocean and the land is noted, one must decide whether the land moved up while the sea level remained stationary or if the land remained fixed and the sea level went down. Of course, if the sea retreated, worldwide evidence of this event should be present. Evidence indicates that emergent coasts typically involve tectonic uplift of the landmass rather than a drop in sea level.

SPITS AND BARS Spits are extensions of the beach across indentations in the shoreline such as bays or estuaries (see Figure 16.9). Spits may continue to grow out into the bay as materials accumulate at its end. With continued growth the spit may completely close the front of the bay. Such features are called *bar barriers* or *baymouth bars*.

An extensive spit may acquire a tip that is recurved, either by further wave refraction or by storm activity. This feature with a claw-shape at its end is simply termed a *recurved spit* or a *hook*. A *tombolo* occurs when a spit connects the mainland to an island. These features are shown in Figure 16.9.

A bar is an offshore, submerged, elongated body of sand built by wave action or by longshore currents. Bars may be partially exposed at low tide and may eventually develop into a barrier beach.

Barrier Islands

Barrier beaches are prominent landforms that occur along the Atlantic Coast from New Jersey to Florida and also prevail along the Gulf Coast from Texas to Florida. Also called barrier islands and offshore beaches, they parallel the shoreline, consist of elongated masses of sand, and are located some distance offshore. Typically they are separated from the mainland by a lagoon and most are cut by one or more tidal inlets. Individual islands may range from 330 ft (100 m) to more than 1000 mi (1600 km) long and generally have a width of less than 2.5 mi (4 km) with an average of only 0.25 to 1 mi (0.4 to 1.6 km). The islands usually lie less than 10 ft (3 m) above sea level although sand dunes may rise as high as 40 ft (13 m).

The barrier islands that form the Outer Banks of North Carolina are shown in Figure 16.10.

Barrier islands are depositional features formed by detrital sediments (sand and gravel) and are therefore distinguished from organic formations such as reefs. They are transitory because they are continuously undergoing changes as a result of natural, dynamic processes, which more recently have been complicated by the activities of humans. These islands are thought to have originated on submergent shorelines through either the emergence of offshore bars, the continual growth of complex spits, the submergence of coastal beach ridges, or some combination of these occurrences. Humans are greatly interested in preserving the barrier islands because of their natural beauty and commercial value, so that some of the real challenges of beach erosion prevention are experienced in these areas. The subject of beach protection structures is considered later in this section.

Dune Ridges

In many areas a dune ridge or dune line is associated with the sand beach along an ocean or major inland lake. In a natural state the dune line migrates forward and back in response to beach erosion and deposition, in keeping with the continuously acting processes of tides and longshore currents and the intermittent processes associated with severe storms. The dune system may take the form of a primary and secondary dune with a trough area between them. This is illustrated in Figure 16.11 along with the backdune, bayshore, and bay areas.

HUMAN ACTIVITY IN DUNAL AREAS   When humans develop a coastal area for housing and recreation, the first tendency is usually to stabilize the dune line. This protects the local residences and provides a roadbed for the primary transportation route. Eventually small towns and villages develop in these recreational areas and the natural changes such as tidal inlet openings and dune migration or beach overwash are no longer acceptable. The dune line must be kept in place, with overwash and inlet formation prevented.

Stabilization of the dune line, however, stops the processes of migration behind the dunes. As the sea

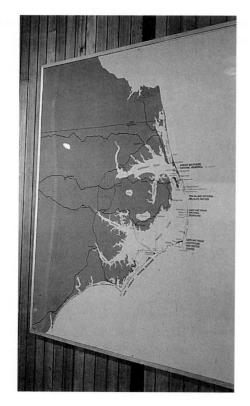

FIGURE **16.10**   Barrier islands forming the Outer Banks of North Carolina.

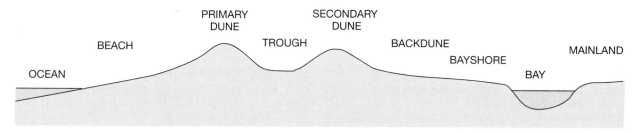

FIGURE **16.11**   Dune features developed on barrier islands.

level rises, which it will intermittently because of storm surges and other phenomena, the beach zone shifts landward. But because the dunes no longer migrate with this shift, the width of the beach and berm is reduced and wave energy is concentrated over a smaller and smaller area of runup. This throws the system out of equilibrium and leads to long-term dune erosion.

Under natural conditions the migrating dunes give up their sand to the local system. That sand is, in turn, redeposited nearby. Dune and beach replenishment occurs when the lower water level is reestablished. However, for stabilized dunes the sands are continually eroded and taken from the system by longshore currents carrying it some distance downcoast. Eventually the wave energy will be dissipated directly against the dune line. Vegetative cover provides erosion protection to the dune surface but does not prevent undercutting. Breaching will eventually occur and with it overwash, both of which greatly concern local residents. The response is to plug the gap as soon as possible and rebuild the dune. Closing the breach by simple sand replenishment is not often a successful long-term solution. Commonly, the next storm breaches the dune in the same area. A more successful procedure is to encourage dune growth with slotted fences (same as snow fences), which have a better chance for permanently reestablishing the dune. In any event, when the dune line is held stationary, the ultimate likelihood of breaching the dune is markedly increased.

DUNE RIDGES ON BARRIER ISLANDS   The dune ridge on a barrier island is affected in yet another way when a continuous dune line is established by humans in order to prevent outwash and tidal delta development. Because of this exclusion of the sea at high tide, the interior portions of the islands, including marshy areas in the lagoons, become progressively lower relative to the rising sea. Closing the natural inlets reduces sediment supply to the marshes and slows the accumulation of organic matter because of this curtailment of fresh sand. The results are twofold: The marsh areas subside, yielding in effect a thinner island, and the barrier loses the buffering effect of the marshes against wave energy dispelled through the inlets. Therefore, the islands are eroded from both sides and are more susceptible to flooding from lower

level surges. With no outlet through the dune to the ocean, surges from the lagoon pile up against the inside of the dune and flood extensive areas.

Hurricanes crossing barrier islands will attack the dune line from both sides as a result of their rotating winds and low atmospheric pressures. The level of the interior will be raised by overwash through breaching of the dune as new sand is added to the backshore and marsh areas. Seaward, the dune line will suffer severe erosion and some beach areas will be nearly devoid of sand. In built-up areas, large sums of money will be required to rebuild dunes and beaches and to remove sand deposited on roads and in residential areas. Flood damage caused by surges from the back bay will also be extensive. In natural areas the damage, measured monetarily, will be less as the dune and back dune area should repair themselves in time through the reestablishment and migration of the dune system.

ESTUARIES

Estuaries are large-volume streams, emptying into the ocean, which flow on very low gradients and experience the ebb and flow of the tides. They are common to drowned coastlines such as the Atlantic seaboard from New England to Washington, DC. Here the land has submerged relative to the sea and submarine canyons mark the location of the downstream end of channels formed during lower ocean levels in the geologic past.

Uses of Estuaries

Estuaries provide excellent harbors because of the shelter of the land. New York City, Boston, Washington, DC, Baltimore, and Norfolk are examples of harbors in estuaries in heavily populated areas. They occur along major rivers, for example, the Chesapeake, Potomac, Delaware, and Hudson Rivers. Estuaries are the center of activity because of this population density and the multiple-use capabilities of the waterways: shipping, fishing, mining, industry, farming, and recreation. Unfortunately, all of these activities cannot coexist without some level of cooperation among these diverse interests.

Pollution of Estuaries

Pollution and its specific causes also comprise an extensive list for estuaries. Major contaminants in-

TABLE 16.2 Contaminants and Their Sources in Estuaries

Contaminants	Major Sources
Pathogenic organisms	
Organic matter	Municipal wastes
Nutrients	
Pesticides	Agriculture
Herbicides	
Heavy metals	
Oil	
Flesh-tainting substances	Industry and shipping
Toxic chemicals	
Heat	Power plants
Sediments	Agriculture and construction

clude municipal wastes, industrial wastes, heat, oil, toxic chemicals, pesticides, insecticides, nutrients, flesh-tainting substances, and sediments. These must be controlled if the multiple uses of the estuary are to be maximized. The primary causes for the contaminants designated above include municipal sewage treatment plants, industry, power-generating stations, farming, construction, and shipping. In Table 16.2 the contaminants are matched with the sources that cause the pollution.

Some of the interactions involving these pollutants and the various uses of estuaries require further discussion. Flesh-tainting substances, those organic liquids contributed by industrial wastes and oil from shipping, discolor both finfish and shellfish, making them unsuitable or unappetizing for human consumption. This can be minimized by proper enforcement of sewage treatment specifications and cessation of the practice by ships of discarding oil and other nondesirable liquids just prior to docking.

Pesticides, insecticides, and nutrients used in farming affect both fishing and recreation. Use of these pollutants should be kept at the lowest level possible. Construction, farming and possibly mining also yield sediments to the estuaries, which on deposition cause a negative effect to shipping, fishing, and recreation. Proper construction, mining, and farming practices can minimize the amount of sediment provided to the streams by erosion.

Heat is supplied by power-generating stations, both fossil fuel and nuclear power plants. Heat loss

into streams should be minimized through the use of cooling ponds with sufficient retention times to preclude the supply of warm water to surface streams. The proper use of cooling towers and mechanical heat exchangers can also lower the temperature of the effluent to an acceptable level.

To a certain extent, several uses of estuaries are potentially at odds with other uses. Mining coexisting with fishing and recreation seems difficult to attain, as is shipping with fishing and recreation. For this reason, it has been proposed that certain areas of a large estuary system be zoned for a specific group of uses much as a county is zoned according to certain types of buildings and purposes: residential, light industrial, heavy industrial, and farming. This arrangement should minimize the conflicts between the various uses and environmental impacts for estuaries.

CLASSIFICATION OF COASTLINES

Coastlines can be classified in a variety of ways depending on the purpose or intended use of the classification. Details of origin and/or physical characteristics are generally involved to some extent. In some cases it is advantageous to apply the classification based on recent emergence, or submergence, of the landmass from the sea. This classification is supplied in Table 16.3. One of its advantages is the easy identification from a topographic map of the features used for classification. Other ways in which

TABLE 16.3 Classification of Coastlines Relative to Ocean Level

Submergent Coastline
   Drowned shoreline
   Rivers become estuaries
   Irregular coastline
      Active erosion with sea cliffs
      Numerous rocky islands
Emergent Coastline
   Terraces
   Few bedrock islands
   Straight coastline with baymouth bars, spits, and
      large beaches
Compound Coastlines
   Contain features of both submergent and
      emergent types or have no distinguishing
      characteristics

TABLE 16.4 Additional Classifications of Coastlines

Lithology of Shoreline
    Igneous and metamorphic rocks (massive rocks)
    Sedimentary rocks (less massive rocks)
        Durable sandstones
        Less resistant, carbonates, weakly cemented
           clastics
Geomorphic Process of Formation
    Glacial
    Littoral
    Tectonic
    Reef rocks, carbonates

---

shorelines are classified are presented in Table 16.4. Two examples of classifications are presented. Both of the classifications in Table 16.4 have some shortcomings because neither includes all types of coasts. They do, however, provide a valuable basis for comparison of coastal features.

SHORELINE PROTECTION STRUCTURES

Several types of shoreline protection structures are used to prevent or reduce the effects of erosion and deposition of sand along a shoreline. These include groins, jetties, breakwaters, seawalls, bulkheads, and revetments. Groins and jetties are typically built perpendicular to the shoreline extending into the breaker zone; breakwaters are built some distance offshore for harbor protection; and seawalls, bulkheads, and revetments are placed parallel to the shoreline separating the land from the water and providing protection against wave attack.

Groins and Jetties

Groins and jetties are structures used to trap or modify the longshore transport of sands. This sand is supplied to a coastal area by two distinct means: rivers transporting sand, which is produced by weathering of quartz-rich rocks, from upstream areas, and erosion of the upcoast region of the shoreline by longshore currents. In the first situation, sand enters the longshore drift at the mouths of rivers. If dams are constructed on the upstream reaches of such rivers, the fluvial sediments (sands) are trapped behind the dams and the reduced sediment supply encourages erosion downcoast from the river's mouth.

The manner of construction of groins and jetties is quite similar. The distinction instead exists in their intended use and placement along the shoreline. A groin is used to extend a beach area or to retard erosion by trapping longshore drift. A jetty, by contrast, extends into the water to direct or confine river or tidal flow in a channel and to prevent or reduce shoaling (sediment buildup) of the extended channel by longshore drift.

Groins are generally placed along continuous sections of shoreline, whereas jetties are built to extend the river banks out into the river mouth, tidal inlet, or harbor entrance. Jetties at the entrance of a channel (see Figure 16.5) also protect the channel from wave action and crosscurrents. These structures are illustrated in Figure 16.12. Photographs of various shoreline protection structures are shown in Figure 16.13.

As previously indicated, groins and jetties are both constructed perpendicular to the shoreline and vary in length between 100 to 300 ft (30 to 90 m). Beach extension will equal about 50% of the structure's length with the structures spaced at one to three times that length apart. Wider spacing is commonly associated with a continuous, straight shoreline. A limiting factor to groin length (and hence for the additional beach width they provide) is that groins extend only to the breaker zone because they are not very effective beyond that point (or to a water depth of about 2 m). Groins may be classed as high or low, long or short, fixed or adjustable, and permeable or impermeable; they may be constructed of timber, steel, stone, or concrete.

Groins and jetties usually induce erosion in the downdrift region because of their interruption of sediment transport. Sand passing these structures will not reach the shoreline for a distance of three to five times the structure's length. To alleviate sediment loss, sand can be pumped from the updrift side of the jetty or groin to the sand-starved beach area below, bypassing inlets where present. Studies should be made to determine the effects of a groin or jetty design for the specific beach in question.

Breakwaters

Breakwaters are structures used to protect shore areas from wave attack by reducing or eliminating wave action. They cause the waves to break some distance offshore, expending their energy in that location rather than immediately adjacent to the

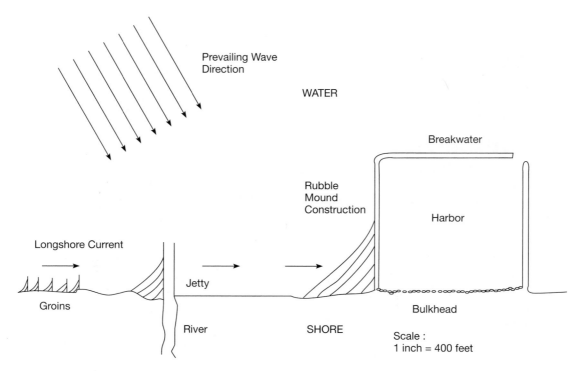

FIGURE 16.12   Shoreline protection devices.

shore. These structures may be shore-connected or offshore breakwaters and may be constructed of a rubble mound or a cellular steel–sheet pile structure. The rubble-mound variety is the stronger of the two structures, is adaptable to almost any water depth, and can be designed to withstand severe wave action. It may contain armour stone rock pieces as heavy as 10 metric tons each. The cellular steel–sheet pile combination structure is used for lower energy conditions. Jetties and groins can be used in combination with various breakwater designs and configurations for shoreline structures.

With the elimination of wave action by breakwaters, longshore transport is greatly reduced and the downdrift beaches are deprived of their normal supply of sediment. In the case of an offshore breakwater, sand is deposited in the quiet area between the structure and the shore. When breakwaters are built updrift of inlets, they impound sand, preventing the shoaling of the inlet channel. Shelter is also provided for dredging operations from which sand is pumped either across the breakwater into the longshore drift or onto the beach.

Seawalls, Bulkheads, and Revetments

Seawalls, bulkheads, and revetments are generally placed parallel to the shoreline to separate the land from the water. Their primary purpose is to provide protection against wave attack. Bulkheads are also used to provide slope stability for the soil behind the structure. Seawalls are generally the most massive and they are designed to resist the full force of the waves. Bulkheads are next in capacity and their function is to hold fill in place, but they are generally not exposed to severe wave action. Revetments, the lightest of the three structures, are designed to protect the shoreline against erosion from currents and light wave action.

Seawalls may have a curved face, stepped, curved and stepped, or rubble-mound construction. The curved face and rubble-mound types are generally used for higher intensity wave action than are stepped structures. An armour of large stone is needed for the curved face structure to reduce the scouring effect of the waves. Sheet piles are generally used in conjunction with the curve face structure to prevent loss of

(a)

(b)

FIGURE **16.13** Shoreline protection structures. (a) Seawall during construction. (b) Breakwater consisting of armor stone.

foundation materials and to reduce uplift and seepage forces.

Bulkheads can be constructed of concrete, steel, or timber. They consist of a thin wall structure imbed-ded into the sediment, with sand backfill behind the wall and a height of water in front. Tie rods and restraining weights at intervals along the wall hold the flexible sections in place.

(c)

FIGURE 16.13   (c) Groins along a sandy beach; longshore drift is from right to left.

Revetments are sloping structures placed against the shoreline and may consist of ridged, cast-in place concrete, flexible articulated concrete, or a riprap structure. In all cases, relief of the hydrostatic uplift pressure generated by wave action is provided through the use of a gravel or crushed stone filter beneath the structure for its total length. The selection of the specific type of structure for shore protection (seawall, bulkhead, or revetment) is based on foundation conditions, exposure to wave action, availability of materials, and cost analysis.

Figure 16.14 shows an outline for solving coastal engineering problems. It includes many of the protective structures and details discussed in this chapter.

Nonstructural Methods of Beach Protection

Beach protection, much like that for floods along major rivers, can be accomplished by either structural or nonstructural means. Structural means for beach protection have been discussed at length with regard to jetties, groins, breakwaters, and seawalls. Nonstructural means, in contrast, entail the use of the natural sand for beach and dune control and repair.

Beach nourishment involves the addition of sand at the upcoast end of the longshore drift system in order for it to be deposited downcoast in an eroded section. Dune reconstruction is accomplished by storing extra sand in dune areas so that, following storm wave erosion, rebuilding of the dune ridge can be performed. Building regulations and zoning of the coastal area are other nonstructural methods for beach and dune line protection.

SENSITIVITY OF SHORELINE FEATURES TO HUMAN ACTIVITIES

Regarding regulations and zoning for the coastal area, reference should be made once more to Figure 16.11. Each portion of the dune system can be considered in regard to its proper use. First, the ocean is tolerant to intensive recreation and subject to pollution controls. The beach is tolerant to intensive recreation but no permanent buildings should be constructed there. The primary dune is truly intolerant. No passage, breaching, or building should be allowed. The natural

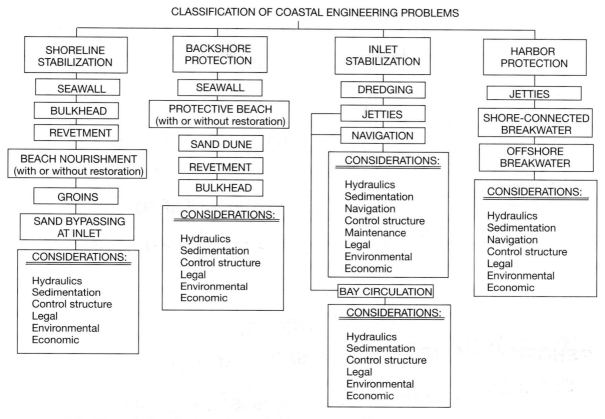

FIGURE 16.14   General classification of coastal engineering problems.

grasses prevent deflation and pathways across the dune destroy this grass. If a passage to the beach is needed, a tunnel should be constructed through the dune or a bridge over it. The secondary dune is equally intolerant. The same restrictions generally hold for it as well.

The trough area between dunes is relatively tolerant. Limited recreation and construction are possible there. It is a proper site for parks and picnic areas with picnic shelters and other day-use facilities. If permanent housing is constructed, it should be widely distributed.

The backdune region is the most tolerant and is suitable for development of housing and other permanent structures. This is also the best location for a highway running parallel to the sea and the dunes. If sufficiently elevated it will provide a view of the sea and tend to negate the need to travel on the

primary dune. If dredging is needed to elevate the roadbed, the sand should be taken from the ocean not the bay. The beach and ocean area are not very rich biological environments, whereas the bay is extremely rich and varied. These biological colonies should be preserved.

The bayshore also is intolerant and no filling of the low-lying areas should be allowed. It is best left untouched. The silts of the bayshore are not suitable for septic tank systems because of their low permeability. Pollution of ground water is a concern for areas with greater permeability. A sewer system and sewage treatment facility are necessary if any degree of population density develops.

Bay areas are tolerant to intense recreation much as is the ocean. Fishing, boating, and shelling are favorite pastimes. If the coastal zone is used carefully, respecting the sensitive environments, then the benefits of this attractive area can be enjoyed without the underlying risk of undue degradation of these landforms.

Erosion of Lake Michigan Shoreline

In 1975–76 the Lake Michigan shoreline experienced extensive erosion primarily caused by the higher than normal lake levels. With the elevated lake surface, wave action against the coastline caused extensive erosion during storm activity. Erosion to the lakeshore was experienced in Illinois adjacent to Chicago, in Indiana along the National Lake Shoreline to Michigan City, and in many areas in the state of Michigan. Losses to property ranging from beach cottages to large mansions and public beach facilities were extensive. Between 1977 and 1982 the lake levels receded, but accelerated erosion continued in some areas for several years, presumably because of the elevated ground-water surface and the response of the slope processes to the initial erosion at lake level. Periodically, lake levels rise in response to climatic conditions and shoreline erosion is reinitiated. Between the times of higher water level the lakeshore becomes more stabilized and generally maintains equilibrium until the next major rise in lake level occurs.

EXERCISES ON COASTAL PROCESSES

Map Reading

Boothbay, Maine; 15-min quadrangle Map

1. What type of shoreline is shown based on the classification of coastlines relative to ocean level?

2. Give two reasons to support your answer to Exercise 1.

3. Why are there swamps in the highlands? (A geologic process not included in the shoreline classification is involved.)

4. Locate by latitude and longitude to the nearest minute a tidal flat.

5. List two highway construction or maintenance problems for this area and indicate how they might be solved.

Oceanside, California; 15-min quadrangle Map

6. What type of shoreline is shown based on the classification of coastlines relative to ocean level?

7. Give two reasons to support your answer to Exercise 6.

8. Why doesn't the San Luis River outlet to the ocean? What happens to the water?

9. Why is there no tidal marsh in Canyon de Los Encinas?

10. What are the relatively flat areas parallel and adjacent to the seacoast? How were they formed?

Tamalpais, California; 15-min quadrangle Map

11. Locate by name or general location:
    (a) tidal flats
    (b) a baymouth bar
    (c) a raised terrace (between what elevations?)
    (d) headland
    (e) sea stack(s)
    (f) a cove

12. How was Lake Lagunitas formed?

Sandy Hook, New Jersey; (Use the Engineering Map Edition)

13. What is Sandy Hook?

14. In view of your answer to Exercise 13, how did Spermaceti Cove form? What has caused the interesting "stairstep" pattern in the vicinity of Normandie (SE part of map)?

15. Why the offset (indention) of the east coast at the end of the piling in the vicinity of C tower (base of Sandy Hook)? What is this structure called?

16. What direction is the littoral sand drift from the town of Highlands to East Keansburg? On what information do you base your decision?

17. Why is the seawall necessary at Atlantic Highlands (along railroad) when there is abundant sand at Water Witch and Atlantic Beach Park?

18. Is there any apparent reason in the contrasting topography and shorelines of the Atlantic Highlands and Belform areas? Explain.

19. What are the straight dark blue lines in the Belford march area called? What is their function?

20. One inch on the map equals how many miles on the ground? Is this a large-scale or small-scale map? Explain.

Oberlin, Ohio; 15-min quadrangle Map

21. What are North Ridge, Middle Ridge, and Butternut Ridge? Did they form at the same time or were they

formed separately? Back up your answer and if you decide they formed at different times, what was the order of formation?

22. What kind of material would you expect to find in the ridges mentioned in Exercise 21?

23. Are there good bathing beaches in this region? Why or why not?

Written Questions

24. If the tides in a certain coastal area have a period of 8000 sec, what is the frequency of the wave in wavelets per day? What is its period in hours? At this rate approximately how many of these long wavelengths occur during the time from high tide to low tide?

25. If an oscillatory wave begins to feel the bottom at a depth of 3 m as it approaches the shore and it has a period of 10 sec, what is its wavelength and wave velocity? What is the wave height at the wave's peak? At what water depth does the wave break?

26. This refers to the example on orbital particle velocity of waves discussed under Wave Motion in the text. The example wave had $L = 100$ m, $H = 5$ m, velocity of the wave trace = 45 km/hr, and the orbital particle velocity = 7 km/hr. The text indicates that the circular orbit with a diameter of H is traveled in the time T. This yields for V_p the particle velocity:

$$V_p = \frac{\pi H}{T}$$

(a) Calculate T from the equation $V = L/T$.
(b) Calculate V_p from the equation $V_p = \pi H/T$. Check this with the value given in the problem.
(c) For a second example wave, make a similar calculation using $L = 380$ m, $H = 10$ m, $V = 90$ km/hr, $V_p = 7.4$ km/hr.

27. The following equation relates wave velocity to water depth and wavelength:

$$V^2 = \frac{gL}{2\pi} \tanh \frac{2\pi d}{L}$$

where g, the acceleration of gravity, is 9.81 m/sec². For shallow water, or $d = L/20$, it reduces to $V^2 = gd$ and, for deep water $d = L/2$, it reduces to $V^2 = gL/2$.

(a) If $d = 1$ m and $L = 30$ m, calculate V using the full equation and the reduced equation. Compare answers.
(b) If $d = 20$ m and $L = 30$ m, calculate V using the full equation and the reduced equation. Compare answers.

28. Why are the tides, which should theoretically vary only about 50 cm on Earth, much more complicated than this? How is this indicated? What factors on Earth affect the tidal variation?

29. What complications related to the tides would be involved if a sea level canal were built in the general vicinity of the Panama Canal? Would the two levels eventually equalize? Explain.

30. (a) Calculate the relative attractive force between the Sun and Earth and between Earth and the Moon. Use the equation:

$$F = G\frac{m_1 m_2}{d^2}$$

where G is Newton's gravitational constant = 6.67×10^{-8} in cgs units. For the Earth-Moon system use d equal to 320,000 km and for the Sun-Earth system 150×10^6 km. For mass use mass of the Earth, $M_E = 1$, mass of the Moon, $M_m = 1/82\ M_E$, and mass of the Sun, $M_s = 3.32 \times 10^5\ M_E$. Is the value of G actually needed to calculate the relative attractive forces? Explain.

(b) What is the difference in the force for each if 12,800 km (Earth's diameter) is added to the distance between the celestial bodies in question? Explain.

31. Tidal power generation had been considered for the Bay of Fundy about 15 years ago but preliminary design considerations concluded that the winter was too severe to accomplish the construction. What is the latitude for the Bay of Fundy? How does it compare to the northern coast of France? Explain the difference in weather conditions for the two. More recently, construction of this power-generating facility was considered again and the decision was made to build the power plant. Why do you think the decision was reversed in the 15-year period?

32. Give the name of the estuaries associated with New York City, Boston, Washington, DC, and Norfolk, respectively. What is the relationship between estuaries, harbors, and tide differential? Explain. What type of mining is likely to be associated with estuary areas? Explain.

33. In a beach area the groins are spaced 60 m apart. How much beach widening is possible if the groin extends outward 45 m but goes 8 m beyond the breakers at low tide?

34. Erosion occurs beyond the down beach side of jetties. If a jetty extends 150 m into the water, for what distance would beach erosion likely occur down beach? How can this be alleviated or minimized?

35. Breakwaters are commonly constructed of rubble mounds with sizable pieces of armour stone for outside

protection. What would be the dimensions of a 10 metric ton cube of granite? Why is it difficult to find proper armour stone today for Lake Michigan's breakwaters? Explain.

36. The terms *structure, process,* and *stage* have been used relative to the development of landscapes or physiographic provinces. Indicate how this relates to the development of the shoreline of equilibrium discussed in the text.

37. Refer to the classification of coastlines relative to ocean level, Table 16.3. Is it likely that some certain combinations of the first and second groups could occur in the same areas such as active sea cliffs and sea terraces? What about irregular coastlines and numerous spits? Discuss in detail.

38. Breakwaters are sometimes blamed for causing beach erosion in areas adjacent to their location. What effect would the breakwater shown below have on downbeach erosion? Explain. What distances might be involved?

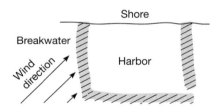

39. What is meant by "beach protection using nonstructural methods"? What is beach nourishment?

40. Why is breaching of the primary dune such a poor practice relative to beach preservation? How can traffic to the beach be accomplished otherwise?

41. Why must filling of the bayshore area be prevented? What is the origin of the compressible silts of the bayshore area?

42. In the erosion of Lake Michigan's shoreline in the mid 1970s some of the affected areas consisted of glacial till with interbedded outwash sands. Why would this sequence be particularly prone to failures by undercutting and landslides? Explain.

Additional Readings

BASCOM, W., 1959, "Ocean Waves," *Scientific American,* Vol. 201, No. 2, pp. 74–84.

BASCOM, W., 1964, *Waves and Beaches,* Anchor Books (Doubleday and Co.), New York.

Oceanography, 1971, W.H. Freeman and Co., San Francisco.

WEYL, P.K., 1970, *Oceanography, An Introduction to the Marine Environment,* John Wiley & Sons, New York.

17

Arid Environments and Wind

CHAPTER OUTLINE

Arid Climates
Geologic Processes in Arid Regions
Wind

ARID CLIMATES

The climate in a region has a significant influence on the geologic processes that prevail and on the appearance of the resulting landscape. The dominance of chemical over mechanical weathering is a consequence of abundant, unfrozen water, whereas the nature of streams or the presence of glaciers is a direct result of rainfall and temperature, that is, of climate.

In arid regions, the action of streams is entirely different from that of humid areas. Yet the effects of running water are commonly most profound in arid areas because of scarcity of vegetation, greater erodibility of the soil, and localized heavy rainfall.

In dry areas, wind is also an important geologic agent because of its contribution to erosion and deposition. In fact, wind was once thought to be the dominant agent in shaping desert landforms, but today only secondary details of landscape are attributed to its role. Nevertheless, the effects of wind are highly visible in arid environments.

Deserts or arid lands in low to temperate latitudes are typified by low rainfall, high temperatures (at least seasonally), and a rather higher evaporation level than precipitation. Typically, deserts receive less than 10 in. (25 mm) of rainfall per year and experience daytime temperatures well in excess of 100°F (38°C) during the summer months. Death Valley in the Basin and Range Province of southeastern California has the record in the United States for these climatic extremes—an average of less than 2 in. (5 cm) of rainfall per year and a maximum, recorded tempera-

FIGURE 17.1 Desert vegetation and topographic features, Mojave Desert, California. A Joshua tree is shown in the foreground.

ture in excess of 112°F (44°C). Because of the extreme heat and evaporation rate during the day many motorists elect to drive through the desert at night. Features known as cool deserts are also found throughout the world. They occur at high latitudes and/or high elevations and typically have low rainfall but much lower temperatures than those occurring in low to temperate latitudes. The Atacama Desert in Chile and the Gobi Desert of Tibet are examples of cool deserts.

Most deserts also experience frequent, high-velocity winds. This is a result of the convection that develops over these hot areas as the heated air rises and other air moves in as a replacement. Convection also provides much of the precipitation for deserts. As the rising air is cooled, it becomes saturated and delivers its moisture as localized and commonly heavy rainfall. Such cloudbursts are a regular event in desert areas and the geologist performing field work in arid and semiarid regions is well advised to watch for the buildup of such storms. Dry, scorched, drainage basins can become a torrent of runoff on short notice.

Vegetation is scant at best in arid regions but the actual amount is a direct function of the available rainfall. Sagebrush will grow in the semidesert areas of the United States as will juniper trees but not when the rainfall is below 10 in. (25 cm) per year. Cactus, Joshua trees, yucca, and small hearty bushes may be the only vegetation in these drier regions. The low bushes grow some distance apart leaving extensive, bare areas between them. This setting is conducive to erosion by both wind and running water. Desert vegetation and topographic features in the Mojave Desert of California are presented in Figure 17.1.

GEOLOGIC PROCESSES IN ARID REGIONS

In arid climates the unconsolidated material (regolith) is thin and coarse textured and is primarily a product of mechanical weathering. Slopes are generally more steep than in humid regions and, as discussed in Chapter 8 on rock weathering, sandstones and limestones form cliffs. Rock units tend to break along joints, yielding rugged cliffs known as buttes and mesas in flat-lying sedimentary rock terrains.

In humid areas the regolith is relatively fine textured, formed mostly by chemical weathering and transported downslope by creep. It is typically covered by vegetation and sandstones are the only prominent cliff-formers in sedimentary terrains. The contrast between slope development in humid and arid regions is illustrated in Figure 17.2.

Deserts are characterized by well-entrenched stream channels, which are dry much of the time. The water that flows in them shortly after heavy rainfall soon disappears by infiltration or evaporation and is unable to flow to the ocean. This condition is known as internal drainage. The Colorado River is an exception to this situation because the discharge from its source in the mountains is so great that the river continues to flow despite the losses under desert and

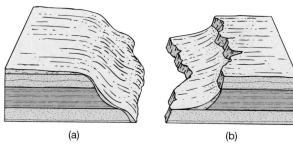

FIGURE 17.2 Effects of climate on slope weathering: (a) humid climate and (b) arid climate. Rock sequence for both diagrams, top to bottom: sandstone, shale, sandstone.

semidesert conditions. Ground water contributes to the discharge in the Colorado River in some areas as well.

Massive erosion of alluvial floodplains occurs when heavy rains result in flash floods. Banks may be undercut and slopes may cave in to form steep-sided channels, forming features called box canyons. Erosional and depositional effects may cause considerable damage to alluvial areas; bridge abutments and center span footings may be undermined while transverse roads are buried by debris. A flash flood east of San Diego, California, in 1978 covered a major highway in the Imperial Valley with 5 ft (1.5 m) of deposits, requiring the construction of a new section of road at the higher elevation.

Clay-rich units with their attendant low permeability provide a terrain where little infiltration occurs

and, therefore, runoff is abundant. In the absence of vegetation in arid climates, very closely dissected topography develops, yielding landforms referred to as badlands. The Painted Desert of Arizona and the Badlands of South Dakota are well-known examples of this feature. The South Dakota Badlands are shown in Figure 17.3.

A somewhat similar landscape has developed in the old strip mine areas of the Midwest where unreclaimed coal mine refuse is closely dissected. Vegetation is curtailed by the high acidity of the soil, caused by the weathering of pyrite from shale units. These area strip mines predate the enactment of state reclamation laws, which took effect in the mid-1950s or early 1960s. An abandoned mine lands area of Indiana is shown in Figure 17.4.

In the deserts of the southwestern United States, basins formed by block faulting are prevalent. This is the Basin and Range Province described in Chapter 13. Streams flowing from the highlands (ranges) seldom persist until they reach the center of the adjacent basin. After high-intensity storms, however, water may accumulate in the basin to form a shallow lake that persists for a few days or weeks. These are known as playa lakes or when dry the lake bed is called a *playa*. Figure 17.5 is a playa located in the southwestern United States.

The Great Salt Lake in Utah contains a playa lake today, although the basin itself has a complex history of block faulting and subsequent glacial deposition. Because of the internal drainage, dissolved salts are not carried away to the sea but are concentrated in

FIGURE 17.3 South Dakota Badlands.

FIGURE **17.4** Abandoned mine lands area, southwest Indiana. Severe erosion developed on coarse coal mine refuse.

the playa lakes instead. On evaporation, salts accumulate on the surface of the playa or, if the water infiltrates, brackish water is carried downward into the subsurface. The nature of the salts deposited is a function of the composition of rocks in the adjacent ranges. Deposits may be rich in sodium, chloride, alkalis (sodium and potassium carbonates), bittern salts (sulfates), or borates (borax and related minerals). The rocks of the Basin and Range Province have a complex history of igneous and metamorphic origin so that all of the ions for the salts just mentioned are available in one area or another.

A prominent depositional feature in a desert basin is the alluvial fan that forms near the change in slope between the sloping basin and the highland. As erosion proceeds, the highlands recede and additional alluvial fans are formed. The nearly flat area downslope from an active alluvial fan consisting of coalescing fans is called a *bajada.*

Another feature of desert terrain consists of a sloping rock surface at the base of the highland and is erosional in nature. Cutting across bedrock, this landform, a *pediment,* has a general shape that is concave-up, much like the long profile of a stream. At the downslope extreme of the pediment, alluvium covers the rock surface. These various landforms are depicted in Figure 17.6.

WIND

Action of the Wind

Winds in arid regions yield both erosional and depositional features. As previously noted, the effects of running water are most significant in shaping desert topography but wind plays an important secondary role. The contribution of both of these geologic agents regarding desert landscapes is shown in Table 17.1. The features associated with running water have been discussed previously.

Researchers working in deserts have observed that two layers of particles are transported in the air when strong winds persist. The lower layer contains sand grains and extends only 4 in. (10 cm) to perhaps 1.5 ft (0.5 m) above the ground. The upper layer carries silt and clay particles and it commonly extends to signifi-

FIGURE **17.5** Playa, southwestern United States, showing a gentle slope to the left.

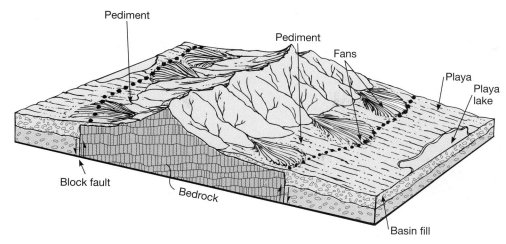

FIGURE **17.6** Landform features in arid climates; block faulted highlands and adjacent basin.

cant heights—330 ft (100 m) or more high. The two layers are equivalent to the bedload and suspended load of a stream.

Sand grains move by saltation in a series of long jumps. These grains usually rise only about 4 in. (10 cm) off the ground as indicated by wind tunnel experiments. Evidence in deserts suggest that 18 in. (45 cm) is the highest level to which telephone poles are sandblasted. Diameters of sand grains transported by the wind as bedload range from 0.15 to 0.3 mm, which is a medium-size sand. Air velocities of 11 mph (5 m/sec) are needed to lift sand grains from the ground.

Most sediment finer than sand remains at low altitudes. Gravity pulls these particles downward at velocities greater than particles in water because of that liquid's greater density and viscosity. A compari-son of water versus air as the transporting medium is shown in Figure 17.7. This indicates that a coarse silt particle has a settling velocity of 1 m/sec in air and 0.250 cm/sec in water. An upward velocity of 1 m/sec (2.2 mi/hr) therefore is necessary to maintain this coarse silt particle in suspension.

Turbulent eddies in air (as in water) move in all directions. On average, the velocities of upward eddies are about one-fifth the forward velocity. Therefore, a wind at 5 m/sec, which is required to lift sand grains from the ground, would keep silt and finer sizes in suspension as well.

A dead air layer persists immediately above the ground. Its thickness is about one-thirtieth that of an obstruction, and the common obstacle on a soil surface is an average-sized sand grain about 1 mm in

TABLE **17.1** Landforms and Landform Features for Arid Regions

Primary Agent	Depositional Features	Erosional Features
Running water	Alluvial fan	Mesa
	Bajada	Butte
	Playa	Box canyon
		Pediment
Wind	Dunes	Deflation basin
	Loess	Wind caves and arches
	Tephra deposits	Deflation armor
		Ventifacts
		Desert varnish

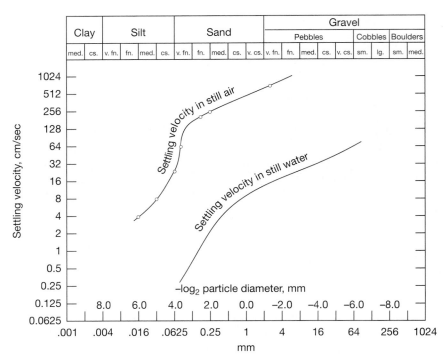

FIGURE **17.7**   Diameter of sand grains versus settling velocity in air and in still water.

diameter. This yields a dead air layer some 0.033 mm thick, which is equivalent to coarse silt size. Silt and finer particles are protected by this dead air layer and they will not be picked up by the wind until saltating sand grains or some other disturbance breaks the smooth surface. A dry gravel road in the country serves as an example. The wind will generate little or no dust from the road until a car comes by, which breaks up the smooth surface and allows the silt particles to become airborne. Wetting the road will reduce dust from traffic by increasing the cohesion of the soil, which in turn resists the uplifting force of the wind. This practice is common on construction sites where water trucks wet down the roads to control dust generation.

Most silt- and clay-sized particles settle within several tens of kilometers from the source. Loess or wind-deposited silt is formed in this way. Both average grain size and thickness of the deposit decrease in the downwind direction. This is discussed in more detail in a later section.

In some instances fine particles are carried long distances by travel through the stratosphere. Updrafts in the atmosphere lift the material, and downdrafts, related to major air circulation, deposit the particles a great distance away. Falls of tephra from distant volcanoes are of this nature and the average diameter of particles carried does not necessarily decrease downwind.

Eruption of the Hekla Volcano in Iceland in 1947 provides an example. Some tephra was raised 29,500 ft (9000 m) into the atmosphere and traveled northwestward to Finland some 1875 mi (3000 km) away. The 1980 eruption of Mount Saint Helens in the Cascade Range of Washington deposited tephra in five adjoining states. Dust accumulations in the stratosphere were noted as far east as the Ohio River.

Wind Erosion

The wind erodes surface materials in two major ways, by deflation and by abrasion. Deflation is the removal of fine, loose particles by the wind and accounts for most of volume of material eroded by it. Abrasion is the natural sand-blasting effect of the wind, which cuts and polishes bedrock surfaces and rock pieces of pebble size and larger.

Deflation can completely lower the surface or can work in localized areas to yield deflation basins. When vegetation is absent in arid regions, which occurs during periods of extended drought, large-scale deflation will develop. Several feet of the landsurface may be removed in a period of a few years under such conditions.

Deflation basins occur in a large number in semiarid regions such as the Great Plains of the United States and Canada. They are typically less than a mile long and have depths of a few feet. In wet years they are covered with grass and may contain shallow lakes. In the past such features were known as buffalo wallows. Some, in western Nebraska, have been mined for potash or borax.

Deflation of a soil mass consisting of sand, silt, and pebbles will eventually form a protective coating, which limits further deflation. The stages of development are shown in Figure 17.8. The sand and silt are carried away by the wind and the pebbles remain

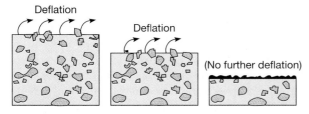

FIGURE **17.8** Three stages in the development of deflation armor, known also as desert pavement or lag deposit.

behind. When a surface develops, consisting of pebbles tightly fitted together, the surface will stabilize. This feature is referred to by several terms: *deflation armour, desert pavement,* or *lag deposit.* It is defined as a surface of coarse particles developed primarily by deflation. A photograph of a desert pavement is shown in Figure 17.9.

Abrasion in desert areas is concentrated on rock at the surface that is not removed by deflation. A *ventifact* is a rock fragment cut and faceted by wind action. They are characterized by polished, often pitted or fluted surfaces, with facets separated from each other by sharp edges. Most of the cutting and polishing is accomplished by saltating sand grains. Studies have shown that the facet actively being abraded faces the wind. If the rock is rotated by undermining or shifts for some other reason, two or three facets can be cut on the rock. The German terms *einkanter, zweikanter,* and *dreikanter* are sometimes applied to these rocks; *kanter* means "edge" or "face." Figure 17.10 shows the development of ventifacts.

Fine-grained massive rocks that undergo abrasion by the wind also take on a dark brown to purple color known as desert varnish. This coating is thought to consist of iron oxide or manganese oxide imparted by the sand-blasting effects. Desert varnish tends to reduce the contrasting appearance between different rock types in outcrop in the desert. Care must be taken to break a fresh surface on a rock cliff or hand sample before examining the lithologies involved.

FIGURE **17.9** A desert pavement or lag deposit.

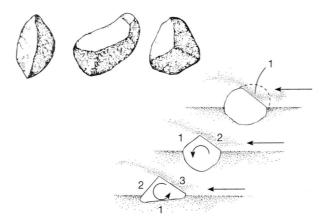

FIGURE 17.10 Ventifacts, pebbles shaped and polished by the wind.

Wind Deposits

Wind-deposited landforms are of two basic types: dunes of sand formed close to the source and blankets of silt or clay formed much farther away. A dune is a mound of sand piled up by the wind. Generally it is initiated by an obstacle that disrupts the flow of air. Sand dunes are not symmetrical in shape but have a steep lee slope with a more gentle slope on the windward side. Sand grains are rolled or pushed up the windward slope and drop over the crest of the dune to the lee slope, coming to rest at their own angle of repose. For a loose sand this repose angle is about 34°. As more sand is pushed over the crest, the sand slides or slips downward, maintaining the angle of repose. Hence the straight, lee slope of a dune is sometimes referred to as the slip face. The angle for the windward slope is determined primarily by the wind velocity and it ranges from about 5° to 15°. This is illustrated in Figure 17.11.

Sand dunes may reach heights of from 100 to 300 ft (30 to 90 m) and in some extreme cases in deserts may be up to 600 ft (180 m) high. Movement of sand from the windward to the lee side of a nonvegetated dune can cause this landform to migrate slowly downwind. Rates of migration in deserts may be as great as 35 to 70 ft/yr (10 to 20 m/yr). Adjacent to sandy beaches associated with the oceans or the Great Lakes, migration of dunes can bury houses, cover roads, and generally disrupt inhabited areas. Encroachment is best curtailed by establishing vegetation that can thrive in the dry sand soil. Beach grasses also prevent deflation by holding the sand grains in place.

The stratification of sand in dunes becomes fairly complicated as migration occurs. Originally, the cross strata formed on the lee side yield a series of beds truncated by the windward face of the dune (Figure 17.12). With shifting wind direction and velocity, dune migration changes direction. Erosion may be replaced by deposition, causing a truncation of one series of cross beds and establishment of another. This more complicated feature is also depicted in Figure 17.12. In any event, the steep beds indicate the lee side of the dune during the migration of a specific episode, with the wind coming from the direction opposite the dip direction of these steep beds.

The shapes of dunes or dune fields are quite diverse relative to both geometric form and orientation with regard to the wind. Availability of sand, wind velocity, constancy of wind direction, general terrain, and vegetative cover are factors that determine the nature of the dunes. Sand dunes can develop in a variety of geologic settings that have a supply of sandy materials at the surface. Beaches, coastal plains, floodplains, outwash plains, and extensive outcrops of sandstone can provide the sand needed to generate dunes.

Most sand dunes consist of quartz grains, primarily because of the abundance and weathering resistance

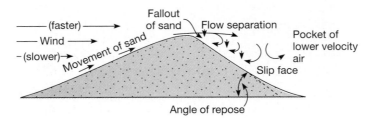

FIGURE 17.11 Development of the windward and lee slopes of a bare sand dune.

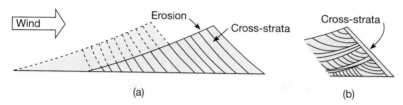

FIGURE 17.12 Cross section view of migrating sand dune: (a) simple parallel layers and (b) complex cross strata caused by variations in wind direction.

of that mineral. These grains may have experienced several cycles of lithification and erosion before coming to rest in a present-day dune. In certain areas of the world other minerals are abundant as sand-sized pieces that accumulate into dune forms.

On the island of Bermuda, well off the South Carolina coast in the Atlantic Ocean, the dunes consist of calcite grains derived from the limestone that forms the island. In New Mexico near Alamogardo, at White Sands National Monument, the area is covered by dunes consisting of gypsum grains. These are derived from the rock gypsum exposed at the surface in this arid environment.

It is usually possible to distinguish sandstones of water-deposited origin from those formed by lithification of wind-deposited dunes. The eolian deposits (wind-derived) have larger dune forms and more extensive cross bedding. They also have frosted, more rounded grains than water-deposited sands. Water cushions the grains, making them less subject to rounding by abrasion. The frosting results from the mutual impact of sand grains during saltation by the wind. Complications can occur in such analysis, however, because some water-deposited sands are derived from wind-deposited sandstones. In this case the frosted surface of grains may persist in the second sandstone. However, not all of the grains would be frosted because some of the sand undoubtedly would be derived from other sources.

Silt Deposits

Loess is wind-deposited silt, usually accompanied by some clay and fine sand. Typically it is not stratified because of the small range of grain size deposited together and the contribution of plant roots and worms, which work on the sediment during and after its deposition. (Loess is also discussed in Chapter 12.)

Commonly, loess will stand in a vertical fashion when exposed. A loess cut near Vicksburg, Mississippi, is shown in Figure 17.13. This is likely the result of the fine particles present, because the silt grains are surrounded by a clay surface coating that

FIGURE 17.13 Loess cut near Vicksburg, Mississippi, showing vertical excavation.

provides a cohesiveness to the material. Yet porosity and permeability are high because of the angular nature of the silt and the open network of pores between them. For this reason also, loess holds water, which is in turn available to plants, and thereby yields a productive, agricultural soil.

Loess is typically yellow in color and consists mostly of quartz, feldspar, mica, and calcite. The color is caused by slight oxidation of iron present in minor amounts. It closely resembles the fine product known as *rock flour* that is formed by glacial action. In the United States, loess is found immediately leeward of major glacial outwash and valley train deposits. Loess was deposited in glacial times in areas just beyond the glacial ice. At this location, the cold, windy climate and denuded valleys choked with glacial outwash provided the source areas for the silt. Deflation in the valleys moved sand grains and provided silt particles for the suspended load of the wind. The silt settled, forming thick blankets up to 100 ft (30 m) thick on the leeward side (eastern side mostly) of major stream valleys and trailing out to thicknesses of 5 ft (1.5 m) or less within about 50 mi (80 km).

The silt came to rest within the zone and was not transmitted further downwind. This is probably because the fine size was not as easily picked up again in the absence of sand grains to initiate its suspension and because obstructions protected it from the wind. Grass lands and forested areas were likely present on the upland areas where the loess came to rest.

Five major areas of loess deposits exist in the United States: 1) the Lower Mississippi River section of the Gulf Embayment, 2) the central area of the United States in the Interior Lowlands including parts of eastern Nebraska and Kansas, Iowa, South Dakota, Missouri, Minnesota, Wisconsin, Illinois, and Indiana, 3) the plains of eastern Colorado and western Nebraska and Kansas, 4) the Snake River Plains of southern Idaho, and 5) the Palouse area of eastern Washington and north-central Oregon. These areas are shown in Figure 17.14.

Loess in the Lower Mississippi River section extends along the east bank of the river floodplain from New Orleans to the Ohio River. The silt deposit is thickest near the river, 100 ft (30 m), and gradually thins within 50 mi (80 km) or so. The northern portion of the deposit is shown in Figure 17.15 and depicts the area of confluence of the Missouri, Mis-

sissippi, and Ohio Rivers. Shown also is Crawleys Ridge of southern Missouri and northern Arkansas, which is capped with loess and lies west of the river. Floodplain sediments lie east of the ridge, which were likely the origin of that loess deposit. Rainfall in this region is about 50 in/yr (127 cm/yr).

Loess in the Interior Lowlands is generally underlain by glacial drift, which in many cases is Illinoian or Kansan in age. A portion of the eastern segment of this loess deposit is also shown in Figure 17.5. On that drawing, loess blankets the eastern part of Illinois, covering Wisconsin drift on the north and Illinoian drift on the south. The deposit is thickest on the uplands east of the Mississippi River, and the Illinois River, another glacial sluiceway, contributes more loess on its leeward side. In Indiana, not included on the map, loess is found on the east side of the Wabash River. It is more prevalent in the southern half of the state between Terre Haute and Evansville. Rainfall in this extensive area ranges from 30 in./yr (76 cm/yr) in eastern Kansas to 40 in./yr (102 cm/yr) at its eastern boundary in Indiana.

Loess of the plains lies on a large treeless region sloping gently eastward from the Rocky Mountains. In eastern Colorado and western Kansas and Nebraska, the loess overlies Rocky Mountain outwash materials. In central Kansas it overlies chalky limestones and shales, whereas in eastern Nebraska and Kansas the underlying material is glacial drift. Because of erosion, this glacial drift is frequently exposed in small isolated areas, particularly along valley slopes. The rainfall ranges from about 20 in./yr (51 cm/yr) in the western portion to nearly 30 in./yr (76 cm/yr) in the eastern part.

In the Snake River Plain of southern Idaho, the loess deposits cover an extensive area of about 20,000 mi^2 (51,000 km^2) and lie at an altitude between 4000 and 6000 ft (1200 to 1800 m). Rainfall is 10 in./yr (25 cm/yr) or less and natural vegetation is primarily desert shrub and sagebrush. Irrigation from the Snake River or from ground-water supplies for agricultural purposes is prevalent in the area. The underlying bedrock is Columbia River basalt, which extends through the loess in some areas yielding a terrain formed on basalt flows known as *scablands*.

The Palouse area of eastern Washington and north-central Oregon is a loess-covered upland on the Columbia Plateau with an average elevation of about 2000 ft (600 m). Annual rainfall is 20 in. (51 cm) or

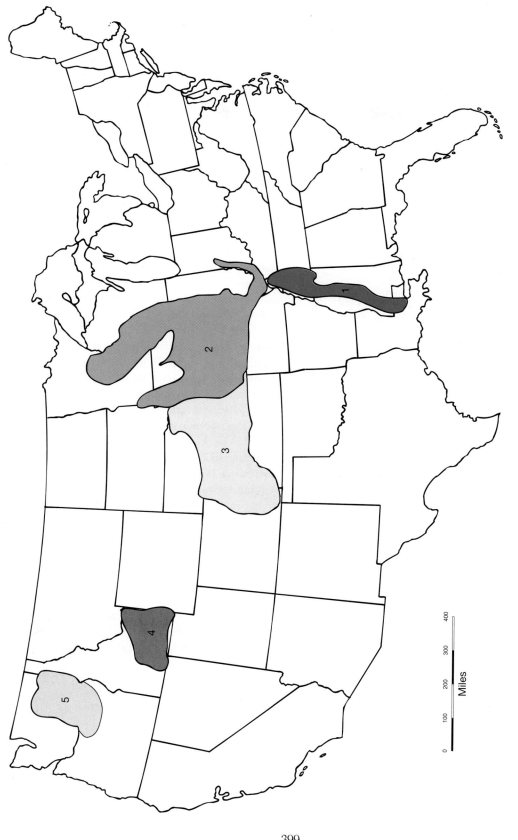

1. Lower Mississippi River Section
2. Central U.S., Interior Lowlands
3. Great Plains, Nebraska, Kansas, & Eastern Colorado
4. Snake River Plain, Southern Idaho
5. Palouse Area, Eastern Washington
   & Northeast Oregon

FIGURE 17.14 Map of United States showing the five major areas of loess deposition.

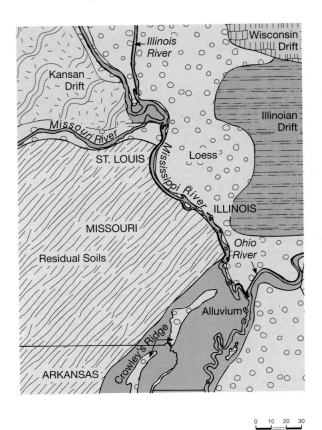

FIGURE **17.15** Surficial geology, area of confluence, the Missouri, Mississippi, and Ohio Rivers.

the adjacent cities and states of the northwest. The Palouse loess area is used extensively for farming. In western Washington some areas can supply wheat with a minimum of irrigation. In other areas, irrigation is required for various grain crops.

The engineering problems associated with loess or indeed with other silt deposits are outlined in Chapters 8 and 12. Some of these characteristics are presented here as well. Loess is generally uniform in texture with 50% to 90% in the silt-size fraction. It can usually be identified by visual inspection because of the yellow color and striking, vertical exposure. For this soil, the liquid limit is typically about 30 with a plasticity index of about 6. It requires care during compaction because silts have a narrow range of moisture for maximum density and because they dry readily. Silts are subject to cracking in earth dams if compacted on the dry side of the optimum moisture content. Refer to the discussion in Chapter 7 for details on soil compaction.

Slopes in loess tend to be stable when cut vertically provided that the loess will remain relatively dry, well below saturation. When saturated, however, or when the phreatic zone is intercepted in a deep loess cut, the back slope (cut slope) must be reduced markedly to a 2 : 1 or 3 : 1 relationship (that is, 2 vertical to 1 horizontal). Trees with deep roots and vegetation supplying thick ground cover will aid in drying out loess from a saturated condition.

In some regions of the world, loess is found in large areas that lie in the leeward direction from deserts. Mechanical weathering obviously provides the silty material that accumulates as alluvium on the desert floor adjacent to the mountains. The loess of western China derived from the great desert basins of central Asia is up to 230 ft (70 m) thick. An earthquake in this material caused the massive landslide and great loss of life described in Chapter 14 on slope stability and landslides.

less and native vegetation is primarily bunch grass. The silt can be 75 ft (23 m) thick or more but basalt exposures can occur abruptly in the otherwise deep loess areas. The Palouse loess is thought to be partly derived by wind from residual soil of the basalt and from volcanic ash. This latter contribution seems quite likely in light of the 1980 eruption of Mount Saint Helens, which supplied considerable tephra to

EXERCISES ON ARID ENVIRONMENTS AND WIND

Map Reading

The following questions pertain to two maps from arid regions in the southwestern United States.

Furnace Creek, California; 7.5-min quadrangle

This area represents a youthful stage in the cycle of erosion in an arid region. It lies within the Basin and Range physiographic province, which is characterized by a series of north-south trending fault block mountains and valleys.

1. Draw an east-west profile across the map area. Note the different topography of the mountains, the valley sides, and the valley bottom. Sketch in the surface area that is probably bedrock.

2. What is the gradient (typical) of a stream in the mountains, along Trail Canyon, of Salt Creek in the center of the valley?

3. What do the arc-shaped contour lines adjacent to the mountains outline?

4. Why do some streams end abruptly?

5. What do the numerous nicks on the surface of the alluvial fan contours represent? What does this suggest about local relief?

6. What is the base level of the map area?

7. In what direction is Salt Creek flowing?

8. Drainage in this quadrangle is described as internal. Why?

9. How will the valley shape change with time?

10. Explain the green strips radiating from Furnace Creek Ranch.

11. List engineering problems that might be associated with the area.

Antelope Peak, Arizona; 7.5-min quadrangle

This area represents the erosion cycle in an arid climate. It is also located in the Basin and Range physiographic province.

12. Sketch a profile from Antelope Peak to the northeast corner of the map. Sketch the bedrock surface.

13. Contrast topographic features on this map with those on the Furnace Creek Quadrangle.

14. What do the large, gently sloping areas on this map represent?

15. Why are alluvial fans on this map not as prominent as on the Furnace Creek Quadrangle?

16. What is the dominant agent of erosion in this area? Cite evidence.

Written Questions

17. The Colorado Plateau physiographic province in the southwestern United States consists primarily of flat-lying sedimentary rocks of the Paleozoic age. Which of the sedimentary rock types would likely be cliff-formers and which would be slope-formers in this arid environment? What is the significance in this regard of the formation named Redwall Limestone, which crops out in the Grand Canyon, Arizona?

18. The Appalachian Plateau of West Virginia also consists primarily of flat-lying sedimentary rocks. Which of these would be cliff-formers and which slope-formers in this humid area?

19. How does the vegetative cover differ between the Colorado Plateau and the Appalachian Plateau area considered in Exercises 17 and 18? In what way does this effect the rock exposures? Explain.

20. Why is runoff and stream erosion in a desert region credited with a greater extent of erosion than is wind action, despite the infrequent rainfall that occurs in such regions? What is the effect of vegetation in this regard?

21. Outcrops of coarse-grained granite in the desert because of mechanical disintegration do not develop desert varnish to the extent that basalts or fine gneisses would. Why is this likely to be the case? Explain.

22. The settling velocity of sand in air is given as 1 m/sec. For how long would the sand be airborne if it fell from a height of 45 cm? This would require that wind gusts of what velocity persist for that period of time to sandblast posts in the desert up to 45 cm high?

23. In the past, road oil has been placed on gravel roads to prevent dust from billowing as automobiles pass. Why is such practice effective in reducing dust? Explain.

24. What is the origin of sand grains that form a sandstone? Why is it likely that several cycles of weathering and lithification could take place if the quartz was formed during the Precambrian era?

The following questions pertain to Figure 17.15.

25. Note that the northern portion of St. Louis, Missouri, has a loess covering at the surface but the southern part of the city has residual soils instead. Much of this soil was derived from limestone.

    (a) Why is this deposit of loess in northern St. Louis located on the west side of the Mississippi River? Note that it is not in a valley area but on an upland surface.

    (b) What compaction problems likely occur when constructing fills or embankments on such loess material? Why? What should be done to alleviate this?

    (c) What problems likely occur during construction of buildings in the residual soils in southern St. Louis? Explain.

26. Note the three different ages of glacial drift shown in Figure 17.15. Arrange these three in chronological

order from younger to older. How would the topography differ in the three areas of differently aged drift? Explain.

27. Note that the loess extends farther eastward from the Mississippi River in the area of the Illinois River than it does to the south from there. Give a likely reason why this is so.

Additional Readings

BAGNOLD, R.A., 1941, *The Physics of Blown Sand and Desert Dunes,* Methuen Publishing Co., London.

BRIAN, K., 1932, *Erosion and Sedimentation in Papago County, Arizona,* Bulletin 730, U.S. Geological Survey, Washington, DC.

DENNY, C.S., 1967, "Fans and Pediments," *American Journal of Science,* Vol. 265, pp. 81–105.

DOEHRING, D.O., Ed., 1977, "Geomorphology in Arid Regions," in *Proc. Eighth Annual Geomorphology Sympo.,* Binghampton, New York, State University of New York.

18

Earthquakes and Geophysics

E arthquakes are shock waves that travel through the Earth. They develop apparently as a result of the elastic rebound of rocks when strain energy is suddenly released. The energy builds slowly in a fault zone until the stored stresses become greater than the rocks can withstand and the earthquake occurs.

Geophysics involves the application of physical laws to study the Earth and earth materials. It is used in purely scientific studies but also in subsurface exploration to search for valuable minerals or to determine subsurface characteristics. The connection between earthquakes and geophysics lies in the common concern for waves propagating through the Earth and what they tell about the nature of the propagating medium. Earthquakes are examined first.

EARTHQUAKE STUDIES

Types of Waves

Earthquake waves can be divided into two types, body waves and surface waves. The body waves travel through the volume of the mass or body, and the surface waves occur at the boundary of the mass. Body waves are of two types, primary waves and secondary waves. Primary (P) or longitudinal waves are compressional in nature; they travel by a series of compressions and rarefactions of the wave form in the solid. They are a push-pull–type wave. Sound waves are compressional waves. These waves vibrate parallel to the direction of propagation and have the greatest velocity of the earthquake waves. Consequently, they are the first waves to arrive at the Earth's surface from the source and hence are termed *first arrivals*.

Secondary waves (S) or shear waves are transverse in nature and vibrate perpendicular to the direction of propagation. They push out into the medium and that medium must push back in order to keep the wave moving. For this to occur, the material must possess some shear strength, or the wave will not propagate. Materials without shear strength such as liquids cannot supply the return push and S waves will not travel through them. On this basis the conclusion has been made that the outer core of the Earth is a liquid because shear waves do not penetrate the Earth from one side to another. This subject is discussed in more detail in a later section on the Earth's interior. Secondary waves travel slower than primary waves but have a greater velocity than surface waves.

When body waves contact the Earth's surface, new waves, called surface waves, are generated. There are several types of surface waves including Love waves and Rayleigh waves but they all are characterized by long wavelengths and high amplitudes. Surface waves are collectively referred to as L waves (for long wavelength). Their velocities are slower than S waves, so they are the third wave to arrive at a location following an earthquake. The velocities can be summarized as follows:

$$V_p > V_s > V_L$$

The primary and secondary wave velocities can be defined in terms of several materials properties of rocks. The properties of interest are E, the modulus of elasticity or Young's modulus; K, the bulk modulus; S, the shear modulus; ρ, the density; and μ, Poisson's ratio. Modulus of elasticity E is the familiar compressive stress/strain relationship for a solid ($E = \sigma/\epsilon$), and K is the stress divided by volumetric strain $= \dfrac{\sigma}{\Delta v/v}$. The shear modulus S for most materials is numerically somewhat less than half that of E.

The shear modulus provides the relationship between shear stress and deformation in that

$$S = \tau/\delta$$

where

S = shear modulus
τ = shear stress
δ = deformation.

Density ρ is the mass per unit volume and Poisson's ratio is the lateral unit strain divided by the vertical unit strain under load:

$$\mu = \frac{\Delta B/B}{\Delta L/L}$$

Using established relationships

$$V_p = \sqrt{\frac{K + 4/3S}{\rho}} \quad \text{and} \quad V_s = \sqrt{\frac{S}{\rho}}$$

By inspecting these equations we observe that when S, the shear modulus, is zero or when τ the shear stress is zero, V_s is zero. This is in agreement with the understanding that liquids do not propagate shear waves.

Also well established is the fact that

$$S = \frac{E}{2(1 + \mu)} \quad \text{and} \quad K = \frac{E}{3(1 - 2\mu)}$$

Substituting these relationships for K and S in the earlier equation for V_p yields:

$$V_p = \sqrt{\frac{1}{\rho}[\frac{E}{3(1 - 2\mu)} + \frac{4/3E}{2(1 + \mu)}]}$$

$$= \sqrt{\frac{E}{\rho}\frac{1 - \mu}{(1 - 2\mu)(1 + \mu)}}$$

In like fashion

$$V_s = \sqrt{\frac{S}{\rho}} = \sqrt{\frac{E}{\rho}\frac{1}{2(1 + \mu)}}$$

Therefore,

$$\frac{V_p}{V_s} = \sqrt{\frac{K}{S} + \frac{4}{3}} = \sqrt{\frac{1 - \mu}{1/2 - \mu}} > 1 \quad \text{so } V_p > V_s$$

This shows that the velocity of the primary wave is always greater than that of the secondary wave.

Several other details are also of interest. One of these is the so-called "density paradox" for V_p. The equation

$$V_p = \sqrt{\frac{K + 4/3S}{\rho}}$$

suggests that as ρ increases, V_p decreases. This is not actually the case because as ρ increases, K also increases but at a rate greater than that of ρ. Therefore, in general, as the density increases, so does the velocity of the primary wave. Hence the paradox.

The equations also show that the shear wave velocity can be determined from E, ρ, and μ, parameters that are easily determined in the laboratory. These values are typical of the static case; dynamic values differ somewhat. For this reason in studies of earthquake resistance for many major structures, the shear wave velocity and shear modulus are typically measured under field conditions to obtain dynamic values.

EXAMPLE PROBLEM 18.1

(a) A rock has a Poisson's ratio of $\mu = 0.25$. What is the ratio of V_p/V_s?

(b) If $E = 6.5 \times 10^6$ psi, what are the values of S and K?

(c) If V_p is 14,500 ft/sec what is V_s in ft/sec and km/sec?

ANSWERS

(a) $\dfrac{V_p}{V_s} = \sqrt{\dfrac{1-\mu}{1/2-\mu}} = \sqrt{\dfrac{1-0.25}{0.50-0.25}} = \sqrt{3} = 1.732$

(b) $S = \dfrac{E}{2(1+\mu)} = \dfrac{6.5 \times 10^6 \text{ psi}}{2(1+0.25)} = 2.6 \times 10^6 \text{ psi}$

$K = \dfrac{E}{3(1-2\mu)} = \dfrac{6.5 \times 10^6 \text{ psi}}{3(1-0.50)} = 4.33 \times 10^6 \text{ psi}$

(c)

$V_p = 14{,}500 \text{ ft/sec} = 14{,}500 \times 1/3.28$

$\qquad = 4421 \text{ m/sec} = 4.421 \text{ km/sec}$

$V_s = \dfrac{V_p}{1.732} = 8372 \text{ ft/sec} = 8372 \times \dfrac{1}{3.28}$

$\qquad = 2552 \text{ m/sec} = 2.552 \text{ km/sec}$

Earthquake Scales

Two distinct scales are used to measure the level of earthquakes. These are the magnitude scale, a measure related to the amount of energy released, and the intensity scale, which measures the damage the earthquake causes.

Magnitude Scale

The Richter scale (M_L) is used today to measure magnitude. Presented in the current form in 1957 by Charles Richter, it is based on the maximum displacement or amplitude of the earthquake wave measured on a standard seismograph 100 km from the epicenter. That reference point, the epicenter, lies on the Earth's surface directly above the focus or place of origin for the earthquake. The Richter value is determined by measuring this maximum amplitude in thousandths of a millimeter (or in microns = micrometers = 10^{-6} m) and taking the common logarithm of this number. Hence a 0.100-mm displacement would yield a Richter value of 2 and 1.000 mm would yield a 3. In like manner 0.500 mm becomes 2.698. A factor of 10 times the displacement is involved from one integer value to another. This shows that a Richter value of 6 has a displacement 1000 times greater than a value of 3.

Because the Richter scale is based on a linear measurement of length, there is no upper limit to the scale as far as physical measurements are concerned. A common misconception creeps in, however, when the news media or others on occasion allude to some maximum, usually 10, for this scale. This is simply not the case. The Earth does, however, seem to impose its own limit of 8.8 or possibly 9.0; this appears to be about the maximum of equivalent stress that the Earth can store before releasing its load. The highest Richter magnitude ever assigned to an earthquake as of early 1994 is 8.7. Earthquakes above 6 are important and those of 7 or more are major earthquakes. The truly massive, devastating earthquakes in history have been in the range of 8 or more, and fortunately, have occurred quite infrequently in the recorded past. Some of the major earthquakes in recent history are listed in Table 18.1.

In recent years additional studies of earthquake magnitudes have lead to the development of modifications to the Richter scale. The designation used for the local magnitude, that is, the Richter magnitude, is M_L. In addition, there is the surface wave magnitude M_s, body wave magnitude m_b, duration magnitude M_D, and moment magnitude, M_w.

A detailed discussion of these other magnitude scales is beyond the scope of this text but it is important to realize that M_L is not an adequate measure for all the magnitudes of earthquakes ranging from less than 3 to greater than 7.5. This is because M_L is not particularly sensitive to earthquakes that are well below magnitude 3 and the M_L scale approaches a saturation point at about 7 and does not distinguish very well between earthquakes

TABLE 18.1 Richter Magnitudes of Selected Significant and Major Earthquakes in the World

Year	Location	Magnitude
1755	Lisbon, Portugal	8.7?
1811–1812	New Madrid, Missouri	8.7?
1886	Charleston, South Carolina	8.6
1899	Alaska	8.6
1906	Columbia	8.6
1906	San Francisco, California	8.2
1906	Chile	8.4
1911	Soviet Union–China	8.4
1914	Quebec	5.5
1920	China	8.5
1925	Santa Barbara, California	6.5
1933	Long Beach, California	6.3
1933	Japan	8.5
1940	El Centro, California	7.1
1950	Pakistan–Tibet–Burma	8.6
1952	Kern County, California	7.7
1960	Chile	8.3
1964	Anchorage, Alaska	8.4
1965	Puget Sound, Washington	6.5
1970	Peru	7.8
1971	San Fernando, California	6.6
1977	Indonesia	8.0
1979	El Centro, California	6.5
1980	Italy	7.2
1981	Iran	7.3
1983	Japan	7.7
1983	Turkey	6.9
1985	Chile	7.8
1985	Mexico	8.1
1986	El Salvador	5.4
1987	Columbia–Ecuador	7.0
1988	Burma–China	7.0
1988	Armenia	7.0
1989	Loma Prieta, California	7.0
1989	Australia	5.6
1990	Iran	7.7
1990	Phillipines	7.8
1992	Landers, California	7.5
1994	Los Angeles (Northfield), California	6.6

above that value. Because of this situation, different scales are suggested for various ranges of magnitude:

M_D or M_L magnitudes less than 3
M_L or m_b magnitudes between 3 and 7
M_s magnitudes between 5 and 7.5
M_w magnitudes greater than 7.5

A comparison of the magnitude scale values for some of the great earthquakes of the twentieth century is shown in Table 18.2. It is interesting to note that the greatest M_s value is 8.7, whereas the greatest M_w value is 9.5.

As discussed earlier, the Richter scale is based on the log of the amplitude of displacement and each integer value has 10 times greater displacement than the preceding one. The energy release between each unit is even greater, more in the order of 30 times between integer values. Examples are shown in Table 18.3 based on several West Coast earthquakes.

TABLE 18.2 Comparative Magnitudes of Some Great Earthquakes of the Twentieth Century

Date	Region	M_s	M_w
July 9, 1905	Mongolia	8.2	8.4
January 31, 1906	Ecuador	8.6	8.8
April 18, 1906	San Francisco	8.2	7.9
January 3, 1911	Turkestan	8.4	7.7
December 16, 1920	Kansu, China	8.5	7.8
September 1, 1923	Kanto, Japan	8.2	7.9
March 2, 1933	Sanrika	8.5	8.4
May 24, 1940	Peru	8.0	8.2
April 6, 1943	Chile	7.9	8.2
August 15, 1950	Assam, India	8.6	8.6
November 4, 1952	Kamchatka	8.0	9.0
March 9, 1957	Aleutian Islands	8.0	9.1
November 6, 1958	Kurile Islands	8.7	8.3
May 22, 1960	Chile	8.3	9.5
March 28, 1964	Alaska	8.4	9.2
October 17, 1966	Peru	7.5	8.1
August 11, 1969	Kurile Islands	7.8	8.2
October 3, 1974	Peru	7.6	8.1
July 27, 1976	China	8.0	7.5
August 16, 1976	Mindanao	8.2	8.1
March 3, 1985	Chile	7.8	7.5
September 19, 1985	Mexico	8.1	8.0

TABLE **18.3**   Energies of Some Major Earthquakes in the Western United States

Location	Date	Richter Magnitude	Energy in Tons of TNT
San Francisco, California	1957	5.3	500
Long Beach, California	1933	6.3	15,800
El Centro, California	1940	7.1	250,500
Kern County, California	1952	7.7	1,990,000
San Francisco, California	1906	8.2	12,550,000
Anchorage, Alaska	1964	8.5	31,550,000

One ton of TNT $= 10^9$ calories; one calorie $= 4.186 \times 10^7$ ergs.

An empirical relationship between the amount of energy released and the Richter magnitude scale has been developed:

$$\log E = 11.4 + 1.5 M_L$$

with E in ergs and M_L on the Richter scale. The two constants 11.4 and 1.5 are subject to modification as more data are acquired; previously a value of 11.8 was used for the intercept value.

The average number of earthquakes per year including all those that can be felt near their epicenters is estimated at more than 150,000. The total number including those earthquakes imperceptible to humans is probably about one million per year. Despite this large number, much of the total energy released in a given year will be associated with just a small number of major earthquakes. A single earthquake of magnitude 8.4 can yield about the same amount of energy as the average energy released each year throughout the first half of this century. Table 18.4 illustrates information on earthquake frequency and magnitude worldwide.

TABLE **18.4**   Annual Number of Earthquakes Relative to Magnitude Worldwide

	Magnitude	Average Number	Release of Energy (Approximate Explosive Equivalent)
Actually observed:			
Great	7.7–8.6	2	50,000 1-megaton (hydrogen) bombs
Major	7.0–7.6	12	
Potentially destructive	6.0–6.9	108	
Estimates based on sampling special regions	5.0–5.9	800	1-megaton bomb 1 small atom bomb (20,000 kg of TNT)
	4.0–4.9	6200	
	3.0–3.9	49,000	
	2.5–2.9	100,000	
			0.5 kg of TNT
	2.5	700,000	

Intensity Scale

Earthquake intensity is a measure of damage to human-made structures that occurs as a result of the ground shaking. It describes the damaging effect that the earthquake motions inflict on buildings and other engineering works. It is not measured by instruments, but is determined instead from reports of damage effects supplied by trained observers. Because intensity is based on actual observations of earthquake damage at specific locations, the evaluation is perhaps more meaningful to the layman than is the magnitude. Considerable time, weeks or months, may be required to assemble an intensity map for a particular earthquake. Another interesting detail is that the intensities for earthquakes that occurred several hundred years ago—well before seismographs monitored such phenomena—can be constructed on the basis of damage reports provided by historians, whereas an accurate measure of the magnitude is more difficult to compile.

The modified Mercalli scale is the measure of intensity used today. The first scale involving earthquake intensities was based on work by de Rossi of Italy and Forei of Switzerland in 1883. This scale, with values from I to X, was used for about 20 years before a more refined measure was needed. With advances in earthquake studies, in 1902, L. Mercalli, an Italian seismologist, developed a new scale with a I to XII range. In 1931, the Mercalli scale was modified by American scientists Harry O. Wood and Frank Neumann to incorporate the modern structural features of buildings. In use in the United States today, the modified Mercalli scale is shown in an abbreviated form in Table 18.5.

There is always some interest in comparing the intensity scale to the magnitude scale. Approximate correlations exist between the two scales, but difficulties are involved because the intensity depends on several factors other than earthquake energy. Geologic conditions and specifics concerning the structures are also involved. These factors are discussed in the next section. The ground acceleration induced by the earthquake, designated in terms of the equivalent acceleration of gravity, is also of interest because structural engineers commonly work in these units when analyzing building safety. Figure 18.1 shows the general correlation among the three measures associated with earthquakes.

TABLE 18.5 Earthquake Intensity—Modified Mercalli Scale (1931)

Intensity	Characteristics
I	Not felt; animals uneasy
II	Barely felt by people at rest
III	Felt indoors; like a passing truck; some hanging objects swing
IV	Like passing heavy trucks; hanging objects swing; autos rock; windows, dishes, etc., rattle; walls creak
V	Felt outdoors; sleepers awakened; small objects upset; doors swing open or close
VI	Felt by all; people frightened and run outdoors; dishes and glassware broken. Furniture moved, overturned; weak masonry cracked; small bells ring
VII	Difficult to stand; noticed in moving autos; furniture and masonry broken; chimneys break at roof line; ponds become turbid
VIII	Auto steering affected; masonry damaged, some collapse; frame houses move on foundations; twisting and falling of large chimneys; branches broken from trees
IX	General panic; masonry damaged or destroyed; frame buildings cracked, underground pipes broken; sand boils, soil cracks
X	Most masonry buildings destroyed; wood frame buildings destroyed; landslides; railroad rails bent
XI	Railroad rails badly bent; underground piping destroyed
XII	Total damage; objects thrown into air

Earthquake Damage

Earthquake damage to human-made structures is influenced by five principal elements: 1) strength of the earthquake wave, 2) length of earthquake motion (and effects of aftershocks), 3) proximity to a fault zone, 4) type of geologic foundation, and 5) building design.

The strength of the earthquake at the location in question is obviously important. This is indicated by the magnitude of the event and the distance from the

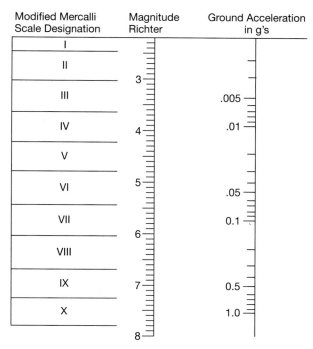

Intensities XI and XII
not included

FIGURE **18.1** Modified Mercalli intensity scale and
approximate relationship with magnitude and
ground acceleration.

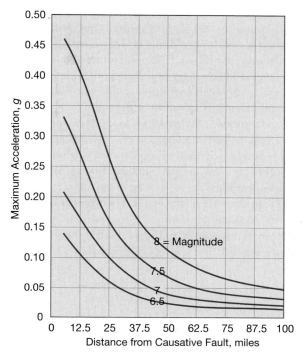

FIGURE **18.2** Relationship between peak bedrock
acceleration, earthquake magnitude, and fault
distance.

epicenter. The ground motion falls off rapidly with
distance as seen in Figure 18.2 where the distance is
related to the maximum acceleration for various
magnitude earthquakes.

The magnitude of a potential earthquake is related
to the length of rupture of the triggering fault.
Obviously the greater the rupture length, the greater
the energy produced and the higher the magnitude of
the earthquake event. The length of rupture is estimated for active faults in order to evaluate the
potential for occurrence of a major earthquake. Table
18.6 shows the relationship between earthquake magnitude and fault rupture length.

The length of earthquake motion is significant
because structural cracking may propagate while
shaking continues so that weakened members may
eventually fail completely. For the most significant
earthquakes, the motion persists for about 10 sec to
1 min. In the Anchorage earthquake of 1964, how-

ever, reports indicate that the shaking lasted for 4.5
min. This, the high magnitude (8.5), and the poor
geological conditions account for the extensive destruction of that major earthquake.

Aftershocks are also important. These occur for a
period of months after the major earthquake as the

TABLE **18.6** Earthquake Magnitude Versus Fault
Rupture Length

Magnitude (Richter)	Rupture Length (mi)	(km)
5.5	3–6	5–10
6.0	6–11	10–15
6.5	11–19	15–30
7.0	19–38	30–60
7.5	38–62	60–100
8.0	62–125	100–200
8.5	125–250	200–400

Earth readjusts following the sudden stress release. For example, the Kern County, California, earthquake ($M_L = 7.7$) of 1952 had an aftershock of 5.8 about one month later. The aftershock caused more damage to Bakersfield than did the main earthquake because the aftershock caused previously damaged, yet still standing, structures to collapse. Illustrated here is a concern for civil engineers: Damaged public buildings should be examined immediately and closed to the public following the main shock, if justified, to ensure that additional effects are minimized if aftershocks develop.

Proximity to a fault zone is important because the energy dissipation seems to concentrate in a fault area, yielding more shaking of the buildings than for adjacent areas. Therefore, if a building is on one fault zone and an earthquake occurs along another fault zone some distance away, the damage is likely to be greater along the nonconnected fault zone than in adjacent nonfaulted material. However, in an area where very strong ground movement occurs, the presence of the fault may be less important than the actual nature of the foundation material.

The geologic materials on which a building is founded play a primary role with regard to earthquake damage. Massive bedrock provides the best foundation because it passes wave motions on with minimum attenuation, resulting in less vibration to the structure. Soil or unconsolidated material tends to amplify rock acceleration, and soft soil may increase this level by several times. Low-density or poorly compacted soils will fare the worst because they supply much greater ground motion to the structure.

In addition to this increased vibration level, low-density soils yield other problems. They may be subject to densification, yielding a volume reduction and loss of support. Low-density, unsaturated sands will settle by reorienting their grains as the ground shaking occurs. Saturated fine sands and silts are subject to liquefaction caused when pore pressures build, thus drastically reducing the effective stress. When in a loose condition ($N = 10$ to 15 blows/foot; see Chapter 19 for details on the standard penetration test), these saturated fine soils are prone to liquefaction and the soil flows downhill except for very gentle slopes. Alluvial areas along major streams are prime locations for such failures.

Human-made fills not properly densified are also

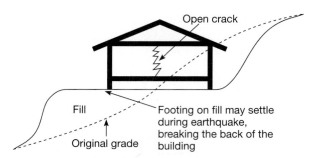

FIGURE 18.3 Effect of earthquakes on structures built on poorly compacted side hill cuts.

subject to settlement during shaking. Graded lots on a hillside can supply major residential failures if the fills are not properly constructed, as shown in Figure 18.3. At the site, soil and weathered rock were pushed over the edge of the hillside during construction and compacted only by overburden weight and that of a bulldozer. Hence the footing on the right is founded on rock and the one on the left on poorly compacted soil. Major settlement of the downhill footing results in failure of the house. Also shown in place is the original topsoil left at the base of the fill. This should have been removed prior to fill-placement because this material can cause slope failure.

Building design is the last of the five factors that influence damage. For residential buildings the object is to build a structure that can overcome inertia and move as a unit with ground motion without the individual parts moving independently of each other. This requires integrity of the structure and is obtained by adequate bracing, with secure anchoring and also bonding of all elements. The whole building—foundation, frame, walls, floors, and roof—must be balanced and well tied together.

Conventional wood frame construction with stud supports and wood veneer or aluminum siding has a high degree of earthquake resistance provided sound workmanship is involved. Adequate lateral bracing is supplied by such elements as the frame, roof, wall sheathing, interior walls, and ceiling panels. Lateral supports can be improved if the sill is bolted to the foundation, all studs are toe-nailed to the sill from both sides, and cross-wall bracing is included. This bracing should be installed in opposing pairs either in an X or V shape. Figure 18.4 shows these aspects in detail.

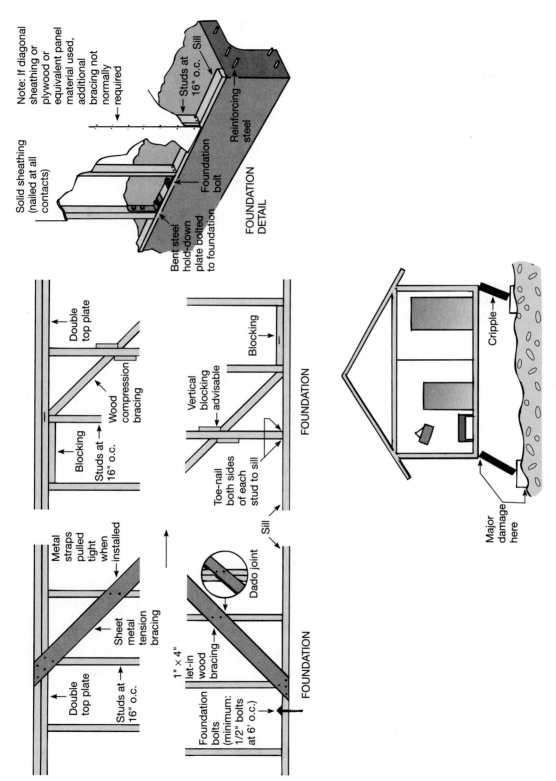

Solid sheathing (nailed at all contacts)

Note: If diagonal sheathing or plywood or equivalent panel material used, additional bracing not normally required

Studs at 16" o.c.   Sill

Reinforcing steel

Foundation bolt

Bent steel hold-down plate bolted to foundation

FOUNDATION DETAIL

Double top plate

Wood compression bracing

Blocking

Studs at 16" o.c.

Blocking

Vertical blocking advisable

Toe-nail both sides of each stud to sill

Sill

FOUNDATION

FOUNDATION

Metal straps pulled tight when installed

Sheet metal tension bracing

Double top plate

Studs at 16" o.c.

1" × 4" let-in wood bracing

Dado joint

Foundation bolts (minimum: 1/2" bolts at 6' o.c.)

FOUNDATION

Cripple

Major damage here

FIGURE 18.4   Structural damage to residential construction during earthquakes, and construction techniques to improve earthquake resistance.

Ordinary masonry structures may be particularly vulnerable to earthquake shaking. This is a consequence of their greater mass than wood frame houses and, commonly, their inability to move as a unit. If good brick or concrete block is laid with strong Portland cement mortar, properly reinforced with steel rods through the masonry, and tied in well with the foundation and inner walls, masonry buildings can resist shaking quite well. Concrete slabs on well-compacted soil and secure footings should perform without undue problems. Nonreinforced bearing walls of adobe or hollow concrete block will be seriously damaged in even moderate earthquakes. This is a problem of large and small buildings alike and for block walls and masonry veneer for buildings where reinforcing is omitted.

Tall buildings must be resistant to transverse motion and to the frequency at which the building vibrates during earthquakes. If a resonant frequency for the building develops, severe damage will be induced. This is known as *site resonance* and involves the relation between the natural period of the ground and the natural period of the structure. Other design details are involved, but these are discussed in a review of the San Fernando earthquake of 1975 presented in a later section.

The secondary wave is typically the most destructive wave form of an earthquake because it induces transverse or horizontal movement in a structure. Increased lateral stiffening of the structure is sometimes required to provide sufficient resistance to transverse shaking. Vertical movement or forces can be more easily resisted because these act in the same direction as gravity or the weight of the building.

Field measurements of the secondary wave velocities of the soil, V_s, commonly provide valuable information concerning earth shaking. Figure 18.5 shows how secondary waves are generated in the field. A sledgehammer is used to strike the end of a wooden plank held against the ground by the weight of a field vehicle. As hammering continues, the sonde is lowered down the hole and V_s is determined at various depths in the soil column.

The period of the soil column P is calculated from the equation

$$P = \frac{4T}{V_s}$$

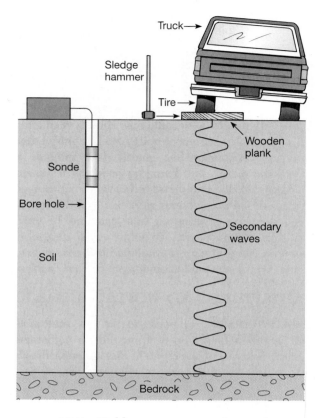

FIGURE 18.5 Field measurements of secondary wave velocity.

where T is the thickness of the soil column and V_s is the secondary wave velocity. For example, if $V_s = 250$ m/sec and the soil or unconsolidated column is 32 m thick, then

$$P = \frac{4(32)}{250}$$

$$= 0.51 \text{ sec for the period of the soil column}$$

A critical condition develops when the period of the soil column matches the period of vibration of a building founded on the soil. These two periods should not coincide because if they do, the building will likely be destroyed during a major earthquake. The rule of thumb for buildings is shown by the following equation:

$$P_{bldg} = \frac{N}{10}$$

where N is the number of floors in the building. So for a five-story building,

$$P_{\text{bldg}} = \frac{5}{10} = 0.5 \text{ sec}$$

Also, some bridges have a period of about 0.5 sec.

Recall the earlier example calculation for the 32-m-thick soil column that had a period of 0.51 sec. A comparison of this 0.51 value to the 0.5 sec for a five-story building leads us to conclude that ground shaking would cause severe damage to this five-story building.

Geologic Investigations of Earthquake Prone Areas

Geologic investigation by necessity is extensive in earthquake-prone areas where new construction is planned or the safety of an existing structure is to be determined. The following are typically required in an engineering geology investigation:

1. A structural geology map of the region indicating recent tectonic movements.
2. A compilation of active faults including the nature of displacements. Field work will likely be necessary and would involve trenching across suspected fault zones. Geologic criteria showing fault movement in the Holocene time (the past 10,000 years) are important; these include displacements in recent soils, dating of organic materials by hydrocarbon methods, and evaluation of landforms to determine relative ages of formation.
3. Mapping of the structural geology for the site, including scarps in the bedrock, differential erosion, and offsets in sedimentary deposits. Included should be rock type, surface features, and local faults. The latter should include probable length, continuity, and sense of movement.
4. Use of geophysical exploration on through-going faults near the site to define recent fault ruptures. Techniques include electrical resistivity and gravity measurements in a direction normal to the fault. Examine evidence for segmentation of the fault, looking for stepover of fault stands and changes in strike.
5. Evaluation of reports involving landslides, major settlements, or effects of flooding at the site.
6. Investigation of ground-water barriers in the area that are associated with faults or may affect the soil response to earthquake shaking.
7. Borings, trenches, excavation, and sampling to locate and describe the presence of sand layers, which may be subject to liquefaction.
8. Measurement of the physical properties of the soil (density, water content, shear strength, behavior under cyclic loading, attenuation) *in situ* or on borehole samples.
9. Determination of *P* and *S* wave velocities and attenuation values in the soil layers using geophysical exploration methods.

This summary addresses particularly critical construction in populated areas. Both the density of population and the nature of the proposed engineering structure impact the level of risk involved in an earthquake-prone area.

Locating Earthquakes Using Seismology

Seismograph stations in the world network receive the wave trace from a strong earthquake starting with the arrival of the *P* waves and followed subsequently by the *S* waves and then the *L* waves. This is illustrated in Figure 18.6. If the velocities of the *P* and *S* waves are known, the distance to the epicenter can be calculated based on the time delay between the *P* wave and *S* wave first arrivals. This is somewhat of an oversimplification, however, because the *P* and *S* wave velocities are not constant with distance from the epicenter but both increase in velocity with greater distances. An example problem serves to

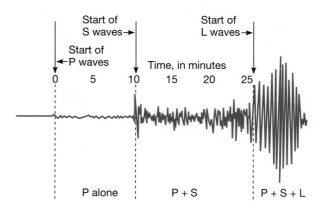

FIGURE 18.6   Earthquake wave traces on a seismogram.

illustrate this ability to calculate the distance, however.

EXAMPLE PROBLEM 18.2

If the time interval at a seismic station between the first arrival of the P wave and the S wave is 250 sec and the velocity of the P wave is 7.89 km/sec and that of the S wave is 4.33 km/sec, what is the distance from the station to the epicenter?

ANSWER

Assume x = distance from seismic station to the epicenter in kilometers. Then

$$\frac{x}{V_p} = \text{travel time for } P \text{ wave} = t_p = \frac{x}{7.89 \text{ km/sec}}$$

$$\frac{x}{V_s} = \text{travel time for } S \text{ wave} = t_s = \frac{x}{4.33 \text{ km/sec}}$$

Therefore, the time interval between the first arrival of the P and S waves or $\Delta t = t_s - t_p = 250$ sec.

$$\frac{x}{4.33} - \frac{x}{7.89} = 250$$

$$7.89x - 4.33x = 250(7.89)(4.33)$$

$$3.56x = 8541$$

$$x = 2400 \text{ km}$$

An accurate plot of travel time versus distance for earthquake waves is shown in Figure 18.7. The slope of these curves (its first derivative) equals $1/v$ where v is the velocity. Observe that the slope decreases or the velocity increases with distance. This occurs because these waves have wave paths located deeper in the Earth than do waves traveling shorter distances. An increase in seismic velocity with greater depth provides this greater velocity.

Also shown in the plot of Figure 18.7 is the cessation of the S wave (corresponding to a depth of 2900 km) and an offset for the P wave. This occurs at the boundary between the mantle and outer core, where S waves are attenuated and the P waves are refracted. These topics are discussed further in a later section concerning the nature of the Earth's interior.

In actual practice, the distance from the seismic station to the epicenter is determined from precise plots of travel-time curves of the nature shown in Figure 18.7. The travel time difference between the P and S waves is compared to the two sloping curves

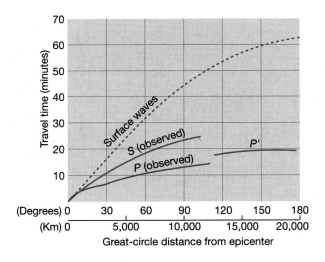

FIGURE 18.7 Travel-time curves for seismic waves in the Earth.

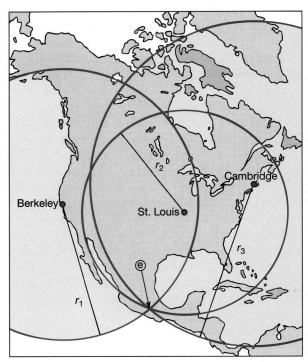

FIGURE 18.8 Locating an earthquake epicenter using three seismic stations.

until a fit is obtained. The distance is indicated by the distance values designated along the abscissa.

After the distance between the epicenter and seismic station is determined, a circle with that radius can be drawn using the station as the center. Three such stations are required to locate the earthquake, and the triple intersection point marks the location of the epicenter. Details are shown in Figure 18.8.

The depth of focus for the earthquake can also be determined. If the focus is only a few kilometers deep, then the P and S waves arrive at the epicenter sooner than they reach points 100 km or so away. By contrast, if the focus is 150 to 300 km deep, the waves arrive almost simultaneously at all points within a large circle around the epicenter. Therefore, the surface area in which the shock is felt simultaneously increases rapidly as the depth of focus increases. The delays in arrival times for the S and L waves relative to the P wave are also indicative of the depth of focus and can be compared from a number of seismic stations.

Based on these data, earthquakes can be classified according to depth. They are subdivided into shallow-focus earthquakes, intermediate-focus earthquakes, and deep-focus earthquakes. Information concerning the depth of focus is presented in Table 18.7.

In recent years the low-velocity zone of the upper mantle (from about 40 to 300 km deep) was discov-

ered through seismological studies. In this zone the velocity of the P wave is reduced as much as 10% and the S wave about 2%. It is now thought that many shallow-focus earthquakes occur above the low-velocity zone, many of the intermediate-focus variety within it, and deep-focus earthquakes below it.

Causes of Earthquakes

Following studies of the San Francisco earthquake of 1906, it was soon generally accepted that earthquakes were caused by faulting as a result of rock rupture in the Earth's crust. This is known as the elastic rebound theory. It was based on observations that the San Andreas fault associated with the 1906 earthquake had shown distortion in years prior to the event and considerable displacement along the fault afterward. This suggested a sudden release in stress during rupture, followed by elastic recovery.

Today this mechanism is questioned for earthquakes that originate at a depth of 20 to 30 km or more. The confining pressures at that depth because of the overburden load prevent sudden extensive ruptures from occurring, but instead the rocks deform plastically. During this deformation some buildup of stress is possible because of rock viscosity so that sudden jerks or ruptures capable of causing earthquakes could interrupt the continuous plastic deformation that occurs. This mechanism seems more acceptable in light of current knowledge about the Earth's interior.

Another item of evidence is the fact that in some cases surface fault movement occurs after an earthquake rather than during it. In the El Centro earthquake of 1979, creep movement across a bituminous road continued several weeks after the road had been repaired following the earthquake. The surface fault displacement was apparently a result of the earthquake-causing mechanism and not the cause for originating the strong ground motions. This suggests that such faulting is a consequence of earthquakes, not a cause for their occurrence.

Earthquake Prediction

The details of earthquake prediction that receive the greatest publicity are location, magnitude, and time of occurrence. For many interested citizens, earthquake prediction simply means predicting the time of occurrence of a major earthquake. In reality, the ability to predict the specifics of the strong ground

TABLE 18.7 Classification of Earthquakes by Depth of Focus

Type	Range of Depth (km)	Relative Abundance
Shallow-focus	70	Most earthquakes occur in this range, more than 75% of total.
Intermediate-focus	70–300	Most of the remaining earthquakes occur in this range.
Deep-focus	300	About 3% of all earthquakes; no earthquakes below about 700 km ever recorded.

motion at a specific site are more important for mitigation of earthquake hazards. Strong motion data are now used for building design, evaluation of existing structures, and zoning for new construction.

In 1990, great concern was experienced about the occurrence of a major earthquake in the lower Midwest of the United States. This concern, which is more typical in certain areas of California and Alaska, was elevated to fear and near panic in the minds of some citizens—to the extent that classes in primary and secondary schools were cancelled on the day of the predicted earthquake.

Iben Browning, a climatologist from New Mexico, predicted that on December 3, 1990, a major earthquake would occur on the New Madrid fault, located in eastern Tennessee and southern Missouri. This was the location of a series of major earthquakes in 1811–1812 that had an estimated Richter magnitude of as high as 8.7, causing major soil ruptures and liquefaction, and vibrations that were felt up to 1000 miles away. No major earthquakes have occurred on the New Madrid fault since that time and the repeat interval for a major event has been estimated as between 200 and 500 years.

Browning's prediction was apparently based on the fact that the gravitational pull on the Earth would be maximized during this particular spring tide event. As discussed in Chapter 16 on coastal processes, a spring tide occurs twice each lunar month during the full and new moons when the Sun and Moon lie in a straight line and their pull on the Earth is additive. December 3, 1990, was a time of full moon, but another celestial condition prevailed, making this a rare occurrence. This event was called a triple junction. In addition to the Sun and Moon being aligned (spring tide situation), the Moon was in perigee, or at a point in its orbit when closest to the Earth. This triple occurrence happens only once about every 61 years.

Selection of December 3, 1990, by Browning as a rare celestial occurrence can certainly be affirmed, but the conclusion he reached (or jumped to)—that this would trigger a major earthquake on the New Madrid fault system—was highly questionable. Other researchers in the past have attempted to correlate earthquake occurrence with celestial gravitational attraction and have found no consistent relationship to exist.

Because of Browning's prediction, people of the lower Midwest have become more conscious of the possibility of a major earthquake at sometime in the not too distant future. Certainly people in such cities as Memphis, Tennessee; St. Louis, Missouri; and Evansville, Indiana; have become more concerned about earthquake damage. This greater awareness is a positive contribution to what was otherwise an unwarranted scare based on a flawed, scientific evaluation by Iben Browning. The news media treated Browning's theory as a hot news item, which helped to elevate the concern to fear and near panic for many individuals in the affected areas of the lower Midwest.

A number of methods for predicting the time and size of an earthquake have been tried. Earthquake locations of course are related to the presence of existing, active faults and overall seismic activity. As discussed previously, fault rupture length is used to provide a rough estimate of the size or magnitude of the expected earthquake.

Physical clues used for predicting the time of occurrence are 1) increased seismicity (foreshocks) and animal restlessness; 2) ground uplift; 3) reduction in P wave velocity; 4) increased radon emission; 5) decreased electrical resistivity; 6) temporal and spatial gaps in the seismicity of a region suggesting that a portion of a fault zone has locked-in stress, which eventually will release yielding an earthquake; 7) determination of a periodicity of earthquake occurrence; and 8) on monitored fault zones, changes in the ground-water level and fault slippage or displacement. All of these physical determinations have encountered complications when applied to a series of earthquake situations. Continued application of these procedures to monitored fault zones should show which, if any, will prove to be reliable prediction techniques for the time of earthquake occurrence.

Distribution of Earthquakes

No portion of the Earth's surface is completely free from earthquakes, yet earthquake occurrence and magnitude are by no means evenly distributed. For most areas only infrequent shocks of small or moderate intensity have occurred. By contrast some zones of the Earth are subject to frequent earthquakes, which range from weak tremors to strong motion events. These comprise the major earthquake belts of the world.

The most prominent seismic areas are as follows: 1) the Circum-Pacific Belt, 2) the Mediterranean and Trans-Asiatic Belt, 3) the Mid-Ocean Ridges of the Atlantic and Indian Oceans, and 4) the East-African Rift Zone.

The Circum-Pacific Belt accounts for some 80% of the earthquakes each year. It extends from the tip of South America at Cape Horn along the Andes Mountains to Central America and along the western area of North America to Alaska, across the Aleutians to Asia and southward down the Asian coast to the island arcs extending southeastward to New Zealand. This area is also known as the "Ring of Fire" because volcanoes are commonly associated with this zone. This extensive belt marks the location where many of the plates of the earth are actively opposing one another. (Refer to the discussion on plate tectonics in Chapter 1 for more details.)

Approximately 15% of all earthquakes occur in the Mediterranean and Trans-Asiatic Belt. It extends from Gibraltar on the west end of the Mediterranean Sea across Italy and Greece through the mountainous regions of the Middle East to the high mountain ranges of southern Asia. Many major earthquakes with severe levels of damage have occurred along this belt during historic times.

The midocean belts follow the midocean ridges, which mark the centers from which seafloor spreading originates. High rates of heat flow are also associated with these highs in the ocean basins. In East Africa the fault zone lies along a line from the Red Sea on the north through Lakes Rudolph, Victoria, Tanganyika, and Nyasa to the south. This area is sometimes marked by a prominent fault escarpment along the earthquake zone and is known as the rift valley.

Today, we also recognize through studies of plate tectonics, that most earthquakes occur along belts marking the boundaries of crustal plates. Earthquakes occur where these plates collide, separate, or simply slip past each other. They coincide with chains of volcanoes, zones of crustal disturbance, active strike-slip faults, deep-sea trenches along continental borders, young mountain belts, and midocean ridges. These are the areas of greatest geologic activity on Earth today. Of the four major earthquake belts described here, only the East African Rift Zone does not lie along a plate boundary.

Earthquakes of the Midcontinent Region

In 1967 and 1968 the U.S. Coast and Geodetic Survey studied approximately 28,000 earthquakes that occurred in the contiguous United States. One form of the data compiled was cumulative strain release from earthquakes between 1900 and 1965. The strain release is assumed to be proportional to the square root of the energy release and this in turn involves the Richter magnitude.

Strain release was determined for 10,000-km^2 areas of the United States and represented in map form using equivalent magnitude 4 earthquakes for each of these blocks. For the Midwest, several locations were designated as high strain-release zones. This map is presented as Figure 18.9. Note that an area in western Ohio has 64 to 256 equivalent magnitude 4 earthquakes, whereas southern Illinois has from 16 to 64 of these equivalents.

Based on these strain-release data and past earthquake magnitudes and epicenters, the 1969 Seismic Risk Map of the United States was developed by Algermissen of the U.S. Coast and Geodetic Survey (Figure 18.10). A seismic zone 3 region surrounded by a ring of zone 2 is centered on the southern Illinois, southwest Missouri and eastern Kentucky border. This is also the site for the massive earthquakes of 1811–1812 in New Madrid, Missouri. Zone 2 corresponds to a "moderate damage" zone of earthquake intensity VII and zone 3 relates to "major damage" with values of VIII and higher.

A midwestern earthquake in recent years, which had considerable effect, occurred on November 9, 1968, in south-central Illinois. It was felt in an area of approximately 580,000 mi^2 (1,485,000 km^2) with a maximum intensity of VII and a Richter magnitude equal to 5.5. An isoseismal map of this earthquake area is shown in Figure 18.11. An isoseismal map uses contour lines to depict those locations with equal seismic intensity. As expected, the intensity decreases with distance from the epicenter, but there are isolated pockets of lower or higher values. Also, contours travel up the stream valleys (Missouri, Mississippi, and Ohio Rivers clearly shown; Wabash River in west-central Indiana also involved), which indicates how the higher intensities are associated with the unconsolidated materials of major stream valleys.

Using information from the 1968 earthquake along

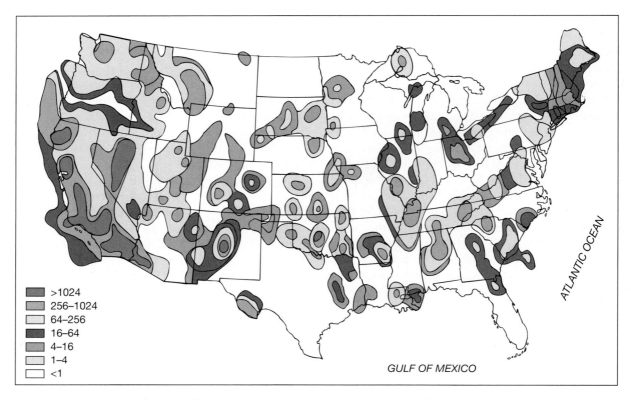

FIGURE 18.9   Strain release in the United States, 1900 to 1965, expressed as
equivalent number of magnitude for earthquakes.

with other shocks that have occurred in the Midwest, we can develop a map of maximum historical intensity. Such a map for Indiana and contiguous states is shown in Figure 18.12. In general, the values increase toward the southwest with a maximum of IX occurring near the confluence of the Ohio and Mississippi Rivers, again near the historic site of the New Madrid earthquakes of 1812–1814. A potential earthquake problem in some areas outlined on Figure 18.12 is the liquefaction of loose, saturated fine sands and silts. This would occur most likely in the alluvial areas in the zones IX and VIII and possibly also VII as well. Homes, industrial and commercial building, bridges, and fill embankments would likely be affected.

A number of lessons can be learned from the analysis of damage in the San Fernando earthquake of 1971 ($M = 6.5$). This earthquake caused considerable damage (65 deaths and more than one-half billion dollars property loss) despite the fact that it was not a so-called "major" earthquake ($M > 7$) and the Los Angeles area had some 40 years of development under seismic building codes prior to its occurrence. Of course, codes are primarily intended to prevent structural collapse during an earthquake; they do not guarantee the usefulness of the structure after the shock.

Highway bridges and embankments were particularly prone to damage in the San Fernando earthquake. Design weaknesses included the following: 1) Concrete columns did not take twisting and grinding; increased strength and closer spiral reinforcing was needed. 2) Expansion joints did not hold bridge decks in place. Movement could be reduced by tension retaining rods. 3) Inadequate bond where concrete shattered; addition of bars at top of block footings and bottom of column caps needed. 4) Weak connection between pile caps and footings; anchor system needed that can take tension. 5) Shear keys did not

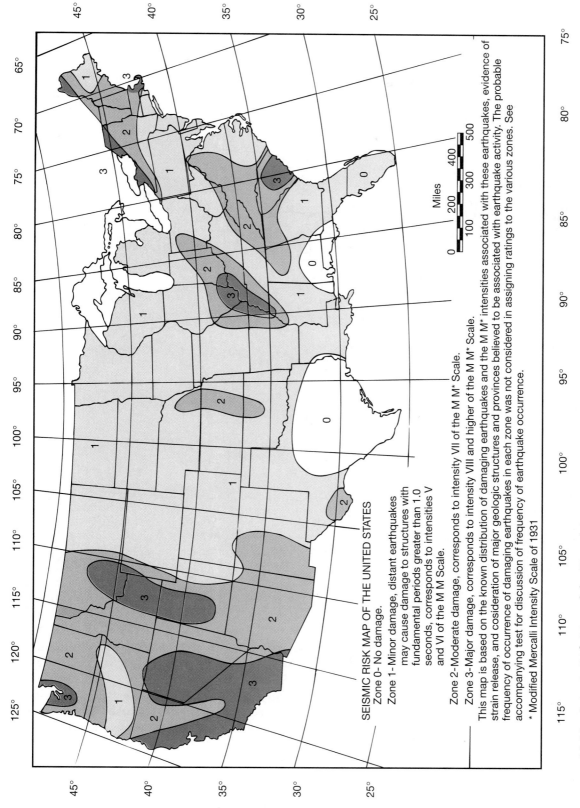

SEISMIC RISK MAP OF THE UNITED STATES

Zone 0- No damage.

Zone 1- Minor damage, distant earthquakes may cause damage to structures with fundamental periods greater than 1.0 seconds, corresponds to intensities V and VI of the M M Scale.

Zone 2-Moderate damage, corresponds to intensity VII of the M M* Scale.

Zone 3-Major damage, corresponds to intensity VIII and higher of the M M* Scale.

This map is based on the known distribution of damaging earthquakes and the M M* intensities associated with these earthquakes, evidence of strain release, and cosideration of major geologic structures and provinces believed to be associated with earthquake activity. The probable frequency of occurrence of damaging earthquakes in each zone was not considered in assigning ratings to the various zones. See accompanying test for discussion of frequency of earthquake occurrence.

* Modified Mercalli Intensity Scale of 1931

Miles

0   100   200   300   400   500

FIGURE **18.10**   Seismic Risk Map of the United States.

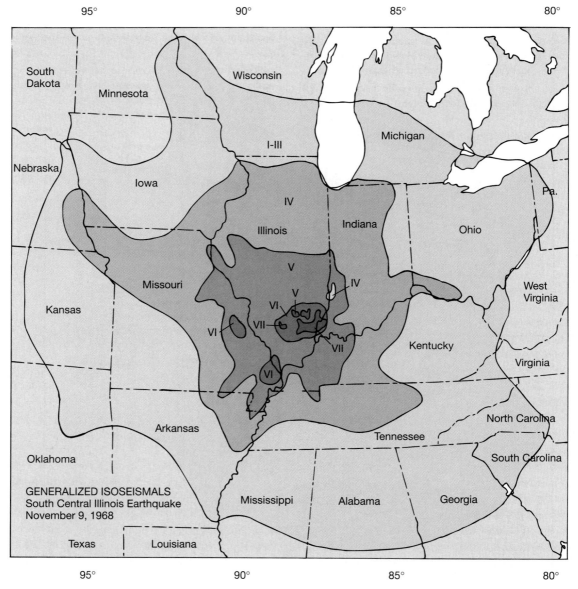

FIGURE 18.11   Generalized isoseismal map of the south-central Illinois earthquake,
November 9, 1968. Intensities refer to the modified Mercalli scale.

provide lateral restraint of bridge abutments; larger
shear keys and longitudinal ties needed. 6) Inclusion
of dynamic earthquake forces in the analysis of a
structure is needed rather than using "equivalent"
static forces.

It is not warranted in the midwestern United
States to build structures that will not sustain some
damage in a severe earthquake. The cost to provide
this is simply too great relative to the level of risk
involved. A structure should be strengthened to prevent collapse and loss of life but funds are unwisely
spent if used for total prevention with expenditures
significantly greater than the cost to repair damage
that would occur.

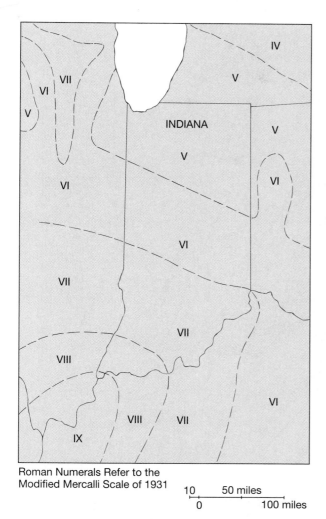

Roman Numerals Refer to the
Modified Mercalli Scale of 1931

10   50 miles
0   100 miles

FIGURE 18.12 Maximum historical intensity map of Indiana, 1811 to 1974.

Loma Prieta, California, Earthquake

On October 17, 1989, at 5:04 P.M., Pacific Daylight Time, a magnitude 7.1 earthquake occurred on the San Andreas fault 10 mi (16 km) northeast of Santa Cruz, California. The epicentral location is shown on Figure 18.13 with San Jose 21 miles (33 km) to the northeast and the cities of San Francisco and Oakland about 55 miles (90 km) to the northwest. This major event, the Loma Prieta earthquake, takes its name from the Loma Prieta mountain peak located less than 10 km from the epicenter.

It was one of the largest earthquakes to occur in California since the catastrophic San Francisco earthquake of 1906. The event caused 63 deaths and at least $5.9 billion in damage, making it the largest dollar-loss natural disaster to that time in U.S. history. It injured 3,757 people and left more than 12,000 homeless.

The focal depth of the earthquake was 17.6 km, thus qualifying it as a shallow-focus earthquake. It registered the following magnitude values: $M_s = 7.1$, $m_b = 6.5$, $M_L = 7.0$, and $M_w = 6.9$. (Refer to the previous discussion in this chapter for the distinction between these different earthquake magnitude measurements.)

Intensities from the earthquakes reached VIII on the modified Mercalli intensity scale (MM) throughout much of the epicentral area. Isolated values of MM IX were assigned to the collapse of the elevated section of Interstate 880 (Nimitz Freeway) in Oakland, and to the damaged Highway 480 (Embarcadero Freeway), and the Marina district in San Francisco. Some areas of San Francisco underlain by thick deposits of Quaternary bay mud and sand dunes experienced intensities of one to two MM units higher than the central part of the city underlain by bedrock.

The area shaken at intensity VII or greater (shaking that causes significant structural damage) measured 1680 mi^2 (4300 km^2) on land. As a comparison, the area of MM VII or greater in the 1906 San Francisco earthquake was 18,750 mi^2 (48,000 km^2) on land or about 11 times greater.

Numerous ground failures occurred over an area 62 miles long by 25 miles wide (100 km by 40 km) extending from San Gregorio to Hollister, with additional coastal bluff landslides occurring along the coast to Marin County north of San Francisco. Ground failures included several types of landslides, most commonly rock falls, rock slides, soil falls, and soil slides, of less than 100 m^3. Liquefaction failures were widespread, occurring from Oakland to Salinas, and produced numerous sand boils and mud volcanoes.

Most people in the earthquake's vicinity experienced about 10 to 15 sec of strong ground shaking stronger than 0.1g. Felt by millions of people, the P wave was experienced first, followed by the strong shaking associated with the S waves and surface waves. The strongest ground shaking recorded in an

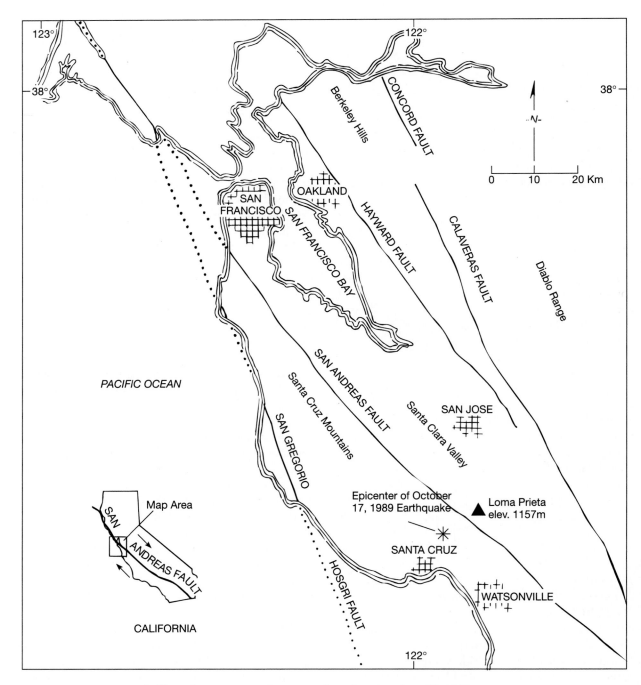

FIGURE 18.13   Map of affected area, Loma Prieta earthquake, October 17, 1989.

area independent of structures (which amplify the motion) was 0.64g horizontal. It yielded little damage, whereas damage nearby, particularly on ridges, was extensive. This suggests that shaking on the ridges and in much of the epicentral area was stronger than 0.64g.

Damage caused by the Loma Prieta earthquake can be considered under two categories: 1) damage near the epicenter and 2) damage located a far distance from it (more than 50 miles). The area near the epicenter includes the Santa Cruz Mountains plus the cities of Santa Cruz and Watsonville.

Severe ground shaking in the Santa Cruz Mountains caused major damage—local peak accelerations may have approached 1.0g. Eyewitness reports indicate household items flew across the room and water tanks became airborne. Nearly 5000 homes in unincorporated areas of Santa Cruz County experienced earthquake-related damage and some houses were thrown off their foundations. A massive landslide closed the northbound lanes of Highway 17 for 33 days and large slumps severely damaged dozens of homes in several housing subdivisions.

In flatter terrains near the Santa Cruz Mountains, ground shaking displaced sidewalks and curbs. In the town of Los Gatos about 31 commercial buildings and 318 residences experienced $240 million of damage. Much of the damage was localized, involving masonry structures or older residences. On the Stanford University campus near Palo Alto, 25 of the 60 damaged buildings required structural reinforcement or repair for an estimated cost of $160 million.

In Santa Cruz damage to commercial properties, residential structures, and municipal properties was about $51 million, $50 million, and $12 million, respectively. On the nearby University of California campus about $5.8 million was incurred, of which about $4.8 million was structural damage. In the downtown area, Pacific Garden Mall and the adjacent residential blocks of mostly older homes experienced the greatest damage. These areas are underlain by recent alluvium of the San Lorenzo River and ground shaking was more damaging than in neighboring areas underlain by older, more consolidated marine terrace deposits. Older brick buildings were extensively damaged and several people were killed when buried under rubble. About 375,000 ft^2 of commercial prop-

erty was demolished. Residential areas experienced collapsed chimneys and porches plus foundation failures. One building burned down, presumably because of a gas line rupture. Some residences, recently upgraded structurally, showed little earthquake damage whereas many houses that had not been retrofitted were severely damaged.

Similar damage occurred in the town of Watsonville. Brick buildings were seriously damaged and many older houses suffered partial collapse from foundation failure. Liquefaction of unconsolidated, saturated, sandy deposits occurred along alluvial plains and near the coast from Santa Cruz to Salinas. A dramatic example of damage was the collapse of a portion of Highway 1 spanning Struve Slough where strong shaking caused liquefaction of the subgrade soils.

Damage far from the epicenter occurred some 60 mi (about 100 km) away in San Francisco and Oakland. The intensity of ground shaking typically decreases with distance from the epicenter for sites underlain by bedrock or well-consolidated soil. In contrast, sites on bay mud can experience high levels of shaking because of amplification of the wave energy by the soft soil. Consequently, areas underlain by bay mud and soft artificial fill near San Francisco underwent liquefaction during the Loma Prieta earthquake. Most of the $2 billion damage for the city occurred in the Marina District underlain by artificial fill. For the city as a whole, 22 buildings suffered so much damage that they had to be demolished, 363 were declared unsafe, and 1250 were only suitable for limited access pending repairs.

The hard hit Marina District is located on an old lagoon filled in to accommodate construction of the Panama-Pacific Exhibition, which celebrated the rebuilding of the city after the devastating 1906 earthquake and fire. Sand boils and evidence of lateral spreading indicate that liquefaction of the sandy fill led to extensive damage. Many of the affected structures were three- and four-story apartments with parking garages on the first floor. The limited shearing resistance of the open, first floor makes these buildings particularly prone to collapse during shaking or lateral spread of the underlying soil. Therefore, a combination of ground deformation and susceptible building design led to the collapse. Lateral spreading of fills also apparently damaged water mains, hampering initial efforts to fight fires resulting from the earthquake. Many unreinforced masonry buildings were severely damaged and six fatalities occurred when a brick structure collapsed.

Across the bay in Oakland, damage was concentrated in areas underlain by bay mud and sandy alluvium. The 1.5-mi-long collapsed portion of Interstate 880 (Cypress Structure) was underlain by thin bay mud and artificial fill, whereas the portion that did not fail was underlain by older alluvium. Locally, enhanced ground motion likely contributed to the disastrous pancaking of the freeway. Similarly, sections of Highway 280 and Highway 480 (Embarcadero Freeway) in San Francisco were closed because of earthquake damage. Other areas of the East Bay margins underlain by sandy fill and bay mud showed damage caused by liquefaction and lateral spreading. The Port of Oakland was damaged by lateral spreading. Liquefaction of sandy fill caused damage to the approaches to the San Francisco–Oakland Bay Bridge, at the Oakland International Airport, and at the Almeda Naval Air Station.

In Oakland, about 1400 residential, 200 commercial, and 12 public buildings sustained damages of about $29 million, $750 million, and $250 million, respectively. In addition, 13 commercial buildings were destroyed, with an estimated loss of $234 million, yielding an estimated total of about $1.3 billion. This does not include the demolition and reconstruction costs of the Cypress Structure. Also, significant structural damage occurred in areas underlain by older alluvium where the ground shaking presumably was less than for bay mud and artificial fill areas. These buildings in West Oakland, which partly collapsed or were thrown off their foundations, were either old Victorian-style or masonry structures that had not been seismically upgraded. In all, much of the damage in downtown Oakland, where the structures are underlain by older alluvium, occurred to masonry buildings.

The nature of the damage and its distribution caused by the Loma Prieta earthquake illustrates that local geology and building construction details greatly affect the extent of structural damage during seismic shaking. Bay mud and loose sandy, human-made fills yield increased seismic hazards for structures built on them. Also, the fatalities in San Francisco and Santa Cruz occurred mostly in unreinforced, masonry buildings. A modification of land use practices to avoid failure-prone areas and reinforcement of existing susceptible buildings is required to reduce future earthquake damage effects.

Human-Induced Earthquakes

Earthquakes are induced in a number of ways because of human activity. There is, of course, the obvious contribution of shock waves by rock blasting for quarries and construction plus the major effects of underground nuclear explosions. In the latter case, distinctions can be made using sensitive seismographs, which sense the unique patterns of such blasts. Details of construction blasting are considered in a later section under geophysics.

In addition, many disturbances at the Earth's surface generate small seismic waves. Included are the wind, auto and truck traffic, railroad trains, and vibrating industrial machinery. These waves can be measured by seismographs and collectively provide part of the microseisms that supply the background displacements on a seismograph.

Earthquakes can be induced by other human activities that have a less obvious connection. The filling of reservoirs after dam construction can yield low-magnitude earthquakes and there is always concern that rapid filling may provide a disruptive shock. We also know that disposal of liquids can increase pore pressures along incipient fault zones. The Rocky Mountain Arsenal in Denver, Colorado, experienced this phenomenon in 1961. It was later established that low-magnitude earthquakes in an area not previously prone to earthquakes (see the seismic risk map, Figure 18.10) were directly related to injections of disposal liquids in the deep well. This deep-level disposal has been discontinued since the relationship was proven by a local Denver geologist.

This suggests a mechanism for releasing smaller magnitude earthquakes to prevent the build up of strain that would ultimately produce a major earthquake. Water injection into a fault zone might trigger the release of a magnitude 5 earthquake when the strain builds to that level. Also the length of a fault rupture might be reduced by pumping water from a fault zone at specific locations, thereby breaking the fault into segments. To do this work with an acceptable level of certainty of the outcome, more research is required leading to specific knowledge about pore pressures and ground-water movements in fault zones.

There is also the question of liability for the damages caused by a 5.0 magnitude earthquake that is triggered by a governmental agency or government contractor. The government could be held responsible for this damage. Also it may be difficult to prove that a natural earthquake of greater magnitude would have eventually occurred to yield greater damage than that caused by the induced earthquake.

THE EARTH'S INTERIOR

Studies of travel paths and wave velocities in the Earth have revealed much information concerning its interior structure. Wave velocities increase in denser materials, waves are refracted and reflected at boundaries, and specific waves (S waves) are not transmitted beyond a depth of 2900 km. This latter fact leads to the shadow zone on the side of the Earth opposite from an earthquake, a zone where no S waves are received.

The refraction of waves reveals major discontinuities or changes in physical properties. From this, four major zones within the Earth are distinguished: the crust, mantle, outer core, and inner core, as depicted on Figure 18.14.

Crust

The boundary between the crust and mantle is the Mohorovicic discontinuity known more commonly as the "Moho" or "M" discontinuity. It was first recognized by the Yugoslavian seismologist after which it is named in a study of a 1909 earthquake in Croatia.

Depth to the Moho is quite different below the continents than below the ocean basins. On the landmass the crust averages about 33 km thick, but it varies between 20 and 60 km, with this greatest value occurring under mountain ranges. This is depicted in Figure 18.15. Under the oceans the crust averages about 5 km thick. Geophysicists have developed plans in the past to drill through the crust in the ocean basins to reach the M discontinuity and the mantle. Project Mohole was proposed in the 1960s and plans made to drill into the mantle, but the project did not proceed to the drilling stage before being cancelled.

The average velocity of waves on the upper portion of the continental crust is about 6 km/sec for the P wave and 3.5 km/sec for the S wave. This is the typical velocity for a rock similar in composition to granite (also granodiorite or granite gneiss). These rocks are rich in silica and aluminum. A transitional layer, likely of intermediate composition, occurs below this zone and at the base of the crust the zone of basaltic composition occurs as indicated by the seismic velocity of the layer. Representative values for these three zones of the crust are given in Table 18.8.

The basaltic layer of the crust apparently extends below the ocean basins as shown in Figure 18.15. The P wave velocity is somewhat lower for this zone below the ocean basins, 6.7 km/sec, as compared to the 7.0 to 7.2 km/sec for that zone under the continents.

Mantle

The zone between the crust and the outer core, about 2900 km thick, is known as the mantle. The mantle transmits both P and S waves and, therefore, must consist of solid materials. It also comprises a major portion of the Earth's volume, some 80% of it.

As mentioned in an earlier section, recent studies have shown that a low-velocity zone exists within the upper mantle. Its upper boundary occurs at a depth of 70 to 80 km below the oceans and at about 100 km below the continents. This zone extends to a depth of

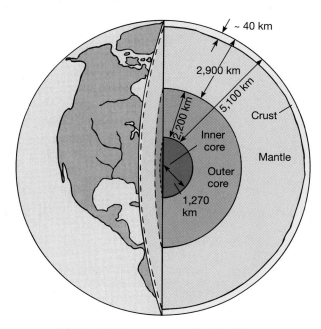

FIGURE 18.14 Major zones in the Earth's interior.

nearly 300 km (see Figure 18.15). The upper boundary apparently marks the contact between the upper, colder portion of the mantle and the lower, hotter part.

This boundary also forms the base of the lithosphere, with the asthenosphere occurring below. In plate tectonics discussions, the plates comprise the lithosphere, which moves slowly over the asthenosphere. Plates consist of crust and upper mantle although they are commonly referred to simply as crustal plates. The upper mantle is in motion, carrying the oceanic and continental crusts above it.

Core

The last major zone of the Earth is the core, which extends from a depth of 2900 km to the center of the Earth at 6370 km. Analysis of earthquake waves that have traveled a great distance shows that two parts of the core exist, the outer core 2200 km thick and the inner core with a radius of 1270 km. Below the discontinuity at the base of the mantle, S waves are no longer propagated and P waves undergo a reduction in velocity. The most widely accepted explanation for this drastic change is that the outer core is liquid. The P waves increase in velocity markedly at a depth of about 5150 km. This marks the boundary with the inner core and the velocity increase suggests that the inner core is solid.

TABLE **18.8**   Continental Crust Thickness, New England Region

Zone	Thickness (km)	Velocity (km/sec) P	S	Rock Composition
Layer 1	16	6.1	3.5	Granitic
Layer 2 (transition)	13	6.8	3.9	Intermediate
Layer 3	7	7.2	4.3	Basaltic
Moho				
Top of mantle		8.4	4.6	

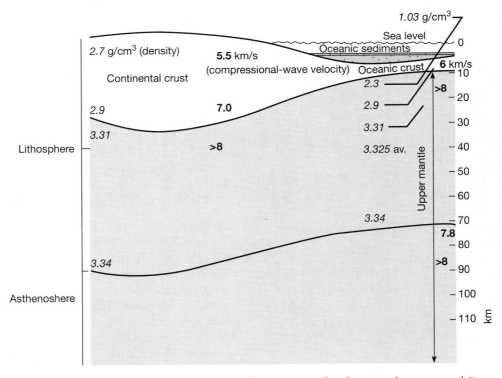

FIGURE **18.15**   Dimensions of the crust and upper mantle, showing densities and P wave velocities.

As the Earth revolves around the Sun it behaves in a manner consistent with a sphere of specific gravity 5.5. It is known that crustal rocks have only a specific gravity of 2.7 for the granitic zone and perhaps 2.9 in the basaltic layer. Rocks of similar composition compressed under the extreme pressures at 2900 km depth or more would increase the specific gravity to about 5.7. Therefore, to produce the average specific gravity of 5.5, the core must measure about 15.0. To accomplish this, the core must be composed primarily of iron with perhaps about 8% nickel and some cobalt in a manner similar to the proportions shown in metallic meteorites.

ENGINEERING GEOPHYSICS

In the study of geophysics, the application of physical laws is used to discern details about the nature of the Earth. As previously stated, this may be applied to studies of basic science or to subsurface investigation involving mineral exploration or engineering construction. In engineering geology the objective is typically related to construction but may involve mineral exploration as well. The application of geophysics to engineering works or construction projects is sometimes referred to simply as engineering geophysics.

Five primary groups of geophysical methods are commonly considered when applied to geologic studies as a whole. These include seismic, gravity, magnetic, electrical, and well logging methods. The seismic methods involve the propagation of waves through earth materials and are a direct extension of earthquake wave studies. They are considered in greater detail in the next section.

The gravity and magnetic methods deal with the strength of the fields of gravity or magnetism generated between a mass of rock and the Earth. From any point on or above the Earth's surface the gravity or magnetic field is measured and compared to that of an adjacent area. Changes in values, known as anomalies, are sought that will supply information about the size, nature, and location of a high- or low-gravity or magnetic source within the Earth. This provides information concerning mineral exploration or details of the subsurface of interest in construction. The relationships have the form:

$$F = G\frac{m_1 m_2}{r^2}$$

for the gravity force. In the equation, m_1 and m_2 are masses, m_1 usually being that of the Earth concentrated at the Earth's center and m_2 that of the instrument; r is the distance between them; and G is the universal gravity constant $= 6.670 \times 10^{-8}$ in cgs units. The measure of F or gravity force is the "gal," named after Galileo and equal to 1 cm/sec^2, with the Earth's gravity force approximately equal to 980 gals.

For the magnetic method:

$$F = \frac{1}{\mu}\frac{P_0 P}{r^2}$$

where F is the magnetic attraction or repulsion depending on the sign, P_0 and P are poles separated by distance r, and μ is the magnetic permeability, which depends on the magnetic properties of the medium in which the poles are located. The unit of magnetic field strength used for geophysical studies is the gamma, defined as 10^{-5} oersted. The total magnetic field of the Earth is normally approximately 0.5 oersted.

Both the gravity and magnetic methods have been used in engineering geology studies, but they are usually applied in rather specialized situations. Also they are typically performed by geophysicists rather than by engineering geologists, geological engineers, or civil engineers.

Electrical methods involve the measurement of electrical properties of earth materials. In general, they consist of two varieties: the measurement of natural earth currents and the resistance to induced electrical flow. Natural earth-current flow is generated under some geologic conditions in which an anode and cathode develop naturally in the subsurface and electron flow ensues. Measurement of the strength and extent of this current is useful in establishing the nature of the geologic conditions.

Electrical resistivity is related to the resistance to electrical flow through earth materials. In this method, a current is induced into the Earth, the resistivity is measured, and information about a basic property of the material under study is thereby determined. Electrical resistivity is the resistance between opposite faces of a unit cube of material. For a conducting cylinder with length L and cross-sectional area A, the resistivity ρ is:

$$\rho = \frac{RA}{L}$$

with R in ohms, A in cm^2, and L in cm to yield ρ in ohm-cm. The conductivity of a material, $K = 1/\rho$, has units in reciprocal ohm-cm. Because electrical resistivity is a primary method of interest in subsurface exploration for engineering construction, it is developed in more detail in a later section.

Well logging methods include a variety of techniques that involve the lowering of instruments into a well or boring and generating data on subsurface lithologies as the instrument is pulled from the hole. These techniques were developed for oil exploration and for 50 years or more have been standard methods for obtaining subsurface information from exploration wells drilled for oil. A number of measurements can be made using well logging methods, the details of which will be more meaningful after further discussion of several geophysical methods of interest. Well logging is of particular significance today because these methods are now commonly used in exploration programs for major engineering construction. Much information can be obtained through the transfer of expertise from standard oil exploration procedures to applications in engineering construction.

As a final detail in this introduction to geophysics, it is worth noting that the methods of exploration we have discussed are indirect. In each of the methods a property of the material is measured (that is, gravitational force, magnetic attraction, seismic velocity, electrical resistivity). However, none of these yield as direct a physical picture as a rock core, rock sample, excavation, or borehole. To infer a rock name or description from geophysical data requires at least one step beyond the basic information obtained, with several assumptions involved. This limitation of the methods should always be realized in geophysical exploration.

Refraction Seismic Method

Two specific seismic methods are available for exploration purposes: reflection and refraction. In the reflection seismic method, waves are reflected from rock unit boundaries in the subsurface and returned to the Earth's surface. In this situation, the angle of incidence, measured from the vertical, is equal to the angle of reflection. This technique is used by oil company seismic crews to explore for oil and gas. Actually they are looking for useful geologic structures or "oil traps" that could accumulate petroleum. The method has the advantage of penetrating thousands of meters into the subsurface but it also involves lengths of seismic cable measured in kilometers and numerous geophones to cover the distance along which the signal is returned. It is a major effort and expense for oil companies to support a seismic exploration crew in the field for extended periods of time.

The refraction seismic method is used for exploration at shallower depths than reflection seismic and therefore is more appropriate for construction-related projects. A depth up to 100 ft or possibly 200 ft (30 to 60 m) is the usual limit attempted for projects of a routine nature using this method.

At major lithologic boundaries, seismic waves are refracted or bent depending on the contrast in velocity between the two layers. This refracted wave has the opportunity, as is shown subsequently in this chapter, to generate waves that eventually travel to the surface before direct waves can arrive traveling just below the surface. Primary (or P) waves are the fastest seismic waves, hence they provide the first arrivals and are used in refraction seismic studies. Therefore, two types of information are supplied by the refraction seismic method: seismic velocities and depth to those lithologic boundaries where the velocities suddenly change.

To initiate primary waves in the ground, energy must be released or generated. This can be accomplished by detonating a blasting cap or larger explosive in the ground or possibly by hitting a sledgehammer against a metal plate on the Earth's surface. The depth of penetration desired for the shock wave, the sensitivity of the seismograph, and the background vibration noise will dictate how much energy must be supplied to obtain the needed detail.

When an explosion is detonated, a wide frequency range of waves is initiated up to 1000 Hz (or cycles per second) but the Earth soon filters out the high-frequency portion, allowing the lower frequencies from 15 to 70 Hz to propagate. Hence the instrument (seismograph) measuring the P wave arrivals should be sensitive to this range of frequencies.

A seismograph for engineering geophysics applications involves an accurate timing device. This device notes the instant of energy generation and of the first wave arrival at a specific location. The shock wave is picked up by a geophone, which is inserted into the soil below the root zone. It consists of a coil in a magnetic field so that when the shock wave arrives the coil moves and a small electrical current is generated, which is registered as an offset on the recording device.

Seismographs can be single-channel or multiple-channel devices. For a 10-channel seismograph, for example, a string of 10 geophones is used to record a single energy-release event. Travel times and distances to each geophone are recorded, which become the basis for velocity measurements and depths for lithologic layers. Smaller equipment consists of a single-channel, single-geophone arrangement and the energy is usually input by sledgehammer on a metal plate. The geophone is located at a specific distance and the travel time is determined. Then the geophone is moved to the next station at a specific distance and the procedure is repeated. This is continued until the specified total distance is covered.

The seismic velocity (*P* wave velocity) has major importance in seismic refraction work. Table 18.9 provides information on seismic velocities. From this table we can observe that soils below the groundwater table can be discerned from nonsaturated soil above the water table. Bedrock, whether above or below the water table, typically has a seismic velocity

above 5000 ft/sec (1520 m/sec) and hence its degree of saturation is not directly involved in determining the velocity. In general, as bedrock becomes more dense with depth the seismic velocity will increase. Velocity also increases with reduced porosity or increased cementation of sedimentary rocks. Porous, poorly cemented sandstones have low velocities but strongly cemented ones have velocities near that of quartzite. Calcareous shales (cementation shales) will have higher velocities than other varieties (compaction shales). Changes in jointing, dipping beds, and overall cementation will affect the seismic velocity as well.

Seismic velocities have been used for many years to predict the ease of excavation of earth materials. Some bedrock varieties require blasting, whereas others can be ripped using a bulldozer with a rock-ripper attachment. With this equipment, ripper teeth are forced hydraulically into the rock and the bulldozer is pulled forward. Thinly bedded or foliated rocks can commonly be removed by this method, which generally turns out to be cheaper than blasting the rock. The diagram of Figure 18.16 provides specific information based on the ability of a Caterpillar D-9 dozer to remove rock materials by ripping. A rule of thumb, that rocks can be ripped using a D-9 dozer if the seismic velocity is below 7000 ft/sec, should be applied with reasonable caution.

To obtain information on depths to lithologic boundaries, a consideration of wave refraction must be included. This is governed by Snell's law. Figure 18.17 illustrates refraction for two cases: 1) a higher velocity layer below the surface layer and 2) a lower velocity layer below the surface layer. When energy is released in the ground, waves move out in all directions from that point. This can be indicated in the form of a ray or as a wavefront. A selected ray is shown in Figure 18.17.

Snell's law indicates that $\sin i / \sin r = V_1/V_2 = n$ where *n* is the index of refraction. In case I, the ray is bent away from the normal and toward the interface. As described next, this can generate a wave that will return to the Earth's surface. For case II the wave is bent downward and no opportunity exists for it to return to the surface. From this it is clear that lithologic units must increase in velocity with depth to ensure that a refracted wave can return to the Earth's surface.

TABLE 18.9 Seismic Velocities for Engineering-Related Studies

Material	Velocity	
Soil to Rock, Nonsaturated	ft/sec	m/sec
Sand	650–6500	200–2000
Loess	1000–2000	300–600
Alluvium	1650–6500	500–2000
Loam	2600–5900	800–1800
Clay	3300–9200	1000–2800
Marl	2600–12,500	1800–3800
Saturated soil	5000	1520
Sandstone	5000–14,000	1500–4300
Limestone	5600–21,000	1700–6400
Slate and shale	6000–15,100	1800–4600
Granite	13,000–18,700	4000–5700
Quartzite	20,000	6100

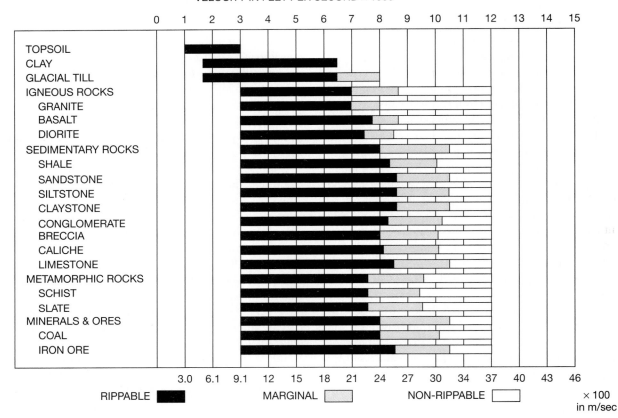

FIGURE 18.16   Ripper performance as related to seismic wave velocities.

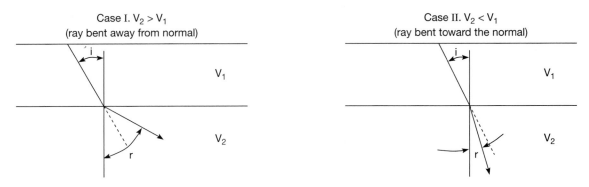

FIGURE 18.17   Illustration of Snell's law of wave refraction.

Referring to case I again, as i values increase for different rays, r will increase until a value of $r = 90°$ is reached. When $r = 90°$, $\sin r = 1$ and as $\sin i = V_1 \sin r / V_2$,

$$\sin i_{cr} = \frac{V_1}{V_2} (1) = V_1/V_2$$

where i_{cr} is the critical incident angle. This provides a ray that moves along the interface of the two layers at a velocity of V_2. As the ray proceeds, it serves as a point source for new primary waves, which emanate from the boundary as the wave proceeds. This is illustrated in Figure 18.18.

Waves move back through layer 1 at V_1 velocity after being carried along the interface at V_2 velocity. Eventually a wave refracted along the interface will initiate a second-generation wave, which reaches the surface before a direct wave can travel immediately below the ground surface. It will travel back along a wave direction also related to i as shown in Figure 18.19. Also shown is a time-distance curve, which indicates when the V_2 velocity begins to exert its influence on the first arrivals. The refracted wave arrives first, and will continue as the first arrival as shown in the diagram. In the same fashion, if a third layer exists with V_3 velocity such that $V_3 > V_2 > V_1$, eventually the wave from layer 3 will control the first arrivals as shown in Figure 18.19.

The velocities of the three layers are determined from the slopes of the time-distance plots. Because velocity = distance/time, velocities are the reciprocals of the slopes of these straight lines. The relationships for the thickness of the lithologic layers can be calculated based on the geometry involved for the ray paths.

For the Z_1 depth, the relationship is as follows:

$$Z_1 = \frac{X_{12}}{2} \sqrt{\frac{V_2 - V_1}{V_2 + V_1}}$$

where X_{12} = the critical distance for the arrival of the wave influenced by V_2; or

$$Z_1 = \frac{T_2 V_1}{2 \cos i_{12}}$$

where T_2 is the time intercept for the $1/V_2$ slope and $\cos i_{12}$ is the cos of the critical angle where $\sin i_{12} = V_1/V_2$.

For the Z_2 thickness

$$Z_2 = \frac{X_{23}(\sin i_{12} - \sin i_{13}) - Z_1 \cdot 2(\cos i_{13} - \cos i_{12})}{2 \sin i_{12} \cos i_{23}}$$

where X_{23} is the critical distance for the wave influenced by V_3 and the angles are indicated in Figure 18.17. Also

$$Z_2 = \left(T_3 - \frac{2Z_1 \cos i_{13}}{V_1}\right) \frac{V_2}{2 \cos i_{23}}$$

where T_3 is the time intercept for the $1/V_3$ slope.

Several approximations have been proposed for the two-layer and three-layer thickness calculations. They are as follows:

$$\frac{2Z_1}{V_1} + \frac{X_{12}}{V_2} \cong T_{12} \quad \text{(two layer)}$$

$$Z_1 + Z_2 \cong 0.85Z_1 + \frac{X_{23}}{2} \sqrt{\frac{V_3 - V_2}{V_3 + V_2}} \quad \text{(three layer)}$$

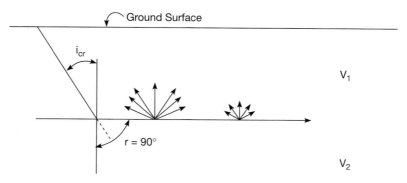

FIGURE 18.18 Refraction and critical wave propagation.

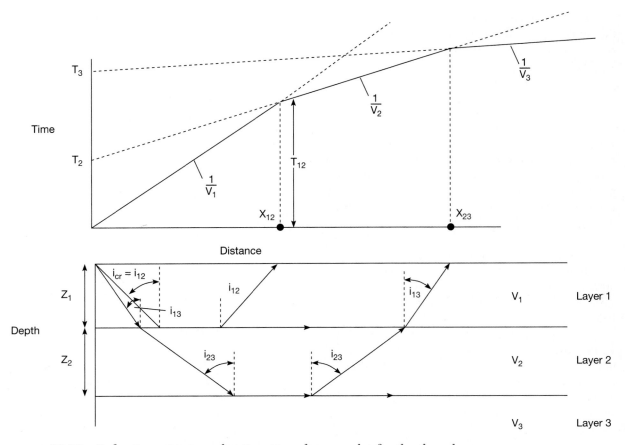

FIGURE 18.19   Refraction seismic exploration, time-distance plot for the three-layer case.

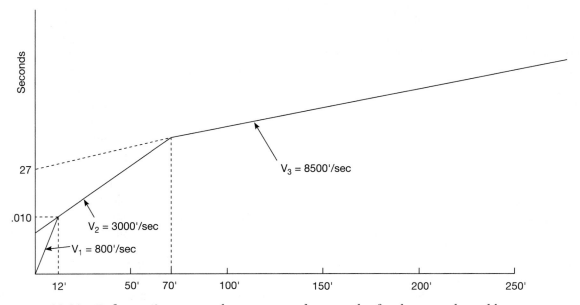

FIGURE 18.20   Refraction seismic exploration, time-distance plot for the example problem.

EXAMPLE PROBLEM 18.3

The time-distance plot is shown in Figure 18.20. See also Figure 18.19 for a similar three-layer cross section. Problem: Calculate Z_1 and Z_2 based on Figure 18.20. First use both the critical distance method and the time intercept method to find both depths. Then use the approximate methods for comparison.

ANSWER

1. $$Z_1 = \frac{12}{2} \sqrt{\frac{3000 - 800}{3000 + 800}} = 4.6 \text{ ft}$$

2. $$Z_1 = \frac{T_2 V_1}{2 \cos i_{12}}$$

    $$\sin i_{12} = \frac{V_1}{V_2} = 0.2666$$

    $$i_{12} = 15.5°$$

    $$\cos i_{12} = \cos 15.5° = 0.9636$$

    $$Z_1 = \frac{0.01(800)}{2(0.9636)}$$

    $$= 4.1 \text{ ft}$$

3. $$Z_2 = \frac{70 \left(\sin \frac{V_1}{V_2} - \sin \frac{V_1}{V_3} \right) - Z_1 \cdot 2(\cos i_{13} - \cos i_{12})}{2 \sin i_{12} \cos i_{23}}$$

    $$\sin i_{13} = \frac{V_1}{V_3} = 0.09411, i_{13}\ 5.4°, \cos 5.4° = 0.9956$$

    $$\sin i_{23} = \frac{V_2}{V_3} = \frac{3000}{8500} = 0.3529, i_{23} = 20.65$$

    $$\cos 20.65 = 0.9357$$

    $$Z_2 = \frac{70(0.266 - 0.09411) - 4.6(2)(0.9956 - 0.9636)}{2(0.2666)(0.9357)}$$

    $$= \frac{12.075 - 0.2944}{0.4989} = 23.6 \text{ ft}$$

4. $$Z_2 = \left(T_3 - \frac{2Z_1 \cos i_{13}}{V_1} \right) \frac{V_2}{2 \cos i_{23}}$$

    $$= \left[0.027 - \frac{2(4.1)(0.9956)}{800} \right] \frac{3000}{2(0.9357)}$$

    $$= 26.9 \text{ ft}$$

5. $$\frac{2Z_1}{V_1} + \frac{X_{12}}{V_2} \cong T_{12}, T_{12} = 0.016 \text{ sec}$$

    $$\frac{2Z_1}{800} + \frac{12}{3000} \cong 0.016 \text{ sec}$$

    $$\frac{2Z_1}{2} + \frac{12}{7.5} \cong 6.4$$

    $$15Z_1 + 24 \cong 96$$

    $$15Z_1 \cong 72$$

    $$Z_1 \cong 4.8'$$

6. $$Z_1 + Z_2 \cong 0.85Z_1 + \frac{X_{23}}{2} \sqrt{\frac{V_3 - V_2}{V_3 + V_2}}$$

    $$\cong 0.85(4.6) + \frac{70}{2} \sqrt{\frac{8500 - 3000}{8500 + 3000}}$$

    $$\cong 28.1 \text{ ft}$$

    $$Z_2 \cong 28.1 - Z_1 \cong 28.1 - 4.6 \cong 23.5 \text{ ft}$$

Blast Damage Effects

A major concern associated with the use of explosives for rock excavation is the potential damage caused by waves generated during blasting. Two concerns are involved: damage by ground vibrations and damage by airblasts. Both are discussed later.

The situation is further complicated by the fact that humans can feel vibration levels much below that required to cause damage, and transient vibrations are inherently alarming to them. Hence complaints will come from citizens even for relatively low levels of blast vibrations. As developed later, particle velocity, measured in inches per second, is the commonly used criteria for relating blast vibrations to possible damage. The value of 2 in./sec particle velocity is commonly considered as the lower boundary of onset for possible damage to a sound, residential structure. Humans can notice transient motions as low as 0.06 in./sec (Figure 18.21).

The response of humans without accompanying noise (or sound effects) is shown in Figure 18.21(a). As indicated, vibrations become disturbing at 0.4 in./sec and feel severe at 1.2 in./sec. If the sound of the blast accompanies the vibration, people react with even greater concern [shown by Figure 18.21(b)]. For this combined situation, blasting is noticeable at 0.02 in./sec and will be judged severe, yielding complaints at particle velocities as low as 0.2 in./sec. This is why good

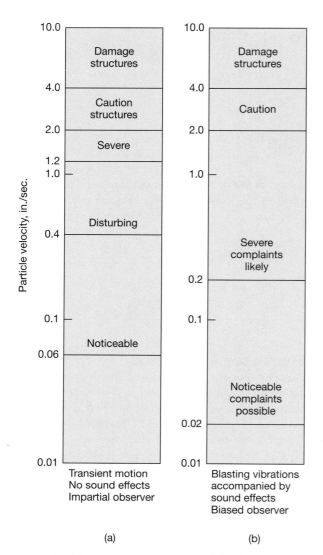

FIGURE **18.21**   Human response to blasting
vibrations.

between the blast site and the observation point, and
the characteristics of the earth material transmitting
the vibration. The term *delay* refers to the short time
intervals, measured in milliseconds (0.001 sec) between detonations of different portions of an explosive charge. Up to 10 delays at several millisecond
intervals can be detonated at one time without the
blast vibrations being superimposed on each other.

The effects of blast vibration fall off rapidly with
distance. The association between weight of explosives and distance has been developed in a form
known as the scale distance relationship. Devine
(1966) of the U.S. Bureau of Mines proposed the
following relationship:

$$V = 121.1 \text{ in./sec } \frac{1(\text{ft/lb})^{1/2}}{R/W^{1/2} \, (\text{ft/lb})^{1/2}}$$

This equation comprises the square root law for blast
damage.

A conservative approach can be used to estimate
the allowable maximum charge using Devine's suggested scale factor of $R/W = 50 \text{ lb/ft}^{1/2}$. This is
included in column 1 of Table 18.10. For purposes of
comparison, a scale factor of 20 lb/ft$^{1/2}$ is shown as
well.

A scale distance of 50 lb/ft$^{1/2}$ yields a particle
velocity of less than 2 in./sec for even the most
undesirable transmitting media in the Earth. Soft,
organic soils yield one of the most destructive vibration levels possible to buildings as contrasted to
massive rock yielding the least effects for a given
blast. If more than a 2 in./sec particle velocity is to be
allowed when a seismograph is employed for moni-

communications with the public, educational re-
sponses, and blast monitoring are needed when blasting in urban or other populated areas is conducted. A
lower criterion that has been developed in some cases
to preclude damage and to gain more tolerance from
the general public is 1.0 in./sec particle velocity.

Blast-Induced Ground Vibrations

Ground vibrations experienced at a certain point
adjacent to a blasting site are dependent on the
weight of explosive detonated per delay, the distance

TABLE **18.10**   Blast Damage Related to Scale
Factors

Distance R (ft)	Allowable Weight of Explosive per Delay	
	50 lb/ft$^{1/2}$	20 lb/ft$^{1/2}$
100	4	25
500	100	6,625
1000	400	2,500
2000	1600	10,100

Based on a scale distance (SD) of $\dfrac{R}{W^{1/2}}$. Therefore, $W = (R/SD)^2$.

toring or if a value less than 50 lb/ft$^{1/2}$ is employed for the scale distance without measuring particle velocity directly, it is suggested that a well-documented study of the site conditions be made in advance. Such a study would be accomplished by a vibrations expert or geophysicist. Additional details on blast damage criteria are available in the literature (Hendron, 1977).

EXAMPLE PROBLEM 18.4 _____

Blasting for a tunnel excavation will take place 130 ft away from a building that needs to be protected from blast damage. What is the maximum weight of dynamite charge that can be detonated based on a scale distance of 50 lb/ft$^{1/2}$? Based on 20 lb/ft$^{1/2}$?

ANSWER

$$SD = R/W^{1/2}$$

where SD = scale distance, R = distance in feet, and W = weight of dynamite charge in pounds. Or

$$W = (R/SD)^2$$

If R = 130 ft and SD = 50,

$$W = (130/50)^2 = 6.76 \text{ lb of dynamite per delay}$$

If R = 130 ft and SD = 20, then

$$W = (130/20)^2 = 42.2 \text{ lbs of dynamite per delay}$$

Effect of Air Blasts

Air blasts occur when surface charges are detonated or when demolition occurs above ground. An overpressure of 1 psi typically will break windows and cause cracks in new plaster. Windows less than 60 ft^2 in area are safe from breakage if the air blast is less than 0.1 psi overpressure. It turns out that overpressures of 0.1 psi for air blast damage correspond fairly well with ground vibrations of 2 in./sec. The overall effect generally is that air blast damage can be prevented by using a scale factor that will preclude damage from ground vibration. Precautions should be taken nonetheless to reduce or minimize air blast effects by proper stemming of holes and the use of blasting mats or barricades if surface blasting is involved.

Blast Damage Criteria

The effects of ground vibration on engineering structures have been studied in detail since the early 1940s (Thoenen and Windes, 1942; Crandell, 1949; Devine, 1966). As one would expect, the level of ground motion needed to cause damage to a structure is not the same for all types of buildings. Also the degree of damage acceptable in one building based on its specific use may not be the same as that accepted in another. Cracking of plaster in a residence was selected as the onset level for damage and this has been used as the lower bound for damage occurrence. In the early work, acceleration a of the ground coupled with wave frequency n was used as the damage criterion. Using conditions of harmonic motion (sinusoidal waves) the energy ratio a^2/n^2 was used to estimate damage effects. It was later found that damage could be correlated with displacement and frequency. Because particle velocity is proportional to a product of displacement and frequency, this velocity became the criteria for assessing blast vibration damage. Displacement seismographs are simpler and less expensive than are accelerometers; this provides a major incentive for use of a displacement-related criteria.

A value of 2 in./sec particle velocity is the vibration level at which damage to plaster was noted (Edwards and Northwood, 1960). On the basis of this work and that previously mentioned, the criteria of 2 in./sec particle velocity has become accepted as the limit of safe vibration for blasting vibration. As mentioned previously, in some situations this acceptable level is reduced to 1 in./sec to ensure that no damage will occur, even to older structures in less than perfect condition.

Earth Resistivity Method

Derivation

The apparent resistivity of earth materials is used to discern details about the subsurface. It, like the seismic method, is an indirect procedure and the data should be compared to boring logs or to an outcrop description to ensure that the lithology is properly identified. Many unnecessary mistakes have been made in the applications of geophysics to engineering exploration that could have been precluded if this simple rule were followed.

To begin the development of this method, a review

of the relationship for resistivity ρ is required. Recall that:

$$\rho = RA/L$$

where R can be expressed in terms of the Ohm's law relationship, $E = IR$ where $E = V$ = electromotive force or voltage, I = current flow in amps. So $R = E/I$ and, therefore,

$$\rho = \frac{EA}{IL} = \frac{VA}{IL} \quad \text{and} \quad V = \frac{I\rho L}{A}$$

In Figure 18.22 a diagram of the earth resistivity array is shown. If materials of constant resistivity are assumed, with current electrodes located at points A and B, plus potential electrodes at C and D, the resistivity can be determined. Because of this configuration and the volume of material through which the current flows, the voltage or potential at electrode C will be:

$$V_C = \frac{I\rho}{2\pi} \left(\frac{1}{r_1} - \frac{1}{r_2} \right)$$

and in a similar fashion the potential at electrode D will be:

$$V_D = \frac{I\rho}{2\pi} \left(\frac{1}{r_3} - \frac{1}{r_4} \right)$$

Referring to Figure 18.22, the potential difference V measured between C and D is $V_C - V_D$. Subtracting these and solving in terms of ρ yields the following:

$$\rho = \frac{2\pi V}{I} \frac{1}{\left(\dfrac{1}{r_1} - \dfrac{1}{r_2} \right) - \left(\dfrac{1}{r_3} - \dfrac{1}{r_4} \right)}$$

which is the fundamental equation for the resistivity method.

To this point, uniform resistivity has been assumed for the material under study. When this is the case, the resistivity value will remain the same, independent of the volume of material measured. However, for natural earth materials the resistivity is not constant throughout the cross section, so a change in the volume of material measured will likely yield a change in the resistivity value. In the case of nonuniform materials, the value of ρ is called the *apparent resistivity*.

For most engineering projects in the United States, the Wenner configuration of electrode spacing is used. In this arrangement the four electrodes are equally spaced along a line, with the two outside ones being current electrodes and the two on the inside being potential electrodes, as shown in Figure 18.23. The distance between electrodes is designated as a, which is increased as successive readings are taken, keeping the same central point for the spread. With an equal distance between the electrodes (refer to Figure 18.22) the equation for resistivity is simplified as shown below. In Figure 18.22, $r_1 = r_4 = a$ and $r_2 = r_3 = 2a$. Therefore, for the Wenner configuration:

$$\frac{1}{\dfrac{1}{r_1} - \dfrac{1}{r_2} - \dfrac{1}{r_3} + \dfrac{1}{r_4}} = \frac{1}{\dfrac{1}{a} - \dfrac{1}{2a} - \dfrac{1}{2a} + \dfrac{1}{a}}$$

$$= \frac{1}{\dfrac{2}{a} - \dfrac{2}{2a}} = \frac{1}{\dfrac{1}{a}} = a$$

and

$$\rho = 2\pi a V/I$$

It is also assumed that the depth of penetration of the current flow lines is approximately equal to a, the

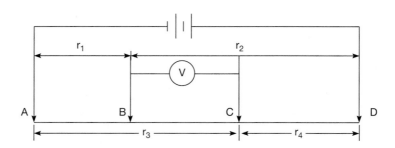

FIGURE 18.22   Elevation view, electrical resistivity survey.

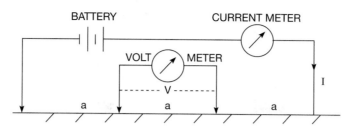

FIGURE **18.23** Wenner configuration, electrical resistivity survey.

electrode spacing. Hence, by increasing a, the depth of exploration is increased.

Application of Method

In practice, when using the Wenner configuration, each potential electrode (P_1, P_2) is a distance $a/2$ from the center position and each current electrode (C_1, C_2) is a distance $3a/2$ from the center. The electrode spacing is increased from one reading to the next, yielding data for ρ versus a spacing for the investigation. If we want to increase the electrode spacing (a spacing) by 10 ft from one reading to the next, each potential electrode must be moved 5 ft further from one reading to the next, and each current electrode must be moved 15 ft. This simple procedure for moving electrodes makes the work proceed faster in the field.

Several suggestions have been proposed for electrode spacing in electrical resistivity surveys or *soundings* as they are sometimes called. For rapid reconnaissance, the "leap-frog" sequence is quite useful. With this procedure the first reading is taken at the largest electrode separation that will be required or is anticipated; refer to this as a_0. It may be, for example, equal to 90 ft. The next spacing will be $a_0/3$ and third $a_0/9$. The location of the potential electrodes of one

reading becomes the location of the current electrode for the next reading. Usually four or five successive readings will cover the range of a spacing desired. This is illustrated in Figure 18.24. For our example, the first a spacing would be 90 ft, then 30, 10, 3.33, and 1. Intermediate points from 30 to 90 and 10 to 30 would be required to complete a detailed survey but the reconnaissance work is quickly accomplished.

For some interpretations of the data, equal interval values for the a spacing make that phase of the study more manageable. These interpretations include the Barnes layer method and the Moore cumulative method. Therefore, increases of 5 ft may be convenient for detailed work. Use of the Wenner configuration, as indicated before, requires that the potential electrodes each be moved $a/2$ or 2.5 ft and the current electrodes moved $3a/2$ or 7.5 ft between successive readings. Intervals involving 0.5 ft are manageable but any other fractions or decimal values of feet are too cumbersome to work with conveniently in the field and still allow for reasonable productivity.

For scientific work and some general geologic studies, the optimum electrode spacing would be that which is equally spaced on a logarithmic scale. For analyses, the ρ versus a spacing values are plotted on double logarithmic paper where the electrode intervals will appear to be equally spaced.

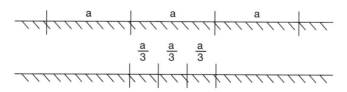

FIGURE **18.24** Leap-frog sequence of electrical resistivity survey readings.

One disadvantage of the Wenner configuration is that variations laterally in the subsurface can be misinterpreted as being caused by variations in the material with depth. As the *a* spacing is increased, so is the depth of penetration, and the lateral extent of the survey is widened as well. If a contrasting material is intercepted laterally, it shows up as a change in ρ, which more than likely is assumed to be caused by a change in a layer at depth. This confusing detail can be discerned by either taking measurements along two perpendicular lines or by using the Lee electrode modification procedure.

The Lee electrode configuration is shown in Figure 18.25. An additional electrode P_0 is driven into the ground at the center of the spread. Three readings are taken for each electrode spacing, the first being the total value using normal potential electrodes P_1 and P_2 in the conventional way. For the other two, the voltage is read between P_1 and P_0 and P_2 and P_0 in succession. If the readings are similar for both [plots overlap on the diagram as shown in Figure 18.25(a)], lateral variations in resistivity are not great. If they differ considerably [diverge as shown in Figure 18.25(b)] the lateral change is significant. A new centerpoint away from the contrasting material may have to be selected or the traverse rotated 180° to obtain a better measure of subsurface changes. Several guidelines are suggested for electrical resistiv-ity soundings in addition to those supplied in the previous discussion:

1. The largest electrode spacing should be at least 3 times and preferably 5 to 10 times the maximum depth of interest.
2. The smallest electrode separation should be less than one-half the minimum depth at which a change in material is expected. However, electrode separations of less than 2 ft are seldom needed.
3. Reading intervals should be more closely spaced at small electrode separations than at large electrode separations. This, of course, is in contrast to the use of equally spaced *a* spacings, which is a convenient procedure for certain methods of interpretation (Barnes layer method and Moore cumulative method). To accomplish this increase in reading interval with increasing spacing, readings should be equally spaced according to the log scale of distances. This procedure is mostly appropriate to scientific investigations rather than engineering site investigation studies.

Interpretation
Interpretation of apparent resistivity data is the most difficult part of this procedure. A number of different methods are used for this purpose and they are summarized in Table 18.11.

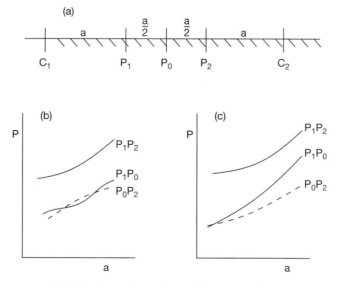

FIGURE 18.25   Lee electrode configuration for an electrical resistivity survey.

TABLE 18.11 Interpretation Methods for Apparent Resistivity Data

Method and Description
1. Plot *a* spacing versus ρ, arithmetic scales. Note extremely high or low values in the plot.
2. Plot *a* spacing versus cumulative ρ, arithmetic scales (Moore cumulative resistivity method). Changes in slope suggest changes of material.
3. Plot log of ρ versus log of *a* spacing. Compare to standard curves.
4. Barnes layer method. Calculate "layer" resistivities based on the Barnes method.

BARNES LAYER METHOD The Barnes layer method was developed because of a perplexing problem that occurs with earth resistivity studies. If a layer at depth is to influence the resistivity value when this layer is included in the volume, the resistivity contrast has to be extremely large. The reason for this is illustrated in Figure 18.26. Each time the electrode spacing is increased, a new incremental volume is added to the total volume to be measured. As the number of readings grows, the increment added to the volume represents a smaller and smaller contribution to that total being measured. A small contrasting resistivity near the surface can affect the ρ measurements as much or more than a large contrasting portion at depth. Stated directly, the ability of near-surface materials to overwhelm the effects of deeper lying

deposits is an inherent problem with the resistivity method.

The Barnes layer method attempts to delineate the resistivity of layers within the sample volume studied. The thickness of the layer is assumed to equal the increment in the electrode spacing. Equal increases of increments work best if for no other reason than it allows the Moore cumulative method to be used for comparison. The Barnes layer method is illustrated in Figure 18.27. The equations are developed as follows:

$$\rho = 2\pi a \frac{E}{I}$$

for the Wenner configuration but $R = E/I$ or $1/R = I/E$:

$$\frac{I}{E} = \frac{2\pi a}{\rho}, \quad \text{or} \quad \frac{1}{R} = 2\pi \frac{a}{\rho}.$$

If[1]

$$\frac{1}{R_n} = \frac{1}{R_n} - \frac{1}{R_{n-1}}$$

where $1/R_n$ is the conductance of a given layer increment when $\dfrac{1}{R_n}$ is the total conductance between the ground surface and the bottom of a given increment, and $\dfrac{1}{R_{n-1}}$ is the total conductance between the ground surface and the bottom of the increment directly above the increment under consideration.

---

[1] This equation assumes that the layers act electrically as though they comprise a parallel circuit.

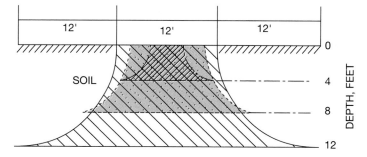

FIGURE 18.26 The increase in volume of soil measured that occurs with increasing *a* spacing for the Wenner configuration, electrical resistivity survey. The cross-hatched volume was measured using a 4-ft electrode separation; the dotted volume with an 8-ft separation; and the total volume with a 12-ft separation.

TABLE **18.12**   Typical Earth Resistivity Values for Selected Geologic Materials

Material	ohm-ft	ohm-cm	2π ohm-cm (for Barnes method)
Soils			
Wet to moist clayey soil	5–10	160–320	1000–2000
Wet to moist silty clay and silty soils	10–50	320–15000	2000–10,000
Moist to dry silty and sandy soils	50–500	1500–15,000	10,000–95,000
Clay	3–35	90–1100	560–7000
Sand	100–6500	3000–200,00	20,000–1,250,000
Sand and gravel with layers of silt	1000–8000	30,000–250,000	190,000–1,500,000
Coarse dry sand and gravel	8000+	250,000+	1,500,000+
Soft shale	2–33	50–1000	300–6300
Bedrock			
Well fractured to slightly fractured with moist filled cracks	500–1000		
Slightly fractured bedrock with dry soil-filled cracks	1000–8000	30,000–250,000	190,000–1,500,000
Massive bedded and hard bedrock	8000	300,000	
Sedimentary Rocks			
Hard shale		800–60,000	
Sandstone		3000–300,000	
Porous limestone		9000–900,000	
Dense limestone		200,000–1,000,000	
Chattanooga shale		2×10^3–1.4×10^5	
Michigan shale		2×10^5	
Calumet and Hecla conglomerates		2×10^5–1.3×10^6	
Muschelkalk sandstone		7×10^3	
Ferrugenous sandstone		7×10^5	
Muschelkalk limestone		1.8×10^4	
Marl		7×10^3	
Metamorphic Rocks			
Group values		5×10^3–1×10^9	
Garnet gneiss		2×10^7	
Mica schist		1.3×10^5	
Biotite gneiss		10^8–6×10^8	
Slate		6.4×10^4–6.5×10^6	
Igneous Rocks			
Group values		9×10^3–2×10^9	
Granite		5×10^5–10^8	
Diorite		10^6	
Gabbro		10^7–1.4×10^9	
Diabase		3.1×10^5	

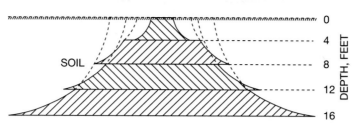

FIGURE **18.27** Barnes layer method: isolating layers at depth from the total volume of soil measured.

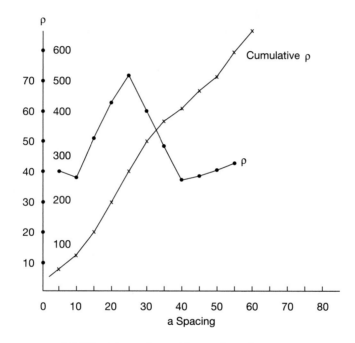

FIGURE **18.28** Example problem, plot of ρ versus *a* spacing and cumulative ρ versus *a* spacing.

$$1/R = 2\pi a/\rho \quad \rho = \frac{2\pi a}{1/R}$$

where ρ is in ohm-feet and *a* is in feet.

$$\rho_L = \frac{2\pi \times 2.54 \times 12A_L}{1/R_n}$$

where ρ_L is the layer resistivity in ohm-centimeters.

$$\rho_L = \frac{191A_L}{1/R_n}$$

where

191 = constant, collecting, 2π and unit conversions

A_L = thickness of given layer in feet
$1/R_n$ = layer conductance of a given increment.

Typical values of resistivity for various earth materials are presented in Table 18.12. Note that no hard and fast rules apply between resistivity values and material types. This is indicated to some extent by the wide range of values shown. Actually the changes and contrasts in resistivity at a specific site may be more useful than the absolute values themselves in determining the nature of the materials present. An example problem is presented as follows.

EXAMPLE PROBLEM 18.5 _____

Data from a hypothetical earth resistivity survey, based presumably on the Wenner configuration, is supplied in the table below. This was constructed with the cross section shown below in mind but not adhered to in a strict fashion. Refer to Figure 18.28 on page 441 for a plot of ρ.

ANSWER

The data recorded are as follows:

a spacing (ft)	ρ (ohm-ft)
5	40
10	37
15	51
20	63
25	72
30	60
35	47
40	37
45	39
50	41
55	44
60	45

The data sheet calculations for the Barnes layer method are as follows (refer also to Figure 18.29):

The data sheet calculations for the cumulative ρ method are as follows:

a	ρ	Cumulative ρ
5	40	40
10	37	77
15	51	128
20	63	191
25	72	263
30	60	323
35	47	370
40	37	407
45	39	446
50	41	487
55	44	531
60	45	576

ρ, ohm-ft		depth, ft
45	Clay	
		10
900	Sand & Gravel	
		25
60	Hard Clay	
		40
1000	Bedrock	

a (ft)	ρ (ohm-ft)	$2\pi a$	$\dfrac{1}{R_n}=\dfrac{2\pi a}{\rho}$	$\dfrac{1}{R_n}=\dfrac{1}{R_n}-\dfrac{1}{R_n-1}$	$\rho_L=\dfrac{2\pi a}{1/R_n}$	Layer Involved (ft)
5	40	31.4	0.78	0.78	40	0–5
10	37	62.8	1.697	0.917	68	5–10
15	51	94.2	1.847	0.150	628	10–15
20	63	125.6	1.994	0.147	854	15–20
25	72	157.1	2.182	0.188	836	20–25
30	60	188.5	3.142	0.960	196	25–30
35	47	219.9	4.679	1.537	143	30–35
40	37	251.3	6.792	2.113	119	35–40
45	39	282.7	7.244	0.452	625	40–45
50	41	314.2	7.656	0.412	763	45–50
55	44	354.4	8.053	0.397	893	50–55
60	45	377.0	8.370	0.317	1189	55–60

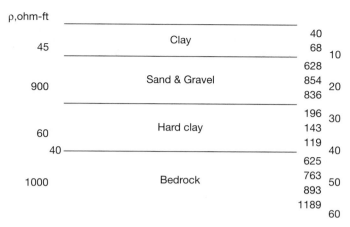

FIGURE 18.29 Results of Barnes layer method calculation related to geologic cross section.

Selection of Geophysical Exploration Methods

The two geophysical methods discussed in detail in the previous sections are refraction seismic and electrical resistivity. As explained these two techniques are most applicable to engineering geology studies because they are the best methods available for exploring shallow depths of the subsurface to 100 ft (30 m) deep or so. Sometimes the two methods are used in combination and the results can augment each other. However, in other instances one method or the other is preferred because of the specific details of the subsurface. These features are examined in the following discussion.

For the refraction seismic method to provide useful data, the subsurface zones or layers must increase in velocity with depth. This is necessary for waves to be refracted upward to travel along the layer boundary generating waves that return to the surface. (Refer to Figure 18.18 for details.) Therefore, if a low-velocity layer lies below a high-velocity layer, the refraction seismic method will not work properly. An example of this would be a sand and gravel layer lying below a clay layer. The gravel has a lower seismic velocity than does the clay, so waves refracted by the boundary between them would not return to the ground surface (see Figure 18.17, case II). Hence, the refraction seismic method is not a proper procedure to explore for sand and gravel deposits within glacial till or clay.

By contrast, the electrical resistivity method should work well in this situation. The contrast in resistivity between the clay (lower resistivity) and the sand and gravel (higher resistivity) should be sufficient that when the *a* spacing or depth of penetration of the electrical flow path reaches the sand and gravel layer, a measurable increase in resistivity is encountered.

The refraction seismic method commonly works well in an area where soil overlies bedrock. The seismic velocity of the bedrock is typically much greater than that of the overlying soil. The electrical resistivity method would also likely work well here, because bedrock typically has a considerably higher resistivity than the soil. However, as the depth to bedrock increases, the resistivity contrast must be greater for the bedrock to provide a significant enough difference to influence the collective resistivity value.

In areas of the Midwest where dense glacial till overlies shale bedrock, the contrast in resistivity typically is not great enough to yield the necessary difference for the resistivity method to delineate the bedrock surface. However, the seismic velocity of the shale can be twice or more that of the till, and the refraction boundary would be discerned. Yet, if the glacial till is extremely dense, its seismic velocity will approach that of the shale and then it would not be possible for the refraction seismic method to delineate the bedrock surface.

The electrical resistivity method can be used to locate the ground-water table in a sand and gravel deposit. The saturated sand and gravel would have a lower resistivity, yielding a contrast that the method could measure. In this case, refraction seismic would likely also work, because the saturated sand and gravel layer has a higher seismic velocity than does the nonsaturated material. The contrast of saturated clay or till versus its nonsaturated counterpart would likely not be great enough, with regard to either resistivity or seismic velocity, for either method to pick up the saturated zone in a clay deposit.

The electrical resistivity method has been used to locate plumes of contaminated ground water in a

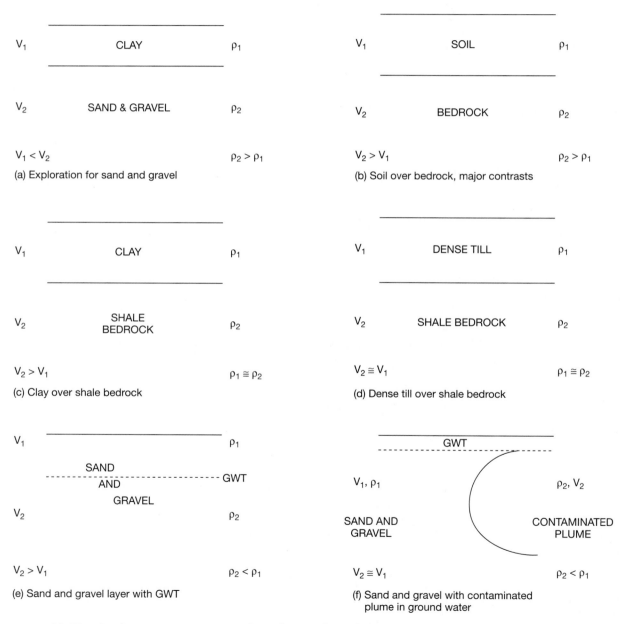

FIGURE **18.30**   Geologic cross sections evaluated using the refraction seismic and electrical resistivity methods.

sand and gravel aquifer. The contaminated water contains a higher concentration of dissolved solids, both anions and cations, which decrease its electrical resistivity. Hence the contaminated plume shows up as a low-resistivity layer. Electrical resistivity surveys, performed periodically, can be used to plot the rate of contaminant transport.

Figure 18.30 shows a series of geologic cross sections that are evaluated in terms of the refraction seis-

mic and electrical resistivity methods. Details of these applications were presented in an earlier discussion.

Well Logging Methods

Well logging or borehole geophysical methods have been used for many years in oil exploration. One of the current developments in site investigation for important engineering structures is the inclusion of well logging methods as part of the investigative

TABLE **18.13**   Common Borehole Geophysical Methods

Method	Description and Uses	Conditions Needed
Electric Logging		
Electric resistivity	Several methods with electrode spacings from 16 in. to 19 ft apart. Measures resistivity between beds. Rock types and effective porosity can be determined.	Fluid-filled uncased hole. Fresh drilling mud usually required.
Spontaneous potential	Potential difference between single electrode in hole and another at ground surface. Values remain relatively constant against shale, reduced values for nonshaley layers.	Fluid-filled uncased hole. Fresh mud.
Radiation Logging		
Gamma ray	Measures inherent radioactivity of rock units. Shale yields higher activity than sandstone.	Fluid-filled or dry, cased or uncased hole.
Neutron	Measures hydrogen-rich aspect of rock or presence of water. Indicates filled porosity of rock. Discerns oil-filled pores from water-filled ones.	Fluid-filled or dry, cased or uncased hole.
Sonic Logging	Measures seismic velocity, can correlate with specific lithologies.	Not affected by fluid, hole size, or mud.
Temperature Logging	Determines thermal gradient, presence of cement behind casing, zones of active gas flow or of lost circulation. In uncased holes, locates fissures and solution openings.	Cased or uncased hole. If no fluid present, logged at slow speed. Fluid should be undisturbed.
Caliper Survey (Section Gauge)	Determines hole or casing diameters. Locates fractures, solution openings, and other cavities. Correlation of formations, used to select zone for packer placement, placement of gravel pack, evaluation of explosive efficiencies, finding details of abandoned wells.	Fluid-filled or dry, cased or uncased hole.

procedures. When a core hole has already been obtained for sampling purposes, the additional cost of well logging with geophysical equipment is quite small compared to the total exploration costs.

In well logging techniques, various probes or sondes are lowered down a completed hole to obtain information on rock units and rock boundaries. These methods are useful to confirm those lithologic boundaries, particularly in areas where core recovery was poor. The methods are also used to obtain data in rock zones that were not cored; this is a common occurrence in deep holes drilled for oil exploration.

The common borehole geophysical methods can be summarized in the following way as shown in Table 18.13. Electrical resistivity well logging is an essential part of the geophysical analysis performed on wells drilled by oil companies. The electric log shows the apparent resistivity of the subsurface formation, along with the spontaneous potential (SP) curves. Shallow wells in freshwater zones show featureless SP curves, whereas brackish and saltwater zones have a negative SP deflection in contrast to clay-rich zones. These show permeable strata in the potential oil-bearing zones in the subsurface.

Gamma-ray logging is performed in soil borings and in bedrock. In this procedure the natural radiation of gamma rays that emanate from certain radioactive elements in the subsurface is measured. Materials containing higher concentrations of radioactive elements, such as uranium, thorium, and radioactive isotopes of potassium, yield greater gamma radiation. Typically, clay and shale contain more radioactive materials than do limestone, sandstone, or sand. Gamma-ray logs therefore are used to indicate clay versus sand units in soils and the clayey and shaly beds versus limestone or sandstone in bedrock. Figure 18.31 shows the relative response of different types of geological intervals to gamma-ray activity.

Gamma-ray logging is commonly performed on borings in glacial material in the Midwest and in other areas. Logging is performed through the casing, which can be composed of either steel or PVC. Borings for foundation studies, water supplies, or environmental studies can be logged in this fashion. The method has proved particularly useful in studies for sanitary landfill sitings, general stratigraphic studies, and the analysis of earthquake potential.

It is commonly important to identify sandy zones

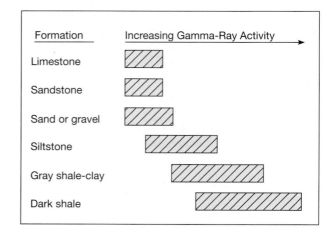

FIGURE 18.31   Relative response of geologic materials with regard to gamma-ray activity.

versus silts or clays in the subsurface (Figure 18.32). In addition, a thick sequence can provide a diagnostic signature curve that can be used to correlate unconsolidated sequences from one area to another. For example, the presence of sand at the base of a unit with higher amounts of clay above it can provide a

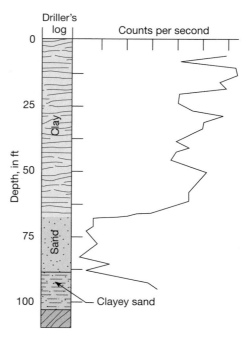

FIGURE 18.32   Gamma-ray log versus drillers log for unconsolidated materials.

diagnostic curve for a unit that has a designation of fining upward (or getting finer in grain size in the upward direction). By contrast, increasing amounts of sand yield a sequence that is getting coarser upward.

Gamma-ray logging of soil sections adjacent to existing sanitary landfills provides details on the migration of contaminant plumes. Water-bearing zones only a foot or so in thickness can be easily missed during subsurface drilling and sampling. Yet migrating ground-water contamination is transmitted by these zones because they control the pathways of leachate migration.

Additional details on geophysical logging methods can be obtained in the literature (see Johnson, 1968). Cross-hole seismic techniques are also available. In that method, a shock wave is generated in one hole and its travel time to an adjacent hole is measured. Seismic velocities are determined based on distances between the boreholes.

EXERCISES ON EARTHQUAKES AND GEOPHYSICS

1. Refer to the appropriate equations involving the shear modulus, bulk modulus, and modulus of elasticity. Given a granite with $E = 5.5 \times 10^6$ psi, $q_u = 22000$ psi, and $\mu = 0.23$, what is the K and S value for the rock? What is likely its τ_0 or shear strength at zero normal stress? What deformation would develop just prior to shearing failure?

2. Given the empirical relationship $\log E = 11.4 + 1.5M$ where E is the energy released by an earthquake in ergs and M is the magnitude of the Richter scale, calculate the increase in energy per integer value of the magnitude scale. Use the tabular form shown below.

Richter Magnitude	$\log E$	E (ergs)	Multiple from Previous Value
1	12.9	8×10^{12}	—
2			
3			
4			
5			
6			
7			
8			

3. Using the relationship from Exercise 2, $\log E = 11.4 + 1.5M$, calculate the energy release for the six earthquakes listed in Table 18.3 of the text. How well does this coincide with the values given for energy release in Table 18.3? One ton of TNT $= 10^9$ calories and one calorie $= 4.186 \times 10^7$ ergs.

4. The average energy released worldwide by earthquakes in a year is about equivalent to that of an 8.4 magnitude event. How much energy is this in ergs, tons of TNT, and in megaton bomb-equivalents?

5. Refer to Figure 18.1 in the text. If an earthquake registers 6.5 on the Richter scale, what intensities are likely associated with it near the epicenter? How extensive would the damage be? What level of ground acceleration would likely occur near the epicenter? (See also Figure 18.2). At a distance of 50 mi what would be the equivalent magnitude, intensity, and acceleration? Explain as needed. If ρ for the rock $= 2.65$ gm/cm^3, what are the V_p and V_s values in meters/sec, ft/sec, and km/sec?

6. Why is the damage so great for adobe buildings in earthquake-prone areas? How could masonry walls in nuclear power plants be made more resistant to earthquake damage?

7. The value of V_p in the sial layer is 6.1 km/sec and V_s is 3.5 km/sec. For a seismograph station 1200 km away from the epicenter, what would Δt or $t_s - t_p$ be equal to at this location using the velocities given?

8. The depth to the focus for a certain earthquake is 20 km.
   (a) If a value of 6.0 km/sec for V_p is used, how long does it take for the P wave to travel from the focus to the epicenter directly above it? How long does it take for the P wave to travel from the focus to a point on the Earth's surface 100 km away from the epicenter? What is the difference in elapsed time for the two?
   (b) For another earthquake the depth to the focus is 250 km. Using $V_p = 6.0$ km/sec again, how long does it take to travel from the focus to the epicenter and how long does it take to travel from the focus to a point on the Earth's surface 100 km from the epicenter? What is the difference in elapsed time for the two?

(c) For both the earthquake with a 20-km depth of focus and the one with a 250-km depth of focus, find the area of the circle located around the epicenter in which the arrival time difference is 1 sec or less. How many times greater is the area for the deep-focused earthquake?

9. At a 30-km depth, what is the overburden pressure on the rock? Assume the density of the overburden to be 2.8 g/cm^3. Calculate this in kg/m^2, lb/ft^2, psi, and kN/m^2.

10. On what is the Seismic Risk Map of the United States (Algermissen's map) based? What specific earthquakes caused the concentric effect observed around southeastern Missouri, around eastern South Carolina, and around eastern Massachusetts on Figure 18.9 in the text? Refer also to Table 18.1 for information.

11. Refer to Figure 18.11. Note that the zone IV area extends eastward in an elongation along the Ohio River. What is the cause for this extension?

12. In what seismic risk zone is Denver, Colorado, located? Refer now to Figure 18.10. How many equivalent magnitude 4 earthquakes are indicated for the Denver area? Based on this, is it likely that small natural earthquakes would suddenly occur in this area? What was the cause of these earthquakes?

13. Examine Figure 18.15. How deep would a hole have to be drilled to reach the M discontinuity if done so in the oceans? Was this ever planned or attempted? Explain.

14. Examine the equation for the gravity method given in the text. What effect does elevation on the Earth's surface have on the F value obtained? Why is this so? Why would a mass of granite yield a higher value than a mass of sandstone below the Earth's surface, all other factors being equal?

15. What limitations and restrictions are imposed when it is noted that geophysical methods are of an indirect nature? How can geophysical information be checked by direct information in the field? Explain.

16. What two types of information are supplied by the refraction seismic method? What type of wave is used in these studies? Why?

17. Why must a seismograph for engineering geophysics studies be an accurate timing device and be sensitive to frequencies in the 15- to 70-Hz range? How does a geophone work?

18. Examine Table 18.9 which provides seismic velocities for common soils and rocks. What is the seismic velocity listed for granite, measured in m/sec? How does this compare for the velocity in the upper crust discussed previously (see also Table 18.8).

19. It is typically not possible to distinguish the groundwater table in a continuous mass of bedrock using the refraction seismic method. Why is this the case? Why then can the water table be discerned in soil materials in some instances?

20. Refer to Figure 18.16 on seismic velocity versus ripper performance. Note that both rock type and ripper size are factors in this figure. Why is this significant? What is the lowest value for marginal ripping shown on the chart and what material is it for? Why then is 7000 ft/sec likely used as the rule of thumb for rippability? Explain.

21. For a horizontally bedded two-layer situation, $V_1 = 150$ ft/sec and $V_2 = 8000$ ft/sec. What types of materials could make up layer 1 and layer 2? Calculate the value of i_{cr} for this sequence.

22. Information from a refraction seismic survey is supplied below for a three-layer case in which V_3, V_2, and V_1 is involved.

Draw the time-distance curve similar to that in Example Problem 18.3 and in Figure 18.19. Find V_1, V_2, and V_3 based on the slope of the lines in the plot. Calculate Z_1 and Z_2 using both the critical distance method and the time intercept method. Explain any discrepancies in your answers.

23. A quarry owner decided to set off a large blast in the quarry on a weekend when few people were in the vicinity. An equivalent of 10 tons of dynamite was set off at one time. An office building is located 2150 ft away from the quarry. What was the scale distance factor for this blast? Was it likely to cause vibration damage to the office building? Explain.

24. Referring to Exercise 23, if the office building has several large windows of 100 ft^2 in size, would air blast damage be a problem if a direct line of sight occurred between the quarry blast and the building? Explain in detail.

25. Relative to the earth resistivity method what is meant by (a) Wenner configuration, (b) leap-frog sequence, and (c) Lee configuration.

Distance from blast to geophone (ft)	10	20	30	40	50	60	70	80
Travel time of wave (sec)	0.008	0.016	0.018	0.020	0.022	0.023	0.024	0.025

26. Under what geologic conditions could the earth resistivity method give useful results whereas for the same conditions the refraction seismic method would not? Explain and then give a geologic cross section to illustrate this situation.

27. Refer to the Example Problem 18.15 in the text. Given the *a* and ρ values compute the columns for log *a* and log ρ.

a	log *a*	ρ	log ρ
5	0.70	40	1.60
10	1.00	37	
15	1.18	51	
20	1.30	63	
25	1.40	72	
30	1.48	60	
35	1.54	47	
40	1.60	37	
45	1.65	39	
50		41	
55		44	
60		45	

28. Make a plot of log*a* (absissa) versus logρ (ordinate). Compare this plot to that of *a* versus ρ and *a* versus cumulative ρ in the Example Problem 18.5. For this problem, which curve seems to be most useful? Why is that?

29. Complete the following Barnes layer calculation table for another electrical resistivity survey:
(a) Make a plot of *a* versus ρ for the data.
(b) Compute cumulative ρ. Make a plot of *a* versus cumulative ρ.

30. Refer to Table 18.13 in the text on common borehole geophysical methods. For the electrical and radiation logging types, which logs could be run in a cased hole through soil? What useful information could be obtained from these well log data? Explain.

Additional Readings

ALGERMISSEN, S.T., 1969, "Seismic Risk Studies in the United States," Proceedings, Fourth World Conference on Earthquake Engineering, Vol. I, pp. 14–27, Santiago, Chile.

CRANDELL, F.J., 1949, "Ground Vibrations Due to Blasting and Its Effect Upon Structures," *Journal Boston Society of Civil Engineers*, Vol. 36, No. 2, pp. 206–229.

DEVINE, J.R., 1966, "Avoiding Damage to Residences from Blasting Vibrations," Highway Research Record No. 135, Highway Research Board, National Academy of Sciences, Washington, D.C.

DOBRIN, M.B., 1960, *Introduction to Geophysical Prospecting*, 2nd ed., McGraw-Hill, New York.

EDWARDS, A.T., and NORTHWOOD, T.D., 1960, "Experimental Studies on the Effects of Blasting on Structures," *The Engineer*, Vol. 210, September 30.

GRIFFITHS, D.H., and KING, R.F., 1965, *Applied Geophysics for Engineers and Geologists*, Pergamon Press, New York.

HENDRON, A.J., Jr., 1977, "Engineering of Rock Blasting on Civil Projects," in *Structural and Geotechnical Mechanics*, Edited by W.J. Hall, pp. 242–277, Prentice Hall, Englewood Cliffs, NJ.

IACOPI, R., 1964, *Earthquake Country*, Sunset Books, Lane Book Co., Menlo Park, CA.

JOHNSON, A.I., 1968, "An Outline of Geophysical Logging Methods and Their Uses in Hydrologic Studies," Water Supply Paper 1892, pp. 158–164, U.S. Geological Survey.

a (ft)	ρ (ohm-ft)	$2\pi a$	$\frac{1}{R_n} = \frac{2\pi a}{\rho}$	$\frac{1}{R_n} = \frac{1}{R_n} - \frac{1}{R_{n-1}}$	$\rho_L = \frac{2\pi a}{1/R_n}$
5	60	31.4	0.523	.523	60
10	45	62.8	1.395	.872	72
15	50	94.2	1.884	.489	193
20	60	125.6	2.093	.209	193
25	71	157.1			
30	80				
35	89				
40	98				

KOVACS, W.D., 1972, "The Seismicity of Indiana and its Relationship to Civil Engineering Structures," Joint Highway Research Project, Publication No. 44, Purdue University.

MALOTT, D.F., 1967, "Shallow Geophysical Explorations by the Michigan Department of State Highways," Proceedings, 18th Annual Highway Geology Symposium, Purdue University, Engineering Bulletin Series 127, pp. 104–134.

McNUTT, S.R., and SYDNOR, R.H., Eds., 1990, "The Loma Prieta (Santa Cruz Mountains), California Earthquake of 17 October 1989," Special Publication 104, California Division of Mines and Geology.

THOENEN, S.R., and WINDES, S.L., 1942, "Seismic Effects on Quarry Blasting," Bulletin 442, Bureau of Mines, Denver, CO.

19

Subsurface Investigation and Site Selection

CHAPTER OUTLINE

SUBSURFACE MODEL

Detailed knowledge of the subsurface is needed in the design and construction of foundations for engineering structures. The subsurface affected by induced loads from a structure, or earth materials that affect construction of the structure, must be investigated to determine the extent of these effects. Engineering works requiring particular concern for subsurface conditions are earth fill, rock fill, and concrete dams; tunnels; bridges; heavily loaded and certain lightly loaded buildings; highways; levees; docking facilities; and large underground openings such as power-generating stations and subway stations. Information on subsurface conditions is also needed for mining, underground storage of natural gas or liquids,

liquid waste disposal, landfill design, and of course for oil exploration. Whenever the placement of a structure is to be made on or within the Earth, or minerals, rocks, or petroleum are to be removed from it, a clear picture of the subsurface is required.

Almost always, some or even extensive knowledge of the subsurface is known prior to drilling and sampling, or geophysical exploration. Commonly, considerably more knowledge exists than a casual observer would suspect. A general indication of the depth to bedrock is known, as well as the nature of the soil, ground-water conditions, soil strength, and bedrock (if it is to be intercepted). It is extremely difficult to plan a proper subsurface investigation if many of these details are not known prior to initiating the drilling program. The purpose of the subsurface investigation is to confirm or to modify the "model" of the subsurface, which is developed by the individual in charge of the drilling and sampling program prior to subsurface exploration. This individual is typically a soils engineer (geotechnical engineer), engineering geologist, mining geologist or mining engineer, petroleum geologist, or some similarly trained individual.

Depending on the organization and on the nature of the project, varying amounts of effort are expended on subsurface investigation prior to drilling and sampling. This may involve many hours of effort for a major structure such as a large dam, to relatively little effort for a one-story industrial plant or business office. The portion of the total effort accomplished prior to drilling and sampling is typically referred to as *reconnaissance*.

RECONNAISSANCE

Subsurface investigation can be planned more effectively if a review of available information is made prior to the initiation of drilling and sampling. This is commonly accomplished in two phases, office reconnaissance and field reconnaissance. For certain projects, such as a highway bridge or foundations for large buildings, the distinction between the two is clear-cut, office reconnaissance first, followed by field reconnaissance; then the boring plan is developed and drilling and sampling commence. For larger structures such as major dams, tunnels, power plants and docking facilities, several stages of office and field reconnaissance may occur with preliminary drilling between them prior to the major drilling program associated with the predesign stage. The nature and scope of the project determine the need for alternating office and field activity, but the important fact must be underscored that both office and field reconnaissance are needed prior to drilling and sampling. Again, for some small foundation projects, reconnaissance may consist only of a brief mental review of the area by the project director and a brief visit to the site during the sampling program.

Office Reconnaissance

In the office phase several steps or tasks can be accomplished as sources of information are reviewed. Not all the sources discussed here are appropriate or necessary to pursue for all sites or possibly for any single site. However, this can be used as a checklist for sources of available information. It should be kept in mind that gaining information by drilling and sampling is an expensive and unwarranted procedure if that same information is already available in the literature. In addition, a boring program can be extremely difficult and expensive if only very limited data are known in advance. It may be necessary to proceed without such information in remote areas of developing nations or in desolate areas throughout the world. If information is available, however, the job should not be made more difficult and expensive by relying on drilling information alone.

Review of Design Plans or Preliminary Plans

Final plans or at least preliminary ones are available for a structure when the subsurface investigation begins. This may consist of a series of alternative plans being considered or a complete building design. The geotechnical engineer or engineering geologist in charge of the foundation investigation should become familiar with the proposed structure through the study of plans and design data. One should determine the approximate magnitude of loads to be transmitted to the foundation; this information is available from the architect or the structural engineer for the project.

Review of Engineering Reports

Many public agencies and private companies have accumulated information on certain geographic locations and on specific types of projects. For a private company this involves a review of their files on previous projects of a similar nature, along with a review of the information of public record. Federal agencies and state governmental offices are primary sources for public information.

Engineering reports may provide general information on the soil, rock, and ground-water conditions that are likely to prevail in the study area. A study of previous construction activity may also prove helpful for anticipating foundation problems and yielding geologic detail.

Review of Published Information or Open File Geologic Reports

Other information pertaining to the site area can be obtained from published maps, records of railroads and utilities, aerial photographs, water well logs, oil well logs, and other similar information filed with the state government. Abundant site data are commonly available to individuals who know where to seek such information in governmental agencies.

As with all available information, there are limits on the accuracy and extent to which data can be applied for the situation under study. Knowing such limitations is a vital part of the analysis procedure.

TOPOGRAPHIC MAPS   These maps, which depict geographical information as well as surface elevations, are useful in the reconnaissance program. Site location and access to the site can be determined from these maps. Surface elevations for boring locations can also be estimated. Topographic maps supply a general idea of the geology to the experienced engineering geologist, particularly in conjunction with other geologic information. They also serve as base maps for field reconnaissance.

Topographic maps of 7.5-min quadrangles are available from the U.S. Geological Survey (USGS) for most areas of the United States. These maps, with a scale of 1/24,000, are named for the largest city or town contained on the map and are indexed by state according to this quadrangle name. The state geologic survey for most states stocks a supply of "topomaps" for their own state that are available for resale. Fifteen-minute topographic maps covering four times the area of the 7.5-min sheet may also be available and will prove useful if a larger area is to be viewed. These are mapped at the smaller scale of 1/62,500.

GEOLOGIC MAPS   Geologic maps may also be available for the area in question. On a geologic map, the top of the bedrock surface is depicted, which is what would be observed if the soil were stripped away from the rock. Geologic units of differing rock type and age are delineated on such maps. The mapping units are typically individual formations. Specifics concerning geologic maps are presented in Appendix B.

Information on the bedrock below the site is obtained from the geologic map, which includes general descriptions of the rock's lithology and rock properties. Geologic maps are found in a number of sources, for example, USGS folios with several series such as the GQ series (geologic quadrangle), HA (hydrologic atlas), I (investigations), and C (coal series). An index of USGS geologic maps is available in most libraries in the form of a list of published maps. Regional geologic maps for areas of 1° × 2° may also

be available, with a typical scale of 1/125,000. Several counties in the midwestern United States would be included on such a map.

GEOLOGY-RELATED MAPS   Other types of geological maps are available from many state geological surveys. Glacial thickness maps typically covering an entire county or several counties and bedrock-surface contour maps for an equal-sized area are available in many locations with glacial cover. These maps typically prove useful in determining the depth to bedrock when used in conjunction with topographic maps. Environmental geology maps may prove somewhat useful if available for a study site, particularly since they tend to be mapped at rather large scales, say, 1/12,000 to 1/1000. However, some information must be interpreted from these maps, which requires a more extensive knowledge of geology.

AGRICULTURAL SOILS MAPS   These maps, which were discussed in conjunction with ground-water geology (Chapter 15), are available through the U.S. Department of Agriculture (USDA). They cover an entire county and generally are available from the county farm agent or the Soil Conservation Service office in the county. Agricultural units are depicted on these maps but engineering significance can be applied to the mapping units as well, as discussed in Chapter 15.

AERIAL PHOTOGRAPHS   Aerial photographs are obtained from an aircraft that takes successive photos using a vertical view with overlap in the ground image from photo to photo. A 60% overlap from photo to photo with a 20% overlap between successive flight lines is commonly used. This provides two views of the same object from a slightly different location and hence a stereographic image can be obtained by viewing two photos simultaneously. Most commonly, air photos are taken using black-and-white film, but color and color infrared photography are also used.

Geographic details of the ground surface such as roads, houses, and farm fields are recorded on the photo and are useful for purposes of location. The tone and texture of objects and patterns are also captured on the photo and these features are useful in identifying soil types, rock units, geologic structures, and various landforms. Aerial photograph interpretation is an important tool for reconnaissance study of terrain.

In recent years, color photography and color infrared photography (false color IR) have become available for analysis. Color photography is of particular value in mapping various soil types because color differences show quite different distinctions. Bedrock features are also enhanced by color photography. Infrared photography utilizes the near-infrared portion of the spectrum, just beyond the range of the visible spectrum but in a portion still sensitive to certain types of film. Color IR photography is useful for highlighting green vegetation (or lack thereof) and water bodies. Vegetation takes on a magenta (red) color, whereas water is completely black.

Black-and-white aerial photographs are available from several governmental agencies, depending mostly on the area of the United States in which the study site is located. The USDA has photocoverage for much of the primary agricultural lands of the United States, whereas the U.S. Forest Service has photography for much of the western area of the country. These photos can be obtained after consulting the index of available photos and ordering photos by number from the appropriate agency.

Color and color IR photography in most cases must be obtained by scheduling new aerial photography for a site. No large depository of these photos of specific areas is available for sale. If such reconnaissance is needed, an aerial survey company should be contacted to obtain the new photos. Photo scale and other details are designated when ordering the photos.

WELL LOG DATA   Thousands of water wells are drilled each year in the United States. In many places, state law requires that drillers record the logs for such wells with a state agency, usually the Department of Natural Resources or some similarly titled agency. Such well data are identified according to location and kept on open file for people to use as needed.

Oil well logs are also recorded with a state agency, typically the Bureau of Oil and Gas or some similarly titled agency. These logs commonly begin their detailed description at the bedrock surface, lumping all material above rock as overburden. In areas with a thick glacial cover such collective terminology overlooks valuable information for engineering construction, but it does indicate the depth to bedrock. In areas of residual soils the well logs provide information from the near-surface downward.

In some cases, the USGS has compiled the water well information into lists of data with an accompany-

ing report. These are published as USGS professional papers and water supply papers. USGS publication lists are available from libraries at major universities or in the libraries of most large cities.

PUBLISHED REPORTS   Geologic information, including maps, geologic columns, and descriptive material, are published in many sources in the United States. These include the major journals of the geological sciences plus other related scientific and engineering fields. A set of reference volumes titled *Bibliography of the Geology of North America* is found on the reference shelf of most university libraries. Referenced by subjects and by author, the publications pertaining to a specific geographical area can be found. Published articles in reference journals can be retrieved from the library to determine whether the information is helpful in the reconnaissance study.

Formulation of the Boring Plan

During the final portion of office reconnaissance, a tentative boring plan should be developed for subsequent review during field reconnaissance. Several factors help determine the number and the layout of borings, some of which are the nature of the construction project, the complexity of the subsurface materials, the personal judgment of the project manager, and the policy of the organizations involved, both the organization doing the investigation and the one paying for the work. The purpose is to learn as much as possible about the subsurface and how it will affect or be affected by the construction by doing a minimum amount of drilling and sampling. Therefore, both engineering geology (or geotechnical engineering) and economics play a major role.

Sometimes, to help determine the number of borings needed, a seismic or an electrical resistivity survey of the site will be performed prior to or in conjunction with the boring program. Frequently the subsurface data can be extended between borings by use of these and other geophysical methods. These techniques are discussed in detail in Chapter 18.

Some general considerations concerning the density of borings and geological conditions are as follows. Limestone units may require more borings than other rock types because of their irregular weathering surface and likelihood of cavities. Igneous and metamorphosed-igneous rocks are typically fairly uniform across a small site and may require a minimum of borings. In alluvial valleys and in glacial deposits, it is difficult to generalize except that inhomogeneity is common. Soils associated with shallow ocean deposits tend to be fairly uniform and typically fewer borings are required. Keep in mind, however, that these rules of thumb have many exceptions. They must be applied only as a guide for formulating the plan, with changes forthcoming as the study progresses.

Procedure for Reconnaissance
It is difficult to generalize for all construction projects a precise reconnaissance procedure. Obviously the level of concern is greater for a major dam than for a conventional small building and this is reflected in the extent and detail of the respective reconnaissance programs. For major structures such as large dams, tunnels, or docking facilities, for example, the procedure may include a review of the regional features followed by an analysis of local considerations of the project.

In the regional review for a major project, analysis of geologic details of a significantly large portion of the Earth's surface is involved, perhaps an area the size of a state or physiographic division of the United States. A general view of the broad structural geology features, the geologic column involved, the nature of the rock, general topography, major drainages, climatic details, and other pertinent information is included. This allows the investigator to determine the overall nature of the geologic factors that will affect the total area, making it possible to determine the relationship between the site and the region as a whole.

With this accomplished, the local factors of the site are considered. The geologic structure, rock types, topography, ground-water conditions, soil types, depth to bedrock, and other information of this nature are accumulated for the specific site. This makes it possible to view the local conditions in light of the overall regional concerns.

Again for large constructions, the drilling and sampling program may consist of several phases rather than a single effort. One phase may involve widely spaced borings to determine the general nature of the site in terms of specific data at these locations. The final design of the project may be accomplished following this preliminary drilling program. Prior to construction, a more detailed drilling program would ensue to provide information for

construction details and for bidding information for prospective construction contractors. The plans and specifications for the project would be developed based on all reconnaissance and fieldwork including this last drilling program.

As a final step in the construction of a major project, the geologic data encountered or learned during construction are recorded along with specific construction details. The purpose is to document the construction record and to provide geologic detail should subsequent problems develop during operation of the project. This makes correction much easier. It also provides a learning process by which planners and designers can improve their techniques through experience gained on previous projects.

Field Reconnaissance
After the office reconnaissance has been accomplished the field phase should commence with a visit to the site. Preferably the project manager should make this field inspection but if not possible, a field representative or the project geologist should be involved. For major projects, the field inspection party may include a group of specialists consisting of the designers, project geologist, project engineer, and the construction inspection people. This allows for constructive discussion of the project in terms of all the factors that will have an impact on the total procedure. Site conditions commonly relate to design limitations and construction details. If these are known and discussed early in the project, many subsequent problems can be prevented.

Field reconnaissance has two major purposes regardless of the size of the construction project: 1) To allow an experienced observer to view the site and record information that will affect either the foundation design or boring plan and 2) to gather information needed for the drill crew to accomplish the boring program.

There are a number of items to observe or to look for in the field during reconnaissance, and field notes should be taken for subsequent reference. These are outlined next.

Items to Observe During Field Reconnaissance
PROPOSED LOCATION OF STRUCTURE   The designer's proposed location of the building or structure on the tract of land should be observed on the ground. If the proposed location appears to be in poor material and

better locations are available on the tract of land a recommendation to shift the location may be in order. Swampy or low marshy areas or unstable slopes are examples of locations that could likely be improved in most situations. A knowledge of the proposed structure (from the review of plans during office reconnaissance) makes it possible to view the needs of the structure and the field location collectively.

TOPOGRAPHY AND VEGETATION   In nearly all cases a topographic map is available for the site, either a USGS topographic quadrangle sheet or a larger scale topographic map specifically prepared for the site. During reconnaissance the field observer should walk over the tract of land while studying the topographic map. If stereographic coverage of aerial photographs is available, this also should be referred to as the site is traversed.

Notations concerning vegetation and topographic relief should be made as the traverse continues. Notes on topography are useful to provide indirect evidence of the subsurface. For example, narrow steep stream channels are suggestive that bedrock is close to the surface and wide alluvial channels suggest a much thicker soil cover. The notes on topography are valuable to the drill crew as well, allowing them to plan access to the site, and to indicate the type of equipment needed, such as truck-mounted, skid-mounted, or all-terrain vehicles or winched equipment.

Vegetation, to some extent, is indicative of surface soils and ground-water conditions. Growth is likely to be dark and lush in areas of ground-water seepage. In some parts of the United States, the growth of certain kinds of trees or shrubs is linked to locations of specific rock types or even rock units. Other trees are indicative of slow slope movement and of course treeless, scarred areas within a forest indicate a recent snow or rock avalanche or other landslide. Obviously, the use of vegetation to determine detailed information requires much experience on the part of the observer and in the specific part of the country involved.

SURFACE SOILS, GULLYING, AND NATURAL SLOPES   Surface soils are studied by direct observation of the surface and through the use of a shovel, posthole digger, or hand auger. These soils may be indicative of the specific parent material from which they were derived, either the rock type or even the individual formation.

Alluvial soils supply information about the stream from which they were deposited. Boulders in a small stream suggest a high stream velocity and possibly scour problems during high flow levels.

Glacial soils in upland areas, terrace deposits, wind-deposited soils, or residual soils on bedrock uplands can be investigated during the on-site investigation. A knowledge of geomorphology (the study of landforms) is extremely helpful in this situation.

Any surface depression, slope dissection, or other natural or human-made excavation in the ground can provide valuable information about the subsurface. Commonly such cuts into the subsurface provide more information than a borehole because of the larger size and three-dimensional observation available. Soil type, depth to bedrock, and type of bedrock may be indicated at such locations. It should be emphasized that narrow, actively downcutting streams are good sources of subsurface data, which must be explored during the on-site reconnaissance. Countless examples can be quoted in which a proper examination of the stream valleys would have provided a wealth of information for use in developing the drilling program and for correlation of boring data after the drilling is completed.

SURFACE AND SUBSURFACE WATER   The presence of surface or subsurface water is valuable information in the design of foundations for structures and in the planning of the boring plan. All surface flows, surface seeps, and springs should be noted as well as available information on ground-water levels. Natural lake levels may be indicative of the ground-water surface, and major streams commonly set the lowest level of the ground-water table for the area.

GEOLOGY OF THE SITE   Depending on the size and nature of the project and on the site itself, different degrees of geologic field investigation are warranted. For a major project such as a large dam, tunnel, or deep highway cut, a detailed, large-scale geologic map should be prepared. This would involve a careful traversing of the site by an engineering geologist, geological engineer, or field geologist during which time the nature of the rock units and their attitudes (strike and dip) would be determined. Maps would be prepared on the topographic base map with assistance from the aerial photographs for the site. Knowledge of field geology and geomorphology are necessary to accomplish an accurate mapping program. This map will be used, along with the other information gathered (as described earlier), to plan the drilling program so that samples can be obtained from rocks at depth and their physical character determined. This information is used to modify the model of the subsurface that is being formulated during the reconnaissance program.

Information Needed by the Drill Crew

The drilling crew needs to know the following items: how to get to the site, where to drill, what equipment to take for the project, and what difficulties will likely occur. It is important that most of the drill crew's time be spent drilling, with a minimum of time spent moving from point to point and virtually no time spent searching for the site or specific boring locations or traveling back to the shop for more equipment. Generally the following factors are involved.

FINALIZATION OF THE BORING PLAN The proposed boring locations should be checked for accessibility and for suitability relative to the needs for drilling information to complete the subsurface model. Deletions, relocations, and additions should be made based on the accessibility and need to supply the necessary details.

TYPE OF EQUIPMENT REQUIRED During reconnaissance, notes should be taken that provide information concerning which type of drilling equipment is most suitable (rotary, auger, auger type). The most applicable method should be used based on field reconnaissance and accessibility. If a continuous flight auger is to be used, will the soil samples from the auger flights be suitable or if a three-flight auger is used will the hole stay open when the auger is withdrawn? When caving of the hole is anticipated, a rotary drill rig is required with casing to hold the hole open, or drilling mud will need to be used instead. Wash boring methods, if proposed, require a water supply for the drilling fluid. A rotary coring rig will also require a water supply as do hollow stem augers when certain sandy strata are encountered. The size of pump, length of hose, and similar details must be determined for this situation in addition to the location of a water supply.

Any borings to be drilled in a steam bed require special consideration. The size of barge, its anchorage, size of drill rig to penetrate the required depth, and whether special drilling permits are needed must be determined prior to sending the drillers to the site. The type of equipment might be influenced by the time of the year in which the borings will be made. The winter season may be a better time for drilling some sites if frozen ground will facilitate access.

BENCHMARKS AND LOCATION OF BORINGS Reconnaissance should determine if benchmarks or other reference points are in place and whether they are adjacent to or on the site. These benchmarks should be properly referenced on the plans. The boring locations may be laid out during reconnaissance if close tolerance location is not critical. If surveying is required to locate borings, a survey crew should be used to accomplish this task.

If a large-scale topographic map is available for the site, boring locations can be made directly from this plot without careful location. In this case, the boring locations should be laid out (staked) during reconnaissance.

PERMISSION OF PROPERTY OWNERS Drilling to be accomplished on land not yet purchased or if access through property owned by others is involved, permission to proceed should be accomplished by the reconnaissance party or a special team of property specialists if this portion of the job is a major one. The use of time by the drill crew to accomplish this is a poor and generally expensive practice.

LOCATION OF UTILITIES Underground or overhead utilities on the site should be accurately shown on the plans, or their locations staked on the ground, or both. The names of the agencies or people to contact before work begins should be supplied on the plans for the drilling foreman.

The use of site maps alone to locate utilities without field verification is not a good practice. Checking with management personnel of the utilities involved is a recommended procedure, which should be accomplished during reconnaissance. A careful record of the names and job titles for the utility should be included with the information supplied regarding utility line locations.

PERTINENT NOTES Notes providing general information for the drill crew should be prepared during reconnaissance. This would include instructions on the best route to the site and routes between borings, as well as other pertinent information. If additional exploration methods, such as geophysical surveys, are applicable to the site, notes on locations and types of exploration techniques desired should be made during reconnaissance. This information, of course, is subject to modification based on the results of the boring program.

SUBSURFACE EXPLORATION: SOUNDING, DRILLING, AND SAMPLING

For most foundation studies, the reconnaissance program is followed by subsurface exploration. This may be limited to finding the depth to refusal by a probing technique or it may involve sophisticated drilling, sampling, and in-hole measuring procedures. (Refusal is the depth at which a probing device is no longer able to penetrate into the subsurface.) Because the means for exploration in soil or unconsolidated material is generally different from exploration in bedrock, these can be used as a twofold subdivision for discussing the various exploration procedures. In this discussion, however, our intention is to emphasize the similarities between soil and rock exploration and to prepare a student for both. Therefore, subsurface exploration is viewed as a single entity. Geophysical techniques and other indirect exploration methods are discussed in Chapter 18.

Soil exploration can be accomplished in three basic ways. These are subdivided generally as follows: 1) probing or sounding of soil depth to reach refusal, 2) drilling and sampling of soil at various depths, and 3) excavation of test pits followed by inspection and sampling. Rock exploration is accomplished by analyzing drill cuttings or rock cores, or with in-hole measuring devices. These are described in turn in the following sections.

Soundings

Sounding, probing, or pricking is a means by which an indication of the depth to bedrock (refusal) is obtained without drilling. A steel bar or hand auger is forced manually into the ground until refusal is reached. This method is most appropriate for soft, soil materials of shallow depths that overlay bedrock. In some cases a minimum soil depth is needed for construction purposes; for example, a gas or oil pipeline, sewer line, or water line. If this minimum depth is reached, then the probing may not extend to refusal but be terminated just beyond the minimum depth. In rough terrain where a drilling rig would have major access problems, probing may be a necessary means of subsurface exploration.

The sounding method has serious shortcomings. No samples are taken and typically no recording of soil type is made, even of near-surface material. Refusal can occur either on bedrock or on any other hard surface including boulders, slightly cemented sands, or strongly consolidated clays. Many assumptions are involved when one concludes that the refusal depth indicates bedrock; this should be checked by a thorough geologic investigation. Soundings may be used to provide information between widely spaced borings. When this is done, the boring information is extended by the soundings but, again, care must be taken to ensure that no false conclusions are reached based on refusal depths.

Methods of Drilling

Two distinct operations are involved in obtaining samples from the subsurface: drilling and sampling. The first process involves the making of a hole to the depth of interest and after that the second, sampling, can proceed. The methods for obtaining these boreholes is discussed first.

Cable Tool Well Drilling

The cable tool or churn drill is used primarily for water well drilling and only rarely for engineering exploration. A chisel bit is pounded and churned into the ground by the lifting and dropping action on the hoisting drum of the drill rig. Only an up-and-down motion (no rotation) is used, which allows the chisel bit to pound the soil or moderately hard rock to small pieces. Rock and soil fragments are removed by a bailer (bailing bucket), which is dropped into the well on the cable to collect the fragments and any water that may have accumulated. Sampling by this technique is obviously limited.

Wash Boring

The drill hole is advanced by the washing action of a water jet in combination with the chopping action of the bit. Steel casing is used to keep the hole open and is advanced as the hole proceeds. Cuttings are returned to the surface in the wash water by way of the annular opening between the casing and the hole. Sampling can proceed at selected intervals.

Boulders and large cobbles are difficult to penetrate with this procedure and they can occur under a number of geological conditions. Drilling is limited to soil materials because bedrock is too hard to penetrate in a reasonable period of time.

Wash borings can also be made with or without an outside casing, using only the water nozzle, that is, omitting the chopping bit. Cuttings are returned in the wash water and no other sampling is possible. This provides only slightly more information than do simple soundings and much care should be used so that proper limitations are put on the information obtained.

Rotary Drilling

In this method, the drill rod is rotated around the drill axis (usually vertical) and various drill bits are used to cut through soil or rock. A more elaborate drilling rig than either the cable tool rig or wash boring equipment is required to deliver the rotational power needed to penetrate soil or rock.

Special roller bits are available for soil and rock that will cut through the material in question. Cuttings are returned to the surface through the wash water. Sampling at selected intervals can be accomplished for most rotary drilling methods but not for all.

Four primary rotary procedures are used for exploration drilling: auger boring with solid augers, auger boring with hollow-stem augers, rotary drilling with roller bits, and rock or soil coring. A truck-mounted rotary drill rig is shown in Figure 19.1. Hollow-stem augers are being advanced into the ground in this photograph.

AUGER BORINGS WITH SOLID AUGER STEMS   These augers can be either continuous-flight, three-flight, or bucket augers. Continuous-flight augers appear similar to a wood screw in that the flutes are rotated into the ground and the soil returns to the surface by way of the rotating flutes. Auger cuttings can be taken from the flutes; the soil is disturbed in this process and a delay is involved between the time when a soil zone is penetrated by the auger tip and the excavated soil reaches the surface. These holes are drilled dry and no casing is needed to keep the hole open. As

FIGURE 19.1   Truck-mounted rotary drilling using hollow-stem augers.

with all augers, drilling is limited to soil or unconsolidated materials.

Three-flight augers and bucket augers must be retracted from the hole when the auger flutes or the bucket is filled with soil. The drill stem is attached to the specific auger in use. For the three-flight augers the auger is rotated quickly after extraction from the hole and the soil spins off in a circular fashion, a short distance from the hole. Sampling can be made from these disturbed piles of soil. Holes can be 4 to 15 in. or more in diameter. Typically the holes are not cased so that drilling is limited to cohesive soil or moist cohesionless soil above the water table. Holes as large as 36 in. in diameter, used for caisson holes, can be drilled with three-flight augers. The base and side walls of these large holes can be inspected by lowering a knowledgeable observer (an engineering geologist or geotechnical engineer) into the hole to describe the subsurface conditions. Caissons are cast in place, concrete piles that are founded on a hard, bearing layer. They are used as column supports for buildings.

Bucket augers have two flights of augers at the base of the stem plus a bucket attached just above the augers. This bucket has an opening at the bottom so that soil will rotate into it but not easily fall out when the stem is pulled to the surface. Disturbed samples can be obtained from the bucket of soil at the Earth's surface. Bucket auger holes are usually 6 in. or more in diameter and typically are not cased. Therefore, they have the same limitations concerning soil type as do the three-flight augers. Only cohesive soils will not cave into the hole when the auger is extracted.

HOLLOW STEM AUGERS   These are large-diameter, usually 8- to 12-in. continuous-flight augers that have a hole 3 in. in diameter inside the augers. These augers are rotated down into the ground and the soil rotates upward carried by the flutes. Instead of sampling this returned, disturbed soil, or pulling the augers to the surface to obtain soil collected near the bottom flights, sampling is accomplished through the hollow portion of the auger. This feature is discussed later. The hollow-stem augers act as a casing for the hole during the sampling process. Because these augers are not extracted from the hole until the exploration is completed, sampling in cohesionless or cohesive soil above or below the ground-water table can be accomplished. These augers cannot penetrate unweathered bedrock.

ROTARY DRILLING WITH ROLLER BITS   In this method of drilling, a tricone bit is used to drill through the rock or soil, grinding the material to coarse sand size or smaller particles, which are returned with the wash water. No other sampling is possible than that obtained through analysis of the cuttings. This procedure is used in soil if the soil unit is not of interest or previously defined or if we want to drill to bedrock as quickly as possible. When this method is used in rock, the purpose is to drill down to a depth of interest where rock cores will be taken. For large portions of the deep oil wells drilled for the petroleum industry, only the major zones of interest are cored, with roller bits used for the remainder. This is done both in the interest of saving time and money. Useful information can be obtained by a trained geologist by studying the rock cuttings returned with the drilling fluid.

CORING OF SOIL OR ROCK   This is actually a sampling technique but in the sampling process the hole is advanced as well. A solid cylinder of soil or rock is obtained in this procedure. More details are presented in the subsequent discussion on sampling.

Percussion Drilling

This type of drilling is accomplished by use of jackhammers or air tracks, which advance a hole by percussion effects. This method is used in bedrock, including both soft and hard materials. The rock is ground to dust and only color and mineralogy can be determined with any degree of assurance. Typically, only the rate of penetration is recorded, whereby soft seams and cavities can be discerned. The holes are usually drilled for purposes other than exploration, for example, for presplit blasting holes, holes for explosives, and holes for grout placement.

Sampling Methods

To complete an adequate subsurface exploration program, it is necessary to obtain samples for identification and classification of soil and rock plus reasonably undisturbed samples for laboratory testing. If the samples are not representative of the materials in the subsurface, misleading results will be obtained.

Disturbed samples are those in which the soil structure has not been maintained and they can only be properly used for description and soil classification and for laboratory tests in which soil structure is not important. In contrast, relatively undisturbed samples are used also for wet and dry density determinations; triaxial, shear or unconfined compression tests; and permeability and consolidation tests.

The degree of disturbance of soil samples depends on the nature of the material being sampled, the core barrel or sampling method used, the drilling equipment, and the skill of the driller. Soil samples, on extended exposure to the atmosphere, will become unusable for testing, mostly because of changes in moisture content. Therefore, proper sealing, transportation, and storage of the relatively undisturbed samples must also be accomplished.

The rock cores of weak shales, coals, and other similar, rather soft materials should not be exposed to the atmosphere for extended periods prior to logging the core (detailed description obtained) or to wrapping the samples for laboratory testing. Commonly, clay-rich shales and mudstones will form thin wafers or disks on drying in the field even when placed in a core box. Cores should be logged as soon as possible after they are obtained. Samples should be selected for testing and properly protected at that time.

The simplest procedures for soil sampling have been mentioned in passing in the preceding discussion on drilling. For continuous-flight augers the disturbed soil is taken from the auger flutes as the soil returns to the surface. For three-flight augers and bucket augers the soil is obtained when the auger is retracted from the hole periodically when either the bucket or the flutes become filled with soil. In the cable tool method, soil or rock cuttings are returned to the surface by the bailer, whereas when using the wash boring or the rotary method with a cutting bit, rock or soil cuttings are returned by the circulating water. Percussion drilling in rock provides only rock dust, which yields little if any specific information about the character of the rock being drilled.

Standard Penetration Test

The most common procedure used for routine soil sampling involves the standard penetration test (SPT) and the split-barrel or split-spoon sampler. This sampler, when disconnected from the drill rod, splits in half vertically to expose the column of soil obtained during the sample-driving operation. This generally disturbed sample can be described and portions stored in glass jars with screw-top lids for future reference. The soil can be used to determine Atterberg limits, specific gravity, moisture content, and gradation.

The standard penetration test (ASTM D1586) is a driven sample test in which the split-spoon sampler, 1⅜-in. ID, 2-in. OD (36-mm ID, 50-mm OD), is attached to the drill rod and advanced into the ground by a series of blows from the drop hammer. This hammer has a standard weight or mass of 140 lb (64 kg) with a height of fall of 30 in. (0.76 m). The sample is advanced in three increments of 6 in. (150 mm) with the number of blows recorded to accomplish each increment. It is assumed that the first increment is used to seat the sampler in the bottom of the hole and the last two increments are the significant ones. The results would be recorded as, for example, 6/7/5, meaning 6 blows for the first 6 in.

(150-mm) increment to seat the sampler and 7 and 5 blows, respectively, for the last two increments. The N value is the sum of the last two increments, in this case, 12, and it represents the number of blows per foot (0.3 m) with the standard equipment. The SPT or N values are used as an indication of density or consistency of the soil. An enormous quantity of data has been amassed, based on this test, during the last 50 years so that useful comparisons of soil density or consistency can be made through the use of this test.

The SPT is a valuable tool in determining soil conditions for the subsurface but the test is not without problems. The N values are subject to variations depending on the free fall of the hammer. The hammer may not fall the full 30 in. (0.76 m) or excessive friction on the hammer stem may occur due to the following reasons: too much friction because of caked oil or soil; hammer may strike the upper stop too sharply on retrieval, pulling the sampler off the bottom between each blow; the boring may not be clean, reducing hammer energy; a rock may partially plug the sampler; in an uncased hole, soft clay may close around the drill rod, causing additional friction; and in deep holes the hammer energy loss may be greater due to the heavy drill stem and rebound within the stem itself.

Standard penetration tests can be taken in conjunction with drilling by the wash boring method, through hollow-stem augers, or in cased holes drilled by three-flight, bucket, or continuous-flight augers. For these latter three methods, the auger is withdrawn from the hole and the split-spoon sampler is inserted. Soils subject to caving or squeezing will yield problems when the augers are withdrawn. Dry cohesionless soils and cohesionless soils below the water table are particularly prone to caving. Soft saturated clays are subject to squeezing. Cased holes are needed in these situations.

Figure 19.2 is a photograph of the standard penetration test in progress. The split-spoon sampler is attached to the drill rod with the standard 140 lb weight shown at the top of the photo. Soil density and strength have been correlated with the standard penetration test results. Table 19.1 shows the relationship between these parameters.

Hollow-stem augers may develop problems in loose, cohesionless soils below the water table. Such

FIGURE 19.2   Standard penetration test in progress.

sands tend to push up inside the augers so that the split-spoon sampler cannot be extended to the base of the auger stem. Washing of the sand under pressure with the water nozzle can remove such sands and sampling can then proceed. This pushing-up action disrupts the soil below the auger, however, and a truly disturbed sample is obtained. Plugs for the auger opening prevent soil from coming into the hollow opening. The plugs are removed after drilling and before each split-spoon sample is taken. Use of such plugs is not problem free because pebbles can lodge

against the plug making it extremely difficult to disengage the plug after drilling and before sampling.

Pushed Tube Samplings

Pushed barrel or thin-walled tube sampling is used for soil or soft rock to obtain a relatively undisturbed sample. A thin-walled tube (Shelby tube) is forced into the soil, using the static force from the weight of the drilling rig, or driven into soft rock. The tubes range from 3 to 50 in. (75 to 1250 mm) in diameter with 4 in. (100 mm) being a common size. The tubes

TABLE 19.1   Comparison of N Value to Density and Strength of Soil

Soil	Penetration Resistance (N blows/ft)	State	Relative Density Percentage	Approximate Unconfined Compressive Strength, q_u (psf)	(kN/m²)
Sand	0–4	Very loose	0–15		
	4–10	Loose	15–25		
	10–30	Medium	35–65		
	30–50	Dense	65–85		
	50	Very dense	85–100		
Clay	2	Very soft		500	25
	2–4	Soft		500–1000	25–50
	4–8	Medium		1000–2000	50–100
	8–15	Stiff		2000–4000	100–200
	15–30	Very stiff		4000–8000	200–400
	30	Hard		8000	400

1 kN/m² ≅ 20.88 psf.

are pushed (or driven as above) into the soil for a length of 10 to 20 diameters, with a 3-ft (1-m) sample length common for the 4 in. (100 mm) diameter sample. The soil or soft rock must have sufficient cohesion to remain in the barrel while the sampler is being withdrawn from the hole. For cohesionless materials, sampling can be facilitated by the use of a piston sampler, which allows the influence of a vacuum effect to aid in keeping the sample in the tube. The piston, which is controlled by a rod, remains stationary while the outer thin-walled tube is forced ahead into the soil. The Osterberg-type sampler activates the piston hydraulically, whereas the Hong-type sampler uses a rachet mechanism. Common tests subsequently performed on thin-walled tube samples include unconfined compression, moisture content, mass unit weight, consolidation, triaxial compression, permeability, and Atterberg limits.

Rotary Coring

This method is used to obtain a cylindrical sample of soil, soft rock, swelling clay or swelling soft rock. An annular hole or kerf is cut into the soil or rock by rotating the outer tube of the sampler with the sample sliding up inside of this tube.

Rotary coring of soil or soft rock is accomplished by a core bit with embedded steel teeth, which cut through the subsurface. Soft materials are protected by a stationary inner barrel, which accepts the sample as the outer core barrel penetrates the soil or rock. Cuttings are flushed upward by the drilling fluid, which comes down through the core barrel and upward on the outside or perimeter of the hole. Adaptations include the Denison sampler, which has a fixed cutter on the outer barrel, and the Pitcher sampler, which has a spring-loaded outer tube. The Acker core barrel uses air and mud as the drilling fluid to support the hole in soft materials. Relatively undisturbed samples, 2 to 8 in. (50 to 200 mm) in diameter and 1 to 5 ft (0.3 to 1.5 m) long can be retained in the inner barrel of the appropriate size. Sampling is effective in firm to stiff cohesive soils and in soft but intact rock; in loose sand below the water table, samples are difficult to obtain and gravels can seldom be sampled by this method.

Coring of rock by the rotary method involves a core bit with embedded diamonds attached to a core barrel, which rotates to cut the annular hole in the rock. The core is protected by a stationary inner barrel and cuttings are flushed upward by the drilling fluid, which is commonly water, but it can be air or drilling mud. The rock cylinder (core) can range from 7/8 to 4 in. (22 to 100 mm) in diameter with perhaps the most common size being $2\frac{1}{8}$ in. (54 mm) known as N_x size. The other common core diameters are listed in Table 19.2 with the 4-in. (100-mm) size prevailing for some projects. Five- or ten-ft lengths of core (1.5 or 3 m) are common, but this can be extended to lengths of 20 ft (6 m) if the rock is extremely sound. Each core barrel full of rock core is referred to as a *core run* or a *run*.

For conventional rock coring the core barrel is attached to the drill rods, which are rotated by the drill rig to advance the hole. The diamond bit is attached to the core barrel and these two rotate with the drill rod. The inner barrel does not rotate and it accepts the core as the bit rotates downward into the rock. Typically rock core holes are not cased, and when the core barrel is full (or the length of the core barrel has been drilled) the entire drill stem is lifted

TABLE 19.2 Core Sizes for Conventional Coring in Rock

Core Designation	Core Barrel Outside Diameter (in.)	Approx. Diameter of Core Hole (in.)	Approx. Diameter of Core (in.)
E_x	$1\frac{7}{16}$	$1\frac{1}{2}$	$\frac{7}{8}$
A_x	$1\frac{27}{32}$	$1\frac{7}{8}$	$1\frac{3}{16}$
B_x	$2\frac{5}{16}$	$2\frac{3}{8}$	$1\frac{5}{8}$
N_x	$2\frac{15}{16}$	3	$2\frac{1}{8}$
3 in.	$4\frac{3}{16}$	$4\frac{1}{4}$	3
4 in.	$5\frac{7}{16}$	$5\frac{1}{2}$	4

from the hole to obtain the core sample. In this procedure, also called a *run*, the core barrel is brought to the surface, the core removed, and the barrel returned to the bottom of the hole. This cycle may take a reasonably long time if the hole is fairly deep because only 30 ft or so of drill stem can be kept intact without needing to break it down (disconnect the stem) as it is extracted from the bore hole. Further details of this procedure are discussed in a subsequent section.

The dip of bedding and of joints is evident in vertical rock cores but the strike cannot be determined from conventional drilling. In broken rock, blockage in the core barrel can prevent drilling from proceeding.

Core recovery is an item of major concern in any drilling program because a continuous sequence of samples is the objective of coring. Core recovery is indicated as the percent of core obtained compared to the length of hole drilled. Lost core creates questions concerning what caused the core loss. Closely spaced rock fractures, weak rock, and changes in lithology are only a few of the common reasons for core loss. The drilling equipment and skill of the driller are also involved. Although 100% recovery is the objective, this may not be physically possible under the geological conditions at the site. An acceptable core recovery using the most appropriate sampling techniques should be stipulated by the engi-

neering geologist or geotechnical engineer. This is the performance level to be expected from the driller or specified in the drilling contract.

In addition to core recovery, another calculation can be made that is based on core length and this is known as the *Rock Quality Designation* (RQD). This is the ratio of the sum of the lengths of core pieces 4 in. (100 mm) or more long to the total length of hole drilled, represented as a percentage. Cores should be N_x size or larger for this value to be meaningful, and the core breaks should be natural ones—not those induced by drilling or drying out of the core. A photograph of rock core placed in a core storage box is shown in Figure 19.3. These are 3-in.-diameter cores obtained from limestone bedrock.

When the orientation (both strike and dip) of bedding planes or joints is desired, oriented rock core can be obtained with special equipment. The method is similar to conventional coring, but in addition continuous grooves are scribed on the rock core relative to a specific compass direction. The core is typically about 2 in. (54 mm) in diameter with a 5-ft (1.5-m) run. In highly fractured rock this method may not work because of core rotation and possible core blockage.

In recent years core drilling using the wire line procedure has become more common in subsurface investigations for engineering purposes. Long used in oil exploration drilling, the method is best suited for

FIGURE 19.3   Rock cores placed in a rock storage box or core box.

deep holes where there is a much faster cycle of core recovery and resumption of drilling than is possible by conventional core drilling. This is possible as the core and stationary inner barrel are retrieved from the outer core barrel by a lifting device suspended on a thin cable or wire line. The lifting device is disengaged from the outer core barrel, and the inner barrel and core are brought to the surface while the outer barrel and core bit remain in place in the hole. Rather than small-diameter drill rods, a large-diameter tube through which the core can be lifted serves as the drill stem for the wire line apparatus. Consequently, the equipment to outfit a wire line operation is different and more expensive than that used for conventional drilling. However, the rate of coring (production) is greatly improved using this method when deep holes (50 ft or more) are involved.

Rock cores 1½ to 3⅛ in. (36.5 to 85 mm) in diameter are possible sizes using the wire line method, with the N_Q size (1⅞ in. or 48 mm) being most common. Core barrel lengths of 5 to 15 ft (1.5 to 4.6 m) are available. Another advantage of the wire line method in addition to increased core production (and an attendant lower cost) is a decreased tendency for core loss. In some instances a third core barrel or inner lining of plastic is used to retain the sample, which yields improved core recovery for soft or weathered materials. This barrel can also be used in the conventional rock coring procedure to improve that recovery as well. Swelling clays or swelling rocks to be sampled commonly require the special triple barrel core sampler to minimize core loss.

Another procedure is available to sample badly fractured or weathered rock. Using rotary coring, an integral or single sample can be obtained by adapting the standard coring method. A central hole ⅞ in. (22 mm) in diameter is drilled through the length of the proposed core and a steel rod is grouted into this hole using a cement slurry. A large-diameter core 4 to 6 in. (100 to 150 mm) is drilled around the grouted rod to obtain the core sample using the conventional rotary equipment.

Preparation of Borehole Logs

An important part of any site exploration program following sampling is the development of the borehole log, which provides the basis for further subsurface evaluation. This log should supply an accurate record of the soil and rock units encountered in the boring along with any other pertinent information obtained during drilling.

Generally, borehole logging is accomplished in the field as the samples are brought to the surface by the drill rig. A more detailed description of the soil or rock samples commonly is provided sometime later in the laboratory but the initial description of the samples on retrieval from the sampler provides valuable information. Ultimately the borehole log is developed from the field and laboratory descriptions. The format of the log and its accuracy depend on the drilling method and on the type of material drilled.

A complete borehole log typically contains much of the following information. An example boring log is shown in Figure 19.4.

1. Name of the project, date, and commonly the weather conditions during drilling.
2. Borehole identification, elevation, and location relative to project grid or accompanying boring plan.
3. Method of drilling and sampling plus details of equipment as needed to identify drilling specifics.
4. Whether boring is vertical or inclined; if inclined, an angle of inclination and the bearing.
5. Rate of drilling progress either as penetration resistance in soil (N values), or for rotary drilling in soil or rock the penetration rate (such as 1 ft in 15 min) and the rpms during the period.
6. Location of coring runs, samples, and *in situ* tests if included; also depths where casing was used and its diameter.
7. Descriptions of ground-water level, changes in standing water level, zones of water loss (when water used as drilling fluid), or zones of water gain (when air is the drilling fluid).
8. Detailed geologic description of the soil or rock samples. For soils this involves a description of the moistness, color, texture (grain size), consistency, and organic content, if any, using proper terminology. For rocks the texture, color, significant mineralogy, type of cementation, if appropriate, rock name, and degree of weathering or alteration are included. A description of the joints, seams, and openings in the rock core is also included, and for joints the degree of openness, spacing, inclination, and infilling material is noted.

PROJECT NAME ___Coal City___
LOCATION _____
CONTRACTOR _____
DRILLING METHOD __HSA & RC__
VERTICAL BORING ____X____

BORING NO. _____
DATE _____
CREW CHIEF _____
BOREHOLE DIA ___8", HQ core___
INCLINED BORING _____

SHEET NO. __1__ of __4__
SURFACE ELEV. _____
WEATHER _____
_____
ANGLE AND BEARING_____

DESCRIPTION	Stratum Depth, ft.	Ground Water	Depth Scale, ft.	Sample No.	Blows/6 in. 3-6" increments	% Recovery	Shelby Tube No.	Notes on Drilling Conditions, Etc.
Ground Surface								
Topsoil	0.7							
Brown, moist, medium-dense Clayey sand (SC) with little gravel	2.5			1	1/3/2	75		
Brown, moist, medium-dense Clayey sand (SC) with little gravel			5	2	4/7/7	50	1	
				3	3/3/2	75	2	
loose	8.5							
Brown, moist, medium-dense, well-graded sand and gravel (SW-GW)			10	4	8/11/14	40		Scale Change
			15	5	13/8/10	10		
			20	6	14/19/22	20		
	23.5		25	7	20/29/50	50		
Brown, moist, very dense, well-graded sand (SW) with trace gravel			30	8	22/50/.4	75		
			35	9	41/50/.4	75		
wet	38.5		40	10	22/33/50/.4	75		
Brown, moist, very dense, well-graded sand and gravel (SW-GW)		▽	45	11	8/19/22	75		
Brown, wet, dense, fine silty sand (SP-SM)	48.5		50	12	8/20/26	75		

BORING METHOD
HSA = Hollow Stem Auger
CFA = Continuous Flight Auger
RC = Rock Coring

SAMPLING METHOD
SPT = Standard Penetration Test
ST = Shelby Tube

GROUND WATER
Noted on rods _46.0_
On completion_____
After_____ hours____

FIGURE **19.4**   Log of test boring.

PROJECT NAME ___Coal City___
LOCATION_____
CONTRACTOR_____
DRILLING METHOD_____
VERTICAL BORING_____

BORING NO. _____
DATE _____
CREW CHIEF _____
BOREHOLE DIA _____
INCLINED BORING _____

SHEET NO. ___2___ of ___4___
SURFACE ELEV._____
WEATHER_____
ANGLE AND BEARING_____

DESCRIPTION	Stratum Depth, ft.	Ground Water	Depth Scale, ft.	Sample No.	Blows/6 in. 3–6" increments	% Recovery	Shelby Tube No.	Notes on Drilling Conditions, Etc.
	53.5							
Brown, wet, dense well-graded sand (SW)			55	13	8/28/42	75		
	58.5							
Brown, wet, dense well-graded sand and gravel (SW-GW)			60	14	6/14/22	75		
	63.5							
Gray, very hard weathered shale with gravel			65	15	50/.4	10		
	67.2		70	16	50/.4	30		
Gray, moist, very hard clayey silt (ML) or highly weathered shale			75	17	50/.3			Cased with HQ casing to 79.5'
	79.5		80		Coring information recorded			
Gray highly to slightly weathered shale	87.0			1	7.5 / 7.5	100		Rock Coring RQD= 60%
			87					
Black shale	90.0		90					
Gray shale	91.5			2	10.0 / 10.0	100		
Black shale	93.0							
Coal	94.5							Run 2 RQD = 55%
Gray shale			97					
highly weathered to slightly weathered			100	3	10.0 / 10.0	100		Run 3 RQD = 45%
			107					
			110	4				

BORING METHOD
HSA = Hollow Stem Auger
CFA = Continuous Flight Auger
RC = Rock Coring

SAMPLING METHOD
SPT = Standard Penetration Test
ST = Shelby Tube

GROUND WATER
Noted on rods_____
On completion_____
After_____ hours____

FIGURE **19.4** Continued

PROJECT NAME ___Coal City___
LOCATION _____
CONTRACTOR _____
DRILLING METHOD _____
VERTICAL BORING _____

BORING NO. _____
DATE _____
CREW CHIEF _____
BOREHOLE DIA _____
INCLINED BORING _____

SHEET NO. ___3___ of ___4___
SURFACE ELEV. _____
WEATHER _____
_____
ANGLE AND BEARING _____

DESCRIPTION	Stratum Depth, ft.	Ground Water	Depth Scale, ft.	Sample No.	Coring Information	% Recovery	Shelby Tube No.	Notes on Drilling Conditions, Etc.
Gray Shale			117	4	10.0 / 10.0	100		Run 4 RQD = 65%
	120.0		120					Run 5, RQD = 20% Coal washed away from 120.5 to 125.0'
Coal	125.0			5	5.5 / 10.0	55		
Gray shale, weathered			127 130					
	133.5			6	10.0 / 10.0	100		Run 6 RQD = 65%
Coal	134.5							
Gray shale, weathered			137					
	141.0		140					
Coal	142.5			7	4.0 / 10.0	40		Run 7 RQD = 20%
Gray shale			147					
	150.0		150					
Sandstone seam	154.0			8	7.0 / 10.0	70		Run 8 RQD = 60%
Gray shale	157.5		157					
Coal, very friable	159.5		160					
Gray shale, highly weathered				9	10.0 / 10.0	100		Run 9 RQD = 70%
			167 170	10				

BORING METHOD
HSA = Hollow Stem Auger
CFA = Continuous Flight Auger
RC = Rock Coring

SAMPLING METHOD
SPT = Standard Penetration Test
ST = Shelby Tube

GROUND WATER
Noted on rods _____
On completion _____
After _____ hours ____

FIGURE **19.4** Continued

PROJECT NAME ___Coal City___   BORING NO. _____   SHEET NO. ___4___ of ___4___
LOCATION _____   DATE _____   SURFACE ELEV. _____
CONTRACTOR _____   CREW CHIEF_____   WEATHER _____
DRILLING METHOD_____   BOREHOLE DIA _____   _____
VERTICAL BORING _____   INCLINED BORING _____   ANGLE AND BEARING _____

DESCRIPTION	Stratum Depth, ft.	Ground Water	Depth Scale, ft.	Sample No.	Coring Information	% Recovery	Shelby Tube No.	Notes on Drilling Conditions, Etc.
			175	10	$\frac{4.8}{8.0}$	60		RQD = 45%
Gray shale, hard 170-175'								
	180.5		180		$\frac{8.0}{10.0}$	80		RQD = 70% 5" of coal washed away
Coal	181.5			11				
Sandstone, hard, thinly bedded			185					
Cross-bedded			190	12	$\frac{6.0}{10.0}$	60		RQD = 40%
			195	13	$\frac{4.0}{5.0}$	80		RQD = 60%
			200					
Bottom of boring 200.0'								

BORING METHOD
HSA = Hollow Stem Auger
CFA = Continuous Flight Auger
RC = Rock Coring

SAMPLING METHOD
SPT = Standard Penetration Test
ST = Shelby Tube

GROUND WATER
Noted on rods _____
On completion_____
After_____ hours____

FIGURE **19.4**   Continued

9. Percent recovery of rock cores or soil sample is included for each run. Core fractures in breaks per foot are also determined for rock in addition to the RQD in the manner described previously. A description of the rock quality can be added, usually in parentheses, to the general description according to Table 19.3.

10. Unusual occurrences should be noted, such as a sudden drop of drill rods or change in color of return wash water.

11. A symbolic log is sometimes included at the extreme left or right of the sheet. It is drawn to scale and consists of standard petrographic (or soil) symbols depicting the stratigraphic layer described adjacent to it. The symbols used for soils and rock are given in Figure 19.5.

Details of Drilling Procedures

An explanation of the drilling procedures is supplied in the following paragraphs in order to emphasize some of the specifics involved. Both split-spoon sampling and rotary rock coring are considered.

In soil materials split-spoon samples can be taken continuously, at specific intervals, or at changes of soil type. For interval sampling for a 25-ft boring, a common practice is to sample at 2.5-ft intervals to depths of 10 ft or 15 ft and at 5-ft intervals for the remaining distance to 25 ft. The split-spoon sampler is driven 1.5 ft, so to end the sample at a depth of 2.5 ft, sampling is begun at a 1-ft depth. The sequence, therefore, is as follows assuming hollow-stem auger equipment: 1) drill to 1-ft depth, insert split-spoon sampler; 2) drive split-spoon to 2.5-ft depth in three 6-in. increments recording the number of blows for each increment, 3) remove split-spoon, open spoon,

describe sample, place sample in a glass jar, and tighten with a screw-top lid, 4) drill to depth of 3.5 ft, 5) put split-spoon sampler back in the hole and drive to depth of 5 ft. This procedure continues until the full depth of the hole is sampled.

For deep borings in soil the procedure becomes more complicated. To obtain a sample at 30 ft, drilling proceeds to 28.5 ft and the split-spoon sampler is inserted. The sample is driven to the 30-ft depth (in three increments) and the sampler is pulled from the hole. This yields an 18-in. sampler on the end of 30 ft of drill rod. The drill rod is hoisted into the air against the derrick of the rig and held in place by a slip ring on the sand line until the split-spoon sampler is removed from the end of the rod. After soil removal the sampler is screwed into place on the rod, and when drilling is completed the rod is returned to the hole with 5 ft of rod added to the top of the drill stem. When depths of more than 30 ft are reached, no more rod can be safely held in place by the slip ring and derrick, so 10-ft lengths must be dismantled in turn as the rod comes out of the hole until the final 30 ft are left and these are pulled into the air as before. Therefore, for a 100-ft-deep hole, seven 10-ft sections must be dismantled in turn as the rod comes out of the hole. This is very time consuming. Some drill rigs utilize two sand lines with slip ring attachments. In this case, for a 100-ft depth, four 10-ft lengths are dismantled in turn as the rod comes out of the hole. Then, a 30-ft section is pulled from the hole and disconnected to be held by the first slip ring and sand line and finally the last 30 ft are pulled out and held by the second line while the sampler is removed. As illustrated here, deep sampling takes much longer to accomplish than does shallow sampling and this must be reflected in the price for taking the samples.

In conventional rock core drilling, the core barrel and diamond core bit are connected to the drill rod [about a 1.5 to 2 in. (37 to 60 mm) in diameter steel rod], which rotates to cut the core. Five- and 10-ft core barrels (1.5 and 3 m) are the most common sizes used for subsurface exploration in engineering geology. These, of course, yield 5- and 10-ft runs of rock core, respectively.

Before rock coring can begin, a casing is set on the bedrock surface and coring proceeds through the casing. Typically, a casing is not placed in the rock section unless that rock is extremely weak and unable

TABLE 19.3   Rock Quality Designation

RQD [% recovery 100-mm lengths (4 in.)]	Description
10–25%	Very poor
25–50%	Poor
50–75%	Fair
75–90%	Good
90–100%	Excellent

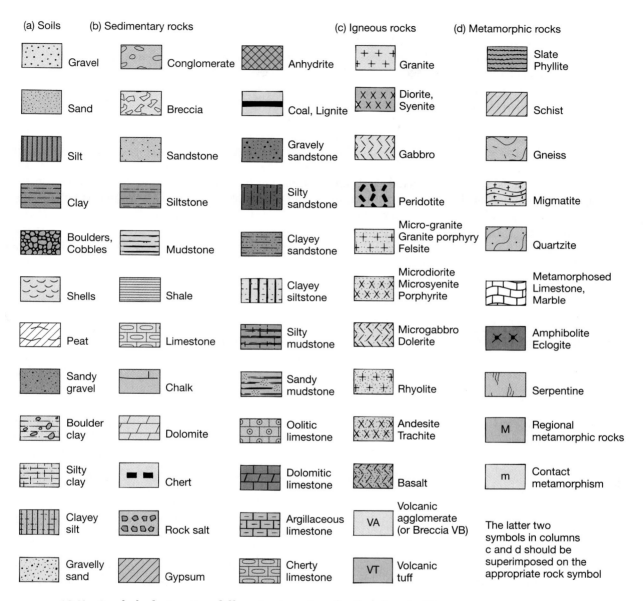

FIGURE 19.5 Symbols designating different categories of soil and rock. (From Attewell, P.B., and Farmer, I.W., 1976, *Principles of Engineering Geology*, pp. 472–473, Chapman & Hall, London.)

to remain open when unsupported. Rock coring can also be accomplished through a hollow-stem auger drilled to refusal on the bedrock surface.

The procedure for rock coring below a thick layer of soil is considered in the following discussion. For a geologic section with 80 ft of soil above bedrock, the soil cross section would be sampled using split-spoon sampling likely at 5-ft intervals except for the upper 15 ft or so where 2.5-ft intervals would prevail. In the last soil sample ending at 80 ft the split-spoon sampler would be retrieved by raising the drill rod to the surface. Thirty feet of rod could be held by the slip ring and sand line. If a second sand line were available on the drill rig, another 30 ft of rod could be held intact by that apparatus. Without the second sand line, the drill string would have to be broken down in 10-ft segments until the lower 30-ft section was reached. This would involve five 10-ft segments. With two sand lines, two 10-ft segments would be separated first, then two 30-ft sections would be lifted and held on the slip rings and sand lines of the rig. This is obviously a time-consuming operation in both cases but considerably greater when only one sand line is available.

To core rock at the 80-ft depth the core barrel and core bit replace the split-spoon sampler at the base of the drill rod stem. The drill stem is put back into the hole and the sections are reassembled to reach the 80-ft depth. Typically, a 10-ft-long core barrel is used so that a 10-ft core run is accomplished. On completion of the rock coring, the sample is brought back to the surface by breaking down the drill stem in the same manner described before. After the rock core is removed from the core barrel, the barrel is put back into the hole for another 10-ft run beginning at the 90-ft depth. Obviously, it takes a long time to core a distance of 50 to 100 ft into bedrock under these circumstances. Therefore, for extensive rock coring projects, below thick sections of soil the conventional coring procedure is very time consuming. In fact, coring rock at depths below the ground surface of 50 ft or more becomes much slower because of the travel time of going in and out of the hole with samples.

Rock coring at these greater depths can be accomplished more quickly using wire line drilling rather than the conventional method. For the same example as before, an 80-ft soil section above rock, the hollow-stem auger would again be drilled to refusal on the bedrock surface. For wire line drilling the drill stem consists not of a small-diameter steel rod but of a tube of sufficient diameter that the core barrel can slip through the inside of it. This drill rod stem or tube must extend from the surface downward to the level of drilling, and this tubing for wire line drilling is more expensive than is the small-diameter drill rod used for conventional drilling. After a 10-ft core run is completed, the core barrel (containing the core) is detached from the drill stem and brought to the surface on a wire line that leaves the remaining hardware behind. The core barrel is lifted to the surface quickly by the hoist, the core removed, and

the barrel returned to the bottom of the hole where the coring is resumed.

In addition to much greater core production, the wire line method typically produces better core recovery than does the conventional method. For wire line drilling in highly fractured, weathered, or altered rock a triple barrel is used.

When the core barrel is brought to the surface the core is removed and stored in wooden core boxes. These boxes are typically 5 ft long and made of 0.5-in.-thick wooden planks. The core is placed in the core box in columns according to the order in which it is taken from the core barrel. Wooden blocks are used to mark the points where the individual core runs start and end. Depths are indicated on these blocks so that the cores are properly located in the core box. Core recovery percentages, RQD, and fractures per foot are determined from measurements made of the cores in the core box.

Test Pits and Trenches

Open pits and trenches are a useful means for obtaining both disturbed and undisturbed soil samples from relatively shallow depths. They also provide a way to observe a large exposure of the subsurface in the side walls of the excavation, which is not possible by other exploration methods. Details of soil stratigraphy, for example, alternating layers of different soil textures, can be fully appreciated by such observation to an extent that could not be discerned by split-spoon samples (even taken continuously) or from pushed thin-walled tube sampling (Shelby tubes).

Open test pits can be dug either by hand or more commonly with backhoes or clamshells on excavating equipment. Typically such pits are in the range of 4 ft by 4 ft (1.2 by 1.2 m) to about 6 by 8 ft (2 by 2.5 m) in size. In some soils dewatering and/or bracing would be required to keep the pit open while excavating and sampling. In the past, pits have been dug by hand to depths as great as 100 ft (30 m) particularly in developing nations where hand labor is inexpensive. In other situations hand-dug pits are restricted to areas inaccessible to heavy equipment. Samples are carefully cut from the side walls and properly protected for laboratory testing. Samples for consolidation testing and uniaxial and triaxial shear tests are obtained. Descriptions of the pit should include the four side walls and possibly the bottom as excavation proceeds.

Pits should be braced, ventilated, and covered as conditions warrant for safety purposes. In addition, if pits are dug in or near a footing location, the depth should not extend below the footing, which prevents any disturbance or loosening of the bearing material.

Test trenches are used to expose soil materials along a given line or section. When dug by hand they are usually about 3 ft (1 m) deep, whereas a bulldozer can extend such trenches to depths of 12 to 15 ft (4 to 5 m). Using a dragline, test trenches can reach depths of 20 to 30 ft (6 to 10 m). Clamshells and backhoes are also used to dig test trenches. Samples typically are removed from the side walls but can be taken from the bottom of the trench during excavation. As with test pits the total exposure of soil should be described for the trench. Color photographs provide a permanent record of the soil details of the excavation.

Trenching is particularly suited for exploration of moderately steep slopes but is adaptable to flat terrain as well. The excavation may consist of a single slot trench down the slope face or it may include a series of short trenches spaced at short intervals along the slope. Test trenches are used extensively to locate fault traces in alluvial materials in order to date the time of fault movements.

MISCELLANEOUS CONSIDERATIONS

Several other items related to drilling and sampling also deserve consideration. One method to gain further information from a borehole is through the use of a borehole camera. This equipment consists of a photographic or television camera that is lowered into an N_x size or 4-in.-diameter hole or larger to yield a circular photograph or scan of the side walls. This is used to view stratification, fractures, and cavities in the walls of the borehole. The method works best above the water table, in a pumped down hole, or when the hole can be stabilized by clear water.

Calyx holes are another means for obtaining subsurface information. These are typically large holes with diameters of 36 in. (0.9 m) or more. Chilled shot are fed into the wash water and become partially imbedded in the soft, coring bit. Rock cuttings are removed by circulating water. When a section has been drilled the core is extracted by drilling a small hole in the top of the core, an eye bolt grouted in, and the core pulled from the hole using the hoist on the drill rig. Problems may develop when calyx holes are drilled through cavernous rock because the shot are lost into the openings.

Calyx holes are particularly useful for first-hand observation of the subsurface conditions by viewing the side walls of the hole. A geologist or foundation engineer can be lowered on a bosun's chair into the hole to describe the lithologies, stratification, fractures, and cavities. This may require dewatering of the hole prior to the investigation. Because of the considerable cost of these calyx holes and because the observations of the side walls of the hole are the most rewarding detail, there is a tendency under current conditions to take no actual core but to cut through the rock with a large-diameter roller bit. The cuttings are removed by the wash water.

Exploration drifts and pilot tunnels are sometimes used to obtain direct information about rock conditions. These are typically 4 to 6 ft wide (1.2 to 2 m) and 4 to 8 ft high (1.2 to 2.5 m). The floor (invert) slopes toward the entrance to facilitate drainage and mucking (rock removal). Descriptions of the tunnel are accomplished and some *in situ* tests are performed at selected locations.

Geophysical logging of boreholes is becoming more common in current exploration of major engineering projects. A long-time practice for exploration in the petroleum field, these procedures are now proving helpful in supplying needed detail for engineering studies. Well logging consists of running various measuring devices into the boreholes to determine a number of physical properties of the rock column that has been drilled. These techniques are discussed in Chapter 18 on geophysical techniques.

EXERCISES ON SUBSURFACE INVESTIGATION AND SITE SELECTION

1. What is the distinction between office reconnaissance and field reconnaissance regarding subsurface investigations for an engineering construction project? List the categories of information to review during office reconnaissance.

2. Regarding published information or open file reports, list the maps and other data that should be reviewed prior to performing subsurface investigation.

3. Give the two major purposes for field reconnaissance prior to subsurface investigation. List the primary items to be observed during field reconnaissance.

4. List the typical information that is needed by the drill crew before they can begin subsurface investigation. Discuss the aspects of (a) permission of property owners and (b) location of utilities.

5. What is meant by the term *sounding* when applied to subsurface investigation? What are the limitations of this method?

6. What is the distinction between drilling and sampling? List the eight different methods given for drilling holes. Discuss the hollow-stem auger method.

7. Describe how the standard penetration test (SPT) is accomplished. What is the length of the split-spoon sampler?

8. Two holes are to be drilled to a depth of 50 ft. Both will involve hollow-stem auger drilling and the SPT method.

   (a) In the first hole, SPT tests are to be taken at 5-ft intervals with the bottom of the sample ending at multiples of 5 ft. Draw this in a cross section. How many SPTs are involved? What percent of the boring is sampled in this case?

   (b) In the second hole, SPT tests are to be taken at 2.5-ft intervals with the bottom of the sample ending at multiples of 2.5 ft. Draw this in a cross section. How many SPTs are involved? What percent of the boring is sampled in this situation?

9. What is a pushed tube soil sample? What are its advantages over a SPT? What are its disadvantages, if any? What laboratory tests can be run on these samples?

10. What drilling and sampling procedure would most likely be used to investigate a 25-ft-thick lake bed on which a three-story building was to be constructed?

11. If a standard penetration test yielded the following results: 6/5/7, what would be the N value? If the soil was a sand, what would be the state and relative density? What would these be if the sample was a cohesive clay?

12. Give an example in which probing would be adequate. Give another geologic setting in which it would not be. Explain why.

13. What would be the nature of the reconnaissance for an interstate highway through a mountainous region? Explain in as much detail as possible.

14. What is meant by *rotary coring*? For rock coring, what is the difference between a core bit and a core barrel? What is the diameter of an N_x core? How big is the N_x core hole?

15. What is meant by *core recovery*? Why is it important to have a high core recovery?

16. Distinguish between conventional core drilling and wire line core drilling. What is the major advantage of wire line drilling?

17. What are bore hole logs? Why is it so important that these be made accurately and completely?

18. When logging core how can natural joints in rock be distinguished from fractures that occurred during the drilling process? Which ones are included in determining the RQD? Why?

19. Prepare a drilling log depicting the following information. Include a lithologic section using proper symbols.

    Vertical hole, 31 ft of soil, 20 ft cored into rock, N_x size. XYZ Project, Boring No. 1. Drilled ten days ago.

    Split-spoon samples at 2.5-ft intervals ending on multiples of 2.5 ft. Values were 2/5/4; 3/3/3; 4/3/5; 5/6/7; 6/6/8; 8/7/9; 6/8/8; 12/12/14; 14/14/13; 15/15/16; 15/17/18; 30/30/28.

    Soils interrupted 0 to 2 ft, topsoil; 2 to 7 ft, sand; 7 to 18 ft, silty clay; 18 to 27 ft, clayey silt; 27 to 31 ft, silt with rock fragments.

    Rock intercepted: 31 to 42 ft, brown friable, quartz sandstone; 42 to 51 ft, gray shale.

    Condition of core: 31 to 41 ft, 9.4 ft of core, all pieces greater than 4 in. except for 1 ft, 3 in. total; 41 to 51 ft, 8.9 ft of core, only 6 ft, 4 in. of core consisting of pieces greater than 4 in. long.

    Water on completion and 48 hr later, 12 ft below the ground surface.

20. What is the RQD or Rock Quality Designation? How is it determined? Why is it significant?

21. What drilling equipment would be needed to explore along the centerline of a proposed earth dam where it crosses an existing stream? What samples would be taken and at what depths?

22. If rock coring costs $17 per foot for N_x size conventional drilling for borings to a depth of 40 ft and $25 per foot for depths of 40 to 75 ft, how closely spaced can borings be placed for a 500 ft × 500 ft area if the borings will be 65 ft deep and the drilling bill cannot exceed $15,000? Make a sketch and locate the borings.

23. What is the advantage of test pits and trenches over rock or soil sampling? What are the limitations for pits and trenches? Explain.

24. What is a calyx hole? How large are they? Can they be viewed directly by human inspectors? Why is ground water a concern in that situation?

25. What would be the purpose of driving a pilot tunnel into a dam abutment prior to dam construction? What information could be supplied?

26. What is meant by *geophysical bore hole logging*? Refer to Chapter 18 and the section on geophysical techniques for further information.

Additional Readings

American Association of State Highway Officials, 1967, *Manual on Foundation Investigations*.

Attewell, P.B., and Farmer, I.W., 1976, "Site Investigation," Chap. 7 in *Engineering Geology*, Chapman and Hall, London.

Krynine, D.P., and Judd, W.R., 1957, *Principles of Engineering and Geotechnics*, Chap. 6, McGraw-Hill, New York.

Legget, R.F., and Karrow, P.F., 1983, *Handbook of Geology in Civil Engineering*, McGraw-Hill, New York.

LeRoy, L.W., LeRoy, D.O. and Raese, J.W., Eds., 1977, *Subsurface Geology, Petroleum, Mining, Construction*, 4th ed., Colorado School of Mines, Golden, CO.

Transportation Research Board, 1978, "Landslides Analysis and Control," Special Report 176, Chap. 4, Field Investigation.

U.S. Department of the Interior, Bureau of Reclamation, "Engineering Geology Field Manual."

20

Engineering Geology and Environmental Geology

CHAPTER OUTLINE

Engineering Geology
Environmental Geology

Engineering geology is the field of study in which geologic factors are related to various types and phases of engineering construction. The purpose is to consider the geologic factors in a proper fashion so that an economic and safe engineering structure is assured. As defined in the glossary of the American Geological Institute (AGI), "engineering geology is the application of the geological sciences to engineering practice for the purposes of assuring that the geologic factors affecting the location, design, construction, operation and maintenance of engineering works are recognized and adequately provided for."

Engineering geology had its formal beginning in the United States following the failure of the St. Francis Dam in California in 1928. Analysis of the failure showed that weak geologic materials in the abutment (side wall) of the dam had been overlooked

and geologic input was desperately needed in such important construction projects. Today nearly all major construction endeavors include engineering geology input from the planning through construction phases of the work.

Environmental geology as a designated specialty developed in the late 1960s. A concern for the environment affected the geological sciences in much the same way as it did other fields of study, and a specific focus on the effects of humans on the geologic environment became a key issue. Discussed in forums at geological society meetings, the subject soon became an area of interest for many geologists. The definition by P. T. Flawn (1970) describes the essence of the subject, "Environmental geology is a branch of ecology . . . that . . . deals with relationships between man and his geologic habitat; it is concerned with the problems that people have in using the earth—and the reaction of the earth to that use." A related term is *urban geology,* which refers to environmental geology of urban or heavily populated areas. In another sense, this can be considered the geology of cities, a title used for a text by Legget (1973). Currently a series of papers on the engineering geology of cities is being published in the *Bulletin of the Association of Engineering Geologists.* (See the Additional Reading list at the end of this chapter.)

In distinguishing between engineering geology and environmental geology, a few general conclusions can be reached. However, even these would not be readily accepted by all concerned. Engineering geology, because it is the longer established field, tends to be somewhat more analytical and detail-oriented in its approach to a subject, whereas environmental geology deals with subjects in a more general and perhaps more descriptive way. Environmental geology has the reaction of the Earth to human use at the heart of the subject, whereas in engineering geology, environmental concerns are one of many parallel details that must be considered relative to geology and construction. A number of geologic specialties are included within environmental geology and one can maintain a primary specialty (such as glacial geology) and still be considered as an environmental geologist. Engineering geology comprises a major specialty involving a number of subspecialties or related areas such as hydrogeology and soils engineering. As a final point, environmental geology encompasses a somewhat larger subject matter: geologic hazards, environmental health, mineral depletion, current land use, and land use planning to name several key subjects. Engineering geology is more tightly focused but more detailed in geology applied to all facets of engineering construction. Also environmental geology may include both the biological and physical factors of nature, whereas engineering geology is concerned primarily with physical geology.

ENGINEERING GEOLOGY

Many features of physical geology become significant in the field of engineering geology. In this text, the most important and most commonly involved subjects of physical geology have been discussed with the specific purpose of showing how they relate to engineering. From minerals and rocks to subsurface investigation and earthquakes, the physical features of geology in engineering have been detailed.

A course in "Geology Applied to Engineering" is the first step in the recognition process by which an engineering student or professional becomes acquainted with the relationship between the two subjects. When the geologic fundamentals have been mastered and the link between geology and engineering is appreciated, the study of engineering geology becomes the next step in the process.

Engineering geology is a specialty within the science of geology. The specialist is a geologist founded in the basics of the subject who is also well aware, through training and experience, of the engineering aspects directly related to geology in construction. These topics would likely include hydraulics and hydrology, soil mechanics, engineering materials, and possibly rock mechanics. In past years a working knowledge of construction obtained through on-the-job experience was sufficient to develop a geologist into an engineering geologist. Today it is more efficient and productive to obtain much of this knowledge through course work in engineering, primarily in the civil engineering specialty and in specific engineering geology studies as well. Engineering and geology course work shape the student into an engineering geologist.

Geological engineering is a sister field of endeavor to engineering geology but here the student is an engineer well versed in geological knowledge rather than vice versa (as is the case in engineering geology). More engineering courses are required and somewhat less geology on the bachelor's degree level. The distinction between engineer and scientist persists in this case; the engineer is involved in design of engineering works and is responsible for their long-term performance. An engineering geologist supplies geologic input for design purposes and ensures that geologic factors are properly considered through all construction phases, but typically does not perform a design function.

For many people the distinction between geological engineer and engineering geologist becomes less clear-cut if they have advanced training in both geology and engineering. In such cases the engineering geologist through additional study and experience likely obtains registration as a professional engineer.

Geological engineering can have three distinct major specialties: petroleum, mining, or engineering construction. For mining and construction the engineering background is similar, calling heavily on civil engineering and on some mining engineering expertise. Petroleum aspects of geological engineering are somewhat different from the others; more involvement with subsurface reservoirs, liquid and gaseous flow in porous media, and oil well systems are prime considerations. Geological engineers in mining and construction are more closely aligned to civil engineering than to oil and gas production.

The concerns of engineering geology are those of construction with and on earth materials, including the origin of these materials in relation to their behavior. The subjects of physical geology involved in engineering geology comprise the list shown in Table 20.1. The applications of geology to engineering can be considered in regard to specific construction projects. Some of the more common examples are listed in Table 20.2 including a few specific activities for each.

The case history approach is sometimes used in the study of engineering geology. In this method of instruction specific construction projects are reviewed in an attempt to evaluate the procedures used for a specific set of conditions. Various phases of the work can be studied: exploration, design, construction, and various subdivisions of each. Care should be taken in this approach to ensure that the specifics of engineering geology are covered rather than broad generalities. One useful method is to discuss the construction procedures in detail first and then examine the case history. For example, for a case history on grouting, discuss types of grouts, mixtures, equipment, procedures and limitations, then review an article on curtain and foundation grouting for a specific dam.

In many ways engineering geology encompasses the elusive part of construction, which involves the uncertainties of geologic materials. Examples are bedding planes and joints yielding rock-block failures in an excavation, seeps in clayey deposits caused by sand layers to yield water into an opening, deeply weathered soil or pinnacles of rock when another condition is expected, solution cavities in limestone in

TABLE 20.1 Physical Geology Subjects Related to Engineering Geology

1. Rock description and identification
2. Engineering properties of rocks, materials for construction
3. Rock weathering and soil development
4. Map reading, both topographic and geologic
5. Structural aspects—bedding planes, joints, and faults
6. Mass movement and landslides
7. Running water—erosion, flood effects, water impoundment
8. Ground water—control during construction, water supply, pollution, subsidence, slope stability
9. Shoreline erosion and protection—engineering structures, jetties, groins, breakwaters, sea walls; finding appropriate rocks for shoreline protection such as riprap and armour stone
10. Earthquakes, seismic risk, slope failure, liquefaction
11. Glacial deposits—construction materials, ground-water supply
12. Arid environments, wind, accelerated erosion
13. Subsurface geology, conditions of stress at depth

a building foundation area or a methane-bearing zone in a shale tunnel. Many times it is the most challenging aspects of construction that involve engineering geology so it should be no mystery that knowledge of geology and of engineering is needed to anticipate such occurrences and to react in the best way possible when they occur.

Engineering Geology Construction Examples

Table 20.2 provides a list of construction projects related to engineering geology. Each of these would require a chapter or a major section for discussion to explore the subject fully. Such detail is beyond the scope of this introductory text and cannot be provided here. As an indication of the nature of the subject matter, however, three construction categories are discussed in the following sections. Included are highways, tunnels, and dams.

TABLE 20.2 Construction Projects Related to Engineering Geology

1. *Highways:* location, alignment, classification of soil and rock borrow, ground-water control, slope stability, materials inventory, and selection
2. *Dams:* site selection, materials selection, subsurface investigation, geologic mapping, foundation support, ground-water control, grouting, instrumentation
3. *Building foundations:* site selection, subsurface investigation, geologic correlation, rock or soil strength determination, ground water, construction monitoring
4. *Tunnels:* Regional geology, site selection, subsurface investigation, geologic correlation, strength determination, estimation of rock behavior during tunneling, blasting, ground-water control, support recommendations, geologic mapping during construction
5. *Rivers:* erosion of banks, dredging, scour effects, channel control
6. *Surface mining:* subsurface investigation, geologic correlation, ground water, slope stability, excavation aspects, blasting, stripping procedures, reclamation
7. *Underground openings and mines:* subsurface investigation, rock strength, stress conditions, ground water, orientation of openings, extraction ratios, subsidence
8. *Sanitary landfills:* site selection, subsurface investigation, geologic correlation, ground water, soil and rock permeabilities, baseline water quality determination
9. *Environmental geology and land use:* preparation of large-scale geologic maps and interpretive maps, determination of current land use, development of acceptable future land uses based on geologic, economic, and to some extent political restraints
10. *Evaluation of materials of construction:* materials inventory and selection, evaluation of masonry construction materials
11. *Slope protection:* along shorelines and lakes, beach erosion, engineering structures
12. *Nuclear and fossil-fuel power plants:* subsurface investigation, geologic correlation, structural geology analysis and faulting evaluation, determine seismicity of site, ground water, and other aspects of building foundations
13. *Blasting and vibration effects on engineering structures:* particle velocity determination and blast design aspects
14. *Environmental impact:* measure baseline information, estimate the changes that will occur if a structure is added

Highways

A brief outline of engineering geology considerations for highway construction is provided as item 1 in Table 20.2. This shows the nature of the work performed by engineering geologists working on highway projects. It is presented in more detail in Table 20.3.

Highway location is dictated in part by the nature of geologic materials present in the area to be traversed. In glaciated regions, care is taken to avoid compressible soils and other poor foundation materials. Existing maps are used to select the highway alignment. River crossing locations are carefully selected and road distances are kept to a minimum compatible with the other constraints. In bedrock areas, attempts are made to minimize the volume of cuts and fills, thereby reducing the amount of earth and rock to be moved. Where troublesome rock

formations exist, road grades and elevations are selected that minimize the exposure of the problem rock units. Swelling shales and landslide-prone rocks are examples of rock formations to be avoided or minimized.

Exploration borings are performed along the proposed highway centerline and for bridge foundations. Standard penetration tests are taken in the soil and rock corings performed in bedrock. Borings are made in both the cut and fill sections. Borings should extend to a minimum depth of 6 ft below the road grade in cut sections. If a large embankment is to be placed, borings should extend to a depth of two-thirds the height of the embankment or to bedrock, whichever is less. As with the cut section, a 6-ft minimum depth below the grade line is required. Atterberg limit and grain size distribution tests are performed

TABLE 20.3 Highway Geology Considerations

1. Highway alignment, locations of right-of-way for the proposed construction
2. Subsurface exploration along highway centerline and for bridge foundations
3. Classification of materials for excavation, rock versus common borrow (soil)
4. Cut and fill volumes determined to minimize the need for offsite borrow pits or rock waste areas; volume changes in both soil and rock from the cut to the fill are estimated
5. Overbreak of rock cuts minimized by presplitting of rock faces
6. Recommend angle of back slope (rock cut slope) based on rock conditions
7. Groundwater aspects related to construction
8. Evaluation of landslide-prone areas
9. Recognition of compressible soil materials
10. Legal aspects, highway effects on adjacent landowners
11. Construction materials, location and inventory
12. Increasing problems with highway location because of human activities

on the soils. Proctor density tests are performed on borrow pit soils used to construct the embankment.

Subsurface investigation for bridge foundations is performed to provide the needed details for the bridge footing design. Appropriate samples are obtained for laboratory testing.

The distinction of rock versus soil or common borrow excavation must be established. This requires that the soil-bedrock interface be determined. The cost of soil excavation is in the range of less than $1 to a few dollars per cubic yard, whereas rock excavation is typically $100 per cubic yard to several times that amount. The soil-rock interface is established in the drilling program and this information may be augmented using engineering geophysics.

During the design process, attempts are made to balance cuts and fills so that offsite borrow or rock waste areas are kept to a minimum. Soil borrow pits are common for interstate highway construction because the grade is typically elevated above low-lying areas to prevent flooding and encourage runoff. Careful calculations are required to ensure that the volumes of cuts and of fills (embankments) are accu-

rately determined. This occurs because rock increases in volume from the cut to the fill, whereas soils decrease in volume during this process. In general, clayey soils tend to decrease more during the compaction process than do sandy soils. By contrast, massive rocks such as well-cemented sandstones, limestones, and hard igneous or metamorphic rocks increase considerably in volume from cut to fill area, whereas weaker rocks such as shales have a smaller volume increase. Rock has a higher volume in the fill because the rock pieces have void spaces between them. This is illustrated by an example problem.

EXAMPLE PROBLEM

A cut section is located adjacent to a fill section for an interstate highway project. In the cut, 15,000 yd^3 of clay and 12,000 yd^3 of sand will be excavated, along with 20,000 yd^3 of hard shale and 13,000 yd^3 of limestone. The shrinkage factor for the clay is 15% and 5% for the sand; the swell factor for the hard shale is 10% and for the limestone is 20%. What will be the volume of the fill constructed from these materials? What volume change from the cut to the fill will occur?

ANSWER

Clay volume in the cut:
15,000 × 0.85 = 12,750   Volume in fill
Sand volume in the cut:
12,000 × 0.95 = 11,400   Volume in fill
Hard shale volume in the cut:
20,000 × 1.10 = 22,000   Volume in fill
Limestone volume in the cut:
$\frac{13,000}{60,000}$ × 1.20 = $\frac{15,600}{61,750}$ Volume in fill
Total

Volume increase is

$$61,750 - 60,000 = 1,750 \text{ yd}^3 \text{ or } \frac{1,750}{60,000}$$

$$= 2.9\% \text{ increase}$$

If the fill could accommodate only 60,000 yd^3 the 1750 yd^3 would have to be disposed of in an offsite waste pile.

Overbreak is the extra volume of rock blasted from a rock cut in excess of that designated in the slope design. It occurs during production blasting in highly jointed or fractured rock. Overbreak can be minimized by performing presplitting prior to production blasting. This consists of drilling closely spaced boreholes 6 to 10 ft apart along the designated boundary of the rock cut. These holes are loaded lightly with explosives and blasted prior to the larger, production blasting. A plane of fracture is established at the designated rock cut boundary so that no overbreak occurs during the production blasting operation.

In horizontal sedimentary rocks, the angle of the cut slope or back slope is largely dependent on the durability of the rock. Massive, durable rock such as limestone, dolomite, and well-cemented sandstone can be cut nearly vertical, ⅙ to 1, for example, whereas soft shale may be cut as shallow as a 1 to 1 slope (45°). Intermediate conditions such as interbedded shale and limestone may be cut on a ½ to 1 slope. There is a major advantage in making the rock slope as steep as safety permits. Steeper slopes require a smaller volume of rock removal and a narrower width right-of-way at the top of the cut. Both of these considerations reduce the cost involved.

Ground-water problems related to highway construction are of two types: 1) ground-water seepage into cut sections and at the base of fill sections and 2) ground-water supplies adjacent to the highway. Seepage into cut sections is controlled by installing underdrains below the pavement to intercept the water. This prevents seepage onto the highway and subsequent icing of the road during freezing conditions. Springs are intercepted at the base of the fills and carried by drains under the road. This prevents saturation of the embankment, which could lead to possible settlement problems.

The water supplies of adjacent landowners may be affected by road cuts that intercept the ground-water table. The water table may be locally depressed in this case. Effects on neighboring wells must be properly documented by a preconstruction and postconstruction survey of these wells. Property owners affected by such road cuts should be compensated by supplying them with a replacement water supply.

Water supplies are also needed at regular intervals along the roadside for rest areas for interstate travel-

ers. The water supply demand is considerable for such facilities and a major water supply, in the hundreds of gallons per minute range, is required. For this reason, locating a water supply for a large rest area can be a major geologic challenge.

Engineering geologists evaluate landslide-prone areas along the proposed highway right-of-way. Geologic information is provided for slope stability analysis through a review of published information, aerial photograph interpretation, field investigation, and in some cases subsurface drilling and sampling. Compressible soils are identified during the planning stages and recommendations made to minimize their effects on the completed highway. Avoidance, excavation, or settlement prior to construction are evaluated.

Highway geologists may be called on to provide expert opinions regarding domestic water supplies including wells, springs, and spring-fed ponds. Evaluation of the effects of highway construction on oil well fields, shale pits, stone quarries, and other facilities are commonly required. Testimony as an expert witness in court or in hearings involving these situations may also be necessary.

A primary job of many highway geologists involves the location and inventory of construction materials for highway construction. Aggregates consisting of gravel, crushed stone, or possibly blast furnace slag comprise the major constituents of a highway pavement. Figure 20.1 shows an idealized cross section of a highway. Aggregates are used in all three portions: pavement, base course, and subbase course.

The subgrade consists of the compacted soil at the base of the highway section with the grade line marking the top of the subgrade. Compacted aggregate forms the base and subbase courses. Subbase material typically is of lower quality than the base course. It is used in the lower portion of flexible pavements where generally a greater total thickness is required than for concrete pavements. High-quality aggregate is used in both the concrete and asphaltic pavements.

Highway geologists and materials engineers are responsible for maintaining an inventory of aggregate sources including gravel pits and stone quarries. This includes the details of the stratigraphy and nature of the pits and quarries as related to the engineering quality of the aggregate. In the western United

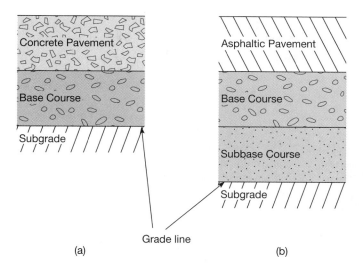

FIGURE 20.1 Schematic diagram of highway pavements: (a) rigid (concrete) pavement and (b) flexible (asphaltic) pavement.

States, gravel supplies vary greatly relative to rock composition and overall quality. Because transportation of aggregates is a major expense, quality aggregates must be located as close as possible to the construction site.

Challenging problems to highway construction have been generated by human activity. Today highways must cross areas of abandoned or active underground mines, former strip mines, landfill areas, subsiding oil well fields, sliding hillsides, compressible floodplain soils, and other challenging situations. Highway geologists are called on to decipher the details of surface and subsurface conditions that affect highway construction in these complex areas. Avoidance of poor areas for highway construction is no longer a possible choice for many projects today.

Tunnels

A brief discussion of tunnels in rock is presented here to provide an introduction to this subject. A much more extensive presentation can be found in advanced books on engineering geology, applied rock mechanics, or tunnel engineering. An overview of the geology of tunnels is provided in Table 20.2 under item 4. This forms the basis for the following discussion.

Site selection and subsurface investigation for rock tunnels can be an extensive endeavor. Tunneling through poor-quality, weak rock is extremely expensive and time consuming, so that great benefits are obtained by positioning tunnels in sound rock.

The investigation is accomplished by geologic field studies and extensive rock coring along the proposed tunnel alignment. Rock cores are described, water conditions noted, the RQD is determined, and selective rock strength tests are performed on rock samples in the laboratory. Geologic mapping is also performed during tunnel construction so that any future problems can be referenced to the geologic details of the existing rock.

Much of a major rock tunnel is commonly located hundreds to thousands of feet below the ground surface. The roof of the tunnel does not, however, typically have to support the total overburden stress. This is because the load is redistributed around the tunnel by a condition called *arching*. Only the load of the immediate roof must be supported by the tunnel, and the stronger the rock the smaller the rock load. The concept of arching is shown in Figure 20.2. Shown also are the dimensions B for the tunnel width, H_t for the tunnel height, H_p for the height of the rock load, and H for the total overburden height. As with most rock tunnels sections $H_p << H$.

Terzaghi in 1943 presented a method for estimating the tunnel loads based on a description of the

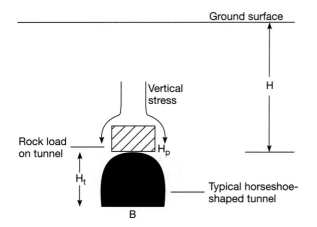

Arching and rock loads on tunnels.

ANSWER

$$H_p = 0.3B = 0.3(12 \text{ ft}) = 3.6 \text{ ft}$$

$$P = H_p \gamma_{rock} = 3.6(165) = 594 \text{ lb/ft}^2$$

This load would need to be carried by the supports placed in the tunnel. If for this tunnel, the following relationship prevailed:

$$H_p = 1.1(B + H_t)$$

$$H_p = 1.1(12 + 16) = 41.8 \text{ ft}$$

Then

$$P = H_p \gamma_{rock} = 41.8(165) = 6900 \text{ lb/ft}^2$$

rock encountered in tunneling. As stated, the better the rock quality, the lower the rock load. The height of the rock load H_p, was represented as a simple function of the width, or width plus height of the tunnel. The following equations show this relationship:

$$H_p = fB$$

for very sound rock, with f varying from 0 to 0.5, and

$$H_p = f(B + H_t)$$

with f varying from 0.35 to 1.1 for most, weaker rocks. For worst case situations, $H_p = H$, the total overburden height. Rock descriptions were provided for different rock conditions that occur in tunnels and f values suggested for each category. This level of detail is beyond the scope of this discussion and is not provided here. An example problem is presented to illustrate the concept of rock load.

EXAMPLE PROBLEM

A horseshoe-shaped tunnel, 12 ft wide and 16 ft high, is driven through a fairly massive rock with a unit weight of 165 lb/ft³. The f factor of Terzaghi yields the following relationship:

$$H_p = 0.3B$$

What is the rock load that must be carried by the tunnel roof?

An important consideration in tunnel construction is known as the *standup time* or *bridge action time*. This is the amount of time that rock (or soil) will stand intact after tunnel mining before it begins to break apart and fall into the tunnel. As would be expected, sound rock has a longer standup time than does poor-quality rock. Obviously, some amount of time is required after mining to erect tunnel supports, so that a rock with little or no standup time is very challenging to tunnel through.

Other techniques for estimating rock loads in tunnels have been developed in recent years. These are based in part on rock and rock mass properties typically determined from rock core samples. Measurements made in the tunnel during construction form the basis for evaluation of the actual loads. These include the use of load cells under the steel supports and borehole extensometers installed in the roof above the tunnel. This latter technique determines the dimensions of the rock load (diagonal symbol) shown in Figure 20.2.

A final method used currently is the New Austrian Tunneling Method (NATM). In this method rock loads in the next tunnel section to be excavated are estimated based on measurements and description of the material currently being mined. Adjustment of tunnel supports is made continuously. Lighter lattice girder supports are commonly used.

Consideration of the major geologic hazards in tunneling is helpful in the understanding of rock tunnels. These are listed as follows:

1. Sudden increase in rock load as the tunnel advances

2. Sudden reduction in standup time as the tunnel advances
3. Rapid inflow of ground water
4. Presence of harmful gases

If the rock load increases suddenly when a new section of the tunnel is mined, the operation for supplying tunnel supports is greatly affected. A brief discussion of tunnel supports is provided later in this section.

A sudden reduction in the standup time for the excavated rock also greatly affects the tunnel construction process. This typically is associated with the sudden increase in rock load and is a consequence of a major change in the rock quality.

A rapid inflow of water into the tunnel can be disastrous. It must be pumped from the workings and requires the proper equipment to do so. Heavy water flows are dangerous to tunnel workers and large volume flows may lead to casualties. Rock units with open fissures and conduits such as cavernous limestones or porous basalts are possible candidates for this problem. Geologic studies must provide the knowledge that such conditions exist so that water can be drained or controlled before tunneling occurs.

The presence of harmful gas in a tunnel is a very dangerous condition and can cause catastrophic effects. Harmful gases include CO_2, CO, H_2S, SO_2, and methane. Methane is also an explosive gas when mixed in the proper proportion with the air in the tunnel. Methane, CO_2, and CO can occur in organic-rich soils and rocks, including black shales and coal-bearing strata. H_2S is a product of black shales and of volcanic rocks. SO_2 is also present in volcanic rocks. Of the various rock types, Holocene volcanics (less than 10,000 years old) can be the most treacherous rocks for tunneling because all four of the major geologic hazards have been known to occur in them.

The two basic procedures for advancing a tunnel in rock are 1) the drill and blast method and 2) the tunnel boring machine method. The drill and blast procedure is sometimes termed the conventional method because it preceded the more recent TBM method.

The drill and blast method is a six-step operation: 1) drilling blast holes, 2) loading these holes with explosives, 3) shooting or blasting the explosives, 4) ventilating gases from the explosion, 5) mucking or removal of the blasted rock, and 6) erecting tunnel supports. The standup time of the rock must be greater than the time elapsed between steps 3 and 6.

DRILL AND BLAST METHOD   The common explosive used in tunnel excavation is ANFO, ammonium nitrate and fuel oil. These ingredients are inert in themselves, which is a safety advantage over the use of dynamite or similar explosives. Blasting caps are used to detonate the ANFO and a delay shooting array is used. Delay shooting is used in lieu of shooting all the explosives simultaneously and involves some millisecond (thousandths of a second) delays between the explosion of different groups of the loaded holes. This allows for a more efficient breaking of the rock, reduces overbreak, and spreads the amplitude of the wave form from the blast, relative to time. This is to prevent blast damage to nearby structures. ANFO can be used only in dry blast holes, or the mixing of ammonium nitrate and fuel oil in proper proportions is not possible. In rocks that produce significant amounts of seepage, dynamite or a similar explosive must be substituted for the ANFO.

The types of temporary supports for tunnels include 1) steel sets, 2) roof bolts, 3) bolt straps and mesh, 4) shotcrete, and 5) spiling. A permanent concrete lining is typically placed over the temporary supports during a later stage of tunnel construction.

Steel sets are I beams or wide flange beams that are deformed into the shape of the tunnel. Purchased from the steel mill prior to construction, they are transported to the job site and then underground, where they are assembled to support the rock load. Considerable labor is involved in erecting large steel sets in a tunnel. Obviously, the larger the rock load, the more steel support required.

Roof bolts are placed in holes drilled into the roof of the tunnel. They typically are 3 ft long but longer bolts can be required. Bolts are tensioned after placement to provide greater support. Several types of bolts and anchorages are used but this level of detail is not considered here. Bolts are typically placed at regular intervals, such as at 3- or 4-ft centers. Bolt straps and mesh can be placed against the roof to hold small, loose rocks in place.

Shotcrete is a relatively dry concrete containing a small (0.5-in.) top size aggregate, which is sprayed on the roof and sides of a tunnel to provide strength by holding surface materials together. Layers of shot-crete are 6 in. or more in thickness. Rock bolts can be placed through the shotcrete to anchor it into the rock above.

Spiling consists of concrete reinforcing bars (deformed bars or rebars) and larger pieces of steel that are driven at an upward angle into the roof of a tunnel prior to excavation. This method is used in poor-quality rock with a very short standup time.

Tunnel Boring Machines The second method for advancing a rock tunnel involves a tunnel boring machine (TBM). A TBM consists of a rotating cutter head system that is rotated against the tunnel face. Rock is sliced and sheared off as the TBM advances. The cutter face is rotated at a few rpms and the muck collected on a conveyor system, which transports it down the tunnel into a muck hauling system. Rock bolts are commonly placed in the roof as the TBM moves forward to expose the newly cut surface.

TBMs up to 35 ft in diameter have been employed to excavate rock tunnels. TBMs are typically most efficient on large tunneling jobs involving tunnels of a mile or more in length. The high cost of setting up a TBM system requires that the project be large enough to amortize this cost and that of the expensive equipment involved.

TBMs cut circular tunnels rather than the tradi-tional horseshoe shape. They also disturb the rock much less than does the drilling and blasting method. Typically, the amount of temporary support in a TBM tunnel is considerably less than that of a drill and blast tunnel of comparable size with the same initial rock conditions. This is, however, somewhat mislead-ing because TBMs would not be used in some of the poor rock materials that provide heavy loads when the drill and blast method is used. The rate of progress with a TBM can equal or exceed that accomplished by drilling and blasting.

As suggested earlier, not all rocks are suitable for TBM excavation. Poor rock conditions lead to too many problems using a TBM. Also very hard, strong rocks can not be economically mined using a TBM because the bit wear on the rock cutters is so extensive. This causes undue amounts of downtime related to repairs and replacements.

Rock with a Mohs' hardness of 6 and above cause greater amounts of cutter bit wear. Also rocks with a high unconfined compressive strength, above 15,000 to 20,000 psi, also cause extensive wear. Because of this situation, the softer, low-strength but intact rocks are the best candidates for tunneling by a TBM. The geologic conditions should be as uniform as possible; the same rock unit with similar overall conditions yields the best situation for the systematic work of the TBM. A TBM is shown in Figure 20.3.

FIGURE **20.3** Tunnel boring machine.

Dams

Dams are engineering structures, usually built across stream channels or stream valleys, that impound water at a considerably higher elevation on the upstream side. Typically they are constructed from large volumes of materials to hold back a much larger volume of water, which in turn possesses an enormous destructive power if released suddenly. Dams are one of the few engineering structures with a capacity for destruction that greatly extends beyond their own constructed area. A simple comparison to highways, buildings, tunnels, mines, and even sanitary landfills points to this conclusion.

The role of a dam is to accept the pressure of the impounded water plus lesser associated loads such as ice, wind, and silt, and transfer the pressure to its base (the foundation) and its side walls (the abutments). Dams are dependent on environmental conditions, particularly on the site geology, perhaps more than any other engineering structure.

The purposes for dam construction are flood control, water supply, irrigation, hydroelectric power generation, and recreation. Most major dams serve more than one of these functions.

In the eastern United States, the U.S. Army Corps of Engineers is responsible for construction of most large dams. Electric utility companies, the Soil Conservation Service of the U.S. Department of Agriculture, and the Tennessee Valley Authority (TVA) have constructed sizable dams as well. In the western United States, the U.S. Bureau of Reclamation is responsible for constructing dams that provide irrigation for the arid and semiarid regions. State governments and electric utility companies also have constructed large dams in the western United States.

Dams built as permanent installations are classified according to their materials of construction. These are concrete dams, earth-fill dams, and rock-fill dams. Temporary dams, of much smaller size may be built of timber or of steel sheet pile.

Site geology and the configuration of the valley greatly influence which of the three major types of dams will be selected. Concrete dams require stronger foundations than do earth-fill or rock-fill dams. Also because of the higher unit cost of concrete, these dams are more appropriate in narrow valleys. Earth-fill and rock-fill dams are selected for wide valleys that commonly extend a thousand or more feet in length. Earth-fill dams are constructed where sufficient, fine textured soils are available. In some mountainous areas where clays are in short supply and coarse-sized deposits prevail, a rock-fill dam may prove to be the most economical type to construct.

CONCRETE DAMS   Concrete dams consist of three basic varieties: gravity dams, buttress dams, and arch dams. Gravity dams are roughly trapezoidal in cross section, which approaches a triangular shape because of its much greater width at the base than at the crest (top) of the dam. Sound rock is the preferred foundation material for gravity dams but they can be constructed on fractured or variable rock units or in some cases on alluvial fill in the stream valley. In the latter case, potential leakage below the dam is a major problem to be solved. Also, intact abutments are required to prevent seepage around the dam, but the strength requirements are less stringent than that required for arch dams. A gravity dam is shown as Figure 20.4.

Buttress dams consist of a sloping upstream slab of reinforced concrete and buttresses or vertical walls on the downstream side that transmit the water load from the slab to the foundation. The buttresses are vertical walls perpendicular to the dam axis with sloping surfaces in the downstream direction. A buttress dam is shown in Figure 20.5.

Buttress dams have the advantage of a smaller volume of concrete than a gravity dam and a decreased volume of foundation excavation because of the smaller size of foundation footings. The area between buttresses can also be used to locate the outlet works and power house. The various parts of a dam are discussed in a following section.

FIGURE 20.4   Gravity concrete dam.

FIGURE **20.5** Buttress concrete dam.

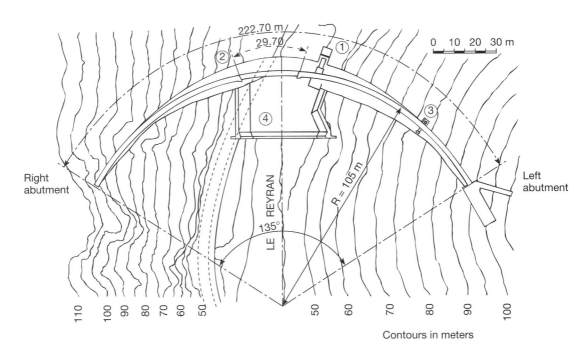

FIGURE **20.6** Schematic views, Malpasset thin arch dam.

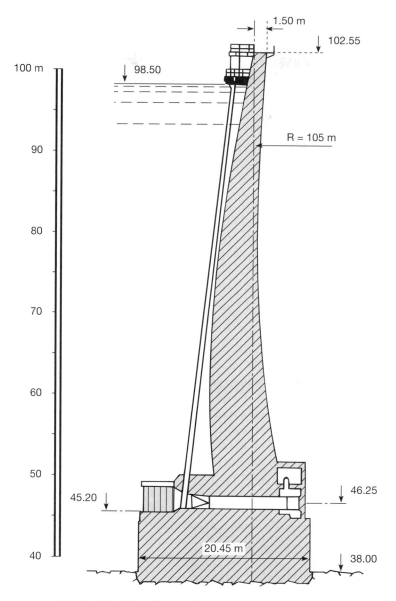

FIGURE **20.6** Continued

Buttress dams typically require more labor than gravity dams because of the added detail of the buttress configuration. Hence there is a trade-off in volume of concrete (less for buttress dams) versus labor (more for buttress dams) that must be weighed in the dam selection process.

Buttress dams also require a good foundation because buttresses transmit greater pressures to the foundation. The space between buttresses is not loaded, so in poorer foundations a buttress may punch through the rock. Gravity dams by contrast are not subject to this concentrated loading condition. In general, a buttress dam requires a sounder foundation than a gravity dam. However, like gravity dams the abutment requirements are less stringent than that needed for arch dams.

Arch dams are composed of a curved concrete wall in plan view with the convex face pointing upstream. A schematic view of an arch dam is shown in Figure 20.6. Part of the water pressure is transmitted to the rock abutments by the arch action so that strong abutments are required to take this thrust. Famous examples of arch dams are the Vaiont Dam in Italy, the Malpasset Dam in France, and the Hoover and Glen Canyon Dams on the Colorado River. The Glen Canyon Dam is shown in Figure 20.7. Recall that the Vaiont reservoir was emptied in 1963 by a major rockslide that caused the dam to be overtopped by a wave of water 330 ft (100 m) high. Two thousand to 3000 people downstream were killed but the dam did not collapse. The reservoir has not been allowed to refill since this failure.

Figure 20.6 shows the plan view (a) and cross section view (b) of the Malpasset Dam, built between 1950 and 1952 on the Reyran River in southern France, only 15 miles upstream from the Mediterranean Sea. As shown in Figure 20.6 (b), the dam was approximately 200 feet high above the foundation, with a free spillway and a spillway apron downstream, plus a dewatering gate and a water intake for irrigation supply. It was not a hydroelectric dam.

On December 2, 1959, during its first filling, the dam failed, and approximately 340 people downstream were killed. Failure of the dam is attributed to excessive uplift pressure from water flowing under the left side of the dam through a fault zone located 100 feet below the dam foundation in a contorted gneiss. This fault zone was hydraulicly connected to a fracture zone or second fault zone below the reservoir, so that the total reservoir pressure (more than

140 feet of water) was applied upward on the foundation and blocks of rock were displaced more than 30 inches. This caused massive failure of the left side of the dam. In the subsurface investigation prior to construction these faults were not detected. However, it is questionable that exploration, no matter how thorough, would locate such faults, particularly the one upstream. The lesson learned from this catastrophe is of the necessity to provide foundation drainage for arch dams to relieve excessive uplift pressures.

Arch dams can be further subdivided based on the ratio of base thickness to height. The following relationship is used

Description	Base Thickness / Height
Thick arch	≥ 0.3
Medium thick arch	0.2 to 0.3
Thin arch	< 0.2

A sample calculation can be used to illustrate this. How is the arch dam shown in Figure 20.6 classified according to the preceding relationship?

$$\frac{\text{Base thickness}}{\text{Height}} = \frac{33.44}{180} = 0.18$$

This would qualify as a thin arch dam.

PARTS OF CONCRETE DAMS Terminology concerning various parts of a dam is discussed in this section. It applies mainly to concrete dams and to concrete overflow spillways for earth-fill and rock-fill dams. A schematic cross section of a dam is shown in Figure 20.8.

The *heel* is the upstream base of the dam in contact with the foundation, whereas the *toe* is its downstream counterpart. The *crest* forms the top of the dam, over which a road or walkway is commonly placed. *Freeboard* is the vertical distance between the maximum pool level and the top of the dam. The *axis* of the dam is an imaginary line through the center of the crest extending from one abutment to the other. *Galleries* are openings through the concrete dam, running longitudinally and located just above the dam foundation. They are used to drain seepage water from the face or the foundation, and to provide

FIGURE 20.7 Glen Canyon Dam, Utah. A thin arch concrete dam.

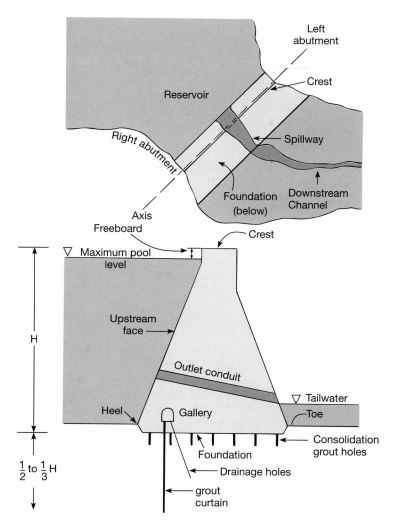

FIGURE 20.8 Schematic cross section of a concrete gravity dam showing various parts of the structure.

openings within the dam to drill grouting and drainage holes and allow access to equipment or instrumentation inside the dam. *Tail water* is the water at the downstream base of the dam discharged from the spillway, outlet works, or power house. The outlet works shown in Figure 20.8 is the conduit type, extended through the dam. An outlet works is used to divert water for irrigation, water supply, or power generation. Spillways, by contrast, are used to convey floodwater from the upstream to the downstream side of the dam where it is released as tail water.

Spillways can be of several types, with the normal or overflow spillway illustrated in Figure 20.8. In this case, water flows over the crest of the dam and is carried downward by a channel or chute, as illustrated in Figure 20.9. This spillway consists of an open channel with a rectangular cross section. Various parts of the normal spillway are labeled in Figure 20.9. The abutments of the spillway and crest are shown, along with the chute over which the water flows down the dam, and the stilling basin with its dentates or energy dissipaters. The concrete walls forming the sides of the chute are called *training walls.*

In a side-channel spillway, water flows from the

FIGURE **20.9**   A chute or normal spillway, dam under
construction.

reservoir along a pathway that is perpendicular or at an angle to the dam axis. It flows around the dam in an open channel through the abutment in an excavation cut perpendicular to the axis. Earth-fill and rock-fill dams commonly employ a side-channel spillway.

Another structure used is the *shaft spillway,* sometimes termed a *morning glory* or *glory hole* spillway. This is illustrated in Figure 20.10. Water is removed through the gates into a vertical or oblique shaft that connects with a gently sloping section that usually extends below rather than through the dam. A variation of the shaft spillway excavated into rock and shown in Figure 20.10 is a tower spillway constructed of concrete, which extends upward into the reservoir.

It is also founded on bedrock and is connected to a gently sloping section that extends below the base of the dam. This type of spillway can be used in conjunction with an earth-fill or rock-fill dam.

Gates are used to vary the amount of water flow allowed to enter the spillway. Gated spillways are known as controlled spillways, whereas an uncontrolled spillway is without gates.

An emergency spillway is an uncontrolled spillway designed to carry floodwater that exceeds the capacity of a normal spillway. They are included in a reservoir design to ensure that an intense rainfall event occurring when the reservoir is nearly full will not allow the dam to be overtopped if the normal spillway capacity

FIGURE **20.10**   Shaft or glory hole spillway located in
reservoir area; photo during construction.

is exceeded. The elevation of the emergency spillway is located slightly above the maximum pool level, within the free board distance, and well below the crest of the dam.

Emergency spillways are located in natural topographic saddles of the reservoir rim and are enlarged by excavation. No concrete construction is included and it is understood that extensive erosion of the base and side walls of the excavation will occur if the spillway is ever used. This would likely be a once in 500- to 1000-year occurrence, but the effects would be catastrophic if instead the dam were overtopped by floodwaters.

An outlet works as mentioned earlier is used to divert water from the reservoir for water distribution or power generation. There are two types of outlets, the conduit type shown in Figure 20.8 or the tunnel type where the water is carried either in lined or unlined tunnels through the abutments. For the power generation case, the outlet works provides water to the hydroelectric power plant. This installation can be aboveground near the toe of the dam or be constructed belowground with an extensive opening excavated into bedrock.

Penstocks are steel pipes of large diameter that carry water under pressure along the ground surface, extending below the dam to the aboveground power station. They supply water to the electrical generators at the pressure of the upstream elevation.

EARTH-FILL DAMS   The design concepts for earth-fill and rock-fill dams are similar. Both involve zones of different soil materials compacted to form a cross section, with the less permeable, lower strength ma-terials in the center to reduce seepage, and the coarser, stronger materials on the outside to increase stability. Because of this similarity, only the earth-fill case is considered in the following discussion.

Earth-fill dams are selected for wide valleys with an absence of firm bedrock. Large volumes of con-struction materials are involved. For example, a 100-ft-high 1000-ft-long earth-fill dam would consist of about 1 million cubic yards (710,000 m^3) of material for the portion above the original ground surface. Zoned sections in the dam provide for the combined requirements of low permeability to reduce seepage and soil strength for stability. A cross section of a typical zoned earth dam is shown in Figure 20.11.

Three major zones are shown: a central clay core and the upstream and downstream shells of silty material. A sand drain on the downstream side of the core collects the seepage after it has passed through the core, which drains away the water and relieves pore pressures. A 2.5 to 1 slope (2.5 horizontal to 1 vertical) is shown for the upstream slope and a 2 to 1 slope for the downstream portion.

The foundation for the dam is depicted as perme-able material, which is typical of alluvial or stream deposits. The clay core has been extended through this permeable material to intersect the low perme-ability material below. This serves as a cutoff wall to minimize underseepage below the dam and to reduce the uplift pressure under the downstream portion of the dam.

Riprap, consisting of large size boulders or large blocks of quarry stone, is used for wave protection along the upstream slope of the dam. Rock pieces must be large enough that wave action will not

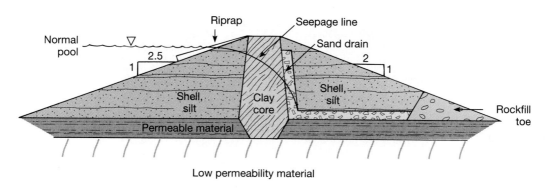

FIGURE 20.11   Cross section through a zoned, earth-fill dam.

dislodge them and durable enough so the pieces do not reduce in size with time as weathering processes prevail.

Along the toe of the dam, a small rock-fill zone has been added. This helps support the downstream slope and allows for drainage both from the sand drain and from the shell. It also provides added strength to the slope.

Earth berms are sometimes added at the base of the downstream slope. Such berms are typically about 50 ft high and about 25 ft wide at the top. The addition of a downstream berm makes it possible to steepen the downstream slope of the dam and may reduce the total volume of fill required.

Rock-fill dams are constructed from coarse rock pieces, gravel size and larger, with an impervious core of silt or sandy silt material and an impervious upstream face. Rock-fill dams tend to be more economical when the following conditions are combined: silt and clay size materials are in short supply, the foundation is not sound enough for a concrete dam, suitable large-sized rock is available in large quantities, cold weather greatly limits the season for rolled fill construction, and considerable earthquake hazards exist. Rock-fill dams, because of their inherent flexible nature, are less prone to damage during earth shaking.

Earth-fill dams are constructed by placement of different soils in the designated zones, followed by compaction with sheep's foot rollers to achieve the necessary density. The soil is transported to the fill by large dump trucks or by large scrapers. Construction of an earth dam is shown in Figure 20.12.

MATERIALS FOR DAM CONSTRUCTION   Large quantities of construction materials are required for the construction of dams. For concrete, the coarse aggregate consisting of gravel or crushed stone must be located near the construction site. Sand for fine aggregate in the concrete is also required. For earth-fill dams, large quantities of concrete are still needed for the spillway and other engineering structures.

Large quantities of clay, silt, and sand are required for an earth-fill dam. The core, shell, sand drain, rock-fill toe, and riprap (Figure 20.11) are composed of materials from the dam site. Considerable geological and engineering investigation is required to locate and select the needed materials for construction.

Special planning is required if excavated rock or soil is to be used for dam construction. Rock excavated from the foundation or in stripping back the abutments, must be processed for use as aggregate, riprap, or soil borrow material. Excavation of rock from spillways, diversion tunnels, or supply tunnels can also provide useful construction materials.

EXPLORATION FOR DAMS   Dam sites are selected using office reconnaissance and geologic field studies. Alternative sites are evaluated and the prime site is selected on the basis of geology, design, construction, and economic considerations. The details of this major undertaking are beyond the scope of this discussion.

FIGURE 20.12   Construction of an earth-fill dam, looking along the dam axis.

Exploration for the selected dam site consists of soil borings and rock coring along the centerline of the dam. Standard penetration tests and Shelby tube samples are obtained in the soils. Soil descriptions are made and detailed cross sections developed. Soils are tested in the laboratory to obtain Atterberg limits, grain size distribution, and strength parameters.

Bedrock is cored to the depth of influence of the dam, considering both the foundation loads and the seepage paths of water after the reservoir is impounded. Typically, N_x rock cores or larger ones, 3 in. diameter for example, are obtained and detailed stratigraphic sections are developed. The RQD is determined on the rock cores.

During the rock coring process, water pressure testing is conducted in the core holes. This is used to locate the areas in the bedrock where permeability is greater and water would travel from the reservoir. This information is later used to determine the zones where grouting will be done. This process is discussed next. Rock cores are obtained from the foundation and the abutments. Vertical and angle holes are commonly drilled. The rock strength and permeability of the rock units are determined.

Grouting is done prior to and during dam construction in the abutments. In this operation, holes are drilled and a mixture of cement and water is pumped into the holes under pressure. The cement sets up to seal the openings in the rock. It is imperative that excess pressures are not used that would lift the rock column and thereby induce fractures in the rock mass. This would weaken the rock and increase fracture permeability.

Two primary locations and types of grouting are performed on the dam foundation. These locations are shown on Figure 20.8. Consolidation grouting is accomplished for the contact area of the foundation for a concrete dam to increase the strength and Young's modulus of the rock. This tends to equalize the amount of deflection of the foundation under the base of the dam.

For concrete dams, curtain grouting is extended below the dam from the galleries to form a cutoff to the flow of water below the dam. This is accomplished when concrete placement is nearly completed, which allows grouting at a higher level of pressure. It is performed upstream of the centerline of the dam. By contrast, if curtain grouting is needed

for an earth dam it must be done prior to construction of the earthen embankment because grouting through the embankment is not recommended. Such practice may erode the earth fill, causing voids to develop in the embankment. For earth dams, consolidation grouting and curtain grouting are accomplished first and then the dam is constructed above the foundation.

As stated, for concrete dams, curtain grouting is performed from the galleries and is accomplished when construction of the concrete work is nearing completion. This is shown in Figure 20.8. The purpose for waiting until this time is because the weight of the dam adds pressure to the foundation and higher grout pressure can be used without lifting the rock column. Cement is forced into rock fractures and bedding planes, which reduces the ability for seepage water to flow through the foundation. This greatly reduces underseepage from the reservoir.

ENVIRONMENTAL GEOLOGY

Many of the same subjects considered in engineering geology are also discussed under the environmental geology category. However, as mentioned previously, the level of detail is generally somewhat less for environmental geology and the scope of the subject is somewhat broader. There is a greater tendency to classify and discuss in environmental geology rather than to calculate specific answers as is commonly done in engineering geology. In a sense, environmental geology points out the problems but the primary solution offered may be omission or relocation. Engineering geology is faced with solving the problem through a direct solution rather than by indicating that a project should not be built or simply that it should be built somewhere else. However, in the mid-1990s environmental studies are taking on a new, more scientific approach promoted by federally funded research. There is now the potential for environmental geology to become more analytical and less descriptive in nature.

Environmental geology typically addresses the subjects listed in Table 20.4. These are usually discussed in terms of statistics and the detrimental factors of the problems are underscored.

Environmental geology also includes the biological aspects of the land. This involves the biosphere or the

TABLE 20.4   General Outline of Environmental Geology

1. Geologic Hazards
   Earthquakes
   Landslides
   Floods
   Tsunamis
   Volcanoes
   Subsidence
   Loss of mineral resources by urbanization
2. Geologic Constraints to Construction
   Slope stability along actively downcutting
      streams
   Siting sanitary landfills
   Septic tank percolation fields
   Shortage of natural construction materials
   Excavation in bedrock areas
   Compressible, organic soils
   Loess and lacustrine deposits
   Karst terrain
   Expansive soils
   Presence of faults
   Ground-water complications
   Reactive concrete aggregates
   Soil erosion and siltation
   Floodplains, flooding, and building restrictions
3. Environmental Health
   Geologic factors and environmental health
   Trace elements and health
   Chronic diseases related to geologic
      environment
4. Human Interaction with the Environment
   Water supply
   Waste disposal
   Effects of urbanization on the landscape
5. Mineral Resources and Depletion
   Mineral resources and population
   Renewable and nonrenewable resources
   Resources versus reserves
   Environmental impact and mineral development
   Recycling of minerals, effects of planned
      obsolescence
6. Current Land Use and Land Use Planning
   Determining current land use
   Landscape analysis and evaluation
   Mineral resource mapping
   Landscape aesthetics
   Land use planning
   Environmental impact

biological portion of the lithosphere. The effects on mineral, elemental, or chemical cycles are a major concern. These include for example the phosphate, nitrogen, carbon, and protein cycles. To ensure that the Earth will continue to support human life in the near and distant future, we must accomplish the following: 1) Regulate the number of humans (unless population stability is achieved, everything else will fail), 2) conserve and recycle the basic materials we use to the greatest extent possible, and 3) ensure that the food supply is adequate for the regulated numbers. It certainly is in the realm of possibility for humans to allow the environment to deteriorate to the extent that their extinction could occur. Far short of that extreme, it seems necessary that the environment be preserved in a state similar to that in which it was inherited, both for aesthetic and simple survival purposes. The answer, however, is not purely preservation but conservation, accomplished by applying the best stewardship of the land.

Large-Scale Interpretive Maps

One feature of environmental geology that can be explored in this basic discussion involves the preparation and use of environmental geology maps. These are large-scale maps prepared at a scale of 1 : 1000 or larger (even as large as $1'' = 1'$) that show extreme detail. They may be strictly geologic maps or interpretive maps derived from geologic maps.

Fault mapping in earthquake-prone areas is a situation in which environmental geology maps prove useful. Fault traces in the subsurface are found by trenching, and fault displacements on the Earth's surface are noted for the study area. This information is transferred to a large-scale map (1 : 500 to 1 : 1000 or so) where a high degree of detail and accuracy can be depicted. Areas where buildings will not be allowed can be accurately designated on these maps so that restrictive zoning is established and made meaningful. Past experience with smaller scale geologic maps, in Nicaragua for example, has shown that fault zones designated on small-scale geologic maps are not sufficient to minimize earthquake damage. Although apparently located a safe distance from the fault zone based on the map, the construction of major buildings directly in an active fault zone has occurred. Both a combination of imprecise location of the fault zone and mapping on too small of a scale led to this dangerous situation.

Many different forms of interpretive maps can be developed from geologic maps or information. Maps depicting the extent of available water in the subsurface are made as are maps of construction materials, swelling potential of soils, rock strength or modulus of elasticity, septic tank acceptability, or landslide susceptibility. Smaller scale maps (1 : 24,000 and less) have been used in engineering geology reports as interpretive maps for many years. One of these maps is presented as an exercise for this discussion topic.

The application of geology to land use planning is another significant criteria for environmental geology. The goal is to introduce geologic factors into the planning process to influence procedures for mineral extraction, development of housing tracts, groundwater development, preservation of agricultural land, and so on. The first step is to obtain a detailed geologic map of the area. For major cities and for counties of approximately 500 mi², an appropriate map scale is 1 : 24,000 or 1 : 62,500 (7.5-min quadrangle and 15-min quadrangle scales, respectively). Topography, surface geology, agricultural soils, bedrock and structural geology, depth to bedrock, and ground-water information are important data that must be compiled. Individual maps are made depicting this information. Current land use information for the study area is ultimately obtained as well. With these combined data, site selection for those areas suitable as housing tracts, sanitary landfills, quarries and gravel pits, ground-water supply or recharge, septic tank percolation fields, and industrial building sites can be outlined. Prime agricultural land can be set aside as well. Economic and political restraints are also considered and in the final analysis these plus geological factors can be used for land use planning.

An environmental study of the type detailed lends itself to a database and computer analysis approach. This is accomplished through the establishment of a grid pattern for the area to which values for the significant geologic factors can be assigned on the basis of a unique address for the different proposed uses of the land. In this way a value for each unit area of the grid can be totaled relative to individual proposed uses. Areas registering the highest sum for a particular proposed use are the most suitable in the study area for that application. A study of this type has been accomplished for a county in central Indiana (Hasan and West, 1982) which includes housing construction, sanitary landfills, ground-water supply, septic tank percolation fields, and several other engineering considerations. Geographic Information Systems (GIS) are being compiled by government and research organizations throughout the United States for data storage and retrieval of this type of information.

Environmental impact studies are required for all projects involving expenditures by the federal government. These include details on the current nature of the environment and the anticipated impacts rendered by a proposed project. Significant impacts must be mitigated, or clearance to proceed is not granted. Several opportunities are also provided for the public to make their views known regarding the proposed project.

EXERCISES ON ENGINEERING GEOLOGY AND ENVIRONMENTAL GEOLOGY

Map Reading

Exercises 1 through 5 pertain to the "Hollidaysburg Quadrangle, PA, Folio for Interpretating Geologic Maps for Engineering Purposes" USGS, 1953 (scale 1 : 62,500). This folio contains a topographic map, general geologic map, and several special-purpose engineering geology maps for the Hollidaysburg Quadrangle. It affords a unique opportunity to compare the different maps for one area. Be sure to examine all maps carefully before completing this exercise.

1. Using a road map of Pennsylvania, give the location of Hollidaysburg with regard to major cities in Pennsylvania and indicate its general location in the state. Using this information and with the aid of a landform map of the United States and the physiographic map in Chapter 13, indicate the physiographic province in which this area is located.

2. Refer to the explanation for the general-purpose geologic map (map 2). Construct a simplified geologic column for the quadrangle based on this information. Use the following degree of detail for the subdivisions.

Abbreviations for periods can be used if needed. Continue on from the portion shown here, beginning with the formations of the Middle Devonian:

Quaternary
  Recent
    Alluvium
UNCONFORMITY
  Carboniferous
    Pennsylvania
      Allegheny fm
      Pottsville fm
UNCONFORMITY
  Mississippian
    Mauch chunk fm
    Loyalhanna ls.
    Pocono fm
  Devonian
    Upper Devonian
      Hampshire fm
      Chemung fm
      Brallier shale
      Harrel shale
    Middle Devonian

3. (a) Sketch a map view of the major structural features of the quadrangle (use map 2). Show anticlinal and synclinal axes, faults, sufficient bedding planes, and geologic symbols to indicate the structural features.

Refer to map 1. How are the mountain regions related to the structural features? What rock types comprise the mountain tops in this area?

(b) Name the maps contained in this folio. How are maps 3 through 6 obtained from maps 1 and 2? How is this useful in land use planning?

(c) The following system is used for reference purposes on the Hollidaysburg Quadrangle:

3	2	1
4	5	6
9	8	7

Why isn't the Township and Range method used? Why isn't a latitude and longitude designation used?

(d) How are the rock units grouped to obtain map 3? How would such a map be useful to highway construction? Foundations for heavy buildings?

(e) How are the rock units grouped to obtain map 4? Why is this of use to planners in this area?

(f) How are the rock units grouped to obtain map 5? Why is this of use to planners and home builders?

4. (a) A tunnel is proposed that will extend from East Sharpsburg to the southern end of Oldtown Run

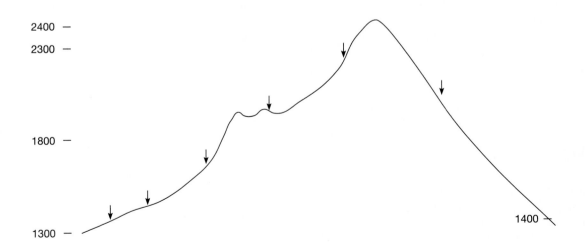

↓ indicate geological contacts

1" = 500' vertical
1" ≈ 2600' horizontal

road. On the topographic cross section on the facing page, draw in the geologic cross section between these two points; include the tunnel grade line on the drawing. A $1'' = 2600'$ map scale for the horizontal dimension has been used in the drawing. For the vertical scale on the cross section, a $1'' = 500'$ scale is employed. What is the vertical exaggeration? Describe any problems that might occur during rock tunneling.

If the load on the roof of the tunnel is expressed by the relationship

Height of rock load carried by tunnel roof

$$= H_p = f(B + H_t)$$

where B = width of tunnel (in feet)
H_t = height of tunnel (in feet)
f = factor dependent on rock conditions.

What would be the pressure on the roof of the tunnel in the Tuscarora Quartzite if the f value of 0.25 is suggested and the tunnel had a height of 20 ft and width 15 ft?

(b) A dam has been proposed for Clover Creek near the town of Henrietta (Rect. 7). What is the gradient of the creek at this general location? Describe completely the geology of the site. What would be the maximum height of the dam? Discuss any problems that might occur during and after construction of the dam. Would you recommend this location as a dam site; explain in detail.

(c) Using a piece of $8\frac{1}{2} \times 11$ tracing paper and map 4 of the folio, outline all the areas in rectangle 8 where underground caverns could likely occur. Indicate also where limestone quarries for concrete aggregate might be located in this rectangle. Explain your choice.

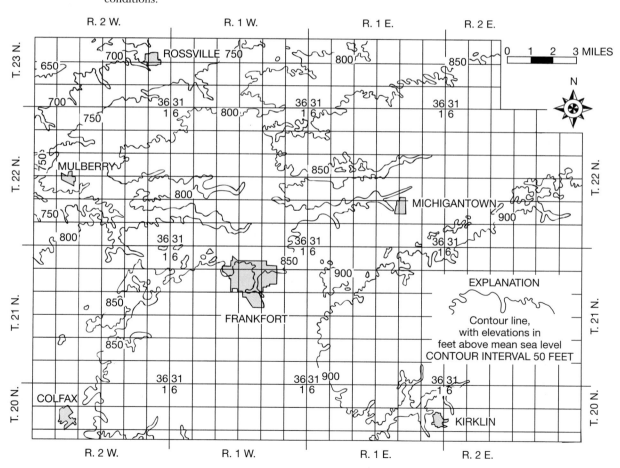

FIGURE 20.13 Generalized topography of Clinton County, Indiana.

5. Environmental geology maps are used today for planning and development. The U.S. Geological Survey has mapped the geology adjoining some large cities with map scales of $1'' = 100'$. Locations with complex geology and rapid growth are mapped first: Denver, Los Angeles, San Francisco, and so on. Why would the Hollidaysburg map fall short in this effort? Explain.

6. The regional geologic map, Danville, Illinois, $1° \times 2°$ map, published by the Indiana Geologic Survey, includes the area of Lafayette, Indiana. What is the map scale? To what extent does this fall short of the $1'' = 100'$ scale mentioned in Exercise 5? What can be done to improve the situation?

7. The environmental geology map of the Saskatoon Alberta, Canada, area is one of the maps presented in the publication available in many university libraries. How is this map an improvement over the Hollidaysburg and Lafayette maps? Explain.

Environmental Geology, Clinton County, Indiana

Given the following maps of Clinton County, answer Exercises 8 through 13.

Figure 20.13   Generalized topography
Figure 20.14   Bedrock geology and bedrock topography
Figure 20.15   Surficial geology and glacial drift thickness
Figure 20.16   Piezometric surface
Figure 20.17   General soils association
Figure 20.18   Generalized glacial geology of Indiana

8. Using the topography map and the bedrock geology map, what is the depth to bedrock at the corner where the boundary between T20N and T21N meets the boundary between R2W and R1W? (*Hint:* This is determined in the following way: Find the surface elevation, then find the elevation for the top of the bedrock; the numerical difference is the depth to the bedrock.)
Check this same point on the surficial geology and glacial thickness map. Are the two values the same? Why or why not?

9. What is the maximum relief in the county based on the topographic map? Why is the area so flat? Refer also to the generalized glacial geology map of the state.

10. Refer to the surficial geology map:
    (a) What material comprises the primary portion of the map area?
    (b) What type of soil texture is this most likely to be? Explain.
    (c) What are the other four mapped materials?

(d) Which is (are) most likely to supply gravel materials for construction? Explain.
(e) Where is the depth to bedrock least in the county? What would be the bedrock material there? Be specific.

11. Examine the piezometric surface map of Clinton County. This is the elevation to which water would rise in a well after an aquifer was intercepted.
    (a) If the piezometric surface is higher than the elevation at which water is intercepted, what type of well is involved?
    (b) At the corner where the boundary between T22N and T23N meets the boundary of R2W and R3W, what is the elevation of the piezometric surface? How far below the ground surface is this? At what depth would a pump intake have to be set in this well? Explain. What is the likely direction of ground-water flow in this area? Explain.

12. Examine the soil associations map of the county. A septic tank system should be located in relatively permeable material. Does soil association 81 qualify for this? Explain. Does association 64 also qualify? Discuss.

13. Section 12, T2IN, R1W is being considered for a sanitary landfill site. Refer to the surficial geology map and the soil associations map. Is this a suitable site for a landfill? Explain in terms of information from both maps.

Written Questions

14. For a highway construction project, a soil cut will be 30 ft deep and an adjacent embankment will be 50 ft high. The cut area will be used as a borrow site for the embankment. The depth to bedrock is more than 100 ft. Describe how typical borings should be taken in the cut and in the fill. Include the depth of the boring, drilling technique, and soil sampling. Include also a list of laboratory tests needed to characterize the material in question.

15. The following volume of soil and rock is to be removed from a highway cut: Soil, 13,000 yd³ of silt and 17,000 yd³ of silty clay; rock, 18,000 yd³ of limestone and 15,000 yd³ of hard shale. The shrinkage factor for the silt is 8% and for the silty clay it is 12%; the swell factor for the limestone is 19% and for the hard shale 9%.
    (a) What will be the volume of the fill constructed from these materials? What is the overall volume change in cubic yards and percent?
    (b) If the volume of the fill is designed to be 58,000 yd³, what would be the needed volume of offsite rock disposal? Next, assuming the rock dumped offsite

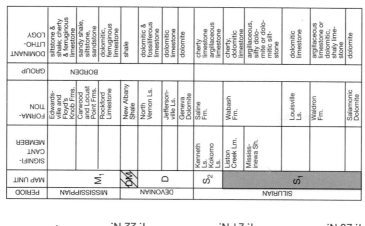

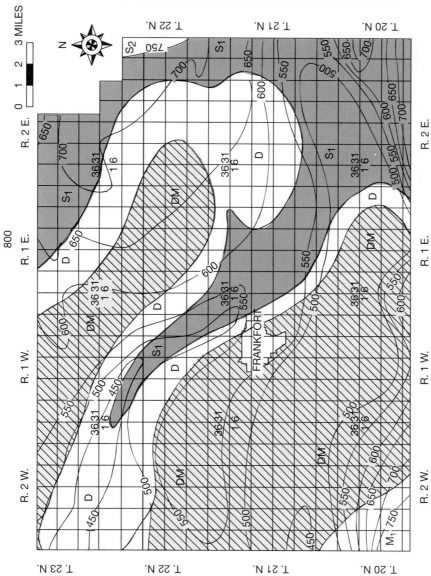

FIGURE 20.14 Bedrock geology and bedrock topography of Clinton County, Indiana.

501

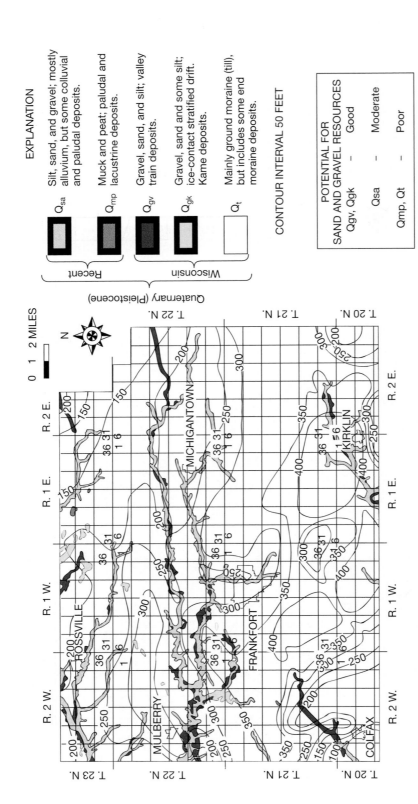

FIGURE 20.15 Surficial geology and glacial drift thickness showing sand and gravel resources of Clinton County, Indiana.

502

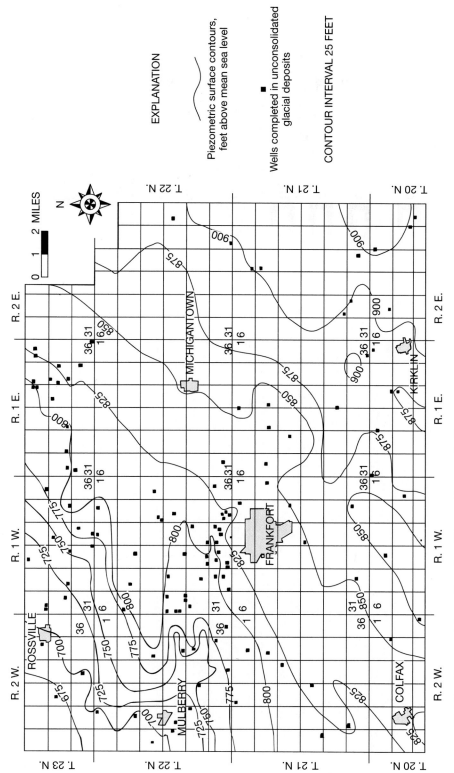

EXPLANATION

⌣⌣⌣ Piezometric surface contours, feet above mean sea level

■ Wells completed in unconsolidated glacial deposits

CONTOUR INTERVAL 25 FEET

FIGURE **20.16** Piezometric surface for Clinton County, Indiana.

503

SOIL ASSOCIATIONS

4. Genesee-Shoals-Eel: Nearly level, well drained, loamy Genesee, moderately well drained, loamy Eel, and somewhat poorly drained, loamy Shoals in alluvial deposits.

64. Crosby-Brookston: Nearly level, somewhat poorly drained, clayey Crosby and very poorly drained, loamy Brookston in glacial till.

66. Fincastle-Ragsdale-Brookston: Nearly level, somewhat poorly drained, silty Fincastle in wind-blown silts and glacial till, very poorly drained, silty Ragsdale in wind-blown silts and loamy Brookston in glacial till.

73. Raub-Ragsdale: Nearly level, somewhat poorly drained, silty Raub in wind-blown silts and glacial till and very poorly drained, silty Ragsdale in wind-blown silts.

81. Miami-Russell-Fincastle: Sloping, well drained, loamy Miami in glacial till, silty Russell in wind-blown silts and glacial till and nearly level, somewhat poorly drained, silty Fincastle in wind-blown silts and glacial till.

83. Miami-Crosby: Sloping, well drained, loamy Miami and nearly level, somewhat poorly drained, clayey Crosby in glacial till.

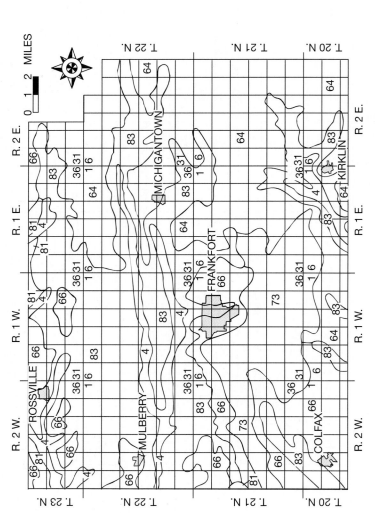

FIGURE 20.17 General soil associations of Clinton County, Indiana.

504

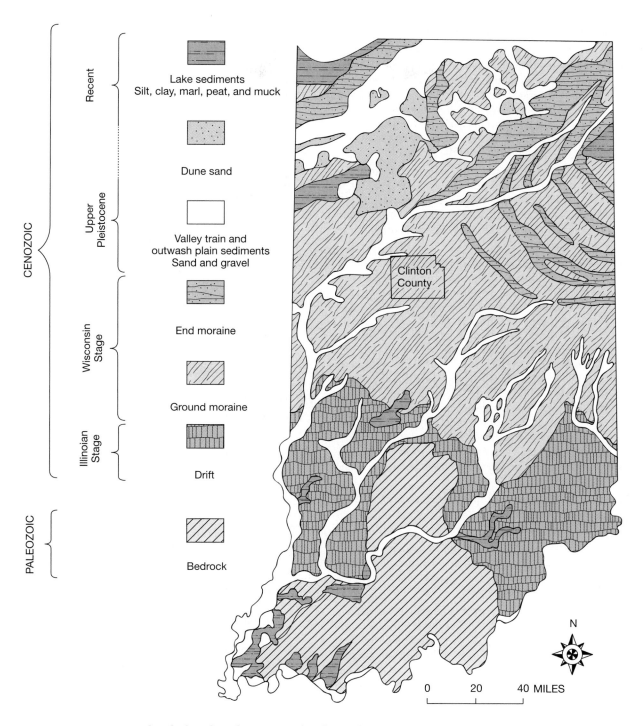

FIGURE 20.18 Generalized glacial geology map of Indiana. (From Wayne, 1958.)

would be shale, how much volume does this represent in the cut section?

16. It is estimated that 12,000 people per day will stop at a rest area along an interstate highway during the maximum tourist period. If the average water use is 3.5 gal/person stopping at the rest area, how much water will be consumed per day? How many minutes per day must a well be pumped if its capacity is 120 gal/min? What percent of the time must the well be pumped?

17. A highway is to be constructed over an old, coal strip mine site that was not reclaimed after mining. It contains elongated spoil piles and a lake, which marks the final strip of coal excavation. List the possible highway construction problems that could occur at the site. State assumptions if any.

18. Under the center of a mountain range, a tunnel is located 1250 ft below the ground surface.
    (a) If the average unit weight of the rock is 165 lb/ft^3, what is the total vertical stress above the tunnel in psf and psi?
    (b) The Terzaghi information for the rock above the tunnel indicates that the equation $H_p = 0.45 (B + H_t)$ would apply. The tunnel is 20 ft wide and 15 ft high. Calculate H_p in feet and the roof pressure in lb/ft^2.

19. (a) List the four major geologic hazards in tunneling. In a fissured limestone, which of these is likely to occur during the tunneling?
    (b) What harmful gases can occur in rocks that contain thin coal beds? How can this problem be minimized?

20. What are the six steps of the drill and blast method for tunneling? What is the preferred explosive? How can blast damage to adjacent structures be minimized?

21. Under what conditions are tunnel boring machines most effective and efficient in rock tunneling? Why might a long tunnel in a fine, massive granite not be a good site for a TBM?

22. What are five primary purposes for dam construction? Where in the United States does the U.S. Army Corps of Engineers have most of its responsibility for dam construction? The U.S. Bureau of Reclamation? Name two other governmental agencies that construct and maintain dams in the United States.

23. List the different types of dams based on construction materials. Include subclasses of the various types. Under what conditions would a thin arch concrete dam be the most feasible? Explain.

24. (a) Why is an emergency spillway a necessary part of a dam construction project, particularly in heavily populated areas?

    (b) List some likely advantages and disadvantages of an underground power generation station compared to an aboveground installation. Consider topography and climate in your analysis.

25. (a) In the past, a major sanitary landfill has been located in Indianapolis, Marion County, Indiana, along the sandy terrace deposits of the White River. Refer to Table 15.11 in the chapter on Ground-Water Geology. Would this site be considered unfavorable based on that table? Explain.

    (b) The White River is immediately downgradient from the landfill and normally leachate would migrate from the landfill to the river. Because the White River has a large discharge, dilution of the leachate would occur. Nevertheless, is this a good practice to allow leachate to migrate offsite in the ground water? What are the long-term consequences of such a practice in the United States?

    (c) Because of heavy pumping of ground water by industry upgradient from the landfill, contaminated water was drawn from the landfill to these industrial wells. What is required to reverse the gradient in this fashion? Does this seem to be a likely occurrence along major flood plain areas in an urban environment? Explain.

26. The Eagle Creek Reservoir is located just to the west of Indianapolis and it supplies a major portion of the drinking water for the city. The Eagle Creek Dam is an earthfill structure that is 85 ft high and 1554 ft long.
    (a) What subsurface exploration procedure would be appropriate for a proposed earthen dam structure in this area? Make a sketch in plan view showing the centerline of the proposed dam, a small existing stream, an adjacent floodplain, and the two valley walls. Locate where borings should be placed along the centerline. What type of borings should be made and approximately how deep should they be?
    (b) The cross section of an earth-fill dam is shown in Figure 20.11. The different sections in the dam are labeled. The materials for construction must be located adjacent to the dam prior to construction. The Eagle Creek dam is located within a Wisconsin age till plain.
       (1) What material would be used for the clay core?
       (2) The silty shell material would likely be obtained from the proposed reservoir area. What is this stream-deposited material called? Why should this not be removed too close to the dam itself? Explain.
       (3) Clean sand is used in the sandy drainage zone. This is used to relieve pore pressures after water has passed through the clay core. This material would likely have to be selectively

removed from deposits in the reservoir area. Why is a clean sand required in this drainage zone? Explain.

(4) Settlement of the foundation and of the dam itself are a concern. Why is it possible that the foundation might settle? What would be the vertical stress induced on the foundation at the center of the dam based on the description provided? What test would be run to determine how much settlement of the foundation is likely to occur? (*Hint:* Refer to the last section in Chapter 7, Elements of Soil Mechanics.)

(5) How is settlement of the dam itself (the embankment) kept at a minimum? How is this accomplished during the construction of the dam?

(6) A concrete spillway is used to control the level of water in the reservoir. What are the materials of construction needed for the spillway? Where would the concrete aggregates be obtained on the construction site?

27. A coal-fired, electric-generating power plant is to be constructed in an area of sedimentary rock strata adjacent to a large stream. The foundation for the power plant and a water intake structure in the river are to be designed by a major engineering company. The power plant will be located on the valley wall on bedrock rather than on the floodplain or terrace of the river.

(a) Why was the power plant probably located on the valley wall rather than in the floodplain? Assume flooding is not the issue, because an area above flood effects was available on the terrace.

(b) Review Table 20.1. Indicate how each of these items would be involved in the design considerations for the project described above. If an item does not apply indicate why this is the case.

28. Why does engineering geology encompass the more elusive parts of construction? Is it to be expected that analytical analysis of geologic phenomenon related to construction does not always provide exact answers? Explain.

29. Geologic hazards generally involve the catastrophic effects related to nature, whereas geologic restraints are more of a long-term effect of geologic agents acting on the landscape. Relative to physiographic units, where do the catastrophic geologic hazards occur and where are the geologic restraints more in control? Consider relative to stable regions (SR) and orogenic belt areas (OB) as presented in Chapter 13 on physiographic provinces.

30. The Fortville fault in central Indiana is a high-angle normal fault that displaces the bedrock surface but does not extend through the Pleistocene glacial deposits above. The bedrock surface displaced ranges in age from Mississippian to Silurian.

(a) From this description, it is concluded that the fault is younger than what age and older than what other age? Hence it is pre- what age and post- what other age?

(b) The fault strikes about N30°E and the south portion is the downthrown side. Draw the fault in map view and cross-section view including the information about the fault provided in this exercise.

(c) If the fault does not extend through the glacial deposits to intersect the ground surface, how likely was the fault's presence determined? Thickness of glacial deposits in this area typically exceeds 100 ft.

(d) There is no increase in earthquakes or microseisms associated with the Fortville fault in Indiana as compared to the surrounding area, which overall is in a low-intensity earthquake zone. What does this indicate about the activity of this fault system? *Capable faults* are those capable of movement because of earthquake activity. Is the Fortville fault capable? Explain. Did movement occur along it within the last 10,000 years? Explain.

Additional Readings

Association of Engineering Geologists Special Publication, "Geology, Seismicity and Environmental Impact," 1973, AEG.

Blythe, F.G.H., and de Freitas, M.H., 1974, *A Geology for Engineers*, 6th ed., Edward Arnold Co., London.

Cargo, D.N., and Mallory, B.F., 1977, *Man and His Geologic Environment*, 2nd ed., Addison Wesley Co., Reading, MA.

Carlier, M., 1974, Causes of the Failure of the Malpasset Dam, in *Foundations for Dams*, American Society of Civil Engineers, New York.

Costa, J.E., and Bilodeau, S.W., 1982, "Geology of Denver, Colorado, U.S.A." *Bulletin of the Association of Engineering Geologists*, Vol. XIX, No. 3, pp. 263–314.

Flawn, P.T., 1970, *Environmental Geology*, Harper & Row, New York.

Hasan, S.E., and West, T.R., West, 1982, "Development of an Environmental Geology Data Base for Land-Use Planning," *Bulletin of Engineering Geologists*, Vol. XIX, No. 2, May.

Legget, R.F., 1973, *Cities and Geology*, McGraw-Hill, New York.

McLeon, A.C., and Gribble, C.D., 1979, *Geology for Civil Engineers*, George Allen & Unwin, London.

Office of Emergency Preparedness, 1969, *Proceedings of a Conference on Geologic Hazards and Public Problems*, U.S. Government Printing Office, Washington, D.C.

Otto, E.E., 1977, "Engineering and Environmental Geology of Clinton County, Indiana," MS Thesis, Purdue University, West Lafayette, IN.

Randall, D.H., et al., 1983, "Geology of the City of Long Beach California, U.S.A.," *Bulletin of the Association of Engineering Geologists*, Vol. XX, No. 1, February.

U.S. Department of Housing and Urban Development, U.S. Geological Survey, Environmental Planning and Geology, 1969, U.S. Government Printing Office, Washington, D.C.

Wayne, W.J., 1956, "Glacial Geology of Indiana, Atlas of Mineral Resources in Indiana, Map 10," Indiana Geological Survey, Bloomington, IN.

West, T.R., and Warder, D.L., 1983, "Geology of Indianapolis, Indiana U.S.A.," *Bulletin of the Association of Engineering Geologists,* Vol. XX, No. 2, May, pp. 105–124.

Appendix A

Identification of Minerals and Rocks

APPENDIX OUTLINE

Minerals
Rocks

MINERALS

Charts used for identification of minerals and rocks are contained in this appendix. The purpose of these charts is to supply the means whereby a beginning student in geology can identify common minerals and common rocks based on the physical characteristics of those materials.

Table A.1 includes the most common rock forming minerals. The chart is so arranged that three groups of minerals are considered with regard to hardness: 1) those softer than steel, 2) those about the same hardness as steel, and 3) those harder than steel. Hardness is used as the primary subdivision because it is a very diagnostic property of minerals and can be readily discerned by a beginning student.

After hardness is determined the properties of cleavage, luster, color, and streak are used to discern the mineral name. Only 19 mineral species are included in Table A.1 because these are the most common rock forming minerals and they comprise the great volume of materials within the Earth's crust.

TABLE A.1 Common Mineral Identification Chart

Hardness	Cleavage or Fracture	Luster	Color	Streak	Remarks	Name and Composition
Group 1: Minerals Softer than Steel						
1–2.5	Not visible	Dull to earthy	White, light brown, other colors due to impurities	White	SpG of pieces about 2, soft compact earthy masses. Greasy feel that becomes plastic when moistened. Clay-like odor to mass.	Clay mineral
1–5	Uneven fracture	Earthy	Yellow	Yellow, brown to black	Commonly in earthy masses SpG 3.5–4	Limonite
1–5	Uneven fracture	Earthy	Deep red	Deep red to brownish red	May be granular or earthy, SpG 5	Hematite
					Iron oxides	
2	One plane	Vitreous silky	White to pastels	White	Fine grained, massive, SpG 2.3	Gypsum (hydrous calcium sulfate)
2–3	Excellent one plane	Pearly to vitreous	Colorless	White	Transparent in thin sheets, yields flexible plates flexible and elastic, SpG 2.8–3.1	Muscovite (Mg silicate)
2.5–3	Excellent one plane	Vitreous	Black to dark brown	Black	Yields plates flexible and elastic, SpG 2.7–3.2	Biotite (K, Mg, Fe, Al, silicate)
					Mica	
3	Three planes, rhomb shape	Vitreous	Colorless or yellow to white	White	SpG 2.7, reacts with hydrochloric acid	Calcite (calcium carbonate)
2.5–4	Rhomb shape	Vitreous to dull	Colorless, white to light brown, pink	White	SpG, powder effervesces slowly in cold, dilute HCl, coarse crystals do not.	Dolomite (Ca, Mg carbonate)
2.5–4		Metallic	Brass		See pyrite	Chalcopyrite (Cu, Fe sulfide)

510

Hardness	Cleavage or Fracture	Luster	Color		Other properties	Mineral
5–6	Two planes at 90°	Vitreous to dull	Black to dark green	Gray greenish gray, brownish gray or blackish gray	SpG 3.1–3.5. Cleavage planes more prominent and more shiny in amphiboles than in pyroxenes (Ca, Mg silicate).	Pyroxene (augite) }{ Pyroboles
	Two planes at 56 and 124°				SpG 3–3.3	Amphibole (hornblende)
6	Uneven fracture	Metallic	Black	Black	Commonly in compact granular masses. Attracted by magnet. SpG 5.2	Magnetite (iron oxide)
6.5–7	Conchoidal fracture	Vitreous	Light green	White to pale green	Granular, sugary masses common, transparent to translucent. SpG 3.2–3.6	Olivine (Fe, Mg silicates)
5.6–7.5	Conchoidal or uneven fracture	Vitreous to resinous	Blood red	White	Transparent to opaque, well-formed crystals common. SpG 3.5–4.3	Garnet group (Ca, Mg, Fe, Mn, Al silicates)
6–6.5	Uneven fracture	Metallic	Brassy yellow	Greenish black	Granular masses and well-formed cubic crystals common. Variety softer than steel nail chalcopyrite. SpG 4.1–5.2	Pyrite (Fe sulfide)

(Continued)

TABLE A.1 Common Mineral Identification Chart (*Continued*)

Group 3: Minerals Harder than Steel

6–6.5	Two planes at about 90°	Pearly to vitreous	White	White, pink	Commonly pink. SpG 2.5–2.7	Feldspar group (Na, Ca, K, silicate) Orthoclase (K feldspar)
				White, gray greenish gray, dark gray	May show striations on cleavage plane. SpG 2.5–2.7 Sodic plagioclase, white to light gray. Calcic plagioclase, dark gray to black	Plagioclase (Na₂Ca feldspar)
7	Conchoidal fracture	Vitreous or greasy	White	White, colorless, pink gray, violet	Massive or well-formed crystals, transparent to translucent. SpG 2.65	Quartz (silicon dioxide)
7	Conchoidal fracture	Dull	White to gray	White to black	Opaque, black variety is flint. SpG 2.65	Chert (silicon dioxide)

Table A.2 is a tabulation sheet for mineral identification. It is so arranged from left to right to prompt the student through the important physical properties that are used to identify a mineral sample. This sheet is used in accordance with Table A.1 for mineral determination.

Table A.3 is comprised of a list of common minerals that have significant economic importance. They can also be considered as ore minerals, or those that when present in sufficient quantities can be mined at a profit. These minerals are arranged alphabetically from top to bottom. Key mineral properties from left to right are supplied. This table can be used for identification of minerals or as a source of information on common economic minerals.

ROCKS

Igneous Rocks

Figure A.1 is a simplified identification chart for igneous rocks. It is based on two primary characteristics of igneous rocks: texture and mineral composition. Texture is designated on the vertical axis on the upper right side. A decrease in grain or crystal size occurs from top to bottom, that is from pegmatite to glassy. Pyroclastic is the igneous texture for rocks whose particles were ejected into the air during volcanic eruptions and form by settling and subsequent lithification.

Mineral composition ranges from high silica (acid) to low silica (basic), which is shown from left to right

TABLE A.2 Mineral Tabulation Sheet

Specimen Number	Hardness	Color	Streak	Type of Cleavage or Fracture	General SpG	Miscellaneous	Mineral Name	General Composition

Identification of Minerals and Rocks

TABLE A.3 Common Minerals with Economic Importance

Mineral	Chemical Composition	Specific Gravity	Hardness	Cleavage or Fracture	Luster	Color	Streak	Form	Other Aspects
Actinolite (an asbestos and an amphibole)	$Ca_2(MgFe)_5Si_8 O_{22}OH_2$	3.0–3.3	5–6	See amphibole, Table A.1	Vitreous	White to light	White	Slender crystals usually	Common metamorphic mineral, asbestos
Apatite	$Ca_5(F,Cl)(PO_4)_3$	3.15–3.2	5	Poor cleavage	Vitreous	Green, also brown or red	White	Massive, granular	Important source of fertilizer (P)
Azurite	$CuCO_3$	3.77	4	Fibrous	Vitreous to dull	Dark blue	Blue	Sometimes radial fibers	Minor ore of copper, effervesces in HCl, commonly found with malachite
Bauxite	Al hydroxides, not a mineral	2–3	1–3	Uneven fracture	Dull	Yellow, brown gray	Colorless	Rounded grains or earthy clay-like masses	Ore of aluminum, weathering product in tropical areas, clay-like odor when wet

Bornite	Cu_5FeS_4	2.5	3	Uneven	Metallic	Bronze on fresh surface, tarnishes to purple	Gray-black	Usually massives, rarely as rough cubes	Important copper ore, associated with chalcocite, chalcopyrite, malachite, and other Cu minerals
Carnotite	$K_2(UO_2)_2(VO_4)_2$	4	Very soft	Uneven fracture	Earthy	Bright yellow	Yellow	Earthy powder	An ore of uranium and vanadium, usually occurs as incrustation on sandstone, radioactive
Cassiterite	SnO_2	6.8–7.1	6–7	Conchoidal fracture	Submetallic to adamantine to dull	Brown or black	White to light brown	Massive and granular	Principal ore of tin
Chalcocite	CuS_2	5.5–5.8	2.5–3	Conchoidal fracture	Metallic	Lead-gray to black	Grayish black	Massive and aphanitic	Important copper ore, associated with bornite, chalcopyrite, and malochite
Chalcopyrite	$CuFeS_2$	4.1–4.3	3.5–4	Uneven fracture	Metallic	Brass-yellow	Greenish-black	Usually massive	Softer than steel whereas pyrite is harder than steel, more brittle than gold
Chromite	$FeCr_2O_4$	4.6	5.5	Uneven fracture	Metallic to submetallic	Black to brown-black	Dark brown	Massive, granular to compact	Only ore of chromium
Copper (native)	Cu	8.9	2.5–3	Hacky fracture	Dull because of tarnish	Copper red	Copper red	Hacky	Minor ore of copper, highly malleable

(Continued)

TABLE A.3 Common Minerals with Economic Importance (*Continued*)

Mineral	Chemical Composition	Specific Gravity	Hardness	Cleavage or Fracture	Luster	Color	Streak	Form	Other Aspects
Corundum	Al_2O_3	4.02	9	Basal parting	Adamantine to vitreous	Usually pink, brown or blue	Colorless	Barrel-shaped crystals or granular	Used as an abrasive, gem forms are ruby (red), sapphire (blue)
Fluorite	CaF_2	3.18	4	Four planes of cleavage yielding octahedron	Vitreous	Bluish, purple, yellow, light green	White	Well-formed cubes or massive	Used as flux in making steel, may show fluorescence
Galena	PbS	7.4–7.6	2.5	Cubic cleavage	Metallic	Lead-gray	Lead-gray	Cube shaped, also massive	Principal ore of lead
Gold (native)	Au	15–19.3	2.5	Uneven	Metallic	Yellow	Yellow	Usually irregular plates or masses	Highly malleable and ductile, this and high specific gravity distinguish it from pyrite, chalcopyrite, and altered mica flakes

Graphite	C	2.3	1	Good in one direction	Metallic or earthy	Black to steel gray	Black to steel gray	Foliated, scaly, radiating or granular	Feels greasy, smudges hands
Gypsum	$CaSO_4 \cdot H_2O$	2.32	2	Good in one direction, flexible	Vitreous, pearly, silky	Colorless, white, gray	Colorless	Prismatic crystals, granular or fibrous	Fibrous variety is satin spar, selenite is transparent, alabaster is massive
Halite	NaCl	2.16	2.5	Perfect cubic cleavage	Glassy to dull	Colorless or white	Colorless	Cubic crystals or massive	Salty taste, table salt
Hematite	Fe_2O_3	5.26	5.5-6.5	Uneven fracture	Metallic to dull	Reddish brown to black	Deep red	Long prismatic crystals or coarse to fine masses	Most important ore of iron
Limonite	Hydrous iron oxide, not a mineral	3.6-4	5-5.5	None	Vitreous to dull	Yellow to dark brown	Yellow-brown	Amorphous, nodular or earthy	Used as a pigment, yellow ocher, found with iron minerals
Magnetite	Fe_3O_4	5.18	6	Some octahedral parting	Metallic	Iron-black	Black	Usually massive granular, aphanitic	Strongly magnetic, primary ore of iron

(Continued)

TABLE A.3 Common Minerals with Economic Importance (*Continued*)

Mineral	Chemical Composition	Specific Gravity	Hardness	Cleavage or Fracture	Luster	Color	Streak	Form	Other Aspects
Malachite	$CuCO_3(OH)_2$	3.7–4.03	3.5–4	Poor	Silky to dull	Green	Green	Radiating fibers and kidney-shaped forms common	Effervesces in HCl, associated with other copper ores
Pyrrhotite	$Fe_{1-x}S$	4.58–4.65	4	Poor	Metallic	Bronze yellow	Black	Massive to granular	Magnetic
Serpentine	$Mg_3Si_2O_5(OH)_4$	2.2–2.65	2–5	Conchoidal fracture	Greasy, waxy, or silky	Variegated shades of green	Colorless	Platy or fibrous	Platy variety is antigorite, fibrous one is crysotile, which is an asbestos
Sphalerite	ZnS	3.9–4.1	3.5–4	Perfect cleavage in six directions	Resinous	Brown to yellow or black	White to yellow brown	Many-sided, distorted crystals common	Most important ore of zinc
Sulfur	S	2.05–2.09	1.5–2.5	Conchoidal to uneven fracture	Resinous	Yellow	Yellow	Irregular masses	Burns with ease, used in chemical industry
Talc	$Mg_3Si_4O_{10}(OH)_2$	2.7–2.8	1	Good cleavage in one direction	Pearly to greasy	Gray, white, apple green	White	Foliated, massive	Greasy feel, metamorphic mineral, used in talcum powder

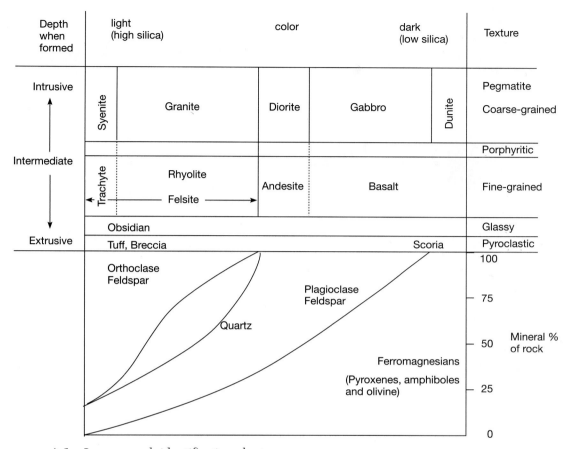

FIGURE A.1   Igneous rock identification chart.

in the diagram. High-silica igneous rocks are also light in color (white, gray, pink), whereas low-silica igneous rocks are dark in color (dark gray, black, green) with intermediate compositions generally having an intermediate color between these two.

Only four groups of minerals are included in the composition chart shown: orthoclase, quartz, plagioclase, and ferromagnesians. The latter group are silicates rich in iron and magnesium; they include pyroxenes, amphiboles, and olivine. In practice, the amount of dark mineral is discerned first (ferromagnesians) then the percentage of quartz and orthoclase are estimated. This approximately defines a vertical line somewhere along the lower diagram. Based on texture this is extended vertically to the upper diagram to obtain a rock name.

For example, assume a dark coarse-grained igneous rock had about two-thirds ferromagnesians and one-third dark feldspar or plagioclase. This vertical line lies about one-quarter of the distance from the right side of the diagram. Extending this upward to coarse-grained texture yields the gabbro category. Therefore, the rock is designated as a gabbro.

Table A.4 consists of a tabulation sheet for igneous rock identification. The texture and composition (or color) are determined first by examination. Then, using Figure A.1 the rock name is obtained by relating the composition to the texture in that chart. Information on origin is provided on the extreme left-hand side of the table. It is related to texture. Extrusive igneous rocks have either pyroclastic or a lava-type origin. Coarse-grained rocks are intrusive in origin, whereas rocks of hypabyssal origin are commonly porphyritic. These concepts are discussed in detail in Chapter 3.

Sedimentary Rocks

Table A.5 is a tabulation sheet for sedimentary rock identification. By this chart the student is directed to

TABLE A.4 Tabulation Sheet for Igneous Rocks

Specimen Number	Texture	Composition	Origin	Rock Name

TABLE A.5 Tabulation Sheet for Sedimentary Rocks

Specimen Number	Texture	Composition	Origin	Rock Name

determine the nature of the texture, specifics of the texture, and the minerals present. Based on this information and Table A.5, the proper rock name can be applied.

Table A.6 is a classification chart for sedimentary rocks. Four subdivisions based on texture are designated in this chart: the clastic, coarsely crystalline, aphanitic, and whole fossil groups. These are determined through careful examination of the rock's texture. Next based on grain size, composition, and other features one ascertains the name of the sedimentary rock.

Metamorphic Rocks

Table A.7 is a simplified classification chart for metamorphic rocks. Two main categories are recognized: rocks that are foliated and those that are not. Determination of the type of foliation and the minerals present leads directly to the name of the metamorphic rock.

Nonfoliated metamorphic rocks are identified according to their mineral composition and their textural detail. Only marble and quartzite are considered in this group of rocks.

Table A.8 is a tabulation sheet for metamorphic rock identification. If the rock is foliated, the type of foliation is discerned. This along with the mineral composition leads directly to the name of the metamorphic rock. If the metamorphic rock is not foliated, the nature of the texture and the mineral composition are used to determine the rock name.

General Considerations

In most courses on rock identification the three genetic groups, igneous, sedimentary, and metamorphic, are studied individually in turn, typically beginning with igneous and finishing with the metamorphics. The student therefore is not concerned with the problem initially of determining to which of these genetic groups the rock specimen belongs. This information is supplied for each group. On completion of the rock study, however, identification of all of these rocks becomes the assignment and suddenly a student is required to identify all rock specimens taken collectively, igneous, sedimentary or metamorphic, presented as a combined group.

A significant problem that must suddenly be addressed is how to determine the genetic group to ensure that the correct rock name is applied to a specific test specimen. Table 5.1 in the chapter on metamorphic rocks is an important aid in making this determination.

Mineral composition, texture, structure, and other features are used to determine the genetic rock group. A careful application of the three classification charts is also used to ensure that the rock belongs in that specific category. Experience, of course, is a major factor because the points of confusion and the diagnostic features for identification become more apparent as more time is spent working with the rock specimens.

TABLE A.6 Classification of Sedimentary Rocks

A. Texture Chiefly Clastic (Composed of Pieces)

Diameter of Pieces	Unconsolidated Sediment	Rock Name	Composition	Descriptive or Special Features
>2 mm	Gravel	Conglomerate	Quartz or varied	Larger pieces rounded, cement commonly calcite
	Rubble	Breccia	Varied	Angular pieces, poorly sorted
1/16 to 2mm	Sand	Quartz sandstone	Quartz grains, calcite or iron oxide cement	Light color (often pink or red) micaceous
		Arkose	Feldspar grains 25–50%	
		Graywacke	Dark rock fragments present	Dark color, cement clay or chlorite
		Clastic limestone	Mostly calcite	Fossil fragments, oolites
1/256 to 1/16 mm	Mostly silt, some clay	Siltstone	Mica, clay minerals, quartz	Laminated, flat-shaped rock
	Silt and Clay	Shale	Clay minerals	Laminated, flat-shaped rock

B. Texture: Coarsely Crystalline (Visible, Interlocking Grains)

Limestone	Mostly interlocking grains of calcite	Soluble in cold HCl
Dolomite	Mostly interlocking grains of dolomite, some calcite	May be porous, effervesces slowly in cold HCl
Evaporites		
Anhydrite rock	Mostly anhydrite	Commonly fine grained
Gypsum rock	Mostly gypsum	
Rocksalt	Mostly halite	Salty taste
Bittern salts	Mostly Mg and K salts	Bitter taste

C. Texture: Aphanitic, Cryptocrystalline, Dense, or Amorphous

Dense limestone (lithographic)	Mostly calcite	Very fine grained, soluble in cold HCl
Dense dolomite	Mostly dolomite mineral	Very fine grained, effervesces slowly in cold HCl
Chert	Very fine quartz, opal, or chalcedony	White, black, green, or brown; conchoidal fracture; hardness of 7 if unweathered

D. Texture Composed of Whole Fossils or Their Alteration Products

Fossiliferous limestone, shell conglomerates	Mostly calcite	Large calcite fossils
Chalk	Mostly calcite	Microscopic calcite fossils (Foraminifera)
Peat	Matted plant remains	Brown; stringy or spongy
Lignite	Mostly hydrocarbons and carbohydrates	Dark brown to dull black
Bituminous coal	Hydrocarbons, carbon and ash	Black; cubical fractures

523

TABLE A.7 Metamorphic Rock Classification

A. Foliated Rocks (Those with Directional Structure)

Type of Foliation	Descriptive or Special Features	Minerals Present	Name	Probable Original Rock
Slaty rock cleavage	Fine-grained, cleavage more pronounced than bedding in shale, mica sheen not obvious	Mica and quartz, but too fine to see	Slate	Shale, tuff
Slaty rock cleavage, surface may have undulations	Cleavage as in slate, mica sheen obvious	Mica and quartz, mica in phyllite larger than in slate and produces sheen	Phyllite	Shale, tuff
Schistosity	Planar directional property due to visible platy or needle-shaped minerals	Muscovite in large, visible sheets or visible needles of hornblende	Schist	Fine-grained igneous and sedimentary rocks
Gneissic	Banding due to segregation of coarse-grained minerals	Quartz, feldspar, and hornblende; lesser amounts of mica than schist	Gneiss	Granite, diorite arkose, graywacke, quartz sandstone

B. Nonfoliated (Without Directional Structure)

Texture	Descriptive or Special Features	Minerals Present	Name	Probable Original Rock
Crystalline	Massive, granular; lack of sedimentary features	Calcite or dolomite	Marble	Limestone, dolomite
Crystalline	Massive, granular; grains cannot be removed as easily as with sandstone	Quartz	Quartzite	Quartz sandstone

TABLE A.8 Tabulation Sheet for Metamorphic Rocks

Specimen Number	Foliated or Nonfoliated	If Foliated, Type of Foliation; if not, Describe Texture	Minerals Present	Rock Name

Appendix B

Topographic Maps

Maps represent a portion of the Earth's surface and provide one of the most useful means for engineers and geologists to record and subsequently present information. They are generally classified into two types, planimetric maps and topographic maps. Planimetric maps depict only the horizontal dimensions showing the location of surface features but not the vertical distance between them. Topographic maps typically show the same features as do planimetric maps, including horizontal measurements, but they also illustrate the third dimension, that is, the configuration of the Earth's surface.

Topography or relief can be indicated in three different ways on a flat map: shading, hatchure lines, or contour lines. Shading can provide only relative information indicating higher versus lower elevations but no accurate vertical distances can be obtained from such maps. Hatchures are lines drawn parallel to the maximum slope direction of the land surface but are inaccurate in depicting topography and again provide no information on actual vertical distances between points on a map.

The most common method used on maps made in the United States to depict topographic information

527 ✓ all

is the contour line. These consist of numerous, fine, brown lines, which provide extensive topographic information to the trained observer, including stream location and direction of flow, shapes of landforms, and specific information on elevation differences.

The United States Geological Survey (USGS) a division of the Department of Interior, has been providing standard topographic maps of the United States for nearly a century. Over the years, three common series of topographic quadrangle maps have been published by the USGS. Early maps included 30 min of latitude and 30 min of longitude, which represents a large area, about 900 mi^2 when located in the middle latitudes (continental United States). Presented on a map approximately 30 in. by 30 in., this yields a small-scale map that cannot provide much detail. Latitude lines (meridians) form the north and south boundaries and longitude (parallels) form the east and west boundaries of the map.

A larger scale map covering 15 min of latitude and longitude was initiated later, representing only one-quarter the area and allowing more detail. At the present time, 7.5-min series maps are being produced, which, of course, depict 7.5-min of latitude and longitude. This large-scale map contains one-quarter the area of the 15-min quadrangle and shows considerably more detail. The publication of 7.5-min quadrangles covering previously unmapped areas provides an active program for the USGS. Their goal is to map the entire nation at that scale. The 15-min and 30-min maps are not available for many areas of the United States and apparently there are no current plans to complete the coverage for those map series.

Quadrangle maps carry as their title the name of the largest city or most important feature that lies within the map's boundaries. Reference maps for each state are available that provide the names and locations of individual quadrangles. Maps are ordered from the supplier and are typically filed in map libraries according to map series and then by the quadrangle name. State geological surveys commonly stock for sale those quadrangle maps available within their state. The USGS provides complete coverage of available maps for the United States and its possessions.

Three major features are shown on topographic maps: 1) relief (topography), which has been discussed in some detail, 2) water features—streams, lakes, ponds, swamps, and canals, and 3) cultural features showing human activity including roads, railroads, bridges, tunnels, land boundaries, various buildings, and so on. The print color is indicative of this threefold division. Topography is shown in brown, water in blue, and cultural features in black. In addition, interstate highways may be printed in red, vegetation in green, and updated information added since a previous map printing is shown in purple. Figure B.1 illustrates some of the standard topographic map symbols.

ELEMENTS OF TOPOGRAPHIC MAPS

Latitude and Longitude

To locate points on the Earth's surface, the system depicting latitude and longitude has been developed. The Earth's surface is divided into units of area by two sets of intersecting lines that form a grid. One set of lines extends in the north-south direction from the true North Pole to the true South Pole. These are longitude lines or meridians of longitude and they are measured up to 180° east and 180° west from the zero degree or Prime Meridian, which passes through Greenwich, England, southeast of London. The meridian or longitude value is determined by the angle between the principal meridian and the meridian in question, measured at the center of the Earth. In the United States all values are west longitude.

The other set of intersecting lines consists of the parallels or the lines of latitude. These circle the Earth in an east-west direction with zero latitude being the equator. Latitude is numbered north and south from the equator to 90° at the poles and is measured by the central angle between that parallel and the equatorial plane.

An important distinction between the two sets of lines is that longitude lines converge at the poles, whereas latitude lines are always parallel. This yields a rectangular shape for quadrangle maps in the middle latitudes (such as the continental United States) and trapezoidal-shaped maps near the poles where rapid convergence occurs.

The unit of measure for the grid system is the degree, which can be subdivided into minutes and seconds. At the equator, one minute of arc is equal to a nautical mile or about 6080 ft in length. A spot on

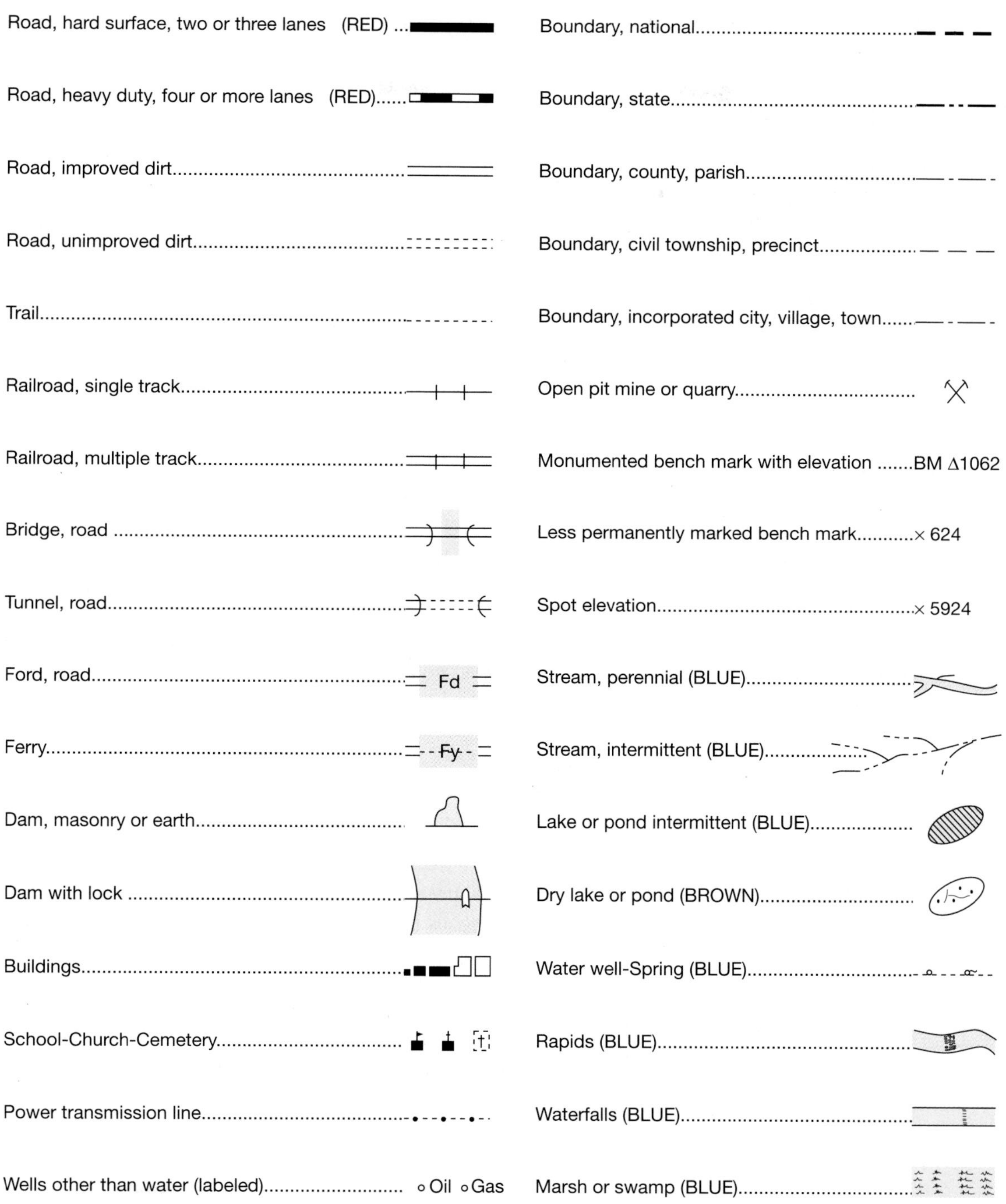

Road, hard surface, two or three lanes (RED) ...▬▬▬

Road, heavy duty, four or more lanes (RED)......▭▭▬

Road, improved dirt..═══

Road, unimproved dirt..----------

Trail...----------

Railroad, single track...─┼┼─

Railroad, multiple track..═╪═

Bridge, road ....⟩╞

Tunnel, road..╪----╞

Ford, road...═ Fd ═

Ferry...═--Fy--═

Dam, masonry or earth..⌂

Dam with lock ....┤▯├

Buildings...▪▬▪▭□

School-Church-Cemetery.......................................▪ ▪ [†]

Power transmission line...•─•─•─•

Wells other than water (labeled)...........................○Oil ○Gas

Boundary, national..── ── ─

Boundary, state...── ·· ─

Boundary, county, parish....................................── · ─ ·

Boundary, civil township, precinct................── ── ─

Boundary, incorporated city, village, town......── · ── ·

Open pit mine or quarry.................................⋊

Monumented bench mark with elevation ....BM Δ1062

Less permanently marked bench mark...........× 624

Spot elevation...× 5924

Stream, perennial (BLUE)...............................≍

Stream, intermittent (BLUE)...................

Lake or pond intermittent (BLUE)...................⬭

Dry lake or pond (BROWN)............................⬭

Water well-Spring (BLUE)...............................─ ○ ── ∽ ──

Rapids (BLUE)...≈

Waterfalls (BLUE)..≣

Marsh or swamp (BLUE)..................................

FIGURE **B.1**   Standard symbols for topographic maps published by the U.S. Geologic Survey.

the Earth is located by giving the latitude first (either north or south) and the longitude (either east or west) both to the degree, minute, and second. One second at the equator is equal to about 101 ft of distance.

Map Scale

The scale of a map indicates the relationship between the distance separating two points on the map and the actual distances between those two points on the Earth's surface. In keeping with the fundamental reason for making maps, map dimensions always represent a greater distance on the Earth's surface than their own actual length.

Three scales are commonly used on topographic maps: 1) fractional, 2) graphic, and 3) verbal. A proper map should have its scale represented in one of these ways and many maps indicate the scale in at least two of them.

The fractional scale or ratio scale is given in the form of a fraction, typically with the numerator equal to 1. The fraction is dimensionless so that one unit on the map is equal to the number of those units on the ground as indicated in the denominator. For example, 1 : 62,500 or 1/62,500 means 1 in. on the map is equivalent to 62,500 in. on the ground. This also holds for any other unit of measure: 1 cm on the map is equivalent to 62,500 cm on the ground, and so on.

The graphic scale or bar scale is a line or bar drawn on the map and is divided into convenient units showing distances on the ground in feet, miles, etc. To determine the distance between two points on the map, the map distance is laid off along the bar scale and the surface distance read directly from it.

The verbal scale is a convenient means for stating the relationship between map distance and ground distance and is commonly used on blueprints and on architectural and engineering drawings. Examples are $\frac{1}{8}'' = 1'$, $1'' = 1$ mi, or $1'' = 100'$.

There are advantages to each of the three scales. The fractional scale can be used for any units and is particularly useful for conversion of metric to English units of length and vice versa. The bar scale lends itself to quick determinations of distance and it has a second advantage. The drawing can be enlarged or reduced and the bar scale information still holds true. This is not the case for fractional scales or verbal scales. In the case of the verbal scale, it is easy to comprehend the relationship between map and ground distances. It is also useful for scaling distances

using an architect's scale or mechanical engineer's scale ("rulers" divided according to different scale relationships).

One concept of importance that commonly leads to confusion is the aspect of large-scale maps versus small-scale maps. A large-scale map shows more detail than a small-scale map and to accomplish this it represents a smaller area on the ground. Using the fractional scale a large-scale map will yield a larger number than will a small-scale map. For example, by comparing the $\frac{1}{24,000}$ scale to the $\frac{1}{62,500}$, the fraction $\frac{1}{24,000}$ is larger than $\frac{1}{62,500}$, and hence that first map has a larger scale. Since $\frac{1}{24,000}$ yields a verbal scale of $1'' = 2000'$ and $\frac{1}{62,500}$ yields $1'' = 5208.3'$, the first map would represent a smaller area in more detail and again be the larger scale map. Refer to Figure B.2 for details.

Converting Scales

We sometimes need to convert from one map scale to another. Conversion from a fractional scale to a verbal scale may be done to provide easier understanding or to a graphic scale for easy distance determination. Only simple arithmetic is involved in any of the scale conversions. The following are examples of these calculations:

EXAMPLE PROBLEM B.1

Convert the fractional scale $\frac{1}{250,000}$ to a verbal scale in terms of inches per mile.

ANSWER

To solve the problem put the fraction in terms of inches per inch and convert to inches per mile.

$1 : 250,000 = 1$ in.: $250,000$ in.

$$1 \text{ in.} = 250,000 \times (\tfrac{1}{12} \times \tfrac{1}{5280}) \text{ mi}$$

$$1 \text{ in} = 250,000/(12 \times 5280) \times \text{mi} = 3.95 \text{ mi}$$

$$1 \text{ in. on the map} = 3.95 \text{ mi on the ground}$$

EXAMPLE PROBLEM B.2

Convert the verbal scale 1 in. = 150 ft to a fractional scale.

ANSWER

To solve the problem convert the ground measurement to inches and cancel the units to obtain a fractional scale.

Square Mile

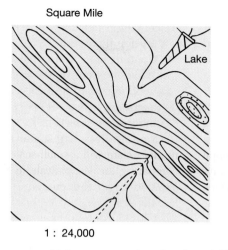

1 : 24,000

Square Mile

1 : 62,500

FIGURE **B.2** Topographic detail at different map scales.

1 in. = 150 ft

1 in. = 105 × 12 in.

1 in. = 7800 in.

1:7800 or 1/7800

EXAMPLE PROBLEM B.3 _____

Convert the verbal scale 1″ = 6000 ft to a graphic scale.

ANSWER

To accomplish this you first decide a convenient, even number to present on the graphic scale or bar graph, which would be most useful to the map reader, these being 1000 ft, 100 ft, 1 mi, and so on, depending on the actual case. In this example, 1000 ft should be selected with subdivisions of this accordingly: 1″ = 6000″ or ⅙″ = 1000′. Mark off distances of ⅙″ apart and label these 1000, 2000, . . ., 6000. Subdivide in half to obtain 500-ft increments.

500

0   5000

It is unacceptable to use ¼″ or some other increment just because it is easy to make into a bar graph. A ¼-in. scale would yield divisions of

1500′ each, which is not that convenient a distance for the map reader to use.

EXAMPLE PROBLEM B.4 _____

Convert the fractional scale ⅟₆₂,₅₀₀ to a graphic scale.

ANSWER

The same procedure is followed as used in Example Problem B.3, but the problem is more challenging:

1 : 62,500

(1 in. : 62,500 in.)

(1 in. : 6208.33 ft)

1 in. : 0:9864 mi

1.014 in. : 1 mi

Lay off distances of 1.014 in. and label them 1 mi, subdivide this distance to obtain ½- and ¼-mile markers.

Topography

Topography refers to the shape or configuration of the Earth's surface. As previously stated topography is most conveniently displayed on a map by means of contour lines, which are imaginary lines of equal elevation. Their elevations are designated with respect to a datum plane, usually mean sea level.

The contour interval, which is indicated on the legend of topographic maps, is the difference in elevation between any two successive contour lines. On standard USGS maps, contour lines are brown and usually every fifth contour line (sometimes every fourth depending on contour interval) is darker in color and it is labeled with the elevation. Typically only one contour interval is used for any map but when exceptions are made, the map legend will so indicate. Contour intervals may be 5 or 10 ft for maps in very flat terrain, 20 to 40 ft in moderately hilly areas, and 50, 80, or 100 ft in mountainous regions. The proper contour interval for a map is one that supplies the maximum detail concerning topography without being so cluttered with closely spaced lines that it becomes unreadable.

Elevations in addition to those indicated by contour lines are commonly supplied on topographic maps. Heights of hill summits, road intersections, and lake surfaces may be indicated for selected locations. These are known as *spot elevations* and their accuracy is given to the nearest foot. Benchmarks, designated BM on the map, are accurate indications of elevations representing permanently fixed points on the ground marked by a brass plate set in concrete. All specific elevations are printed in black on topographic maps.

Relief refers to the difference between the highest and lowest elevations of a feature under consideration. Local relief means the difference between the highest and lowest point in a local area perhaps within a square mile or so. Maximum relief for a quadrangle can be determined by subtracting the lowest elevation (lowest point in the major stream valley) from the highest point of the quadrangle, a hill, or a mountain located by careful examination of the elevation.

Contour lines mark the intersection of horizontal, parallel planes passed through the Earth's surface at specific vertical intervals (the contour interval). Because of this manner of construction, they carry certain requirements or restrictions; they also indicate special information because of their behavior in view of these restrictions. A list is provided below.

1. All portions of a contour line have the same elevation.
2. Contour lines never cross or intersect each other. They can merge as in the case of a vertical or overhanging cliff.
3. Every contour line closes on itself eventually, although this may not be shown on the map portion under study.
4. Contour lines never split or divide.
5. When contour lines cross stream valleys they bend upstream to form a V; that is , the Vs point upstream. They are perpendicular, however, to the water course itself.
6. Uniformly spaced contour lines indicate a uniform slope.
7. Closely spaced contour lines indicate a steep slope.
8. Widely spaced contour lines indicate a gentle slope.
9. Closed contour lines appearing on maps as ellipses represent hills.
10. Closed contour lines with hatchures represent a depression. The hatchures, which are perpendicular to the line, point downslope into the depression.
11. For land surfaces that extend uphill and then into a depression, the first hatchured contour is the same value as the last normal contour.
12. For land surfaces that extend downhill and then into a depression, the first hatchured contour is one interval less than the last normal contour.
13. Maximum contours (ridges) and minimum contours (valleys) always appear in pairs. Therefore, no single lower contour line can lie between two higher ones and no single higher contour line can lie between two lower ones.

Topographic Profiles

A topographic profile is a cross section of the Earth's surface taken along a specific direction. It is the view one would observe of the upper surface of a trench as seen in the side view. This is known as a side, profile, or elevation view as contrasted to the map or plan view, which represents the Earth's surface as seen from vertically above. Topographic profiles can be constructed along any given line on a topographic map. See Figure B.3 as an illustration.

Profiles are easily constructed using common graph paper. For the horizontal scale of the profile, the map scale is selected for convenience. A vertical scale is then arbitrarily chosen that will accentuate the surface features as needed. Because this exaggerates the vertical dimension it is known as vertical exaggeration. Only when the vertical and horizontal

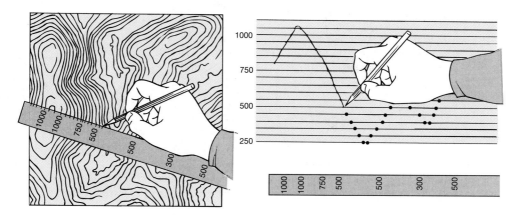

FIGURE **B.3**   Construction of a topographic profile.

scales are the same is the profile a true representation of the cross section. Slope angles of the land surface as depicted on the profile will also be increased if the vertical scale is exaggerated.

Instructions for preparing a topographic profile are as follows:

1. Select the line on the map along which the profile will be made. The student should appreciate that this line can be viewed from either of two sides. Two different profiles, one the mirror image of the other, will be obtained depending on the direction in which the cross section is viewed. Arrows can be placed on the map view to show the direction of sight.

2. Set up a vertical scale on the graph paper beginning with a minimum elevation that is less than the lowest elevation along the profile. A scale of $\frac{1}{10}$ or $\frac{1}{8}$ in. equal to a contour interval may be convenient depending on the divisions of the graph paper itself. Label the horizontal lines appropriately.

3. Place the graph paper along the line of profile. It may be necessary to fold under the margin of the graph paper so the graph subdivisions extend directly to the folded edge. Opposite each contour line which crosses the line of the profile place a small dash or tick mark and label the elevation. Also mark stream locations, depressions, hilltops, and so on, which may be useful in smoothing the final profile. If the contour lines are too closely spaced, only the dark lines should be considered.

4. Extend the points downward on the graph paper until they cross the appropriate lines marking those elevations.

5. Connect the points by a smooth line taking into account the information about stream location and so on noted earlier. Title the profile and indicate the vertical exaggeration after determining it in the manner explained below.

Vertical Exaggeration

Vertical exaggeration is determined by comparing the horizontal scale to the vertical scale. The vertical scale is used as the numerator and the horizontal scale as the denominator to obtain the vertical exaggeration in the following way: Horizontal scale $1'' = 1000$ ft; vertical scale $1'' = 300$ ft

$$\text{Vertical exaggeration} = \frac{\text{Vertical scale}}{\text{Horizontal scale}} \quad \frac{1''/300'}{1''/1000'}$$

The units cancel and VE = 3.33. As another example, consider a horizontal scale $\frac{1}{24000}$, vertical $1'' = 50'$. First put both scales into dimensionless form: vertical $1'' = 50'$ or $1'' = (12 \times 50)''$ or $1'' = 6000''$ or $\frac{1}{6000}$:

$$VA = \frac{1/6000}{1/24000} = 4$$

Figure B.4 illustrates the effect of vertical exaggeration.

Location by the Land Office Grid Method

Much of the United States has been subdivided by a grid method known as the Township and Range system, Land Office grid method, or the Congressional system. It was established by Congress in 1785

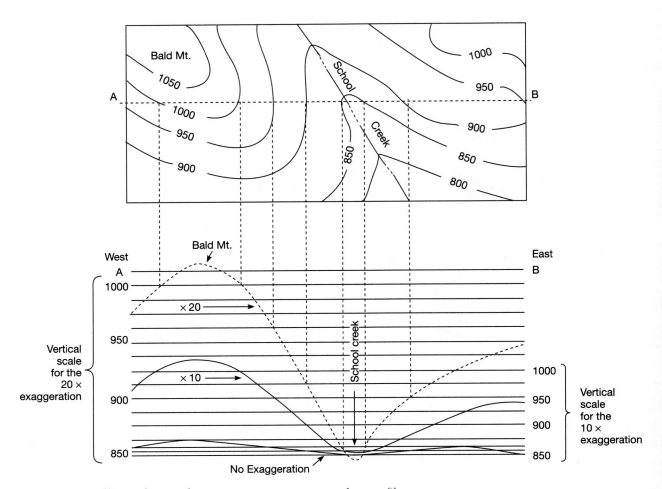

FIGURE **B.4**   Effects of vertical exaggeration on a topographic profile.

to provide a system of land survey that could be applied throughout the developing western territories. The United States has been divided to provide a basic unit, which is a square measuring 6 mi on a side. Not included in the system are the original 13 colonies plus Maine, West Virginia, Kentucky, Tennessee, Texas, and parts of Ohio (Figure B.5).

A principal meridian and baseline are established as the coordinate axes for the grid. Typically, each state of the United States has a designated principal meridian and baseline for surveying reference. Strips or tiers 6 mi wide, termed *townships*, are laid off north and south from the baseline and labeled accordingly: T1N, T2N, . . ., T1S, T2S, . . ., whereas similar strips laid off east and west from the principal meridian, known as *ranges*, are also labeled accord-

ingly: R1W, R2W,. . . . A designated square 6 miles on a side, for example, T10N, R5W, is known as a township and consists of 36 square miles. The townships are subdivided into 36 pieces, each 1 mi^2 in size, known as *sections*, which are numbered consecutively from 1 to 36 beginning in the northeast corner.

The sections in turn are divided into four quarter sections, these quarter sections into sixteenth sections, and they in like manner into sixty-fourth sections. A possible designation at this level would be NE¼ of NW¼ of SE¼, Section 10, T10N, R5W. Refer to Figure B.6 for further details.

The area contained in a section is 1 mi^2 or 640 acres. Therefore, ¹⁄₆₄ of a section as indicated in the example above is 10 acres in size, a relatively small portion of land. An acre is 43,500 ft^2 or a square 208.7

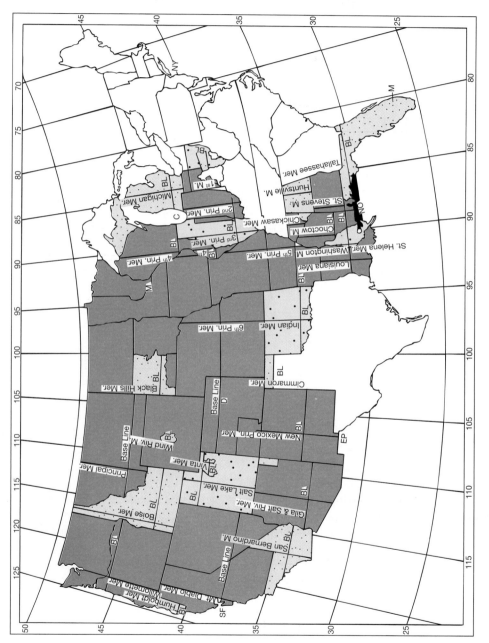

FIGURE **B.5** Principal meridians and baselines for the United States.

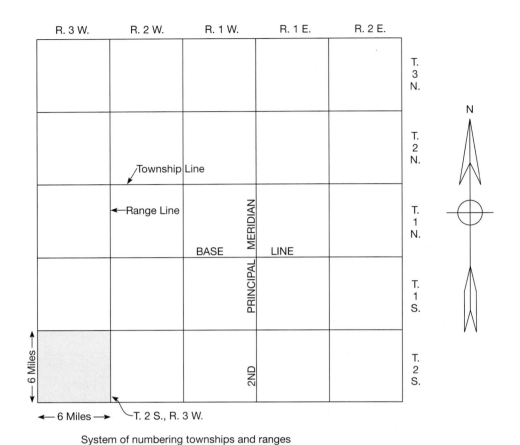

System of numbering townships and ranges

System of numbering sections
of a congressional township

System of subdividing a section
(640 Acres)

FIGURE **B.6**   Land division according to the Township and Range system.

ft on a side. The ability to locate a piece of land contained in a state of the United States to within 10 acres by this simple procedure is a significant accomplishment. Figure B.7 shows the Township and Range system in Indiana.

Gradients or Slopes

The slope of the land surface or of a stream bed is known as the *gradient*. This is equal to the vertical distance divided by the horizontal (or map) distance. For stream gradients, the units of feet per mile are used. For most streams this yields convenient numbers between a tenth of a foot to several tens of feet per mile. Mountainous regions may have steep gradients in excess of 100 ft per mile, whereas the largest rivers may yield several feet or fractions of feet per mile. The profile of a stream along its stream bed is called its *long profile*.

Human-made features extending over long distances typically have their gradients supplied in percent, which equals (vertical distance/horizontal distance) × 100. Some of these features may have a slope of several percent, which would be quite steep compared to stream gradients.

Dips of rock strata are also represented as slopes or gradients. Dipping rock units will not be considered in this section but for purposes of comparison, the dip of such units is commonly expressed in degrees of the angle measured from the horizontal. One degree of dip is equal to about a 2% slope.

Compass Directions

Meridian lines bound a topographic map on its left and right margins. These extend in the true north-south direction so that true north is toward the top of the map. Other maps used as engineering drawings should, along with an indication of scale, contain a north arrow indicating the true north direction.

For most topographic maps, true north will not coincide with the direction determined by a magnetic compass, that is, with magnetic north. The angle between true north and magnetic north measured in the plane of the map is known as magnetic declination.

On standard topographic maps, the magnetic declination is shown along the bottom margin by means of an arrow with an angle measured from true north. The angle is labeled and the direction can be either east or west of true north in various parts of the United States. Magnetic declination listed on such maps depicts the known value at the time of map preparation. Owing to an annual drift in the location of magnetic north, the latest declination must be applied when using compass measurements in a field study. The general details of the magnetic declination in the contiguous states of the United States are shown in Figure B.8.

GEOLOGIC MAPS OF HORIZONTAL ROCKS

The extension from topographic maps to geologic maps that depict horizontal rocks is easily accomplished. Geologic maps show the surface distribution of various types of rock. They also indicate the relative age of the rocks and suggest their manner of association in the subsurface. The plan view of contacts between units are shown on these maps as well. Topographic contour lines may or may not be included on a geologic map.

A geologic contact is the surface of intersection between two different rock units or formations. The term *formation* pertains to a rock unit that is thick enough and sufficiently widespread to be represented on a geologic map. It typically contains a distinctive lithology or rock type that can be traced in the field. On a geologic map, the contact between two adjacent formations will appear as a line.

When the formations are horizontal, their contacts are horizontal as well. Such contacts will occur parallel to the contour lines on a topographic map because contour lines, of course, depict the intersection of the Earth's surface with horizontal planes passing through the Earth. It is an easy matter to trace these horizontal contacts on a topographic map if the contact elevations are known.

Geologic cross sections of horizontal rocks are also an easy extension from topographic cross sections. A geologic cross section shows the distribution of formations in the vertical section. It is simply a topographic profile with the geology of the section added. For horizontal rocks such addition is an easy matter. The contact elevations are added to the cross section and the formations are illustrated by different colors.

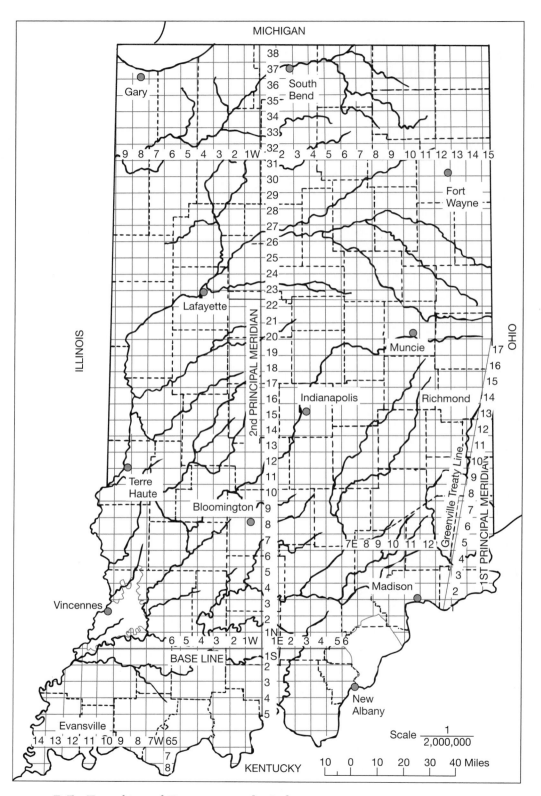

FIGURE B.7 Township and Range system for Indiana.

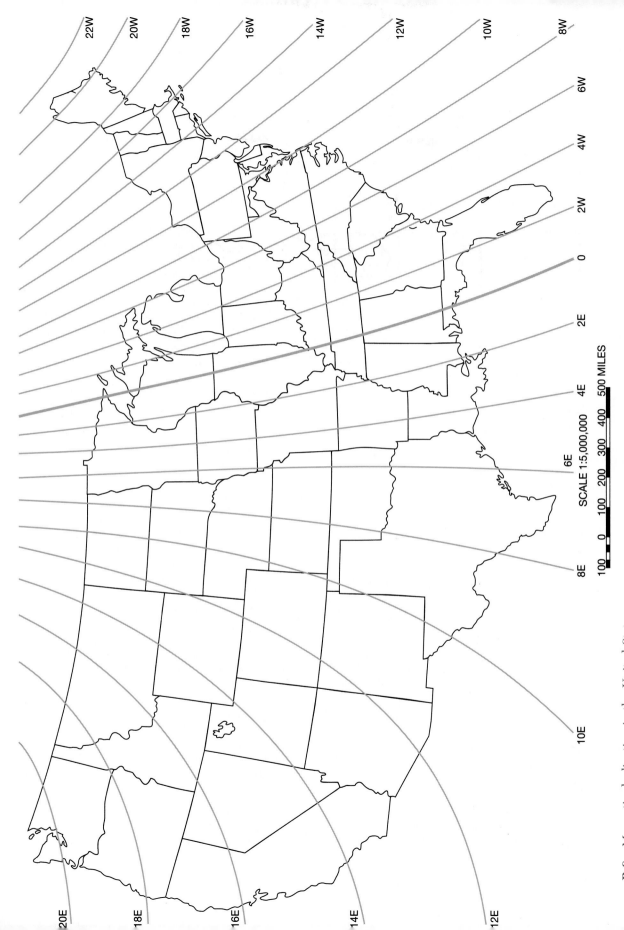

FIGURE **B.8** Magnetic declination in the United States.

EXERCISES ON TOPOGRAPHIC MAPS

1. Given the graphic scale of 1¼ in. on a map represents 1 mile on the ground, compute the fractional scale.

2. Given the fractional scale of 1 : 84,000, convert this to a convenient verbal scale. Convert it also into two graphic scales, one reading in thousands of feet and the second in miles, half miles, and quarter miles.

Bellefonte, Pennsylvania; 15-min Quadrangle Map

This map has been divided into nine rectangles for reference purposes.

$$3 \quad 2 \quad 1$$
$$4 \quad 5 \quad 6$$
$$9 \quad 8 \quad 7$$

3. Why was the Township and Range method not used for this map? What is the approximate area of the map? Give the scale, contour interval, and magnetic declination including direction. What is the area of each rectangle on this map?

4. Using latitude and longitude, determine as accurately as possible the center of the town of Julian in rectangle 4. What is the elevation at this point?

5. Determine the straight line distance between the center of State College in rectangle 8 to the center of Julian in rectangle 4. Give the distance to the nearest tenth of a mile.

6. What is the general direction of flow of Bald Eagle Creek? What is the maximum relief in rectangle 1? Give the location of the maximum and minimum elevations to the nearest ¹⁄₁₆ of a rectangle.

7. The pass through Eagle Mountain just north of Bellefonte is called a *water gap*. What would be the effects of flooding on this narrow portion of Logan Branch Creek?

8. What is the overall gradient of Brower Hollow from its beginning to the confluence with Bald Eagle Creek? Concerning distance, assume it consists of two straight-line segments, one from the stream's beginning to the bend near the 985 elevation and the second from that point to Bald Eagle Creek. Give the gradient in feet per mile.

9. Plot the long profile of Bush Hollow (rectangle 2) from Grindstone Gap to Bald Eagle Creek using the lines provided below. Plot accurately the 100-ft contours only, approximating 50-ft increments by interpolation. For the vertical scale, use $1'' = 400'$; for the horizontal distance use the map distance. (Locate the contour lines of interest on a folded edge of a piece of scrap paper and transfer these points to the cross section.)

What is the shape of the long profile? It is a steady slope, concave down, concave up, or what? Calculate the vertical exaggeration of your drawing.

_____
_____
_____
_____
_____
_____
_____

10. A railroad tunnel is planned that will connect the railroad at Unionville (northern portion of rectangle 5) through the mountain to the Bellefonte Central Railroad on the opposite side. The tunnel would run perpendicular to the trend of the mountain. Determine the following information: (a) length of the tunnel in feet, (b) the gradient of the tunnel in percent, and (c) the maximum height of rock cover over the tunnel. To accomplish these three tasks, sketch the topographic cross section showing the two portals (entrances to the tunnel), the point of maximum elevation of the mountain, and the tunnel gradient. (d) What would be the total overburden stress at the maximum depth of rock cover if the unit weight of rock = 165 lb/ft³? (e) Why would the rock load on the tunnel roof not likely be this great? Explain. (f) If 5 ft of rock load must be carried by the tunnel supports, what stress in lb/ft² would this exert on the tunnel roof? (g) If limestone, sandstone, and shale were present at various locations along the tunnel, what kinds of problems might develop during tunnel excavation?

West Lafayette, Indiana; 7.5-min Quadrangle Map

11. What is the contour interval of this map? Indicate in latitude and longitude the area covered by the map. What is the map scale and the magnetic declination?

12. What is the difference in elevation between the top of Happy Hollow Creek at Grant and Salisbury Streets and its confluence with the Wabash River? Using the straight-line distance between these two points, determine the average gradient.

13. Locate the pond in section 13 near the center of the map using the Township and Range method. What is the approximate area of the lake as shown?

14. Note the change in relief about one mile north of the Wabash River near the southern boundary of the map. The bluff marks the intersection of glacial outwash on

the upper terrace of the Wabash River system with the upland surface of the glacial till plain.

(a) Trace (mentally, no pencils or pens please) this intersection across the Purdue University campus. What is the general direction of these features?

(b) The terrace consists of sand and gravel deposits, whereas the till plain is mostly silt and clay. What effect would this have for buildings (residence halls) that straddle these two landforms?

(c) Make a cross-sectional drawing in the area of the football stadium. Show the geologic materials in the subsurface.

(d) On which landform is the university water supply wells located? Why? On which landform is the water tower located? Why?

15. Obtain a copy of the Lafayette, East Quadrangle, as well as the Lafayette, West Quadrangle, for this question. Based on topographic features and other geographical constraints explain why West Lafayette is growing northward whereas Lafayette is growing southward.

16. Topographic map and horizontal sedimentary rocks. Refer to Figure B.9. The sedimentary rocks in this area are horizontal and have negligible soil cover. Streams are shown as dotted lines. Given the following information, draw in the contacts between formations and color the formations on the map and on the key to the right of the map. The youngest beds are at the top of the key.

(a) (1) The contact between the Savage Sandstone and the Lydon Conglomerate occurs at elevation 1200 ft.

(2) A well is located at point 1. Savage Sandstone crops out at the surface. At a depth of 50 ft the contact between the Savage Sandstone and the Morgan Dolomite is reached and at 150 ft the contact between the Morgan Dolomite and Churp Limestone is reached.

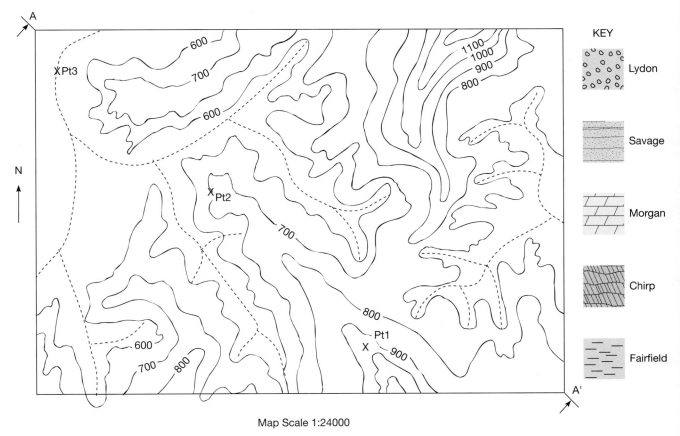

Map Scale 1:24000

FIGURE **B.9**   Map and cross section for area with horizontal sedimentary rocks.

(3) Another well is located at point 2. Churp Limestone crops out at the surface. Fairfield Shale is found at a depth of 100 ft.

(b) Draw a geologic cross section (topographic cross section showing geologic units) along line A-A′ looking in the direction indicated by the arrows. Use a vertical scale of 1″ = 400′ and a horizontal scale equal to the map scale.

(c) A dam is being considered for location at point 3. The local consulting geologist suggested that the reservoir level be maintained below the 600-ft contour level. On what basis do you think he or she made this recommendation?

Additional Readings

U.S. GEOLOGICAL SURVEY, "Topographic Maps," available from USGS, Washington, D.C.

U.S. GEOLOGICAL SURVEY, "Index of Topographic Maps for Individual States," Washington, D.C.

Index

547

Maps (*continued*)
geological, 496
glacial, 191
glacial-thickness, 453
interpretive, 192, 480, 496
landslide susceptibility, 497
large-scale, 496
legend, 191
rock strength and modulus, 497
septic tank acceptibility, 497
subsurface, 496
swell potential, 497
symbols, 191, 203, 206, 213
water availability, 497
Marble, 74, 76–78, 86, 90, 95, 266, 292, 329, 337, 521, 524
Marcasite, 19
Marginal lakes, 251, 253
Marietta Basin, 276, 277, 279
Marina District, CA, 309, 421, 423
Marine, 59, 199, 309, 323
Marine clays, 118
Marine shale, 159
Marine terraces, 423
Mariotte, 324
Marker bed, 210, 212
Marl, 67, 429, 440
Marsh, 379, 456
Masonry, 410, 423, 424, 480
Mass movement, 258, 283, 285, 479
Mass spectrometer, 185
Materials engineer, 482
Materials inventory, 480–482
Matrix, 54, 94
Matterhorn, Switzerland, 249
Mature topography, 240, 241, 243, 268, 271
Maumee Lacustrine Plain, 272, 274
Mauna Loa, Hawaii, 37
Mayon, Phillipines, 37
Meanders
scars, 237
streams, 229, 230, 237, 238, 241, 242
Mechanical weathering, 389, 390, 400
Mediterranean Sea, 417, 490
Mediterranean & Trans-Asiatic belt, 417
Meinzer unit, 331, 332
Melting, 307
Meltwater, 248, 250–252
Member, 174, 193
Memphis, TN, 416
Mercalli, L, 408
Mercalli scale, 408
Mercury (element), 21
Mercury (planet), 5
Mesas, 390, 393
Mesozoic, 182, 187, 266
Meta-igneous rocks, 454
Metals, 155
pipes, 169
Metamorphic grade, 73, 75, 94
Metamorphic rocks, 58, 71–79, 151, 153, 159, 173–176, 179, 186–188, 199, 214, 236, 247, 252, 264–267, 269, 270, 356, 380, 392, 430, 440, 481, 521, 524, 525
defined, 17, 34, 71
engineering concerns, 77, 78
identification, 76, 77, 524, 525
Meteorites, 190, 427
Methane, 354, 479, 485
Mexican onyx, 67
Mexico City, NM, 289
Miami-Russel-Fincastle, 167
Mica, 26, 73, 74, 78, 105, 188, 398
Michigan, 250, 271, 272, 386
Michigan City, IN, 386
Microcline, 187
Microseisms, 419, 424
Microstructure, 120, 121

Midcontinent, 178
Middle East, 417
Middle Rocky Mountains, 267, 273
Mid-ocean ridges, 12, 14, 417
Midwest, 179, 253, 255, 259, 353, 356, 391, 416–418, 420, 443, 446
Mielenz, R.C., 97
Migmatite, 75
Migration, 158, 292, 396, 447
Milky Way, 1, 2, 4, 7, 8
central mass, 8
Mill Creek Valley, OH, 253
Millisecond delays, 485
Milwaukee, WI, 242, 372
Mine spoil, 314
Mineralization, 303
Mineralogy, 95, 460
Minerals, 151, 153–156, 159, 161, 189, 392, 427, 430, 451, 478, 496, 509, 521, 525
abundance, 18
atomic structure, 19
biological activity, 20
carbonates, 28
clays, 28–31
cleavage, 22, 95
color, 20
composition, 20
crystal form, 22
crystal symmetry, 22, 23
crystallography, 23
crystals, 19, 20
crystallization, 19, 20
defined, 17, 19
depletion, 478, 496
development, 496
diaphanacity, 23
economic, 24, 514–518
extraction, 497
feromagnesians, 26
formation, 19, 20
fracture, 22
fusion, 20
hardness, 21
hydrated, 324
identification, 19, 20–24, 509–513
interfacial angle, 22
luster, 20
magnetism, 22
Mohs' hardness scale, 21, 45, 56, 76
oxides, 27
physical properties, 18, 20–24
precipitation, 20
reaction to HCl, 24, 28
resources, 496
secondary, 174
silicate structure, 24–26
silicates, 24–27
specific gravity, 21
streak, 20, 21
striations, 24, 27
structure, 19, 21, 22
sublimation, 20
sulfates, 28
sulfides, 27
tenacity, 23
Mineralogy, 59, 152, 158, 159, 299, 465, 513
Mines, 487
Mining, 292, 312, 337, 348, 349, 359, 379, 380, 391, 451, 479, 484
engineer, 452
geologist, 452
Minneapolis, MN, 228
Minnesota, 190, 255, 265, 398
Minutes, 528
Miocene, 182, 192
Mississippi River, 183, 223, 225, 228, 236, 256, 374, 398, 400, 418
Mississippi River delta, 238
Mississippi River Valley, 266, 417

Mississippian, 178, 179, 183, 192, 274, 276
Missouri, 161, 228, 241, 268, 398, 400, 416, 417
Missouri River, 227, 228, 236, 256, 398, 400, 417
Mitchell Plain, 272, 274
Mitigation, 497
Mobility, 152
Model (of subsurface), 451
Modified proctor test, 122
Modulus of elasticity, 83, 87, 88, 404, 405, 495
Moho, 425
Mohorovicic, 425
Mohorovicic discontinuity. *see* Moho
Mohr circle, 84, 86, 134–136, 290
Mohr-Coulomb criterion, 136–139, 299, 300
Mohr diagram, 84–86
Mohr envelope, 84–86
Mohs' hardness scale, 21, 45, 56, 76, 486
Moisture, 153, 166
Moisture content. *See* Water content
Mojave Desert, CA, 390
Mollusks, 189
Molybdenum, 292
Moments, 293
Moon, 369, 373, 416
composition, 10
full, 416
new, 416
origin, 10
Monitoring, 480
construction, 480
Monocline, 200
Montana, 156, 267
Monterey Chert, 174
Montmorillonite. *See* Smectite
Moore cumulative method, 437–442
Moraines, 250
end, 250, 252–255, 258, 336, 356
ground, 249, 250, 252, 254, 255, 258, 259, 356
lateral, 250, 251
medial, 249–251
recessional, 249, 251, 274
terminal, 249, 251, 274
Mount Lassen, 267
Mount Pelee, Martinique, 41
Mount Ranier, WA, 267
Mount Saint Helens, 40–41, 267, 270, 394, 400
Mount Shasta, CA, 37, 267
Mount Whitney, CA, 267
Mountain building, 264, 269, 277, 417
Mountain ranges, 211, 425
Mountainous regions, 200, 207, 214, 248, 255, 305, 310
Muck, 259
Mucking, 259, 473, 485, 486
Mud cracks, 59, 62
Mud flow, 285, 286, 305, 307, 308, 312
Mudslide, 286
Mudstone, 55, 460
Mud volcanoes, 421
Municipal waste, 390
Municipalities, 348
Muscatatuck Regional Slope, 272, 274
Muscovite, 26, 73, 74, 76, 155, 157, 187, 188, 510
N value, 116, 167, 168, 259, 410, 461, 462, 465, 470
NQ core, 465
Nx core, 463, 464
Nx hole, 473
National flood insurance, 242
National Guard, 228
Natural gas, 54, 451, 479
Natural levees, 236, 237
Natural water content, 168
Nautical mile, 528

558

COMMON CONVERSIONS USED IN GEOTECHNICAL CALCULATIONS

Quantity	
Length	1 statute mile = 5280 ft
	1 nautical mile = 6076 ft = 1 minute of arc of a great circle on Earth
Area	1 acre = 43560 ft^2
	1 square mile = 640 acres
Volume	1 cubic foot = 7.48 gal
	1 barrel = 42 gal
	1 acre-ft = 43560 ft^3 = 1223 m^3
Unit Weight and Density	Water, 62.4 lb/ft^3 or 1000 kg/m^3 or 1.000 g/cm^3
Discharge	1 cubic foot per second (cfs) = 0.646 million gal per day (MGD)
	1 gal per min = 2.228 × 10^{-3} cfs
Hydraulic Conductivity (and Velocity)	1 cm per sec = 1.03 × 10^6 ft/yr = 864 m/day
	1 ft per year $\cong$ 10^{-6} cm/sec
	1 Meinzer = 1 gal/day/ft^2 = 0.134 ft/day = 4.72 × 10^{-5} cm/sec = 0.04085 m/day
	1 darcy = 18.2 Meinzers, water @ 60°F = 8.58 × 10^{-4} cm/sec, water @ 60°F
	1 millidarcy = 0.001 darcy = 0.858 × 10^{-6} cm/sec, water @ 60°F
Slope	1° = 1.75% = 92.2 ft/mi = 17.45 m/km
Transmissibility	1 gal/day/ft = 0.1337 ft^2/day = 0.01240 m^2/day
Time	1 year = 3.1536 × 10^7 sec
Temperature	F° = C°(1.8) + 32 or C° = (F° − 32) × 1/1.8

F	C
60	15.6
70	21.1
80	26.6
90	32.2
100	37.7

one unit of F° = 0.555 units of C° or about an increase of 1°C for every 2°F for small values

Example: 72°F equals about 22°C

83°F equals about 26.6 + 3(0.555) = 28.3°